T0251359

BIOMASS POWER
for the World

Pan Stanford Series on Renewable Energy

Series Editor
Wolfgang Palz

Vol. 1
Power for the World: The Emergence of Electricity from the Sun
Wolfgang Palz, ed.
2010
978-981-4303-37-8 (Hardcover)
978-981-4303-38-5 (eBook)

Vol. 2
Wind Power for the World: The Rise of Modern Wind Energy
Preben Maegaard, Anna Krenz,
and Wolfgang Palz, eds.
2013
978-981-4364-93-5 (Hardcover)
978-981-4364-94-2 (eBook)

Vol. 3
Wind Power for the World: International Reviews and Developments
Preben Maegaard, Anna Krenz, and
Wolfgang Palz, eds.
2013
978-981-4411-89-9 (Hardcover)
978-981-4411-90-5 (eBook)

Vol. 4
Solar Power for the World: What You Wanted to Know about Photovoltaics
Wolfgang Palz, ed.
2013
978-981-4411-87-5 (Hardcover)
978-981-4411-88-2 (eBook)

Vol. 5
Sun above the Horizon: Meteoric Rise of the Solar Industry
Peter F. Varadi
2014
978-981-4463-80-5 (Hardcover)
978-981-4613-29-3 (Paperback)
978-981-4463-81-2 (eBook)

Vol. 6
Biomass Power for the World: Transformations to Effective Use
Wim van Swaaij, Sascha Kersten,
and Wolfgang Palz, eds.
2015
978-981-4613-88-0 (Hardcover)
978-981-4669-24-5 (Paperback)
978-981-4613-89-7 (eBook)

Pan Stanford Series on Renewable Energy
Volume 6

Transformations to Effective Use

BIOMASS POWER
for the World

edited by

Wim van Swaaij
Sascha Kersten
Wolfgang Palz

PAN STANFORD PUBLISHING

Published by

Pan Stanford Publishing Pte. Ltd.
Penthouse Level, Suntec Tower 3
8 Temasek Boulevard
Singapore 038988

Email: editorial@panstanford.com
Web: www.panstanford.com

British Library Cataloguing-in-Publication Data
A catalogue record for this book is available from the British Library.

Biomass Power for the World: Transformations to Effective Use
Copyright © 2015 by Pan Stanford Publishing Pte. Ltd.
All rights reserved. This book, or parts thereof, may not be reproduced in any form or by any means, electronic or mechanical, including photocopying, recording or any information storage and retrieval system now known or to be invented, without written permission from the publisher.

For photocopying of material in this volume, please pay a copying fee through the Copyright Clearance Center, Inc., 222 Rosewood Drive, Danvers, MA 01923, USA. In this case permission to photocopy is not required from the publisher.

ISBN 978-981-4613-88-0 (Hardcover)
ISBN 978-981-4669-24-5 (Paperback)
ISBN 978-981-4613-89-7 (eBook)

Cover image courtesy: Sumi-e painting by Joke van Swaaij
(www.jokevanswaaij.com)

Printed in the USA

Contents

9. Solvent-Based Biorefinery of Lignocellulosic Biomass 289

*Paulus Johannes de Wild and
Wouter Johannes Joseph Huijgen*

Preface

The books of the Pan Stanford Series on Renewable Energy (series editor Wolfgang Palz) are directed to the laymen or at least to the non-specialist in the discussed subject while the authors are often authorities and pioneers in their fields.

Compared with photovoltaics (PV) and wind, biomass is a rather broad topic. The number of technologies developed for PV and wind are smaller and the sole product is electricity. How different is the situation for biomass? Production of biomass in agriculture and forestry and its conversion by mechanical, thermo-chemical, biological, and bio-chemical processes to a wide range of improved fuels, suitable to be introduced into a large variety of modern energy-conversion equipment, is extremely complex and requests knowledge of many disciplines of technologies and sciences. Moreover biomass flows for energy are connected to those for other uses of biomass like for food, feed, materials, chemicals, etc. Within the arenas of politics, business, science, and of non-governmental organizations, the views on the priorities for the use of biomass vary a lot and are often based on partial or incorrect information. Indeed a complete encyclopedia would be required to even partially cover all the issues involved.

We decided to concentrate in the present volume on the conversion technologies for biomass to fuels and energy carriers fitting in the infrastructure of modern and developing society. Of course a kind of rudimentary sketch is necessary for the background and the scenery where these technologies are going to be fitted in.

We are extremely grateful to and proud of the co-authors that joined us in this effort to produce a text accessible to the non-specialist without losing too much of the essential details on barriers and dilemmas that determine the possibilities for deployment of the various conversion technologies. However, a few chapters are devoted to broader themes. In the first part of the introductory chapter, Chapter 1, the role of biomass in the built-up of our living environment is discussed. The direct and indirect use of biomass over centuries by man for satisfying his needs and expanding his ambitions, together with wind and later with fossil fuel is also pictured. The fossil fuels (coal and later oil and gas) took over the main role as energy sources and energy carriers and became primary sources for electricity generation. Now fossil fuels are seen as a threat to our biosphere because of greenhouse gas emissions, especially CO_2 from fossil carbon. Within the renewable and sustainable energy sources, biomass will probably retake an essential position while fossil fuels will have to fade out.

In a second part of Chapter 1 an overview is given in the form of a simplified description of the conversion processes of biomass, producing essentially all fuel types presently obtained from fossil sources. To be complete some overlap with other chapters could not be avoided. In the third part of Chapter 1 an ultra short summary on future developments in conversion technologies is presented, certainly colored by personal visions. The use of biomass and derived products as stored energy and its position between renewable sources with fluctuating output (solar, wind, etc.) is discussed and the potential of electric power conversion to green carbon–containing fuels is discussed in some detail. The chapter ends with a series of overall conclusions.

In Chapter 2 Wolfgang Palz gives his vision on the subject. In Chapter 3 Heinz Kopetz, chairman of the World Bioenergy Association, presents a highly interesting and condensed view on the future role of bioenergy in the global energy system. He draws our attention to practical and probably realistic targets for the midterm future: 2030. In Chapter 4, Wolfgang Palz together with Henri Zibetta presents an extensive overview on the history of the European support program for biomass energy in which Wolfgang and Giuliano Grassi played important directive and stimulating roles.

Biomass energy cannot be considered in isolation and several aspects like employment, rural, social and economic development play an important role. Activities, stimulated in the past by the European Union and still highly relevant for the present situation, are the LEBEN projects based on a strategy for commercial exploitation of biomass at a regional level and described in Chapter 5 by Giuliano Grassi.

All other chapters of the book are devoted to individual conversion processes of biomass to fuels with the idea to clarify the challenges and obstacles while remaining largely accessible for the non-specialist. For most authors this must have been quite a job and unfortunately some invited authors were not able to deliver a suitable manuscript on time. The readers are to judge whether our efforts were successful. We all hope that this book will be useful in promoting the application of biomass in a responsible way as a renewable and sustainable energy source; after all it is the oldest and presently the largest we have. Although as a renewable source its capacity is not unlimited, we will need all our resources to cope with the giant challenge for the future—providing sustainable energy for our future world and finding an accessible route to reach this aim.

Wim van Swaaij
Sascha Kersten
Wolfgang Palz
Spring 2015

Chapter 1

Overview of Energy from Biomass for Nonexperts

Wim van Swaaij and Sascha Kersten

TNW/SPT, University of Twente, Meander 224, PO Box 217,
7500 AE Enschede, the Netherlands

w.p.m.vanswaaij@utwente.nl

An introduction to biomass for energy for nonexperts is presented. The origin of biomass, its use as an energy carrier, and its position amidst other sources of energy is discussed.

The role of biomass in the balance and recycle of carbon in the biosphere is elucidated and aspects of sustainability are shortly touched upon.

The main emphasis of this chapter is on the introduction of biomass into modern energy conversion equipment and the conversion processes of biomass into improved solid, liquid, and gaseous fuels. The chapter is concluded with some discussions of the status and possible future developments of the role of biomass in interaction with other renewable energy sources and possible synergistic effects.

Biomass Power for the World: Transformations to Effective Use

Edited by Wim van Swaaij, Sascha Kersten, and Wolfgang Palz

Copyright © 2015 Pan Stanford Publishing Pte. Ltd.

ISBN 978-981-4613-88-0 (Hardcover), 978-981-4669-24-5 (Paperback), 978-981-4613-89-7 (eBook)

www.panstanford.com

1.1 The Position of Biomass in the Energy Field

1.1.1 Power from the Sun Collected via Biomass and Historical Use

Life, as we know it, on earth is mainly driven by high-quality energy received from our sun in the form of photons. In the central core of the sun, under the influence of extremely high temperatures and pressure, fusion of hydrogen to helium takes place (see Fig. 1.1), producing a tremendous amount of energy.

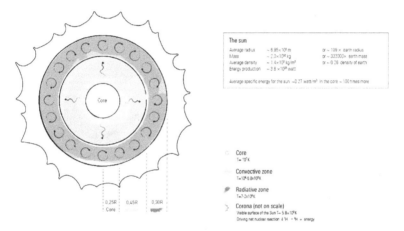

Figure 1.1 The sun, the driving force for life on earth in approximate figures.

This energy is cascaded via radiation in the core and convection in the outer shells into lower-energy photons which form the typical solar radiation spectrum characterized by a surface temperature of about 5800 K [1]. Present-day knowledge places the start of the universe about 13.8 billion years ago with the so-called Big Bang and the subsequent genesis of earlier generations of galaxies and their stars. The formation of the sun, our star, with its solar system took place about 4.6 billion years ago. In the same period our planet came into being, mainly consisting of the heavier elements (heavier than hydrogen and helium), while the sun almost completely consisted of hydrogen and helium. At present the rate of helium production in the sun is 3.1×10^{-19} kg/s per kilogram solar mass and this is accompanied by an energy release of 191×10^{-6} watts/kg of solar material. For comparison a human being produces 5000–10000 times more power per kilogram mass. This may seem a somewhat low rate of the sun's specified heat production but due to the enormous mass of the sun (about 1.99×10^{30} kg) it results in a total energy flow from the sun into space of 3.9×10^{26} watts [2].

A very small fraction of this energy flow is intercepted by our planet earth where the solar flux is about 1361 watts/m^2. Here the incoming radiation is reflected by the earth or transformed into lower-energy photons and emitted to space again. A tiny, tiny fraction is converted in the biological cycles and temporarily stored in biomass before decay processes produce again heat to be radiated as infrared photons to space. Only an extremely small proportion of this biomass fraction gets finally stored as fossil deposits such as coals, gas, and oil.

Life on earth and its produced biomass can thus be considered as driven by a tiny fraction of the outflowing energy stream of the sun, taking advantage of the moderate temperature created on earth by, mainly, physical processes.

The present conditions on the earth and its atmosphere are, however, also created by life itself as it evolved. While the earth is formed mainly from materials heavier than hydrogen and helium, these heavy elements were originally already created in stars of previous generations and spread into space, for example, by supernova explosions and trapped by gravity in the cloud which would result in our solar systems. From this point of view, the earth, life, biomass, man itself, it is all stardust based.

Shortly after the earth was formed its atmosphere had a composition widely different from the present atmosphere. Free oxygen (O_2) was absent and life must have been initially based on energy cycles involving energy-rich components available or created in the environment (see, for example, Ref. [3]). Already in the earlier periods of its existence, life evolved into organisms like the blue-green algae which were able to absorb photons, creating activated electrons in molecular systems which were used in cascades of reactions to remove oxygen from water and to reduce CO_2. Over a prolonged period of time this changed the composition of the atmosphere (see Fig. 1.2) as not all the oxygen produced reacted again with earth surface components or with the reduced organic components created by early life. Some of these latter products were fossilized. The presence of accumulated oxygen in the atmosphere allowed very efficient energy cycles as became present in the eukaryotic cells of later flora and fauna. The eukaryotic cells as we find them today are extremely complicated systems with highly organized structures probably evolved via collaboration and integration and further evolution of originally separate organisms [3].

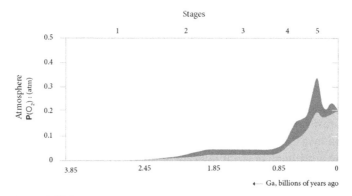

Stage 1	3.85–2.45 Ga	Practically no CO_2 in atmosphere
Stage 2	2.45–1.85 Ga	O_2 produced but absorbed in oceans and seabed rock
Stage 3	1.85–0.85 Ga	O_2 starting to gas out of the oceans but absorbed by land surfaces and formation of ozone layer
Stages 4 and 5	0.85 Ga–present	O_2 sinks filled and gas accumulating

Figure 1.2   Oxygen content of the atmosphere in successive periods of the existence of the earth. The production of oxygen started very early, but it was initially consumed in the oxidation processes of the earth's crust's components (among others, iron). Reproduced with the permission of http://rstb. royalsocietypublishing.org/content/361/1470/903.

Although life nowadays is mainly a closed-cycle process involving creation of biomass with reduced components which are oxidized again in the decay phase of individual organisms, as stated before, a small fraction gets fossilized in carbon-rich deposits finely dispersed in sediments or stored in thick layers of fossilized metamorphosed biomass like coal seams, trapped methane-rich gas, or oil.

These fossilized biomass forms, accumulated over billions of years, are the main source of energy of present-day human society. By combustion of these stocks these reduced components are reoxidized by atmospheric oxygen and release energy in the form of heat which originated from the sun's radiation many millions of years ago.

A similar process of enhanced oxidation by fire can of course be carried out with biomass of recently grown plants. Such processes occur naturally, for example, in fires of forests ignited by lightning or other events or on purpose to clear land. At the dawn of mankind possibly more than a million years ago [4] our ancestors learned to control this process and to use it to their advantage, generating heat at will, modifying materials, cooking to improve food, making better tools and weapons, etc.

At present the use of biomass for these purposes is reduced at the expense of fossil fuels and some other energy sources, but in the future, it will probably, to a certain extent, increase in importance again in refined and modern forms available or to be developed by our civilization.

Like mentioned before, in the early days of mankind biomass was mostly used in heat generation applications for fireplaces to keep warm, for cooking, for tool making, for materials and pottery, etc., and not yet for mechanical energy (e.g., motion) generation.

These needs for mechanical energy generation were fulfilled by human muscle power (including those of slaves), from animal power from horses, oxen, donkeys, etc., and from the ninth century [5] on by wind and water mills. In classical times possibly some power was generated from biomass combustion via steam heat generation, for example, for opening temple doors or blowing a horn. Even some kind of rotating steam engine or turbine based on steam outflow momentum [6] (see Fig. 1.3) was created but these were rather curiosities demonstrated in temples than useful devices.

With the first effective steam engines at the beginning of the 18th century of Thomas Newcomen [7] and later of James Watt [8] and of George Stephenson [9] (see Figs. 1.4, 1.5, and 1.6), generation of power from combustion of fuels, opened the way to large-scale mechanical power generation both in stationary applications and for transport purposes. This was further enhanced by the conversion of mechanical power into electric power by the invention and widely spread application of dynamos and electric motors (see, for example, Fig. 1.7).

Figure 1.3   Early steam turbine or aeolipile described by Hero of Alexandria in the first century AD. Water is heated by a wood fire and steam rises through the vertical tubes. Due to the momentum of the jets of outflowing steam the "ball of the god of the wind" rotates.

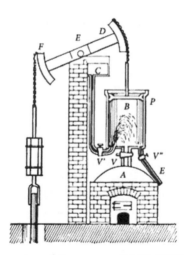

Figure 1.4   Steam engine of Newcomen (1663–1729), an early example of production of mechanical energy from combustion of solid fuel. Steam from boiler A is passed into cylinder B, pushing the piston in the upward direction. Subsequently, cooling water is injected, creating a low pressure under the piston, which is driven back by the atmospheric pressure.

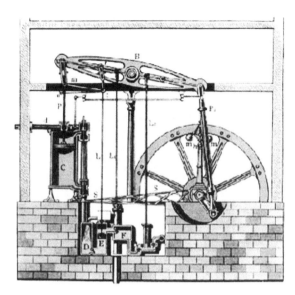

Figure 1.5 Steam engine of Watt (1736–1790). Watt improved the steam engine by introducing, among others, a separate steam condenser. In this way the cooling down of the working steam cylinder could be avoided. The efficiency was thus improved by a factor of 5 or more.

Figure 1.6 The Rocket (1829) was an early and practical example of a steam locomotive, created by Stephenson (1781–1848). Combustion of solid fuel–based mechanical energy generation was successfully introduced in the transport sector.

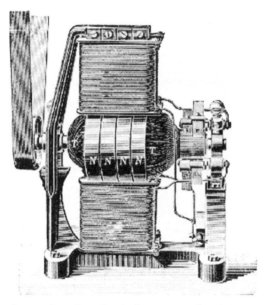

Figure 1.7 Conversion of mechanical energy to electricity got momentum by the development of industrial-size generators (called dynamos) in the second half of the 19th century like the Siemens Hefner–Altenecksche dynamo machine.

Internal combustion engines became more and more efficient in the second half of the 19th century. In 1862 Nikolaus Otto, on the basis of the work of Lenoir [10], produced and sold stationary motors with internal combustion running on lighting gas and later on liquid fuels.

Toward the end of the 19th century Rudolf Diesel introduced his internal combustion engine with compression ignition and after early experiments with biomass, fuels like peanut oil, animal fat, fossil fuels like powdered coal, and petroleum fractions were used to generate mechanical energy by internal combustion with compression ignition at high efficiencies.

This book is devoted to the conversion of biomass into heat, mechanical power, electricity, and fuels. Some attention is given to the production of chemicals. This first chapter will be focused on the present and future role of biomass as a primary energy carrier which is derived from a renewable and almost infinite source: the sun. While biomass can sometimes be used almost without pretreatment in combustion processes, for example, wood, dried dung, or waste combustion, for reasons of transport, storage, and efficient combustion in modern mechanical and/or electrical energy-generating equipment it is often preferred to convert the crude biomass into more convenient solid, gaseous, or liquid energy carriers. These processes will also be shortly discussed in this chapter.

1.1.2 Biomass amidst Other Sources of Energy

It is not easy to answer the question, What is energy? Physicists discovered [11] that the notion of energy provides a very handy yardstick to quantify many things which we want to do with sources in nature to suit our needs, like keep us warm, cook our meals, change materials, etc., and to provide work and power, with the latter muscular power, for the many things we want to realize.

The notion that energy can be present in many interchangeable forms like (sensible) heat, mechanical energy, kinetic and potential energy, radiation, etc., and the law of conservation of energy allow for a proper book keeping of energy flow accumulation and storage in closed and open systems (see the first law of thermodynamics).

The experience showed that in all processes in nature or invented by humans the quality of the energy, as measured by its usefulness and all-roundness for applications, deteriorates until all energy reaches the state of low-temperature heat. In the final stage this is ±2.7 K, the cosmic background temperature of our universe.

This experience has been translated into the second law of thermodynamics and concepts like exergy (see, for example, Ref. [12]). When discussing energy sources we normally mean sources from which we can generate high-value energy like high-temperature heat, electricity, mechanical energy, etc. If we consider, for example, an electricity-driven heat pump for heating our house, we need a large reservoir of low-temperature heat like the soil or the air from which we would like to pump up the heat to, for example, room temperature. We consider with the term "primary energy sources" usually only the high-quality energy sources from which electricity was produced.

In Fig. 1.8 an overview is given from the original sources from which mankind uses its energy.

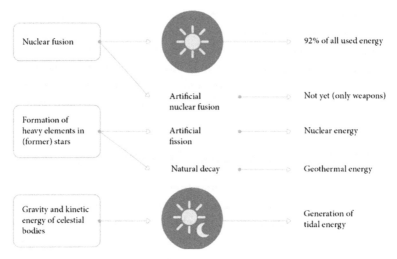

Figure 1.8   Original sources of all our energy.

In the overwhelming majority it is derived from the radiation received from the sun, in its turn based on the fusion of hydrogen nuclei to helium nuclei (see Fig. 1.1). The enormous energy released from this fusion reaction corresponds to a small disappearance of mass from the sun. Mankind has been fascinated by this source of energy, getting so much energy for so little mass. Although fusion energy release in the form of a weapon, the hydrogen bomb, was established within a relatively short time after Einstein published his famous relation $E = mc^2$, getting a controlled, tunable, safe flow of energy from this nuclear reaction was and still is a hard nut to crack. Bringing this process to full use will at least still requires decades of years, despite the brave efforts of a worldwide multinational consortium in the ITER demo unit (500 MW) now under construction (see Ref. [13]).

In 1989 there was a sudden hope on simplification of fusion energy when Pons and Fleischmann [14] supposed to have created fusion-based energy release in a small amount in a simple electrolytic cell.

But these results could not be reproduced systematically and the idea was rejected by the mainstream science community as a mistake and only a small community nowadays is continuing this or similar work under the name of cold fusion or LENR encouraged by unexplained energy release from processes not yet understood. For the time being we have to use the fusion energy which comes or came to us by intermediates of solar radiation.

Another source of nuclear energy is the splitting of heavy nuclei, for example, U^{235} formed in former-generation stars, for example, by supernova, and caught in the formation of, among others, our planet. Also here an enormous amount of energy can be generated, corresponding to a small mass difference between the fuel (e.g., U^{235}) and the fission products. This form of energy has certainly met with large-scale applications. About 5.7% of the primary energy of mankind and 13% of the electrical energy generated are fission based [15]. With the advance of the so-called breeder reactors, producing future fuel from nonfissionable elements looks particularly promising by opening large (almost infinite) stores of energy.

However, the dark side of this source became more and more apparent: long-living radioactive waste and handling and storage of very dangerous materials. Safety (among others, Tsjernobyl), excessive and intensive danger of natural disaster consequences (Japan, 2011), proliferation of nuclear weapons, and targets for terrorist attacks, all are hindering future growth, and some countries have declared to abolish (e.g., Germany) or abstain (Italy) from fission energy.

Nuclear reactions also occur naturally in the earth's mantle and core and these contribute for about 60% (40% is still primordial heat from the formation period of the earth) to the heat flux from the inner core to the surface of the earth where this flux is ±0.087 watts/m^2 [16].

The overall temperature gradient in the earth's crust, normally about 25°C/km, indicates that relatively deep holes have to be drilled to exploit these geothermal energy sources (say 1–4 km, Ref. [17]) but local conditions due to volcanic activity and constitution of the deep layers of the earth's crust may represent special opportunities such as exploited in Iceland where 65% of the national energy consumption is provided by these geothermal sources [18].

Yet another source of energy is the gravity field where the slight displacement of heavily bodies in interaction with their kinetic energy can create accessible, energy for mankind. The gravitation forces of the earth, moon, and sun create tidal waves on the earth, and particularly in favorable places these can create large-flux water flows which represent a high density of kinetic energy. These can be explored as tidal energy plants as is already done, for example, in France in the Rance estuary, which has favorable conditions for tidal energy. The unit has a peak power rating of 250 MW and an annual output of about 600 GWh [19].

1.1.3 Flows of Energy from the Sun

As indicated in Fig. 1.8 the overwhelming part of our energy is passed to the earth in the form of solar radiation. It can be exploited by mankind in different secondary energy flows or stocks. While the flow of solar energy toward the earth is enormous, it is spread out over a large area (average 1361 watts/m^2) and not always available. Generally the human desires are to concentrate the energy flows and to get uninterrupted flow of energy at will. An overview on the different routes to realize this is given in Fig. 1.9.

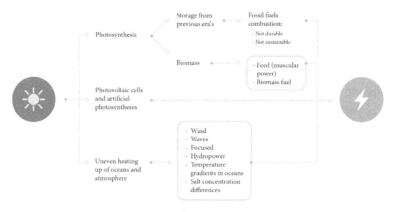

Figure 1.9 Different routes of solar radiation to power production by mankind.

The important key process toward the realization of these human desires is the photosynthesis of plants. Over billions of years photosynthetic life forms have created biomass from mainly CO_2 and water under the influence of solar radiation, while releasing free oxygen and producing reduced compounds such as carbohydrates (reduced relatively to CO_2). Reoxidation via a complex system of internal reactions provide plants (and animals which consume plants) with energy at the required moment. Most of the biomass produced will be reconverted to CO_2 and water, using oxygen, by activity of organisms living on decaying biomass and/or by natural oxidation processes. But as said earlier a relatively small fraction of biomass escaping from this process gets fossilized and is transformed step by step into fossil fuels: grades of coal, oil, and natural gas. Now at a high rate consumed by mankind it represents almost 81% of our energy consumption (2012, Ref. [20]). However, the consumption rate of fossil fuels by far exceeds the rate of formation of new fossil fuels and is therefore not durable or sustainable. This is the more so as the human population on the earth is expanding and there is a rightful desire to increase the wealth and thereby the energy consumption of the poor people.

Instead of using fossilized biomass also recently grown biomass can be combusted as was already done by our ancestors and predecessors who used and manipulated fire. Nowadays about 12% of the worldwide energy production is based on combustion of biomass itself. The overwhelming majority of this biomass is used not in industrial equipment but in stoves and wood fires used by poor people for cooking meals or keeping themselves warm. It is often their principal or only source of energy. Worldwide more than 2.5 billion people depend on it [21]. While also biomass produces CO_2 at combustion the overall effects are different from those of combustion of fossil fuel. In fact the CO_2 from biomass has been absorbed by the plants recently. The uptake and release of CO_2 by biomass, whether combusted or biologically decomposed, are in balance on a relatively short time scale, usually a few years or less. Therefore the growth and use of biomass can, in principle, be considered to be an almost ideal recycle process with the input of solar energy as a driver. Nature provides here the example for mankind. This is in contrast with the role of CO_2 from fossil fuel. Here additional carbon in the form of CO_2 is brought into the atmosphere which does not fit in a recycle stream with a short cycle time and therefore leads to accumulation of CO_2 in the atmosphere. Biomass has of course other functions than fuel for combustion and competition with food and feed is a factor to be considered carefully. This will be discussed later and at other places in this book in more detail.

It should be realized that biomass is also the source of feed for animals like cattle which provide human food and which are also traditionally used to provide mechanical power (like horses, oxen, dogs, goats, etc.).

Solar energy can also be collected and directly converted to electrical energy by photovoltaic (PV) cells and in a separate volume of this series the history of PV and its application and future is vividly described [22]. It does not interfere with the carbon cycles and has a large potential for dematerialization of the energy generation. A high relative growth of this form of energy generation has been sustained now over decades and the obvious advantages of this type of energy generation will promote its further application.

Cost of installed solar cell systems will certainly be brought down and the intermittent nature of the direct solar radiation coped with by an efficient and economic storage cycle of electrical energy (see Table 1.1), a so-called smart electricity grid and possibly by long-distance transport of electrical energy.

Table 1.1   Electric energy storage

Electric energy storage	Remarks
Pumped hydroelectric	In use at suitable locations, large scale
Compressed-air energy storage	Some use in suitable locations, large scale, related to use of turbine
Fly-wheel energy storage	Short time frame, quality control of electricity
Supercapacitors	Short time frame, quality control of electricity
Superconducting energy storage	Short time frame, under development
Batteries: lead acid, sodium sulfur, vanadium redox flow batteries, NiM-hydride, etc.	Can be used on different scales and for off-grid applications
Hydrogen via electrolysis, storage of hydrogen, and power generation, for example, via fuel cells	To be used on different scales and for off-grid applications (emerging).
Introduction of other elements like carbon (next to hydrogen) to produce alcohols and hydrocarbons	Not yet developed, off-grid applications, for example, for the transport sector

Considerable surface areas will have to be covered by solar cells to collect the desired quantities, depending on the average local influx of solar radiation. While, a lot of progress has been made in the efficiency of solar cells, both in research situation and in commercially available units, reduction in capital cost and integration in smart grids will be essential.

Some people dream of an all-electric society but as explained by Groeneveld, liquid fuels have advantages in storage, transport, and utilization (see Ref. [23]). Examples are long-distance truck transport, aviation, shipping on the oceans, etc. See also Table 1.2.

Renewable sources could also work together with a lot of synergy. As is suggested in Fig. 1.10, it could be advantageous to use biomass for the collection of CO_2. If for example agricultural waste would be combusted using pure oxygen almost pure CO_2 can be produced via this oxy-combustion in one step.

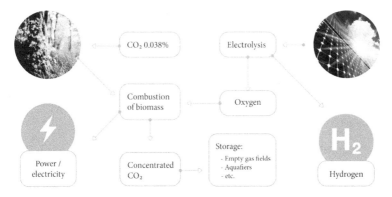

Figure 1.10 Combining renewable energy sources can have important synergetic effects. Electricity from renewable sources such as PV or wind can produce via electrolysis of water hydrogen and pure oxygen. Using the pure oxygen for combustion of biomass in a process called oxycombustion, concentrated CO_2 is obtained, which can be stored, thus effectively withdrawing CO_2 from the atmosphere. Alternatively the CO_2 can be used in combination with the hydrogen for the production of synfuels without the use of fossil carbon. The pure oxygen could also be applied in biomass oxygen–steam gasification delivering syngas (carbon monoxide and hydrogen mixture) for synfuels or chemicals.

The required oxygen would be a by-product of the electrolysis process based on electricity from renewable sources like wind and PV which produced the hydrogen. On top of that the additional solar energy in the form of the biomass heat of combustion becomes available. Similarly via oxygen gasification of biomass also syngas ($CO + H_2$) can be produced instead from which liquid fuels can be synthesized.

Table 1.2 Properties for transport and storage of different power sources

	Energy (MJ/L)	Density (MJ/kg)	Effective storage time
Gasoline	32	43	Years
Methanol	17	21	Years
Ethanol	22	28	Years
Hydrogen liquid (without container)	8.4	120	Weeks
Hydrogen 350 bar (without container)	2.8	120	Years
Hydrogen hydride**	13	6	Years
Pb-acid battery **	0.24	0.12	Weeks
Li-ion battery **	1.2	0.84	Weeks

** Typical

Losses in transport over long distances

	1000 km	5000 km
Oil	<1%	1%
Gas	2%	10%
Electricity (800 V)	6%	21%

Liquid fuels have some advantages over electricity in storage, transport, and transportation.

In transportation this is most clear in aviation, truck transport, shipping, etc. (from Ref. [23]).

In Fig. 1.11 the PV cells are qualitatively compared to biomass energy production and there is a growing interest in going one step further in imitating the natural photosynthesis process by producing not only electrons from photons but also energy-rich molecules in solar cells (see, for example, Ref. [24]).

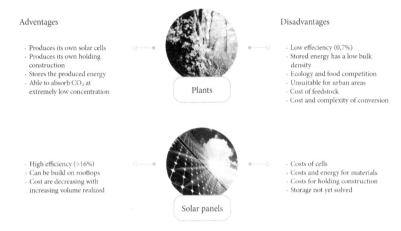

Adventages

- Produces its own solar cells
- Produces its own holding construction
- Stores the produced energy
- Able to absorb CO_2 at extremely low concentration

Plants

Disadvantages

- Low effeciency (0,7%)
- Stored energy has a low bulk density
- Ecology and food competition
- Unsuitable for urban areas
- Cost of feedstock
- Cost and complexity of conversion

- High efficiency (>16%)
- Can be build on rooftops
- Cost are decreasing with increasing volume realized

Solar panels

- Costs of cells
- Costs and energy for materials
- Costs for holding construction
- Storage not yet solved

Figure 1.11 Comparison of biomass production with photovoltaics on several aspects.

At present the production of solar fuels from artificial photosynthesis introducing CO_2 as a carbon source is still facing a lot of technical problems and in any case there would probably be a difficult competition with processes introducing separate electrolysis from PV-based electricity to produce hydrogen from water. This electrolysis process is already available and can have a high efficiency (up to over 85%), although the investment cost is still high. Possibly a similar electrolysis process can be developed to produce syngas ($CO + H_2$) from mixtures of CO_2 and water. Alternatively the required CO in syngas can be produced by reduction of CO_2 by part of the produced hydrogen (so-called reversed shift reaction $H_2 + CO_2 \rightarrow H_2O + CO$).

Syngas can be converted to high-quality fuels, such as methanol, ethanol and hydrocarbons, the latter via well-proven processes such as the Fischer–Tropsch synthesis. These syntheses can be used for hydrocarbon transport fuel production and for chemicals based on renewable energy.

Of course, also in these cases, next to hydrogen, CO_2 will be necessary, which should ultimately be recovered from the atmosphere to close the carbon cycle. This can be done via biomass production or even with chemical sorbents. In recent years some R&D is going on to find processes suitable for direct absorption of atmospheric CO_2 (see, for example, Ref. [25]).

As indicated in Fig. 1.9 uneven heating up of the atmosphere and oceans is the cause of wind and collecting wind energy is an important way to collect solar energy, nowadays a fast-growing business. Separate volumes [26, 27] of this series of books are devoted to the history and technology of wind energy production. As with the other renewable sources the intermitted and highly variable character of the wind requires storage of electrical energy or smart grids backed up with other primary energy sources.

Waves in seas and oceans generated by wind are also indirectly derived from solar radiation (see Fig. 1.9) and can be a source of energy provided that machines can be made at acceptable cost which can collect the energy associated with the surface waves of seas and oceans, especially in "hot spots" where waves are frequently strong and persistent. Figure 1.12 gives an example of such devices. Wave energy is not yet widely applied and still in its infancy. The potential is not yet clear.

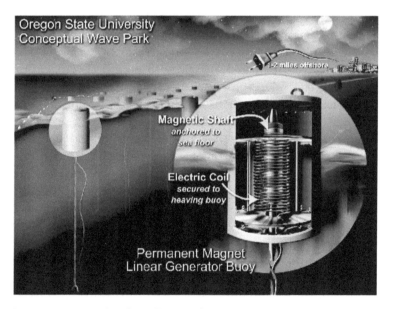

Figure 1.12 Example of a collector of wave energy assembled in a wave park (Oregon State University).

A substantial part of the solar energy used by mankind is heat. This can be passive heating of, for example, houses and stored for later use or it can be applied for driving domestic air conditioning.

Via the uneven heating up of oceans and atmosphere also mechanical energy can be generated via Carnot cycles. This uneven heating up of the atmosphere can of course be strongly enhanced by solar radiation concentrators in the shape of troughs and parabolic mirrors. A very early example is shown in Fig. 1.13. Nowadays large-scale energy generation facilities based on this principle have been realized and for areas with suitable solar radiation the potential can be high if costs can be further brought down and especially if storage of heat carriers or electricity can be provided.

Figure 1.13 Focusing solar power plant producing steam driving a printing press at the World's Fair, Paris, 1878.

Hydropower is based on water evaporated by solar heat from the seas and oceans and precipitated in higher areas and mountains where it represents gravitational potential energy to be exploited on its flow back to the sea. Normally water is collected in reservoirs and electricity is generated in turbines through which the water is passed. Especially in large turbines and generators the efficiency of conversion of potential energy of the water to electricity can be very high (90%). Therefore the system is also used for energy storage by pumping water up to the reservoir again. Turnaround efficiencies can be high and this system is therefore attractive for storage. About 16.4% of the electricity world production is generated via hydropower and the only disadvantages are the lack of suitable places, the cost of realization, environmental concerns, and the use of space for the reservoirs.

Vertical temperature gradients in the ocean can be another source of solar-derived energy. Here evaporation of a suitable solvent in a closed system (e.g., ammonia) can be used to drive a turbine after which the vapor is condensed in deep water with a heat exchanger and pumped back to the evaporator. The efficiency is limited by the Carnot efficiency, which, due to the relatively small naturally occurring temperature differences (about ±25°C), cannot be high (max. 7%). At the moment with the exception of a few small-scale facilities the technology is not yet used and the potential for electricity production is still unclear. Several integrated projects aim at products other than electricity (like freshwater from desalination).

Differences in concentrations of salt in freshwater and seawater, as occurs, for example, where rivers flow into the sea, can be exploited by utilizing the osmotic pressure difference or creating electricity via reversed electrodialysis (see for example Ref. [28]). In both cases large membrane areas are required to create sufficient water flow through the membranes separating the different salt concentrations. Collecting solar energy via this way may become attractive but is still in its infancy.

1.1.4 Future Energy Generation and Consumption

The current, man-manipulated energy consumption, is about 500 EJ/year. This means for each human a flow of 2.3 kW. The future consumption can be expected to be considerably higher.

Today's energy consumption is far from uniform and to a large extent associated with the state of development (Fig. 1.14).

Projections based on scenarios do vary considerably depending on the assumption on which they are based.

Michiel Groeneveld [23] used a method based on the estimates of future consumption of energy in the basic needs of people in the future (see Table 1.3). He considered the sectors food, materials, domestic needs (cooking, heating, communication), and transportation and introduced the 4 kW man of the future.

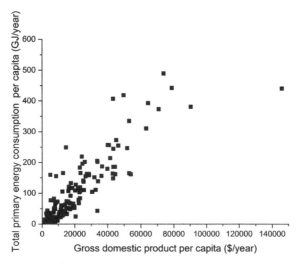

Figure 1.14 Gross domestic product per capita versus energy consumption rate per capita for different countries. http://en.wikipedia. org/wiki/List_of_countries_by_GDP_(PPP)_per_capita; http:// en.wikipedia.org/wiki/List_of_countries_by_total_primary_ energy_consumption_and_production; http://en.wikipedia. org/wiki/List_of_countries_and_dependencies_by_population.

Table 1.3 Estimated future energy needs of an average human person

Need	Energy	Required for
Food	1 kW	• Agriculture (fertilizer, irrigation, tractors, etc.) • Food processing • Food cooking
Domestic	1 kW	• Heating cooling applications • Communication, computer, Internet, telephone, TV
Materials	≥1 kW	• Manufacturing steel, cement, metals • Plastics, chemicals, paper, wood (raw materials and energy)
Transportation	≥1 kW	• People and freights
Total	≥4 kW	

At present the average US individual consumes ±12 kW and the European individual ±6 kW. A world population of 10 billion people consuming 4–5 kW each would result in a yearly energy consumption of about 1300 EJ, which is more than 2.5 times the present consumption rate. When this situation will be reached is uncertain but this could probably not be 2050 if you compare the analysis of Groeneveld with different scenarios based on different assumption methods.

The actual future energy consumption is difficult to predict but this is even more so for the distribution over the sources from which the energy supply will come: especially the role of renewable energy in the spectrum of sources remains in debate and may vary widely from publication to publication. Figure 1.15 is the projection of Shell for their somewhat older scenario "Scramble" [29]. In this scenario, consumption of other renewables and coal are high, but despite this the total energy consumption in 2050 is still much lower than needed for the 4 kW society of Groeneveld. This is the more so for the more modest Shell scenario "Blueprints."

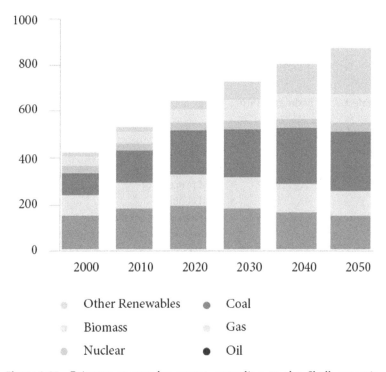

Figure 1.15 Primary energy by source according to the Shell scenario "Scramble" [29].

The distribution of the biomass utilization in different sectors indicated by the Shell scenario is shown in Fig. 1.16.

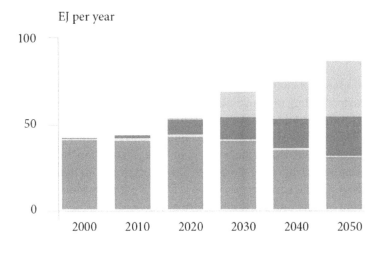

Legend:

- Biofuels - 2nd generation   Electricity
- Biofuels - 1st generation   Traditional

Figure 1.16   Final energy consumption using biomass in the Shell scenario "Scramble." The role of biomass next to traditional is limited in this scenario to biofuels and hardly any biomass is used for electricity generation [29].

In fact the future of the world's energy production and consumption pattern will depend a lot on success or failure of the present technological and socioeconomic developments. Moreover, future development will depend on new sources of energy, yet to be discovered or unexpected breakthroughs in the generation of energy such as for biomass, solar cells, and energy transport like smart grids, electricity highways, etc.

In many scenarios the role of fossil fuels remains large, and although the distribution of fossil fuels over the different kinds like coal, oil, and natural gas may substantially change due to, among others, increased reserves of natural gas related to new recovering methods like fractioning, etc., the overall role of fossil fuels will stay dominant unless radical changes are made.

As stated before for some transport sectors like aviation and long-distance truck transports it is difficult to imagine how the present fossil fuels can be replaced if not by other carbon containing fuels like biofuels or solar-based synfuels.

1.1.5 Fossil Energy and CO_2 Emission

Abundance of energy will be required in the future among others for future recycling of materials and elements. Moreover to protect the environment, additional energy will be required.

Production of energy from fossil fuels represents in itself an environmental threat because it involves extraction from carbon components from the earth which have been stored for many millions of years in the earth's crust to be released within a geological short period of time into the atmosphere as CO_2.

Figure 1.17 shows the increase in atmospheric carbon dioxide with time, indicating that especially since the Industrial Revolution with its associated increasing use of fossil fuels, the CO_2 content of the atmosphere increased by 40%.

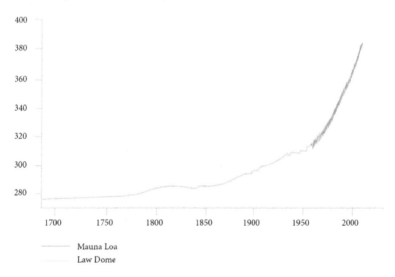

Figure 1.17　Atmospheric concentration of CO_2 derived from ice core measurements obtained at Law Dome, East Antarctica, and from measurements on Mauna Loa, Hawaii. From the start of the Industrial Revolution the CO_2 concentration has risen from 280 to 394 ppm, an increase of 40%.

It is generally assumed that the increase of CO_2 in the atmosphere affects the heat balance of the atmosphere by interference via the infrared radiation absorption/emission behavior, resulting in a slowly increasing averaged atmospheric temperature. More and more people (including those of ICCP) believe that this is happening because of several factors but with a leading role for the increased CO_2 concentration due to fossil fuel combustion and deforestation. UNFCCC assumes that a maximum increase of the average atmospheric temperature of 2 K can possibly be allowed before irreversible and unacceptable effects occur [30]. While these matters are still under debate the precautionary principle will dictate an important action in reducing the extra input of CO_2 in the atmosphere. The situation can best be explained from the carbon balance over the atmosphere.

Figure 1.18 gives a simplified version. The total stock of fossil carbon in the earth is very large, here indicated as 10,000 Gton. Fortunately this is not all readily accessible. Mankind extracts about 8 Gton of carbon in the form of coal, oil, and gas, which will end up as CO_2 in the atmosphere and is about in equilibrium with surface waters of lakes and upper layers of seas and oceans. Biomass on all continents, and in the surface compartments of oceans and seas with an estimated total hold-up of 1000 Gton carbon, is exchanging about 80 Gton carbon in the form of CO_2 (80 Gton carbon corresponds to 293 Gton CO_2) with the atmosphere which is almost in equilibrium with the surface waters with respect to CO_2. These compartments contain together a total of ±1750 Gton carbon. The largest stock of carbon, however, is the deep oceans but the carbon exchange rates with the deep oceans are relatively small. It would take hundreds of years (or maybe thousands) before carbon equilibrium of atmosphere and surface waters with the deep ocean would be established at this low exchange rate. Also carbon withdrawal from the biosphere biomass in the form of fossilization is relatively small compared to the exchange of all biomass with the atmosphere.

Looking at these figures of hold-up and flow we could make a much simplified estimate that at the current rate of absorption of CO_2 by biomass from the atmosphere and surface waters the time to reduce the amount of carbon in the atmosphere to half its present value would be about 15 years, but of course this loss of carbon is in reality compensated for by an equal flow of carbon in the opposite direction from the decay of biomass.

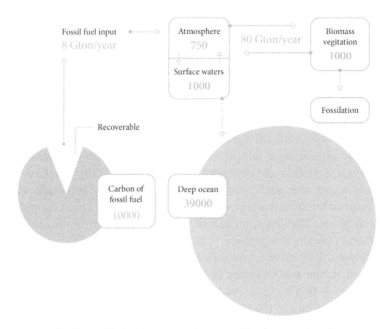

Figure 1.18 Simplified global carbon balance. The flows and exchange rates of carbon are presented here in Gton/year of carbon, which may be present in CO_2, biomass, etc. Stocks for hold-ups of carbon are given in Gton present in fossil fuel, biomass, surface waters, etc. Atmosphere and surface waters are assumed to be (almost) in equilibrium with each other. The carbon flows to the deep oceans and to fossilization are relatively small.

However, the inflow into the biosphere of carbon in the form of CO_2 from combustion of fossil fuels is not compensated for and this 8 Gton/year carbon accumulates in atmosphere and surface water. Neglecting changes in biomass hold-up in the biosphere and other minor sources of CO_2, the whole increase of CO_2 in the atmosphere from the start of the industrialization to present, could be ascribed to the input from fossil fuel combustion of the last 200 years. This would represent a clear mark of the presence of mankind on earth showing an increase of the CO_2 content of the atmosphere by 40%. Although the CO_2 concentration by now is only about 400 ppm (~0.04 vol.%), CO_2 has an important influence on the so-called greenhouse effect which produces the moderate average temperature on earth. Climate models try to explain this greenhouse effect and its enhancement by the increased CO_2 content and that of other greenhouse gases like methane, nitrous oxides, etc. While these models are disputed and debated there is a general concern about the average atmospheric temperature rise in the coming decades. As stated before this temperature increase should be limited to 2 K to avoid large-scale (irreversible) effects like icecap melting, seawater-level rise, extreme weather, etc. The aim is to get this realized by reducing the yearly input of CO_2 of fossil origin into the atmosphere. Climate conferences and agreements on reductive measures have been arranged to this end but the effective emission reductions are relatively small up to now. In view of the increasing future energy consumption and difficulty to switch to renewable energy production on a large scale this CO_2 emission reduction is quite a challenge for mankind. On the long term, drastic reduction of CO_2 emission from fossil fuel utilization will be required and, possibly, even recovering of accumulated CO_2 from historic emission could be requested by some kind of planetary control/engineering.

1.1.6 Why Biomass as a Source of Sustainable Energy

While it is clear that despite attempts to increase the energy efficiency, energy demand may at least double in the coming decades, no single renewable energy source alone can provide sufficient energy to bridge the gap between the need and the provision of energy. Also biomass will play its role in the future. Its advantages and draw backs are summarized below.

Biomass is renewable and can be a sustainable source of harvested solar energy. A special advantage is that biomass is a carbon-based fuel which fits into the present mainly carbon-based (fossil) fuel infrastructure. Because biomass energy is based on a short carbon cycle, it is essentially CO_2 neutral if none or little fossil energy is used at its production. In case the CO_2 would be recovered and stored after combustion of biomass it can even be a sink for atmospheric CO_2.

Unlike some other energy sources the production of biomass is global and not monopolized by a small number of countries. Biomass energy is available in the form of a combustible fuel and so inherently it is in a form suitable for storage, in contrast to other sources like PV or wind electricity for which storage systems have to be provided.

However, biomass also has some shadow sides. It often is primary available in the form of a low bulk density solid, especially in the case of agricultural and forestry waste. It will need some preparation (like pelletizing, chipping, torrefaction, etc.) or conversion (pyrolysis, gasification, liquefaction, etc.) before it can be stored, transported and used in modern energy conversion equipment like motors, turbines, fuel cells, etc. Compared to fossil oil, coal, and natural gas, biomass has a high oxygen content (say 50%) while in most fossil fuels there is little or none oxygen present. The oxygen content of biomass lowers its combustion value to roughly half of its fossil counterparts. Oxygen can be removed in pre-combustion conversion processes and especially for some product applications or fuel blends, some oxygen may be an advantage, for example, to counteract soot formation but it is mostly a neutral diluent. Generally the original high oxygen content of biomass, complicates the application. Another potential disadvantage of production of energy from biomass is the competition of the use for energy of biomass with important other applications as food or feed. Increasing prizes, for example, for maize or sugar or palm oil for food have been already ascribed to increased demands for fuels such as alcohols for gasoline or diesel component from vegetable oils.

An important source of biomass, however, is agricultural or forestry waste and by-products/wastes of food and feed preparation. In these cases there is no competition for food or feed.

In the cultivation of biomass there may be several ecological issues such as degradation of biodiversity and inorganic material balances like for phosphate, bound nitrogen, and other elements. Moreover water requirement and energy used for the cultivation process (tractors and other machinery) have to be taken into account.

1.1.7 Is There Sufficient Biomass Available?

At present 12%–13% of the world's energy production is generated from biomass. This utilization is almost entirely for traditional applications by poor people, like for cooking or heating, and refers to firewood and dried dung combustion or fermentation for biogas production, etc. The technology used is typically outdated and produces health-threatening emissions. The energy efficiency is very low but for more than 2 billion of humans it still represents their sole access to energy carriers. The utilization of biomass for power generation and heating in modern equipment represents only a few percent of present-day biomass use. The potential of biomass for energy generation is still a matter of debate. We can distinguish three different types of sources: agricultural and forestry waste, existing dedicated fuel crops, and potential future crops like algae.

The agricultural and forestry waste represents an important potential, estimated to correspond to about half the world's crude oil production [23]. The advantage of its use is that, as it forms a by-product of human activities to produce food and materials and "nature" there is no need for new land. This does not imply that it is an easy and uncomplicated resource. In contrast to fossil fuels which become available at concentrated spots, biomass (waste) production is distributed over large areas and is generally in its raw form, a low-bulk-density solid difficult and expensive to handle, transport, and store. Some kind of in-field or pretransport treatment is required. Dedicated biomass energy crops form already nowadays a rapidly expanding business. For liquid fuels crops like sugarcane and maize for alcohol production and palm oil for diesel components it is already a multibillion business for countries like Brazil, the United States, Malaysia, etc.

Production forests and wood thinning are for several countries like Canada, Sweden, Finland, Russia, and the Baltic states an intensively exploited source of energy next to high-value materials.

To what extent all these resources can be expanded in the future is yet unclear and will depend on many factors, both promoters and obstacles, summarized in Figs. 1.19 and 1.20.

It will probably have to involve new and high-production species, possibly especially dedicated to energy production or multiple-product plants. To further avoid competition for land, saline land, and marine production systems could be developed and possibly in water cultivation of biomass, for example, with high productive algae. For stable international trading and transport well-developed systems should be created and international partnerships along the whole line of production toward final use. Stable, sustainable, and renewable agricultural production should be in place, allowing for mineral retention or return to the soil. All these factors should be backed up with effective international "green certification" systems

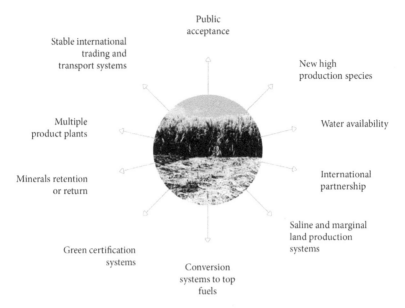

Figure 1.19 Facilitators for large-scale biomass production.

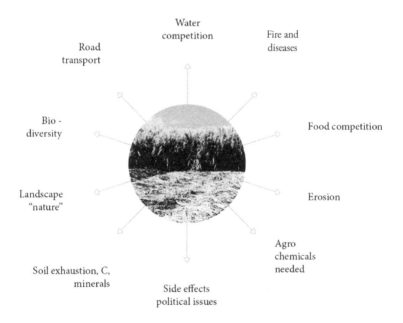

Figure 1.20 Barriers to large-scale biomass production.

to support public acceptance of biofuel chains on the long term. In Chapter 3 a few expectations of worldwide production of biomass for energy, including that from waste, are discussed. Whether high or low expectations will actually be realized will depend on the success of dealing with several obstacles or challenges like biodiversity, questions of landscape or nature, soil exhausting from carbon and minerals, erosion, the agrochemicals needed, water competition, danger of fires and diseases, increased road transport, food competition and several side effects, and political issues. Moreover the technical issues and developments will play a major role.

1.2 Conversion Processes for Different Kinds of Biomass to Advanced Energy Carriers

1.2.1 Introduction

In this section we will present a bird's-eye view of the different processes available or under development to transfer the chemical energy contained in "raw" biomass to other refined energy carriers adapted to modern equipment for storage, transport, and utilization of these carriers.

Within the framework of this chapter we will only touch the principles as the scientific, technical, and commercial literature is covered by a wide range of specialized handbooks and possible hundreds of thousands of relevant publications. Nevertheless we will try to elucidate the most important items, rather in pictures than in numbers or equations, and make it accessible as far as possible for the nonspecialist. Some of the items are treated more in depth in the subsequent chapters by renowned specialists or we will refer to handbooks or other publications.

With the word "biomass" we mean a wide range of materials produced by life and if we restrict ourselves to biomass meant for energy production there is still a large variety of raw materials. Figure 1.21 gives an impression of a few of these raw materials. Only in exceptional cases the biomass can be used without further processing and normally pretreatment and further conversion is required.

Figure 1.21 Typical biomass feedstocks. 1. Wood chips; 2. wood pellets; 3. demolition wood; 4. rice husk; 5. straw; 6. *Miscanthus*, an energy crop, in the field; 7. thick suspension of microalgae; 8. empty palm oil branches; and 9. municipal waste where biomass products are mixed with many other constituents.

From a fuel's point of view the material content of biomass can be classified as organic components (like cellulose, hemicelluloses, lignin, fat, waxes, etc.) (for an overview see Fig. 1.22), ash components (the nonorganic components in the organic structure inside the biomass or dissolved in internal water (attached or included in the bulk), and the water content itself. If compared to fossil fuels the most striking differences with most types of biomass are the much higher oxygen content, typically 50% W and half or less of the combustion value.

The importance of the water content is illustrated in Fig. 1.23, where the lower heating value (LHV) (total or higher heating value minus the heat required to evaporate the water) is given for typical biomasses as a function of the water content. For a water content of more than about 80% the heat released in the combustion process is not sufficient to evaporate all the water and the LHV becomes negative. Of course the wet biomass can be dried first, for example, by solar energy or by recycling some of the heat of combustion or by drying techniques where the heat of evaporation is recovered and after upgrading recycled for drying purposes (vapor compression drying, etc.). However, drying of biomass can still consume a considerable fraction of its heat of combustion and adds to the complexity of any type of wet fuel combustion. Upstream solar and air drying in the production fields are generally to be preferred.

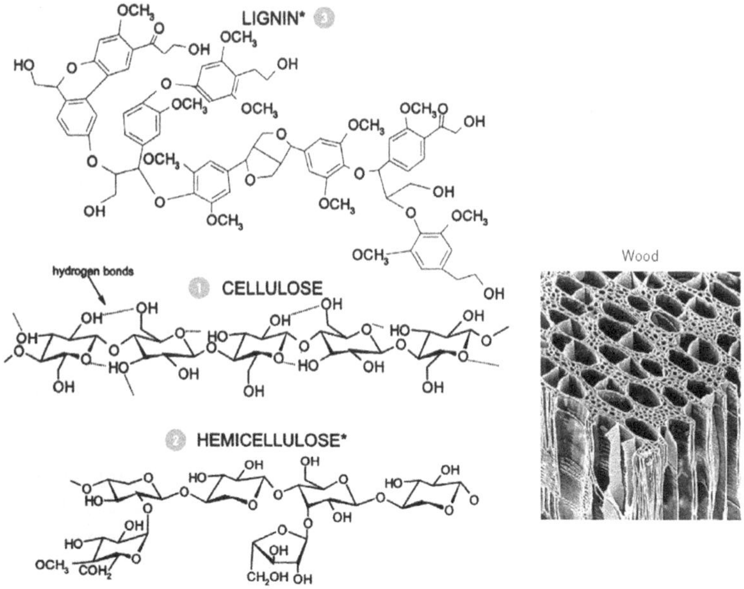

Figure 1.22 Composition of wood as an example of biomass composition. The three most important constituents are sketched here: 1. Cellulose, a polymer of glucose; 2. hemicelluloses, also a polymer of different sugars; and 3. lignin, a complex compound providing strength and stability to the wood structure. Different additional biomass constituents can be proteins, fats, etc.

In many cases it is desired to convert crude biomass to secondary fuels, better adapted to long-distance transport and/or application in modern energy converting systems. To this end many processes are available or under development. As explained in Fig. 1.24, in principle, nearly all products which can be made from fossil fuel can be made from biomass. An important decision aspect in the selection process for conversion to better fuels is whether the biomass feedstock is relatively dry or has a high or very high water content, rendering drying expensive and unpractical.

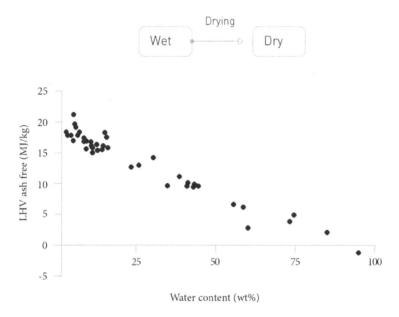

Figure 1.23 Wet biomass lower heating value (WLHV) as a function of the water content. At a water content of more than 80% the wet heating value (WLHV) becomes negative. Water should be removed with high energy efficiency to obtain a combustible fuel at such conditions.

In the case of wet biomass still several conversion routes are available, which avoids the evaporation of most of the water. These routes are schematically indicated in Fig. 1.25.

First there are the *biological routes* which produce secondary fuels in the water phase at moderate conditions with the help of microorganisms and/or enzymes. An example is manure anaerobic digestion, in which manure is partially digested to form a methane-rich gas under anaerobic conditions (absence of oxygen). The methane-rich gas, which separates spontaneously from the water phase, can be directly used as fuel gas or after purification to replace natural gas.

Figure 1.24 From the different generations of biomass feedstocks, in principle, all present energy carriers, chemicals, and carbon-containing materials, now produced from fossil raw materials, can be made. The first generation may be in direct competition with food and feed, the second generation represents waste and mainly lignocelluloses materials, the third generation is new organisms like algae, and the fourth generation uses solar energy and H_2O and CO_2, CO_2 collected from the atmosphere by plants.

Similarly some refined fractions of biomass, like sugar solutions, for example, prepared from sugar reed can be converted into (fuel) ethanol by yeast. Ethanol is separated from the water phase after removal of the yeast (by centrifuge) and further separation of the liquid mixture by distillation. Via azeotropic distillation and or/other dewatering techniques purified alcohol can be recovered. Current developments aim at processes where cellulose and hemicelluloses as present in lignocellulosic feedstocks (like wood or agricultural waste) are converted to sugars to be used in similar processes to produce alcohols or by chemical processes to other liquid fuel components.

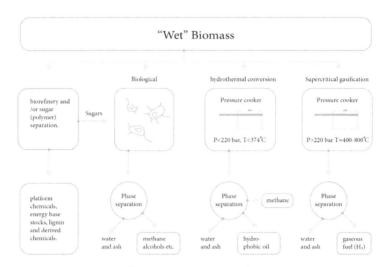

Figure 1.25 Overview of conversion processes for wet biomass.

A different process for producing fuels from wet biomass is *hydrothermal conversion* at high pressure and temperature (indicated in Fig. 1.24 by a high-pressure cooker). Under these high-pressure conditions (up to 220 atm.) and at these temperatures (up to 350°C) the water does not evaporate but remains in the liquid phase. At the reaction conditions biomass is ideally converted into a liquid hydrophobic oil (see Chapter 20) or, especially if a suitable catalyst is available, into methane which is simply separated from the water phase (see Chapter 21).

For a realistic continuous process of this route efficient heat exchange between inlet and outlet streams of the system are vital for this process. A third route operates at still higher temperatures (up to 800°C) and high pressures. Under these severe conditions the desired products are mainly hydrogen, methane, CO_2, and CO. The route is called supercritical gasification as at these conditions water does not form two separate phases (liquid and gas phase—steam) but is in a so-called supercritical state. To obtain a favorable overall heat balance, here also extensive (countercurrent) heat exchange is necessary between inlet and outlet streams of the continuously operated supercritical reactor.

If dry biomass is available, many conversion processes toward improved fuels are available or under development. Figure 1.26 gives a simplified overview.

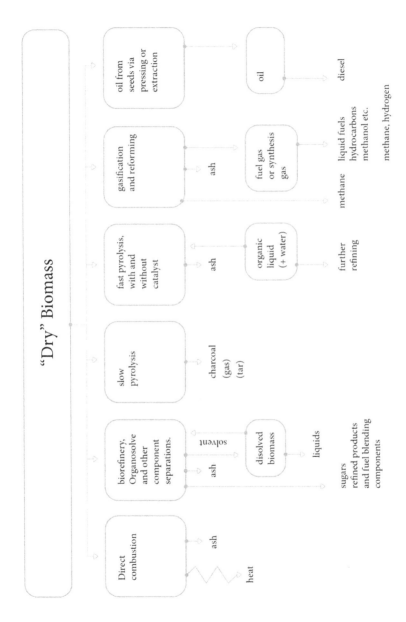

Figure 1.26 Overview of conversion processes for dry biomass for the production of improved fuels.

Direct combustion can be applied to generate heat or power and often some pre-conversion is applied, like pelletizing, size reduction (chips), torrefaction (low-temperature baking), etc., to improve the properties of the solid fuel.

Sometimes the biomass is first splitted up in its constituents: lignin and sugar polymers cellulose and hemicelluloses. This can be done among others with the so-called Organosolv process (see Chapter 9) where biomass is mixed with a solvent and lignin is separated. For liquid fuel production the sugar polymers can be hydrolyzed into the sugars to be converted by fermentation or chemical processes to liquid fuels.

Pyrolysis processes (pyrolysis = decomposition by fire) are applied to convert biomass into liquid, solid, and gaseous fuels. The actual conditions during the pyrolysis process, where biomass is heated up in the absence of air or gas phase molecular oxygen, determines which product type, liquid (pyrolysis oil), gas (CO, H_2, CH_4), or solid (char) is dominant. In Fig. 1.26 fast pyrolysis is indicated, meaning a pyrolysis process dedicated to liquids production but also processes rather directed to char coal (slow pyrolysis) or to fuel gases (high-temperature pyrolysis) are possible by changing the process operation conditions (see Chapters 14 and 15).

With gasification and/or reforming, biomass is converted into gas phase products by partial oxidation with oxygen or air to produce a combustible gas containing CO, H_2, CH_4, etc. The exact composition of the gas produced depends on the type of gasification process used however.

Processes for gasification of biomass are available or are being developed for a wide range of final applications. Products may vary from a simple raw combustion gas suitable for cofeeding in a gas furnace, to a methane-rich gas for substitution of or blending with natural gas. A special case is the production of synthesis gas also called syngas, a purified mixture of CO and H_2 from which liquid fuels like hydrocarbons (e.g., via Fischer–Tropsch synthesis), methanol, and ethanol can be synthesized. These subsequent processes are well established for syngas from fossil fuels and form an additional important future gate way for biomass to top liquid fuels like diesel, gasoline, alcohols, etc.

For completeness in Fig. 1.26 also the utilization of vegetable oils from oil seeds, or the (waste) animal oil and fats are mentioned. With relatively simple processes like transetherification or hydrogenation, excellent diesel fuel (components) can be produced.

All these processes for wet or dry biomass are here discussed separately. However, in future practice these processes may be integrated in so-called biorefineries (see Fig. 1.27) where crude or preprocessed biomass will be fed in. In its widest context such a biorefinery even food and feed leave as products together with a whole range of other products. At present the notion of a biorefinery is still in rapid evolution and diversification and may refer to a wide range of integrated operation on biomass, with biological, biochemical, chemical, and thermal steps.

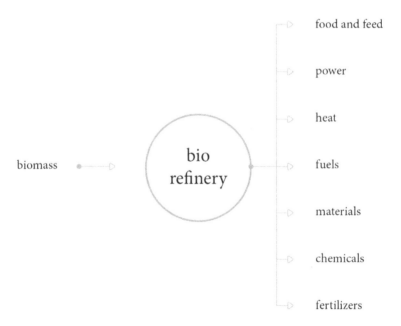

Figure 1.27 General concept of a biorefinery. Processes in the refinery may consist of a combination of physical, chemical, and biological conversion steps, somewhat like a mineral oil refinery.

In the following parts a short description will be given on the different conversion processes with some remarks on the key items in their specific stage of development.

1.2.2 Conversion of Dry Biomass to Advanced Energy Carriers

1.2.2.1 Combustion of solid biomass

While combustion of biomass with air is the oldest process used by man after gaining knowhow on how to control it, which could have occurred maybe more than a million years ago by early *Homo* species, it is still by far the most important biomass conversion process. It comes in a wide variety of technologies from the simple stove fire to highly advanced boilers, generating energy on a hundred megawatt scale. In the development of combustion equipment and integrated furnace/boiler systems, precise control of the mixing and contacting pattern between biomass (or other solid fuel) and the combustion air is optimized. In this way control on the combustion process is gained by manipulation of the throughput, control over the temperature profile, spatial combustion rate distribution and undesired by-product formation (NO_x, CO, tars, obnoxious smells, etc.). Moreover the controllable extraction of heat from the combustion zone and/or from the flue gases is taken into design and operation consideration. For most types of biomass the softening and melting temperature is often lower than for coal and this can be a matter of careful consideration. In many cases also removal or prevention of obnoxious substances like dust, smell, and other environmentally undesired components are integrated into the design. Finally noise abatement is regularly included.

Simple combustion of biomass for cooking or heating of houses using hardly any or very simple equipment is used by maybe more than two billion people and this often forms their only access to additional energy. This type of use of firewood (often collected by women) and dried dung probably represents up to 10% of the world's primary energy consumption. The efficiency of this type of biomass combustion is often low to very low and serious health issues occur related to inside air pollution. Many attempts have been made over several decades to design, operate, and mass-produce more efficient simple wood-fired cookers consuming only 10%–40% of the wood used by the simplest firing of wood under a cooking pot. Generally these efforts are hindered by investment costs, operating and production issues, and cultural barriers between well-intending development enthusiasts and the target group. A recent example of a potential successful wood stove backed by a large industry can be found in Chapter 6.

In some of the countries in Western Europe (like in Austria) direct firing of domestic central heating systems have met with a remarkable success (see, for example, Fig. 1.28). It concerns here modern, fully automated equipment consuming well-defined wood pellets (or chips) while the combustion process is well controlled automatically, among others, by lambda sensors (controlling excess air).

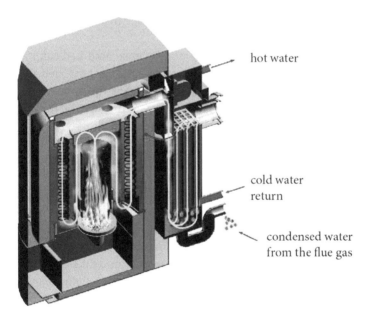

Figure 1.28 Automatically fed high-efficiency small-scale wood pellet boiler. The water in the flue gas is condensed and heat of condensation recovered, improving the overall efficiency. Capacity 8–20 kW. Adapted from the presentation "State-of-the-Art of Small-Scale Biomass Combustion in Boilers" by Prof. Ingwald Obernberger.

For industrial semi-industrial and centralized domestic heat and power generation, modern and older designs of stoker burners, underfeed stokers, and grate boilers are used for a smaller (from a few kilowatts up to a megawatt), intermediate (1–10 MW), or larger scale (>10 MW). Generally in modern and large-scale burners a lot of attention and inventions are directed to increase the efficiency of the combustion, the automation, and safety and to prevent obnoxious emissions to the environment in the form of gas-phase components, microparticles, smell and ash components, sound, etc., often driven by regulations and laws and enforced by local and governmental authorities. In recent decades the progress in combustion efficiency has been impressive for small-scale biomass combustors, from 55% (1980) to over 90% nowadays.

If (electric) power is to be (co-)generated people will, depending, among others, on the scale of operation, select a steam engine (seldom done nowadays) or a steam turbine to convert heat to shaft power.

For the combustion of biomass also dense fluid-bed combustors are selected (see Fig. 1.29). In these combustors a "sand" bed, floating on the up-flowing gases is created which has the appearance of a boiling liquid as much of the gas flow passes through the (expanded) "sand" bed in the form of bubbles.

This creates an intensive axial mixing and allows for a wide range of biomass fuel particle sizes and shapes to remain on the move in the bed without floating too much on the top or sinking and accumulating on the gas distributor plate on the bottom. This would allow the use of biomass fuels, with no or only little pretreatment. These fluid beds became popular for coal combustion with coabsorption of sulfur by limestone used as "sand" in such coal combustors. The overall combustion temperature of ± 800°C–900°C was optimal for combustion and sulfur removal. Sulfur capture is much less important for biomass than for coal, but the principle of in-bed capture of obnoxious substances remains attractive. It also turned out, however, that on a local scale a burning particle may still have a temperature several hundred degrees above the average temperature of the bed and plumes of combustible pyrolysis vapors may rise through the fluid bed, indicating less than ideal mixing. With this respect a dense fluidized bed is not an ideal burner. In most cases both for coal and biomass, the zone above the dense fluid bed, called splash zone or disengaging zone, is used for the final burnout of the gas and solids by injecting secondary air for combustion. These bubbling beds remain attractive for difficult fuels but were for coal more and more replaced by the so-called circulating beds where the gas velocities could be much higher and the solids entrained by the gas flow, is separated by a large cyclone and feed back to the combustion bed (see Fig. 1.30). Similar to the situation in the dense-bed combustors the circulating solids form a heat buffer or "flying wheel" which simplifies control and operation. Compared to the dense bed, the pressure drop over the circulating bed combustor can be lower, which is important because of the large flow rate, and also the operation flexibility (and turn-down ratio) is more favorable. These units can also be made suitable for biomass, usually cofiring with coal because of the high capacities involved for which enough biomass feedstock is not always available.

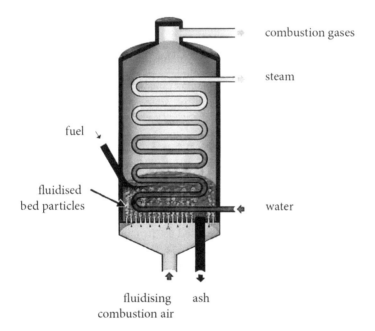

Figure 1.29 Simplified sketch of a dense fluid-bed combustor. The combustion air passes through a distributor and fluidizes the mixture of inert material ("sand") and solid fuel. By action of large gas bubbles formed in the bed, strong mixing occurs while the combustion process takes place, which results in a rather uniform temperature in the bed and strongly increased heat transfer to the water/steam tubes. Adapted from http://www.photomemorabilia.co.uk/FBC.html.

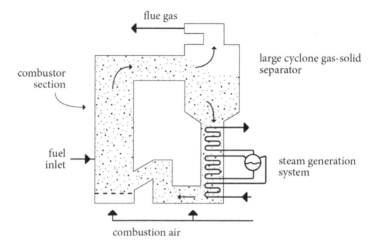

combustion air

Figure 1.30 Simplified sketch of a circulating fluidized bed. Due to the high gas velocity in the combustion section the solids mixture is continuously entrained to the large gas–solid-separating cyclone and via a downer section where steam is generated returned to the combustor. Adapted from Paul A. Berman and Joseph C. Dille, *High Efficiency Pressurized Fluid Bed Systems*, 1982.

Heat generated in the combustion section and from the flue gas is normally used to generate steam from which (electrical) power can be produced via steam turbines and electricity generators.

In the 1980s it was first tried to develop a pressurized fluid-bed combustor which can be considered as the combustion chamber of a gas turbine (expander) (see Fig. 1.31). In this way a certain fraction of the combustion value can be directly extracted in the form of (electrical) power; a kind of open gas turbines running on solid fuels. Future developments will determine whether this operation will meet application for biomass (see, for example, Ref. [31]).

Direct conversion of the heat of combustion of biomass can be realized on a relative small scale by a Sterling engine and by a closed gas turbine cycle where the biomass combustor is only used as a heat source. In this way the combustion process can be optimized in separation from the power generation. The efficiency of such an operation will be on the low side if only power is produced and applications will be much more attractive in combined heat and power generation application.

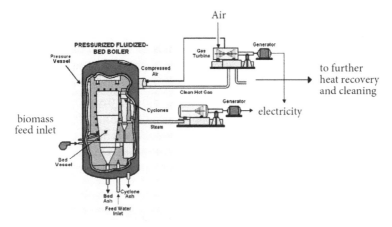

Figure 1.31 Simplified flow scheme for a pressurized fluid-bed biomass combustor. Combustion air is compressed by a flue gas expansion turbine and introduced in the combustor. Electric power is generated both from the expansion turbine and from the steam cycle turbine. The fluid bed is placed inside a pressurized vessel, thus separating the barrier for pressure from the barrier of high temperature. Adapted from NETL (2010c), CCPI/Clean Coal Demonstrations Tidd PFBC Demonstration Project, Project Fact Sheet.

For a very large scale of operation, typically >300 MW, coal-entrained flow combustors, also called pulverized coal combustors, are used. Coal is pulverized to a very fine particle size and injected with combustion air directly into burners in the boiler, not unlike a gas or an atomized liquid. At the operation temperature of such a boiler a flame is formed or a very hot kernel fired by several burners and at these conditions the fuel is rapidly combusted (see Fig. 1.32).

At present this technology is well established with a lot of refinements, optimizing the combusting condition profiles, heat recovery, ash handling, etc. For biomass this technology is specially used for cofiring of biomass into primarily for coal designed boilers. A lot of experience has been gained over the last decade by this type of operation, which is attractive because of the effective way to reduce the CO_2 emission from fossil origin, while maintaining fuel flexibility and use of existing equipment. However, precautions and effective measures should be taken to avoid the pitfalls of fouling of heat exchange equipment, corrosion, changing nature of ash and bottom ash, and many other factors (see Chapter 7). In cofiring of biomass it can be considered to use torrefied (a kind of baked) wood (or other biomass) to be fed to the mills of the coal pulverization. Apart from advantages in storage, transport (which may be in the form of pelletized torrefied wood), and handling in the process of torrefaction (heat treatment at 260°C–300°C) the wood particles become brittle and more suitable for grinding. Apart from that the biodegradation is strongly reduced. These favorable properties are, however, to be balanced by the increased complexity and additional cost. Moreover, torrefaction does not change the ash problem essentially (see Chapter 13).

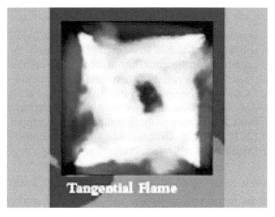

Figure 1.32 By placing the burners of a pulverized coal boiler in a tangential direction a hot kernel is created with a high temperature and combustion rate. (http://koilerxm.info/tangential-fired-boilers/).

1.2.2.1.1 *Oxycombustion*

Because much R&D attention is nowadays directed toward CO_2 recovery and storage from flue gases of fossil fuel combustion this could also be of interest for biomass fired furnaces. Although the CO_2 emission from biomass combustion does not contribute to the longer term increase of the concentration of CO_2 in the atmosphere because it represents a short-term recycle of carbon from atmosphere to plants and back it is still of interest for the CO_2 balance if recovered and stored. In fact in this way CO_2 is taken effectively out of the atmosphere and this may be the only effective way of doing so and reducing the impact of previous CO_2 emission by mankind in the last 150 years (see Fig. 1.10). In combustion of carbon-containing fuels in air, the nitrogen in air (~80%) always forms a diluents in the flue gases, rendering the recovery of CO_2 a costly operation (say ~40 €/ton CO_2 or more). If combustion takes place feeding pure oxygen to the combustor, after cooling of the flue gas, nearly pure CO_2 can be produced (apart from minor other oxidized products of other fuel elements like nitrogen, sulfur, etc.). To avoid extremely high combustion temperatures CO_2 and steam in the form of flue gases is recycled in these systems. For biomass combustion these developments will probably only take place subsequent to those for fossil fuels. Apart from technical difficulties the use of pure oxygen (or strongly enriched air) will increase the overall complexity and/or cost for the oxidant (pure oxygen) and this will have to be balanced by the easier/cheaper recovery of CO_2.

Large-scale production of oxygen is nowadays mainly carried out by liquefied air distillation, but alternative separation techniques (such as by membranes) may further develop and on a long term oxygen produced as a by-product at electrolysis of water for hydrogen production may become an attractive route for oxygen production (see Fig. 1.10). A special form of separation of oxygen from the nitrogen in air is carried out integrated in the combustion process in the so-called cycling combustion. Here a combined oxygen and heat carrier is oxidized with air and this carrier is continuously or intermittently contacted with the carbon containing fuel, which extracts the oxygen from the carrier while forming CO_2 and water. These systems are still in a R&D phase and preferably operated with gaseous fuels, although also solid fuels were tested in a so-called double-zone fluid bed, to be discussed later.

1.2.2.2 Gasification of biomass

1.2.2.2.1 *Introduction*

The aim of gasification is to transfer the heat of combustion of a fuel to a gas phase as much as possible in the form of chemical energy. Especially compared to a solid fuel a gaseous fuel is easier to combust, with a smaller excess of air (more efficient) and a gaseous fuel is better suited to be used in modern energy conversion equipment like internal combustion engines, turbines, furnaces, fuel cells, etc. Moreover gasification allows the removal of undesired components and elements present in the (solid) fuel, like mineral components which become ashes and further other components like sulfur, nitrogen, etc.

A second aim of gasification may be to convert a raw material containing carbon into synthesis gas (also called syngas), a mixture of mainly CO and H_2. From this mixture, after cleaning, a wide range of top-quality liquid fuels like diesel, gasoline, methanol, ethanol, etc., can be made and apart from this methane and a broad range of chemicals can be produced by existing well-proven processes at least for syngas derived from fossil fuel sources. In the case of syngas production also liquid and even gaseous feedstocks are used for conversion and these can also be derived from biomass.

The "cold gas" efficiency of the gasification process, defined as the chemical energy in the product gas over that of feedstock can be up to 80% for large-scale heat-integrated operation. The remaining heat can be partially used for other purposes or is lost to the environment. Gasification takes place at higher temperatures say 600°C–1500°C to get a desirable rate of the different gasification reactions. Heat to achieve these high temperatures can be fed in from the outside, a process which is often also called (steam) reforming, or is generated inside the gasification reactor by partial combustion of the (partially already gasified feedstock which is mostly done. A more descriptive name of this specific process would therefore be gasification by partial oxidation.

Gasification in a wider sense is already an old process which has been used for over 180 years, and while biomass gasification occurred at least already in 1848 most of its development was directed to the conversion of fossil fuels. The status of the technology today is presented in Table 1.4. On the basis of the different feedstocks like natural gas, fossil oil, coals and biomass the status of the production of low-caloric gas, syngas, and methane (as substitute for natural gas) is considered.

Low-caloric-value gas (sometimes called producer gas) is a product gas formed with internal partial oxidation gasification processes which use air as a gasification agent. This use of air means that all the nitrogen in the air fed to the system ends up in the product gas (usually in volume fractions of 40%–60%). In a more complex arrangement of the gasification process to be discussed further on this can be avoided (see also Chapter 11).

Table 1.4 Gasification, status of the technology

Fuel	Low-caloric gas*	Syngas	Methane
Natural gas	Commercial	Commercial	–
Fossil oil	Commercial	Commercial	Via syngas, few plants
Coal	Commercial	Commercial	Via syngas, few plants
Biomass	Dirty: commercial	"Almost commercial,"	PP and lab scale
	Clean: "almost"**	demos	

*Low-caloric gas or producer gas: usually from gasification with air, product is diluted with nitrogen (~50%), 4–7 MJ/Nm^3).
**For small scale, full commercial

While this inert ballast load by nitrogen can be a certain disadvantage for the application as a fuel gas or as diluted syngas the simple gasification to low-caloric gas by air requires no pure oxygen and is less complex. For all fossil fuels this process can be considered as fully commercial and especially in the past as the pure oxygen was prohibitively expensive it had and still has widespread applications. For biomass one should consider two versions of the process. The first one is called "dirty" which means that the gases are directly used, without any cooling or cleaning in a close coupling to a combustor of a boiler, for example, to replace part of the coal fed to the boiler. This can be an efficient way to introduce biomass in an existing boiler, retaining most of the biomass ash in the gasifier and offering a large replacement flexibility (see Fig. 1.33). The second version includes cleaning of the gas from tars, entrained dust, etc. To this end the low-caloric gas is usually passed through a hot cyclone and/or hot filters, cooled down washed with water or a combination of liquids. Mostly the ultimate aim is to use the product gas not in a furnace or boiler but in internal combustion engines, gas turbines or in catalytic processes where high purity is required. Because of the presence of tar vapors and the formation of aerosols gas cleaning is not an easy job which may considerably increase the cost of the gas production via gasification, especially if a very low contaminant concentration is requested (e.g., for application in catalytic chemical reactors).

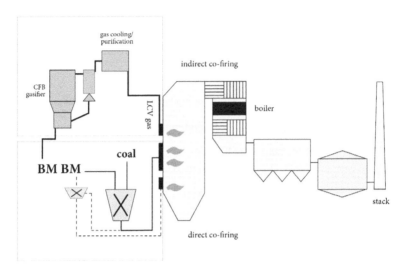

Figure 1.33 Introducing biomass (BM) for cofiring into a boiler can be done directly or via gasification to low-caloric fuel gas. By gasification most of the biomass ash can be kept out of the boiler system. The hot gas from the gasifier can be introduced without cleaning from tars and dust but alternatively the gas can be first cleaned after cooling down. In the latter case it can also be used in a gas furnace, in an internal combustion engine, in turbines, etc., for direct and combined heat and power generation. Reprinted with permission of ECN.

Also for power generation with engines and turbines extensive cleaning to prevent blockage, fouling corrosion is necessary. In World War II, when liquid fuels were scarce in some areas over a million of small-scale gasifiers with relatively simple gas cleaning were constructed to convert auto and tractors, motors, etc. from liquid fuels to solid fuels like wood blocks, char or coals (see Fig. 1.34). From a single type of such gasifiers (Imbert) over 500,000 were produced during the war period. The operation periods were, however, relatively short and frequent repair and cleaning of the equipment was required. Often also an increased wear and tear occurred, which was only acceptable under certain circumstances. As soon as liquid fuels became available again these gasifiers quickly disappeared. For stationary application and for long duration runs similar equipment has been developed in the last decades and applied in small numbers for the market in the developed countries and in developing countries (see Ref. [32] and Chapter 12).

Bundesarchiv, Bild 183-\00670A
Foto: o.Ang. | 1946

Figure 1.34   Woodblock gasifier fueling a motor car.

For the production of synthesis gas from fossil fuels like natural gas, fossil oil, and coal, a wide range of gasification processes is available and these have been applied (see Ref. [33]) for many decades under continuous development toward more feedstock flexibility, increase operation temperature, and higher unit capacity. For syngas from biomass the situation is different and indicated in the Table 1.4 as almost commercial and demos. A certain development of technology and market will still be required to attain wide spread success. This will further be discussed under technologies and challenges for biomass gasification. For a deeper study of the subject biomass gasification and application the reader is referred to the special literature (see, for example, Ref. [32], which is an excellent entry).

Methane (or substitute natural gas) can be produced from synthesis gas in a well-developed catalytic process called methanization. However, due to the high costs and availability of natural gas there are only few production plants based on this principle. One well known example is the Great Plains Synfuel Plant in Beulah, North Dakota (USA) converting coal to methane.

Also for biomass conversion via syngas to methane has been studied and developed (see Chapter 11) as an example. This seems to be especially attractive for remote areas and for special economic circumstances because of competition with natural gas in other areas. For some types of biomass gasification processes, producing a relative high fraction of methane in the synthesis gas the methanization pathway to SNG can probably be more attractive (see Chapter 11). Methanization of syngas is a strongly exothermic (heat producing) process and for coal Exxon [34] developed a process in which the heat released by the methanization is used for the endothermic (heat consuming) catalytically enhanced (steam) gasification process for syngas production in the eighties. Recently this has also been studied for the biomass to methane option [35].

1.2.2.2.2 Technologies and challenges for biomass gasification

Gasification is considered as one of the most attractive options to convert biomass into mechanical energy and electricity in modern equipment or to produce high-quality synthetic liquid and gaseous fuels. However, while a lot of experience is already available for fossil fuels this can only be partially transferred to biomass. The reason is the fact that the scale of possible operation is generally much smaller for biomass compared to fossil fuels due to logistic problems of distributed production of biomass versus mining from concentrated deposits. The smaller scale may lead to higher specific capital and operation cost. Also the composition of biomass plays a role such as the quality and composition of the ash (softening and melting point) and higher oxygen content, etc. For coal, the developments of the last decades lead to pressurized, high-temperature entrained flow reactors for gasification where the feedstock is pulverized and converted to syngas at very high peak temperatures (say 1400°C–1700°C) using mostly high pressure of 10–30 atm. to increase the volumetric conversion rate and to obtain pressurized syngas. At these high temperatures the ashes are completely molten and the final ash product has a glossy appearance for which environmental advantages are claimed in case coal is the feedstock. All desired reactions are not limiting in rate anymore at these high temperatures and tars are completely converted. This operation is suitable for many types of coal and is the present top technology with capacities of up to 2000 ton/day per reactor. However, the technology is complex, very sophisticated control, extensive safety measures and intense heat exchange is required for operation. This drives the economical operation area toward large scales of operation. For biomass tests with biomass have been done on lab scale and pilot plants up to demos where biomass was cogasified in a full-scale operation replacing up to 70% of the coal by biomass with good results [36] (see Fig. 1.35).

Figure 1.35   Cofiring of biomass in a full-scale coal gasification–based power plant (NUON/Vattenfall), Buggenum, the Netherlands. From 2006 onward 10% biomass was cofired to the gasifier (entrained flow, SHELL type) and in later trials extended to 70% biomass (energy basis). In 2013 this still relatively small power plant was closed down due to economic circumstances. Reproduced with permission of ECN (http://www.ieatask33.org/app/webroot/files/file/country_reports/NL_July2013.pdf).

Apart from these efforts, most gasification processes for biomass were directed to less severe gasification conditions: lower temperature and atmospheric or close to atmospheric pressure. It was hoped that this would lead to more economically viable solutions at the scale were biomass is expected to be gasified. Lower temperatures would allow the ashes to remain in the solid state. As indicated in Fig. 1.36 generally the temperature should remain below 1000°C and often below 850°C to avoid (partial) ash melting. At intermediate temperatures, say, 1000°C–1400°C, partial ash melting may occur, rendering the gasification process impossible in technical-scale equipment due to blockage and fouling.

Gasification of solids

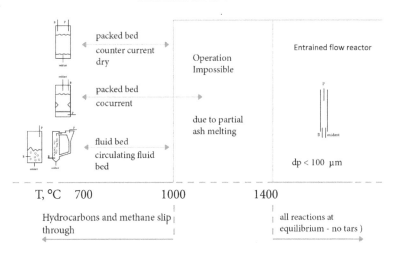

T, °C   700   1000   1400

Hydrocarbons and methane slip   all reactions at
through   equilibrium - no tars)

Figure 1.36 Gasification processes of solid fuels should avoid the temperature range where ashes are partially melted, as this leads to blockage of the reactors. Operation at very high temperatures where all ash is in the molten state is possible, although complex. This would also avoid obnoxious tars in the product gas. At low temperatures the ash remains solid but then the problem of tars in the product gas has to be faced.

However, lower temperatures come with a few problems. Tars are formed at these conditions and are only partially converted (cracked) if no special measures are taken. With a few simplified schemes the processes within a low-temperature gasifier (say 850°C) can be sketched (see Fig. 1.37). The solid biomass breaks down under the influence of heating up due to heat generated by the internal combustion into char, vapors, and permanent gases like CO, CO_2, H_2O, and H_2. As indicated in Fig. 1.37 gas-phase reactions take place between hydrocarbons in the vapors and steam reforming, producing more syngas and modifying the composition of the syngas ($H_2 + CO_2$ <-> $H_2O + CO_2$, called the shift equilibrium reaction). The char can be gasified to syngas by reaction with steam or CO_2 but these reaction rates are relatively low below 900°C. This can be seen in Fig. 1.38. Solid materials (feed and produced char) are rapidly converted in pyrolysis oxidation reactions and much faster if they consist of smaller particles with more contact surface area.

Gasification of solids (char) is relatively slow, however, and is only affected by the particle size for large particles.

Gasification chemistry

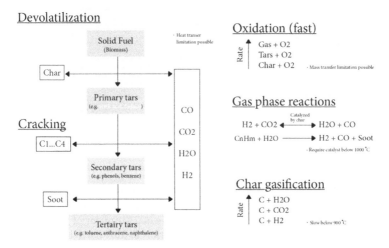

Figure 1.37 Gasification chemistry. Gasification of biomass is a very complex chemical process involving stages like devolatilization of particles, cracking of the vapors (tars), gas phase reactions and char gasification. The driving exothermic (heat producing) reaction is the oxidation of part of the products. Which parts depends much on the reactor design.

To achieve high char conversion in gasification this should preferably be done by the oxidation reactions rather than via the gasification reactions unless sufficient residence time for the char can be realized.

Returning to Fig. 1.37, it is indicated there that the primary tars formed by the pyrolysis reactions are mainly converted by cracking reactions to permanent gases and secondary tars which are more resistant to cracking and without extensive measures like catalytically enhanced cracking these tars end up in the product gas where they cause fouling an blockage in the downstream equipment and in a catalyst bed, etc. At higher temperatures also soot is formed. Although cleaning technologies for dust and tar-loaded gases have been improved in recent years, tars especially remain the "Achilles heel" of low-temperature biomass gasification (see Fig. 1.39).

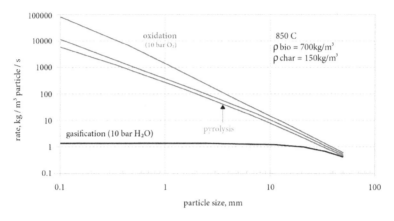

Kinetics of solid conversion by different reactions
at 850 °C and Ptot ~ 30 bar

Figure 1.38 Rate of particle conversion in low-temperature biomass gasification. The rate of solids conversion depends much on the particle size as the rate of the mass transfer of the reacting gases to the surface and interior of the particles depends much on the particle size. This is the more so if the following reaction step is very fast and not limiting the conversion rate at all. This is true for oxidation reactions and to a large extent for devotilization (or pyrolysis) reactions. Char gasification under these conditions is relatively slow and only affected by the particle size for large particles.

To counteract the problems related to the undesired tar production several strategies have been followed. The main tactics followed are given in Table 1.5.

Table 1.5   Strategies for (residual) tar removal for low-temperature biomass gasification processes

---

1. Increase tar-cracking processes inside the gasification reactor (or cascaded to it) by adding tar-cracking catalysts (throw away catalyst or with recovery and recycle).

2. Create areas of very high temperatures through which the vapor-loaded gases are passed and subjected to high cracking rates.

3. Use a cascade of physical cleaning technologies along the cooling-down trajectory of the hot product gas, removing all obnoxious substances: filters, cyclones, oil scrubbers, water scrubbers, electrostatic precipitators, cold plasma separators, etc.

Tar is the "Achilles heel" of
LT biomass gasification

▶ Troublesome mixture of dust, tar, and water
▶ Process water contaminated with tars
▶ Fouling, loss of availability

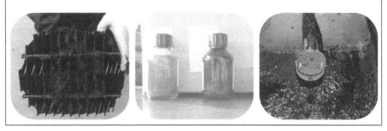

Figure 1.39 Tars, dust, and water are together the "Achilles heel" of low-temperature gasification and these problems are not to be underestimated and careful attention is required in design and operation. Reproduced with permission of ECN.

Catalytic enhancement of the tar cracking reactions can be done by adding various sorts of natural materials (like dolomite) to the gasification system or specially created synthetic catalysts based on Ni or other metals on carrier materials.

A wide variety of catalyst and systems have been developed during decades (for an overview see among others Ref. [32]) but as long as widespread industrial use has not been realized no clear winner can be identified.

A second way to realize increased tar conversion is to create areas of high temperatures in the product gas usually by adding a postgasification section or in a secondary gasification reactor put in series with the primary gasifier. In the past, when applied in the primary reactor outlet the high temperature was reached by additional air or oxygen introduction. The secondary gasification in a separate step has the advantage that, in principle, some high-temperature dust removal can take place avoiding too much ash melting in the secondary gasification step of entrained ash particles. This secondary "gasification" for tar cracking comes at a prize of a higher product gas temperature, an indication that more of the energy of the biomass fuel has been converted to sensible heat rather than chemical energy for combustion. Some of this can be counteracted by heat exchange, for example, between inlet and outlet of the secondary gasification/cracking reactor. Due to the high temperatures this is difficult but not impossible via a system of regenerative heat exchange somewhat like applied in black furnace operation [37].

The most common way to remove tars is to wash these out of the product gases while cooling these down. In the same trajectory several other components have to be removed from the product gases. Dust particles of ash and char (sometimes also some soot) can usually be first removed by cyclones or high-temperature filters (ceramic or at low-temperature bag house filters). Sometimes then a simple water wash is used but due to mist and smoke formation these washers are not always very effective and depending on the application electrostatic filters have to be installed to remove aerosols. In the past oil washing has been tried with some success in special temperature regions. In recent years, the Energy research Centre of the Netherlands (ECN) introduced an integrated systems of syngas and producing gas clean-up incorporating among others an oil wash and subsequent water wash, cleverly positioned along the cooling down trajectory and reported by reaching excellent clean gas figures (see Fig. 1.40).

Apart from this system if the application of the gas is for catalytic conversion to fuels further removal of several trace component would probably still be necessary to meet the specifications of the catalysts used in these processes.

1.2.2.2.3 *Gasification reactors used for biomass*

Figure 1.41 gives an overview of most of the gasifiers used for biomass gasification. All of the types were primary in use for coal, and the principles are 80–160 years old and still built in modern versions today. Also indicated in the table are sketches of the temperature profiles. *The updraft gasifier*, also called the *moving-bed countercurrent gasifier* is the oldest one and still in use (among others in pressurized form of the Lurgi gasifier). Already in the Bischof gasifier for peat and biomass (1839) (see Fig. 1.42) one can recognize this way of operation. Oxidant, air or steam oxygen mixtures are flowing in countercurrent with the solid fuel like biomass. The feed should be in the form of regular blocks to obtain a regular and evenly permeable bed moving in downward direction. Often a rotating grid is used for further stabilization of the gas flow distribution.

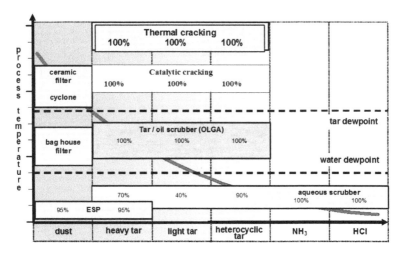

Figure 1.40 Different gas-cleaning technologies for low-temperature gasification are effective at different temperatures (overview chart from the ECN). The ECN developed a cleaning technology called OLGA [38], based on a precise positioning along the cooling-down trajectory of a consecutive oil- and water-scrubbing technology. Reproduced with permission of ECN.

The different profiles of temperature and fuel conversion in the gasifier are positioned as indicated in Fig. 1.41. The heat necessary for gasification is generated by oxidation of the char and the endothermic conversion reactions in the reduction zone and the heating up of the inflowing solid feed further cool down the gases to intermediate values. However, the efficient operation with respect to heat utilization in this countercurrent process also leads to the entraining of the vapors/tars to the product gas outlet giving rise to very high tar contents (say up to 100 g/Nm^3). This requests extensive gas cleaning. The relative simple gasification technique is characteristic by a relatively simple robust construction and operation. It is used in capacities varying from 100 kWth up to tens of MWth.

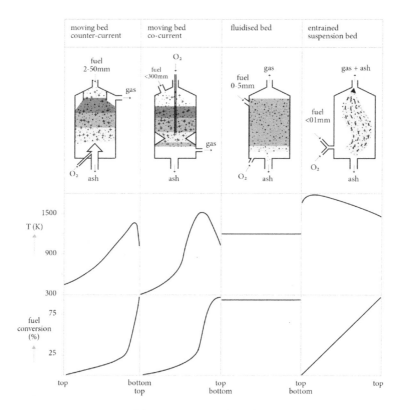

Figure 1.41 Overview of the basic types of gasification reactors with their typical profiles of temperature and fuel conversion.

The moving bed cocurrent gasifier or the *downdraft gasifier* is especially meant for small-scale operations (50 kWth to 3–5 MWth). Here from the inlet of the air (oxygen/steam has only be used in exceptional cases) the gas flow is mainly in down ward direction in concurrent with the partially converted biomass. At the environment of the air inlet a very hot combustion zone is created close to the air inlet and heat is transported to the pyrolysis zone/drying zone by conduction and free convection of the gas phase. In a special variation the gas phase is pumped around by forced convection in a gas loop including an ejector pump and a combustion chamber for vapor combustion and fed back at high temperatures at the air inlet.

Coal, peat or biomass

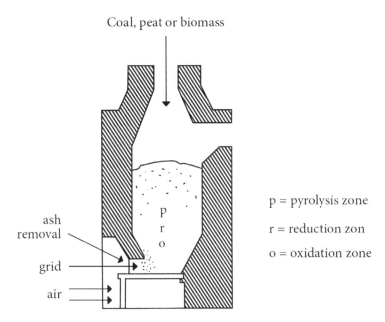

p = pyrolysis zone

ash
removal

r = reduction zon

o = oxidation zone

grid

air

Figure 1.42 Archetype of a reactor for gasification by partial oxidation is this (countercurrent) moving-bed gasifier of Bischof (1839).

The gasification of the char takes place in the lower part of the gasifier and is successful because of the long residence time of the char in this zone. An advantage of this operation is that the level of tar in the product gas is relatively low (say $\sim$200 mg/Nm^3) because all vapors pass through the hottest zone and char acts as a cracking catalyst. The demands on the feedstock dimension and uniformity are severe and relaxing on this invariably leads to deterioration of the operation. Despite this, as stated before over a million of cars and transport vehicles have been fueled by these gasifiers when liquid fuels were scarce. The design of these and stationary gasifiers was based on empirical rules and overall balances. Only in recent years rate-based models could explain the complex flow and transport phenomena of the internal processes in these reactors.

Overall the cocurrent principle is attractive and in many modern design the complex downdraft system has been split up in different parts (like in *Thermoselect*, Ref. [39]) with separately optimized constructions and flows and reactor types. This adds to the robustness of the operation but increases the complexity and the cost.

In the *dense fluid-bed gasifiers* shaped similarly as a fluid-bed combustor but without the heat exchanger (see Fig. 1.29), a finely divided inert solid is forming a so-called fluid bed driven by the up-flowing gases with the appearance of a boiling liquid maintaining a more or less sharp level of fluidized particles. The solid feed (biomass) is introduced in this medium kept at high temperature via partial combustion of chair and part of the product gases: the temperature in the fluid bed is largely uniform and about equal to the product gas temperature. Small particles entrained by the product gas are separated by a cyclone and recycled to the bed. Such a gasifier can accept a relatively wide range of particle sizes and shapes and has an intermediate tar concentration in its product gas ($\sim$5 g/Nm^3). The char gasification rate is not very high (see Fig. 1.38) and because the char hold-up in the bed is usually relatively low this results in a lower char conversion efficiency. Because char is also converted by fast oxidation reactions near the gas distributor at the bottom in some modifications and via the char injection from the cyclone bottom pipe the competition between gas combustion and char combustion by air can be improved. Alternatively, char could be recovered from the ash and used as a separate product. The archetype dense-bed gasifier was the "Winkler" fluid-bed gasifier from the 1920s. Many later modifications, higher temperature, pressurized have been developed and specially modified for biomass. The scale for application of biomass dense-bed gasifiers is from 2–20 MWth.

The *circulating* fluid-bed *gasifier* (see Fig. 1.43) is somewhat similar to the dense fluid but operates at about a factor 10 higher gas velocities (say 4–8 m/sec, while for dense beds 0.4–0.8 m/s is normally applied). This means that hardly any solids level in the fluid bed can be distinguished and the solids inventory (inert carrier solids, ash and char) is completely circulated continuously over the large cyclones where the gas phase is separated and the solids flows back to the bed. Tars in the crude product gas are somewhat similar to dense beds ($\sim$5 g/Nm^3) and also the feedstock flexibility is high. Because of the high gas velocities the bed diameter is less at the same capacity compared to a dense bed (although the circulating bed will be somewhat higher) and it can be built for high capacities up to hundred MWth.

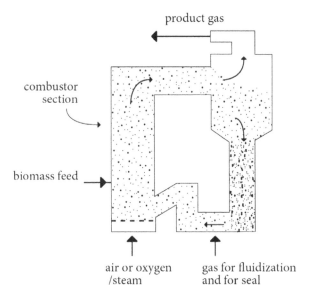

Figure 1.43 Simplified sketch of a circulating fluid-bed biomass gasifier. Details of the layout and seal (here a U bend) may differ in the various designs.

With respect to the char conversion and the selectivity compared to oxidation of vapors and gas in the zone close to the air inlet also the situation is somewhat similar to a dense bed. However, in principle, the back flow of product gas and partially converted feed solids to the air inlet zone can be better controlled or manipulated.

In the *twin-fluid-bed gasifier* (see Fig. 1.44) the gasification zone is completely separated from the combustion zone in which only the char is combusted. In principle, two separate fluid beds are used to this end. Solids from the gasifier containing the product char are introduced in the fluidized oxidizer and there the char is combusted resulting in an increase in the temperature of the carrier solids. These very hot solids are transported to the gasifier where biomass is contacted in a separate fluid bed with steam.

Gasification which takes place here is an endothermic process which cools down the solids which is therefore continuously circulated to the oxidizer again for reheating. This setup has an important advantage.

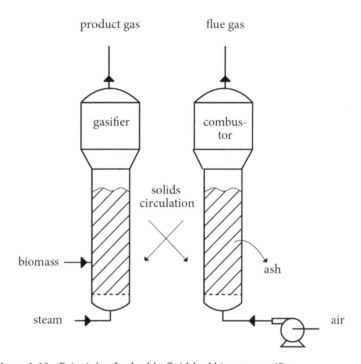

Figure 1.44 Principle of a double-fluid-bed biomass gasifier.

The gasification section is here separated from the oxidation reactions in the combustor where only char coal is combusted. The nitrogen from the air is not mixed and entrained with the product gas but leaves with the flue gas. So effectively nitrogen-free product gas is obtained without separate air separation process. Temperature differences between combustor and gasifier are kept limited by intense solid circulation between these two sections (see also Chapter 11).

Because only solids are exchanged between the two zones there is no inter mixing of the gas phases. So in the oxidizer air can be used for combustion while the nitrogen contained in the air (80%)

ends up in the flue gas of the combustor and does not mix with the gasification product gas which can remain essentially nitrogen free. For syngas ($CO + H_2$) production this is an important advantage. Otherwise to produce a nitrogen free product gas, pure oxygen (mixed with steam) is required as a gasification agent and as stated before, oxygen is relatively expensive.

The product gas of such double bed gasifiers do contain normally about 10% CH_4 (representing over 30% of the combustion value). This requires an additional step (oxygen gasification of the product gas or steam reforming) to arrive at pure synthesis gas. If methane is the desired product the high concentration of methane already in the crude product gas is of course an advantage. The technical solution for such a double bed, were large amounts of solids have to be circulated can vary widely, incorporating circulating dense fluid beds for oxidation zones (examples are the Güssing gasifier based on the work of Hofbauer, see Chapter 11, and the Milena gasifier of the ECN [40]). It has also been tried to realize the principle of partially separated double-fluid beds in one fluid bed shell with stripping fluid beds in between [41].

In the *entrained suspension bed* or *entrained flow gasifier* (see Fig. 1.41) finely powdered carbonaceous material like coal or biomass is contacted in a very hot zone (say 1400°C–1700°C) with the gasification reagents where conversion takes place in a very short time (<1 sec). At these temperatures all ashes are in a liquid form and all tars converted. Normally these units are used for syngas production and operate with oxygen/steam mixtures under pressure, while the solids are different finely divided fuel grades (also for liquid fuels and gases like natural gas similar entrained flow gasifiers exist and are widely applied for syngas production).

In a Shell modification [33] the hot zone around the flame in the entrained flow gasifier is cooled by a membrane wall, producing steam, and protected by a layer of ash, molten at the flame side and flowing in downward direction. The properties of this ash layer can be modified by adding flux materials containing calcium to the feed.

The complexity of such an operation requesting among others a highly sophisticated process control, and the related capital investment, limits the application to large and very large scales (order of a few thousand MWth). Application to biomass may be restricted by logistics and availability of biomass. If wood is used as a feedstock the ash content is very low and the ash protected membrane concept may request large addition of ash-producing components to the feed. In this case it may be better to use another type of entrained flow gasifier, or a modification of an oil entrained flow gasifier. Large-scale tests with cofeeding of biomass in an entrained flow gasifier were reported successful [36]. Also pyrolysis conversion products of biomass have been tested in entrained flow gasifiers (see, for example, Ref. [42] and Chapter 10). A special type of gasification is a plasma flame gasifier used mainly for dangerous waste and municipal waste. This process will not be discussed here. Figure 1.45 shows the scale of operation typical for the different gasification technologies discussed in this section.

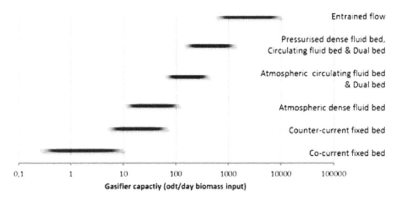

Figure 1.45 Typical capacity range for different gasification reactor technologies. odt: oven dry ton. Data from http://www. ecolateral.org/gasificationnnfc090609.pdf

1.2.2.3 Pyrolysis of biomass

Pyrolysis is decomposition under the influence of heat. It is one of the oldest processes used to convert biomass into other fuels and materials. Man may have discovered its use to produce char coal or hardening of spear points shortly after the discovery how fire can be handled. If biomass like wood is heated up in the absence of air the solid material becomes brown, than black at increasing temperature. Vapors which may condense on cooler surfaces initially contain much water and light organics but at increasing temperatures the condensates becomes dark brown to black and in the gas phase smoke can be observed. Part of the gas phase escaping from the pyrolyzing biomass are permanent gases like CO, CO_2, methane, hydrogen, etc. This simple experiment shows all aspects of pyrolysis used to produce improved fuels. While pyrolysis takes place the main product yield can be steered toward char, liquids, or gases. We will shortly discuss the processes to produce char (also called biochar) and processes with the main emphasis on liquids (also called pyrolysis oil, bio-oil) production. These processes are treated in much more detail in Chapters 14 and 15.

One category of pyrolysis at relatively mild conditions is called *torrefaction* and the product aimed for is called torrefied wood or biomass. It received renewed attention in recent years, especially in relation to cofiring of biomass in large-scale boilers. Torrefied wood has advantages in storage and transport as it is not subject to biodegradation it is brittle and easier to grind to small sizes and can possibly be mixed with the coal feed to the burners. The downside of torrefaction is the loss of some of the combustion value and the cost of torrefaction. The process is described in Chapter 13.

Char coal was produced from biomass by pyrolysis already in prehistoric time because of its special properties. It is easy and convenient to handle in combustion processes where high temperatures can be reached (up to 2700°C), producing little smoke when used for fires by black smiths, for cooking and baking, grilling, barbecues, etc. The low amount and the quality of the ash (sulfur, etc.) in char coal makes a much wanted fuel in metallurgical processes and especially in the past it was used in iron production, where it is of course also a reduction agent. Nowadays with some exception coke produced from coal is used instead.

Apart from use in energy generation char coal found many applications as a carbon source in pyrotechnic mixtures, gunpowder, etc. After activation it is applied in purification, filtration, medical absorptives, and many more, and even as a soil-improving agent (see, for example, "Terra Preta" soils produced in pre-Columbian times, Ref. [43]). The traditional way of producing was the creation of a crude kiln, called a Meiler (see Fig. 1.46), in the following way. Char coal was produced with arranged woodblocks in a conical heap with a central chimney left open. This heap was closed all around, for example, with clay to close the heap from the air, except for the chimney and a few control openings at the base. After ignition the operation depends much on the craftsmanship regulating the air input at different stages of the operator but if well operated efficiencies could reach reportedly from 20%–25% on mass bases (Chapter 14), depending on scale. As the heat is generated by internal combustion this system and the modern versions are called kilns. Modern production methods rather depend on metal containers or retorts, indirectly heated from the outside and in which gases and tar by-products from one retort are separated and also used in the next retort to heat up the next charge (see Fig. 1.47). Also continuously operated rotary retorts are used, wagon retorts and many more variation: even a pressurized version of the retort process aiming at high carbon yield has been studied recently (see Antal) (see Chapter 14 for a detailed overview).

Figure 1.46 Operation of a traditional Meiler for char coal production. (https://verhaal140.wordpress.com/2013/02/20/de-hunneschans-en-de/).

Figure 1.47 Modern char coal production in batch retorts with heat integration between retorts in different phases of the process such as using the gas produced for heating up the next batch retort. The retorts are in position (see dark lids) and the pyrolysis gases are used as fuel for heating (see bright light of the fire through the opening on the front). Photo reproduced with permission of Stassen.

1.2.2.3.1 Pyrolysis of biomass for liquid products

By producing liquid products via pyrolysis several logistical problems with biomass as an energy source in storage, transport over long distances, homogenizing, bacterial decay, etc., can be solved. Figure 1.48 illustrates the potential advantage of decentralized pyrolysis oil production with storage and centralized processing to fuels and products.

Figure 1.49 also shows that production of pyrolysis oil represents a considerable densification of the energy carrier. The pyrolysis process also forms an initial purification step for a fuel, reducing its ash content by removing a very large part of the ash in separating the remaining solids (char and included soil remains). Moreover, a liquid energy carrier is much easier to handle compared to a solid and to feed into modern energy-converting equipment like motors, turbines, and Sterling engines. Also for further refining and production of chemicals, especially in pressurized process equipment a liquid has important advantages over a solid feedstock. That's why processes have been developed specially aiming at the production of liquid products from biomass. It operates at close to atmospheric pressure and usually between 400 and 500°C by fine tuning of the process conditions more and more of the heat of combustion of the biomass ends up in the liquid products. A disadvantage of the simple pyrolysis process can be that while, for example, wood has a composition consisting mainly of sugar polymers (cellulose and hemicelluloses) intimately entangled with lignin, in the pyrolysis oil all decomposition fragments of these components are intensely mixed and/or have reacted with each other. It could be better to first separate these constituents of wood prior to the application of processes to convert these to different fuels or chemical products. These routes are considered in some of the so-called biorefinery concepts and these will not be discussed here.

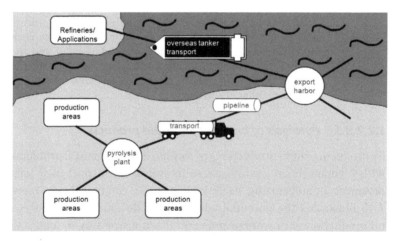

Figure 1.48 Logistics of decentralized pyrolysis oil production. Biomass can be produced at areas remote from where the bio-oil is required. Liquids are easier and cheaper to transport and to process compared to bulky solids. Minerals can remain close to the production areas and brought back to the fields.

In Table 1.6 a typical pyrolysis oil composition is compared to typical fuel oil properties. It can be seen that although a superficial visual resemblance to fuel oil can be observed, in fact pyrolysis oil is a completely different product. It has a highly characteristic and strong smell slightly remembering that of smoked food products. Crude pyrolysis oil has a high water content and about half the heat of combustion of fuel oil. Moreover it has a high oxygen content. The overall elemental composition more resembles that of biomass such as wood than that of mineral oil.

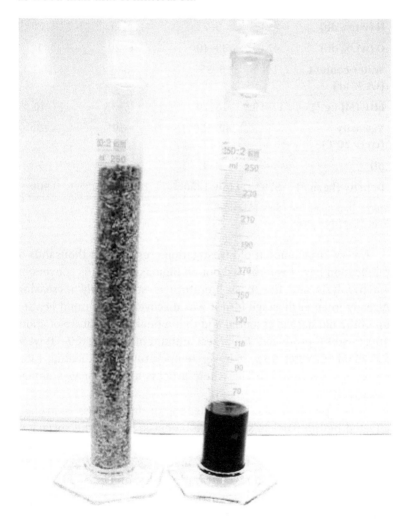

Figure 1.49 Energy densification via pyrolysis. The biomass and the pyrolysis oil represent both 1 MJ heat of combustion. See also Color Insert.

Table 1.6 Elemental composition and some characteristics of pine wood, fast-pyrolysis oil (from pine wood), hydrothermal liquefaction, and heavy fuel oil

	Pine wood	Pyrolysis oil	Hydrothermal liquefaction oil	Heavy fuel oil
C (wt.% db)	46.6	50–64	65–82	85
H (wt.% db)	6.3	5–7	6–9	11
O (wt.% db)	47.0	35–40	6–20	1
Water content (wt.% ar)	9	5–35	3–6	0.1
LHV (MJ·kg⁻¹)	17–19	16–19	25–35	40
Viscosity (cp @ 20°C)	–	40–150	~10^4	180
pH	–	2–3	–	7
Density (kg.m⁻³)	570*	1150–1250	1050–1150	900

Source: Data taken from Refs. [44, 45].
Note: *Particle density

A very large amount of investigations resulting in thousands of publication have been carried out on biomass to liquids conversion via pyrolysis and the mass literature is still rapidly increasing. Already in an early stage [46] it was discovered that rapid heating up of the biomass is essential and that a final temperature of about 400°C–500°C produced the largest amount of oil up to 60–70 wt.% for wood (Chapter 15). At lower temperatures and heating rates more char is produced. The whole process aiming for was named fast pyrolysis.

In fast pyrolysis, the overall heating of the biomass is strongly affected by the particle size of the biomass. Even if the surface of the particles would instantaneously at the right temperature the temperature history of the solid matter is quite different. This results in different product distribution (gas, oil and solid). Figure 1.50 shows a typical biomass particle while it is in a certain stage of conversion (e.g., in fluidized bed reactor, see under gasification).

For engineering purposes simplified reaction schemes have been proposed and used. Under influence of the high temperature in the reactor, heat penetrates in the biomass creating a time dependent temperature profile inside the particle. Gas, vapor and char are formed in a ratio depending on the local temperature. While desorbing, flowing, and diffusing through the outside shell of the particle this primary composition of products is modified due to cracking reactions which may decrease the vapors produced by conversion of these vapors to gases. It has been clearly demonstrated (see, for example, Ref. [47]) that these cracking reactions are strongly catalyzed at pyrolysis temperature by char and alkaline earth minerals (AEMs). That is why it is believed that rapid separation of the vapors from the reacting solid is also important for high liquid yields: the liquid is recovered from the vapor/gas mixture by condensation but this is not easy. While cooling down ultrafine "smoke" type of particles/droplets are formed which are notoriously difficult to separate and require special measures (like electrostatic forces in an e-filter, etc.). Overall the pyrolysis reaction itself seems to have hardly an important heat effect. However, this is still not completely clear. For heating of the feed and evaporation of the water and to compensate for heat losses from the pyrolysis plant energy in the form of heat has to be supplied.

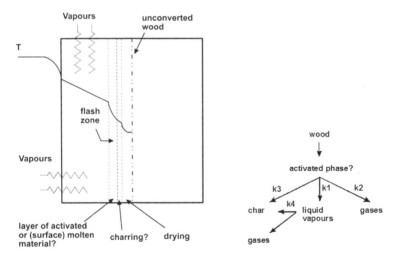

Figure 1.50 Sketch of a larger wood particle during pyrolysis (only one half is shown). Heat penetrates from the surface to the core of the particle and meanwhile pyrolysis takes place at the different local temperatures, and vapors and gases leave the particles through a char layer.

Product char or gases can be combusted for this purpose and different ways of heat exchange with the pyrolysis reactor are practiced.

From the simplified discussion on the mechanisms taken place in pyrolysis of biomass a variety of *process and operation principles* under development today can be understood/explained in Fig. 1.51 presents the most frequently applied reactor types. An often used reactor applied for pyrolysis in process development or process studies is the dense or bubbling fluid bed. It can provide rapid heating up of the particles and heat can be brought efficiently in via wall, heating tubes or via recycling hot gas. Char is recovered from the bed via a gas-solid separator like a cyclone.

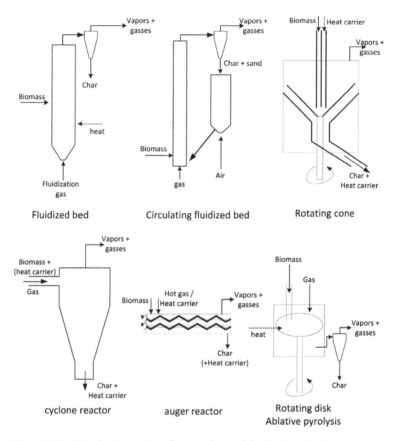

Figure 1.51 Pyrolysis reactors frequently used for fast pyrolysis.

In the circulating fluidized bed, heat carrier sand is transported to a pyrolyzer and is separated together with the char from gas and this solids mixture flows down to a second fluidizing bed where the sand is heated up by combustion of the char. Such a circulating system with a combustion section and a conversion section is used in many shapes and forms (see also the double-fluid-bed gasifier operating though at a much higher temperature) and is in widespread operation in the mineral oil industry (catalytic cracking) and in steam gasification in a dual-fluid bed. Nevertheless it still can present a challenge for operation on a smaller scale and with solids mixtures.

In the ablative reactor it is attempted to press the biomass against a hot zone under a scraping action to remove the char layer as it is formed. The idea is to generate a rapid conversion front (process intensification) of which the vapors are quickly taken away from the char to minimize losses of primary vapors to cracking.

In the vortex reactor and similarly in a cyclone reactor (see Ref. [48]) very small biomass particles can be used, allowing for short conversion times and rapid separation of the vapors to the condenser.

The rotating cone reactor is usually used in a circulating solids set up with a fluid-bed combustor. The hot sand and the biomass are brought into a rapidly rotating cone where the mixture is quickly mixed, while the solids are swept over the edge of the cone and can be circulated back to the char combustor. The vapors escape from the cone without too much secondary contact with the char. BTG BV in the Netherlands has used this idea, originally developed at the University of Twente [49] in a pilot plant, in a demo unit and constructed recently a commercial plant (see Fig. 1.52).

A completely different circulating system is the (twin) screw mixer reactor. Here biomass is mixed with the hot heat carrier material and is subjected to pyrolysis. In this principle the heat transfer which could be augmented by heat transfer through the wall is secured next to the controlled passage of the converting biomass through the reactor section. The vapor cracking may be somewhat more intense but this depends on the details of the design. In some applications where the product envisaged is a bio-oil–char mixture the exact distribution over solids and liquid is less important. Finally, a completely different principle to improve the quality and quantity of liquid is vacuum pyrolysis. Here the residence time and the concentration of the components in the gas phase can be much lower compared to those under atmospheric pressure. The lower pressure will also influence the desorption of the converted biomass and the transport in the porous of the products. This leads to different products. Also the heat transfer and optimal operation temperature will be influenced by the application of vacuum.

Figure 1.52 Picture of the pyrolysis plant of BTL-BTG (5 ton/hr) under construction (2014) in Hengelo, the Netherlands.

A future special version of fast pyrolysis is catalytic pyrolysis. Catalysts aimed at modification of the product slate. This catalyst can be applied as in combination or mixed with the heat carrier inside the process itself or in the space where the gas and vapors still at high temperature are already separated from the solids. At present it is not yet completely clear what direction catalysis should take. Most pyrolysis efforts are still in a lab scale with exception of Ref. [50] (see under application).

1.2.2.3.2 State of development of fast-pyrolysis processes for biomass

Although pyrolysis is an old process and has been used to produce liquids from different kinds of fossil fuels, for example, for energy purposes the fast-pyrolysis process for biomass to liquid is a more recent idea not yet fully commercial (Chapter 15). Strangely enough the only real commercial plant in operation for a long time already produces flavors rather than energy [51]. For a widespread application of fast pyrolysis large-scale demo projects will be necessary, convincingly demonstrating the soundness of the technology and the technical and economical validity of the chains of biomass feed supply and successful commercially valid application of the pyrolysis oil.

1.2.2.3.3 *Application of pyrolysis oil*

An overview of the applications is given in Fig. 1.53. A first class of application is direct use as a fuel for combustion in boilers, directly fired equipment, etc. Several properties of the crude bio-oil like corrosion due to high acidity (low pH, see Table 1.7), instability in storage and transport, polymerization, increasing viscosity with time, gas production, etc., will request special measures. These can be in the form of adjustments in the equipment or pretreatment of the oil. Diesel engines manufacturers are not yet willing to invest on a large scale in the necessary R&D programs as long as the markets for pyrolysis oil fueled diesel engines is not yet visible due to a lack of large quantities of biomass pyrolysis oils. A somewhat similar situation exists in the field of gas turbine. Tests have shown that R&D has to be devoted to turbine development for pyrolysis oil. Deposits in combustion chambers and corrosion and deposits on turbine blades are additional points of attention. By using an indirectly heated turbine cycle, where direct contact of pyrolysis oil combustion products with the turbine blades is avoided, a more simple introduction of pyrolysis oil into the system could be possible. Anyway also here large-scale production of pyrolysis oil will be essential in this kind of chicken/egg situation.

Applications of Fast Pyrolysis Oil

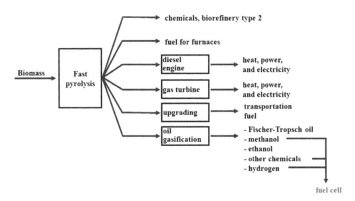

Figure 1.53 Overview of applications of fast-pyrolysis oil.

Table 1.7   Status overview of application of pyrolysis oil

Applications of pyrolysis oils	Remarks
Combustion in boilers and or cofiring in gas-fired furnaces, etc., close to commercial operation	Pretreatment is required. Simple pretreatment and adaption of existing equipment are required.
Fueling diesel engines, despite tests and demos not yet commercially available	This still requires considerable pretreatment, especially for small-bore engines. Mixing/emulsification with mineral oil could improve the chances for application.
Combustion in gas turbines, limited experience and developments yet	For internal combustion turbines probably extensive pretreatment is required. For closed systems less is required.
Gasification/cogasification, limited experience but show promising results	Oil residue gasifiers can be adapted without too much pretreatment requirements.
Refining and corefining with mineral oils, lab-scale and small pilot plants showing possibilities	Extensive hydrotreating seems to be too expensive but mild hydrotreating may be an option.
Raw materials for chemicals via separation and conversion processes, research stage as yet	

Gasification and cogasification of pyrolysis oil to syngas have been tried successfully and were recommended by Higman and van der Burght [33] and this process would open the routes to several top-quality synthesis fuels via the syngas produced: Fischer–Tropsch synthetic diesel, methanol, ethanol, methane, etc. Pyrolysis oil from biomass appears to be a reasonable feedstock for oil residue gasifiers, although some pretreatment like filtering and special treatment to counteract the high acidity with corrosion danger will probably be required. Another potential route toward hydrogen and syngas from pyrolysis oil is steam reforming or coreforming with methane.

Refining and corefining with mineral oils seem a logical step for pyrolysis oil. While in traditional mineral oil refineries the presence of fixed oxygen was usually close to zero nowadays in several fuels a certain amount of oxygen is present or has been added to enhance performance (e.g., octane boosters) in gasoline or as a result of obligation (see EU regulations) to have a minimum content of renewable fuels, realized in the form of, for example, ethanol in gasoline. If the requested amount of biomass-derived fuel could be introduced within the refinery process this could have some advantages. In corefining the much smaller stream of biomass can profit from the economy of scale of the much larger mineral oil process flow. At the downside is the danger of decreasing runtimes due to special problems caused by the bio-oil feeds and products for which the existing equipment was not designed originally. For corefining of pyrolysis oil a start could be the hydrogenation of bio-oil to remove most of the oxygen as water. This would lead, however, to a very high hydrogen consumption, which is too expensive. Still a mild hydrogenation of pyrolysis oil has been advocated to improve the miscibility with mineral oil and to improve the product slate in corefining. See Chapter 16. A suggested route for corefining could be the introduction of the pyrolysis oil into the cat cracker feed. This unit is used in refineries to crack the heavy fraction of the crude oil to lighter molecules. Whether the pyrolysis oil can be introduced directly or after some "light" hydrotreating is still under debate.

Kior is developing a process where catalytic cracking is combined with catalytic pyrolysis. In this setup solid biomass is introduced in a special version of a catalytic cracking unit.

1.2.2.3.4 *Raw materials for chemicals via separation and conversion processes.*

Although the production of chemicals and materials falls somewhat outside the scope of this book, in one of the concepts of the so-called biorefinery, pyrolysis oil is envisaged as a feeding material or important intermediate product (see chapter 15). Next to the stream of different interesting chemicals like resin precursors for partial substitution, fertilizers and soil conditioners can also be recovered from pyrolysis products and have been mentioned in this chapter (e.g., Ref. [51]), but these products fall outside the scope of this book.

By separation commercially interesting side-streams could be generated. Products for flavors (also called liquid smoke) are produced on a substantial scale and applied in the food industry for "smoked" products. Sugar and sugar derivate–rich streams (levoglucosan, levoglucosone), after some modification potential feedstocks for fermentation, organic acids, furfurals, and many more basic chemicals could at least, in principle, be derived from pyrolysis oils.

1.2.2.4 Biofuels from vegetable oils: biodiesel

Vegetable oils and animal oils or fats are excellent feedstocks to produce fossil diesel replacement products. Vegetable oils are produced from pressing or extracting, for example, rapeseed and soybean seeds, sunflower, palm oil, waste vegetable oils, animal fats, oils from algae, and many more produced by living systems. All these oil products have a similar structure (see Fig. 1.54). They can be considered to be components called esters created from glycerol and three long-chain fatty acids after splitting off three molecules of water. Although the vegetable oils can be used as such for application in boilers and heavy diesel engines they can generally not replace petro diesel in modern diesel engines. In a conversion process, called transesterification, they are transformed into mono-alkyl esters and glycerol. Glycerol has been replaced by three monoalcohols, which in most cases will be methanol. The resulting product is then called fatty acid methyl ester (FAME), but also ethanol can be used, giving fatty acid ethyl ester (FAEE). Glycerol, which came as a result of this process as a by-product in increasing quantities on the market, found a new series of applications, even for methanol via syngas production to allow replacement of fossil-based methanol in FAME. These monoalkyl esters have the right properties to allow them to be mixed with modern diesel fuels from mineral oil origin in different proportions or to be used as 100% biodiesel. The long carbon chain of the biodiesel is somewhat similar to the long chain alkanes found in petrodiesels with the exception of the oxygen atoms in the esters. This may change some of the properties like density, heat of combustion (per kg) and flash points. Moreover a different solvent behavior for rubbers and some polymers has to be taken into account. A higher affinity for water increases the danger of corrosion, and ice crystal formation, and fuel crystal formation at lower temperatures (observed in cloud point tests) is to be considered, but not all differences are necessarily negative. Sometimes a cleaner and more complete combustion is claimed and all other points mentioned can be coped with. This type of biodiesel has found widespread use, mixed in different ratios guided by different standards and taxations in different areas and/or helped, among others, to fulfill the European requirements on renewable fuels in transportation.

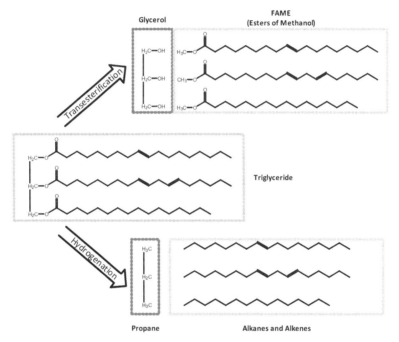

Figure 1.54 Structure of a typical vegetable oil (triglyceride) and its reactions with methanol to FAME, a diesel component with glycerol as a by-product. As an alternative hydrogenation can be applied, producing alkanes and alkenes for diesel together with propane.

A different way to produce renewable diesel from vegetable oil is catalytic hydrogenation (see Fig. 1.37). In this process the triglycerides are converted into alkanes. Instead of glycerol now propane is produced and from the long fatty acid remains the corresponding alkanes are formed. In this way all oxygen has been removed from the diesel product and long chain hydrocarbon molecules identical to much desired diesel components (without aromatics) are obtained. Because it does not contain oxygen, measures to counteract adverse effects such as modifications of engine are not requested. A process based on this type of hydrogenation was developed by Neste and called NExBTL and was realized on a commercial scale (Porvoo, Singapore, and Rotterdam). [52].

Whatever the conversion routes of vegetable oil to diesel are the final products are usually classified as first-generation biofuels as present large-scale production of vegetable oils could be (not always) in competition with markets for human consumption instead of conversion to energy carriers which may cause serious moral dilemmas. Also the effect on the overall carbon emission balance is still under debate, for example, in the case of palm oil and together with the general points on deforestation and other indirect effects. Greenpeace has opposed the use of these kinds of feedstocks. For other feedstocks like waste vegetable oils, algae, jatropha, etc., balances may be quite different.

1.2.3 Wet Biomass Conversion

For biomass containing excessive amounts of water conversion processes producing and separating fuels directly out of the wet biomass, without drying of the biomass may be attractive. Some processes can only be carried out in the aqueous phase and they will also be discussed in this category. As indicated already in Fig. 1.25 these processes can be biological and biochemical (with microorganisms and/or enzymes) or thermochemical.

The main biological processes used are anaerobic digestion of biomass and the production of alcohols, especially ethanol, from sugar solutions. The most important thermochemical processes are the hydrothermal conversion to liquids and/or methane and the supercritical gasification. Nowadays these latter processes are often called aqueous phase reforming of biomass and biomass components.

1.2.3.1 Anaerobic digestion

The anaerobic digestion of biomass occurs spontaneous by in nature in wet environments at places devoid of oxygen. Bubbles collected sediment drew the attention of early chemists in the 17th century, like Robert Boyle and Stephan Hale, who also found that these gases could be ignited [53]. With the improvement of equipment for handling combustible gases, the collection of these types of gases became more attractive for cooking, heating, and later even illumination. Figure 1.55 shows a simple collection/storage system meant for single-farm-scale operations like for cooking, heating, or lighting.

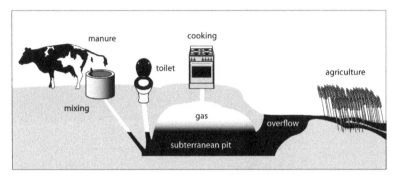

Figure 1.55 Sketch of a very simple biogas-producing system. Copyright Louise van Swaaij.

1.2.3.1.1 *Processes in anaerobic digestion*

Several microorganisms influence the conversion of biomass via enzymatic activity. Different stages can be distinguished in the production of methane from biomass and this also depends to a certain extent on the biomass feedstock, which again depends on the different applications of the anaerobic digestion.

In anaerobic digesters where manures or sewage sludge is converted the biomass conversion has already taken place to a large extent in animals or humans leaving only those parts which can still be converted after the passage through the animal. That's why often codigestion is applied, for example, feeding more convertible high-energy feedstocks (see Table 1.8 for typical examples). Another similar but naturally occurring process of anaerobic digestion takes place in municipal waste belts, where it is often controlled and managed by exhausting the gases formed and after possibly cleaning and removal of obnoxious substances and pressurizing used for energy generation.

Table 1.8 Examples of typical feedstocks for (co-)feeding into anaerobic digesters

Waste products
• Manure from pigs, cows, etc.
• Sludge from wastewater purification systems
• Organic waste from chemical production process
• Waste from food production
• Slaughterhouse waste
• Waste paper
• Grass clippings
Energy crops
• Energy maize
• Water hyacinth
• Miscanthus
• Silage

Feedstocks to be digested are generally complex and may contain both digestible and nondigestible components and may be dissolved or in a slurry form. Generally these contain polymeric carbohydrates like starch and cellulose, and also fats and proteins. Typically lignin type of components and cannot be easily converted by normal anaerobic digesters. The digestion can be described in a simplified form in the following way. Polymeric components will have to be broken down into smaller molecules in a process stage called hydrolysis before they can be further converted. In the hydrolysis process sugars are produced from carbohydrates, from fats fatty acids are produced, and proteins are broken down to amino acids. These components are further converted into carbonic acids, alcohols, some hydrogen, carbon dioxide, and ammonia (acidogenesis) and further broken down to mainly acetic acid (and CO_2 and hydrogen) in the third process step called acetogenesis (see Chapter 18). Finally the methanogenesis takes place by methane-producing organisms, methanogens belonging to the domain of *Archaea*, nowadays considered to be a domain separate from bacteria and eukaryotic domains. The methanogenesis uses the intermediate products produced during the first steps of the process as far as they can be used and mainly in the form of acetic acid and hydrogen.

As not all components in the feedstock can be converted, a digestate will remain, which in some cases can be further processed, used as fertilizer, etc. Generally also a liquid watery effluent will remain to be treated and some of the by-products (like H_2S) will be transferred to the gas phase product which contains mainly CO_2 and CH_4. A very much simplified indication of an overall reaction could be described for glucose:

$$C_6H_{12}O_6 \rightarrow 3CO_2 + 3CH_4$$

Because the feedstock to the digester can vary much and the reaction pathways are much more complex in reality product gas compositions may differ widely. Table 1.9 gives a typical example of the ranges of composition.

Table 1.9   Typical composition of biogas

Component	% Volume
Methane	50–75
CO_2	25–50
N_2	0–10
H_2	0–1
H_2S	0–3
O_2	0–2

With respect to process configurations and operation parameters, digesters have been designed and operated for a wide range of conditions, depending on feedstock, plant capacity, operation temperature, and purpose of the plant. Plant ranging from simple small-scale units, for example, for producing gas for cooking in rural area in developing countries to highly sophisticated, large-scale units for codigestion in modern agricultural areas but also in industrial settings, for example, for generating energy from waste of processing food products.

The operation temperature of the digesters can be close to the environmental temperature in what is called mesophilic digestion (20°C–45°C with an optimum from 30°C–38°C) or in thermophilic operation of 49°C–57°C (even up to 70°C) (Chapter 18).

Different microorganisms will be present under these conditions. Thermophilic conditions promote the specific process rate, reduce pathogen survival but are more difficult to control and less stable.

1.2.3.1.2 *Reactor configurations*

From a reactor engineering point of view the digester system is relatively complex. Generally a two-phase feedstock (solid/liquid) has to be introduced and some products are also in the form of slurry. The nonconverted solid has to be removed from the system, a liquid waste stream and the gaseous product. The entrainment of the microorganisms has to be taken into account and the residence time of the different components has to be regulated for optimal results.

In the simplest systems batch operation is applied in a closed system using microorganisms contained in a previous batch as inoculation material. After the batch is completed and the product gas recovered the batch vessel is emptied. With several batch reactors operating in different process phases in parallel a semicontinuous operation is possible. In a real continuous operation the introduction of the feedstock and the withdrawal of products and waste streams are more or less continuously.

In a continuously operated system the flow and the state of mixing inside the reactor system are of particular interest, especially because of the residence time required in a typical digester is relatively long. For the solids it may be 90 days and for the liquid phase from 1 hour to one day. If a well-stirred reactor would be used several problems would have to be dealt with. For the different stages of the process, optimal conditions may vary. Methane producing microorganisms need a restricted pH range, while acidogenic bacteria, by their products, generate a low pH (Chapter 18).

Also the regulation of the hold-up of the different solids present and the gas separation need attention. However, for all these problems a wide range of design solutions have been found like agglomeration of the microorganisms, control of the internal flows, and distribution of materials inside the single vessel. A famous example is the up-flow anaerobic sludge blanket system (UASB) of Wageningen, which found widespread application (see Chapter 17).

Sometimes more than one stage is applied, usually two. In such a system the methane formation step can be separated and optimized separately at the cost of a higher plant complexity. Complete separation of the process steps will be difficult or impossible.

1.2.3.1.3 *Application*

Anaerobic digestion is used worldwide for cleaning up effluents and sewage treatment. Here often destruction of organic waste is the main purpose and energy generation from the produced methane is welcomed to reduce the energy consumption. Small-scale digesters are used in large numbers and worldwide for local energy production for cooking, etc. In large digester facilities the produced biogas can be burned in reciprocating engines or turbines, preferentially generating both electricity and heat, and can be considered as a renewable (biomass-derived) power source.

Biogas, depending on its production feedstock, may contain a varying fraction of trace components like H_2S. For the removal of these components effective and cost-effective removal processes are available (amine washing, addition of $FeCl_2$, etc.). Also for CO_2 removal if desired, process measures and postproduction washing steps are available (see Chapter 18).

The effluent flows of the digesters in the form of different or mixed solids streams or slurries require special attention. Often further processing such as composting is required to allow its use as compost, fertilizer, or other products.

The liquid effluent (methanoginic digestate) could also be used as a fertilizer in most cases and/or will need further processing (e.g., aerobic oxidation).

1.2.3.2 Ethanol production

Ethanol production is probably one of the oldest processes manipulated by man for the production of alcoholic drinks. Ethanol as liquid energy carrier is produced nowadays in large quantities and used in different blending ratios in transportation fuels (see Chapter 19). The main production route for biomass-based ethanol is fermentation of sugars. These sugars originate from pre-converted products of agriculture. The most important feedstocks are sugarcane in warm climate zones and corn and sugar beet in temperate regions. Other sources for sugars are used like sorghum and cassava. Many of the production routes are based on agricultural products which are in fact important basic components for food or feed application and their application for fuel production has been under intense debate and still is. These products are called first generation biofuels. Utilization of other agricultural and forestry products, not in competition with food and feed like lignocellulosic coproducts or waste are also possible sources of sugars to be fermented to alcohol in which case they fall in the so-called generation II biofuels. If lignocellulosic biomass is specially grown for fuel production still there is debate about the land utilization and the lifecycle analysis and many more issues. However, generally the arguments in favor outweigh the objections in these very complicated issues and this also holds for some specific generation I routes (see, for example, Chapter 20).

For three generation I routes a typical (simplified) production scheme is given in Fig. 1.56, namely, for cane, corn, and beet sugar.

We will discuss especially sugarcane and corn because of the large production presently in operation (e.g., in 2012, 28.3 Mton of sugarcane-based ethanol was produced in Brazil [see Chapter 19] and 39.4 Mton of corn-based ethanol was produced in the United States [54]). For sugarcane if circumstances allow the cane is harvested and cut by mechanical harvesters and brought rapidly to the mill to avoid deterioration. At the mill, roller crushers separate sugarcane juice and the fibrous matter, called bagasse. Bagasse is used in boilers, where it produces steam and electricity. Electricity is generated nowadays in excess and can be delivered to the main grid. To the heated juice several cleaning agents are added (like lime, sulfur, and thickener). After filtration and sieving, the clean juice is sent to the fermentation process. In many cases depending on the market the sugar plants have the option as well of producing various amounts of sugar after evaporation and crystallization/separation.

The alcohol-producing process by fermentation of sugar solutions derived from corn is roughly the same. It is practiced on a huge scale in the United States. Here, however, a substantial amount of fossil energy is introduced in the whole chain of production of ethanol starting in corn production and the conversion process to fuel ethanol. The situation of fuel alcohol from corn will probably be strongly influenced by the increased fossil liquid and gaseous fuels recovered with new techniques like fracking in the United States [55]. The production from sugarcane in Brazil went through a few decades of continuous improvements to reach the present impressive results (see Chapter 19).

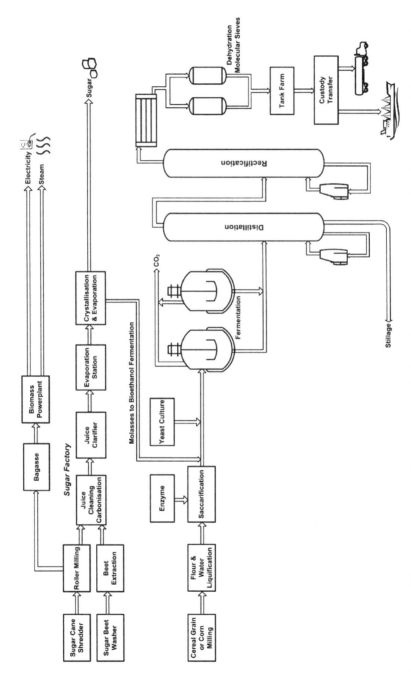

Figure 1.56 Simplified process scheme for production of ethanol from sugarcane, from cereal grain or corn and from sugar beet. Adapted from http://labroots.com/user/activity-group/index/id/350/title/fermentation-technology.

1.2.3.2.1 *Second-generation bioethanol based on lignocellulosic feedstocks*

The sugar used for the ethanol fermentation can originate from a wide range of lignocelluloses biomass. If these feedstocks would actually be used for ethanol production it would increase the capacity for production of fuel ethanol enormously. One reason is that a much larger fraction of the total biomass can be used in the process, and moreover many low-value biomass types can then be converted to ethanol, like agricultural waste, or by-products (such as corn cobs, straw, etc.), forestry products, and forestry waste, high-productive-energy crop species like switch grass, *Miscanthus*, etc. These potential feedstocks are not in direct competition with food and often do not even require additional agriculture land. Of course the sugars from lignocelluloses are not directly available like in sugarcane nor can they be readily produced from starch like from corn. They are much more difficult to be freed from the sugar polymers like in wood from cellulose and hemicelluloses entangled in the protective lignin cages.

It should also be realized that there are also thermochemical routes to liquid fuels using lignocelluloses like via gasification to syngas from which liquid fuels including ethanol can be produced via catalytic routes or biological routes (see, e.g., Chapter 20). Nevertheless producing ethanol via fermentation of sugar, recovered from their polymers (cellulose, hemicelluloses) from lignocelluloses at economic conditions and with low greenhouse gas emissions is considered by many the holy grail for bioethanol production. As explained in Chapter 20 this involves several steps:

1. Size reduction of the feedstock to increase access
2. Pretreatment to remove the lignin and free the sugar polymers
3. Hydrolyzation of these polymers to sugar (e.g., sugars such as glucose and C-5 sugars such as xylose)
4. Fermentation of these sugars to ethanol
5. Production of fuel-grade ethanol via distillation and dehydration (and denaturation), with, of course, some form of wastewater treatment being installed

For each of these steps, several alternative routes are available and moreover from an economical and practical point of view it could be tempting to combine two or more of these steps, although this might as well introduce new challenges due to possible suboptimal operation of the individual steps in the combination. Chapter 20 gives an excellent overview of the world wide efforts to create a viable process which appear to approach the first steps toward mass production of lignocellulosic ethanol. Conversion of sugars of lignocelluloses origin to other fuels than ethanol from fermentation is developed by Virent [56] in their bioforming process setup. Inside the process hydrogen and intermediates of oxygen containing molecules are produced via aqueous phase reforming over a proprietary catalyst at a temperature of 450–475 K and pressure of 10–90 bar. The intermediates are further converted over a catalyst (ZSM-5) to produce gasoline type of products or via alternative routes to kerosene of diesel. If this process reaches full commercial scale under economic conditions it would deliver drop in fuels for gasoline, kerosene, and diesel.

1.2.3.3 Wet thermochemical processes

Several processes have been developed to recover the energy content or to produce fuels from wet biomass and even for diluted suspensions or solutions of biomass components by thermochemical means. A few examples of such wet feedstocks are wood suspensions, sludge from wastewater purification, manure, organic waste from chemical plants or from food processing, etc. See also Chapter 21.

These processes generally operate at temperatures over 100°C and even up to 600°C–800°C and pressures from 10 to 400 atm. Both catalytic and noncatalytic processes are used. The general idea of the high pressure is to prevent the water from evaporating as far as possible at one particular boiling point. Depending on the conditions of pressure and temperature the transition of the liquid water phase to the gas phase (steam) requires a large amount of heat—the heat of evaporation. This heat of evaporation can be recovered, in principle, from the products but in practice at lower temperature where it is less valuable. Preferably the heat required to bring the (wet) biomass to the reaction conditions should be recovered from cooling down the products, as is sketched in Fig. 1.57. In such a countercurrent heat exchanger the feed is heated up, while the products are cooled down. The heat exchanger must be efficient to make such a process viable and this can best be realized if the temperature profiles show a uniform temperature difference between the hot and cold flows. In general the wet processes can be divided into two categories:

- Processes at a subcritical condition of water
- Processes at a supercritical condition

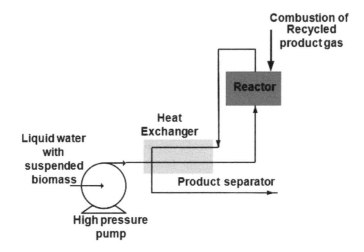

Figure 1.57 The heat exchanger is a very important part of a process for biomass conversion in hot compressed water. The heat exchanger need to recover the heat contained in the products and heat up the feed stream in a countercurrent. Reactions may already start in the heat exchanger and this may lead to fouling.

The supercritical condition of water occurs at the combined condition of $T > 374°C$ and $P > 218$ atm. (see Fig. 1.58). Under these conditions there is no steam or liquid water anymore. There is just a single type of fluid approaching this condition, called supercritical water. Approaching this condition in a transparent vessel where both phases are present (e.g., in a quartz capillary) one can see that the meniscus of the gas/liquid interface suddenly disappears.

The physical and chemical properties of supercritical water differ considerably from those of subcritical water (and from steam), although many properties change gradually around the critical point [57].

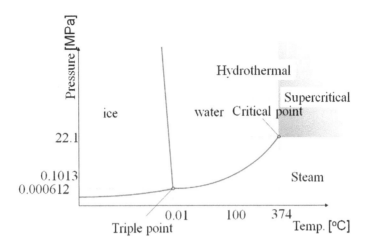

Figure 1.58 Phase diagram of water indicating the phase transition lines between ice, water, and steam. The area where supercritical conditions occur (dark-shaded area) and the area where mostly hydrothermal processes are carried out (light-shaded area) are indicated.

1.2.3.3.1 *Overview of the area of processes in hot compressed water*

To get a global view of what can be expected we will consider the following two equilibrium reactions as in a model component for biomass:

Glucose-reforming reaction with water, producing hydrogen and CO_2

$$C_6H_{12}O_6 + 6H_2O \rightarrow 6CO_2 + 12H_2$$

And methane production from glucose

$$C_6H_{12}O_6 \rightarrow 3CH_4 + 3CO_2$$

These reactions are only an indication for the calculations of thermodynamic equilibrium. It is not suggested that they occur in such a form. Nevertheless they are helpful for the overview. From Fig. 1.59 it is clear that from a single equilibrium calculation it can be expected that for a dilute solution (1% glucose in water) the selectivity of hydrogen formed over the formation of methane is very high and above 600°C almost no methane (as model component representing hydrocarbons) is expected. The situation for higher concentration (25% weight feed, dotted lines) is rather different and at 600°C still a lot of methane is predicted.

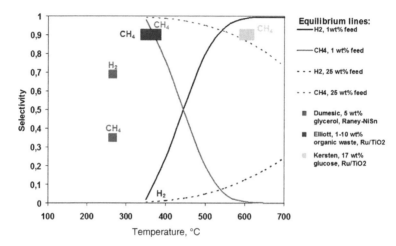

Figure 1.59 Conversion in hot compressed water. Simplified overview of prediction of equilibrium values of CH_4 and H_2 for 1% and 25% glucose solutions in water as a function of temperature and experimental results at various conditions.

Experiments by Kersten et al. have confirmed the presence of much methane here. At lower temperatures, at subcritical conditions, thermodynamics in itself would suggest methane (or hydrocarbons) as the end product.

Elliot (see Chapter 22) showed that with the use of suitable catalyst in the subcritical temperature area, biomass products can be converted to methane and CO_2 with high yields. Also Naber and Goudriaan (see Chapter 21) have shown that at Hydro Thermal Upgrading® (HTU®) conditions, oil-like products with reduced oxygen content are produced, releasing at least part of the oxygen from the biomass as CO_2, somewhat similar to the methane production, but methane is not the desired product in their process. Rather different from the results and this vision on overall equilibrium are the results of Dumesic [60, 61], who showed that at relatively low temperatures with a special catalyst, small oxygenates like methanol, ethylene glycol, and glycerol could produce hydrogen-rich gas streams using a catalyst or Raney–nickel catalyst promoted with Zn. Although with relatively low concentration of glucose also significant amounts of methane were produced, it nevertheless shows the magic of catalysis. The chemical bond C–O cleaving and the methanation are suppressed by the catalyst, while the C–C bond breakage is actively stimulated, and via the water–gas shift reaction $CO+H_2O \rightarrow CO_2 + H_2$ the hydrogen production is promoted. The methane production reactions are thus inhibited and hydrogen ($+CO_2$) produced for selective kinetic reasons far away from equilibrium. This may open the way to the production of several liquid fuels (hydrocarbons or alcohols).

The process step is nowadays called a more general term: "aqueous phase reforming."

1.2.3.3.2 State of affairs and potential of wet thermochemical processes

Starting from the right side of Fig. 1.59 we will first discuss supercritical gasification.

Although supercritical water is nowadays used at a relatively large scale in advanced power stations to increase the highest temperature in the steam cycle for efficiency reasons, its use in a chemical reactor for biomass conversion is considerably more complex. First there is the feeding problem to such a high-pressure and high-temperature reactor, especially complex if biomass slurry has to be processed. Moreover the complicated heat exchanger required; the corrosion occurring at these harsh conditions, requiring special expensive materials; the necessity of a stable catalyst for this demanding conditions; coke formation potentially causing blockage; and salts which are generally hardly soluble in supercritical water and may form depositions, all these points are still far from solved. Therefore these supercritical conversion processes are still in a R&D phase and at lab scale. Technical scale pilot plants were not yet very successfully developed to demo plants. Small-scale pilot plants have been pioneered by Antal (see, for example, Ref. [62]) and somewhat larger plants were operated at the University of Twente (see Fig. 1.60) together with BTG BV and subsidized by NEDO (Japan) and a still larger plant VERENA (100 kg/h) was operated at the KIT-IKFT in Karlsruhe [63].

Figure 1.60 Pilot plant for supercritical gasification at the high-pressure laboratory at the University of Twente. For safety reasons this unit was placed in a concrete safety box.

In the near future the concept will probably be further researched mainly on laboratory scale. At the University of Twente some tiny quartz tubular reactors resisting high temperatures and pressures were applied for continuous operations but mostly for batch operations (see Fig. 1.61). Metal reactors often produced confusing and irreproducible results due to wall catalytic effects further complicated by corrosion. Throw-away quartz capillaries together with a special system of fluid-bed ovens, shakers, and closed capillary crackers and special developed analysis system allowed mapping of a wide range of possible process conditions [58, 59]. Despite the present slow progress in this area the concept of supercritical gasification remains appealing as all kinds of by-products or wastewater streams within a biomass fuel factory or biorefinery could be converted into hydrogen-rich and pressured gases. A technical breakthrough is, however, still awaited.

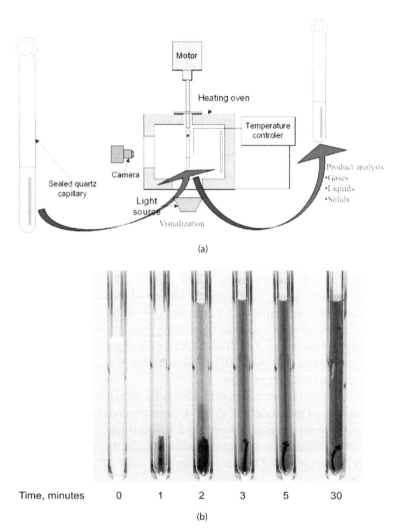

(a)

Time, minutes   0   1   2   3   5   30

(b)

Figure 1.61   (a) At the University of Twente a safe capillary high-throughput experimental technique was developed for reaction under hydrothermal and supercritical conditions. (b) At time $t = 0$ the capillary is cold, but then, after about two minutes the final temperature is reached, as can be seen from the expansion of the water. The wood stick is "dissolving" to a large extent in the water and after 30 minutes only a flint of char is left [58]. The quartz capillaries can stand the high pressures and temperatures, even at supercritical conditions, but the diameter and wall thickness should be adjusted (see Refs. [58, 59]). See also Color Insert.

1.2.3.3.3 Subcritical thermochemical processes for wet biomass streams

For processing of such streams at least four different types of processes can be distinguished, aiming for:

- methane production;
- improved solid and liquid fuels from difficult biomass waste for furnaces and power stations;
- liquid transportation fuels and blending components; and
- production of hydrocarbon liquid fuel components based on aqueous phase reforming.

A typical example of catalytic *methane production* via reforming of biomass in hot compressed water production has been presented by Elliot et al. from the Pacific Northwest National Laboratory in Washington (see Chapter 21). The operating pressure was about 200 atm. and the temperature 350°C. They used different reactors among which were continuously operated fixed-bed catalytic reactors. They were able to demonstrate the continuous production of high-concentration methane containing gases from a wide variety of biomass feedstocks at high conversions [of organics expressed as chemical oxygen demand (COD)] and high values of carbon transfer to gas-phase components, close to 100%.

If such a process can be made sufficiently robust and economical, it could represent a generally applicable conversion step for different biomass streams, waste streams, and by-products of biomass fuel production, chemicals, and food production. Catalysis is essential in this field and Chapter 21 is presenting an in-depth discussion of this subject.

Heating up wet biomass under pressure may serve also other purposes than methane production. It opens the possibility of the *production for improved solid, liquid, and gaseous fuels from difficult-to-handle feedstocks*. A wide range of processes has been developed in this category. To give an example EnerTech Environmental Inc. (USA) [64] developed a process for converting sewage sludge into an energy carrier called "slurrycarb process," the final product being called "E-Fuel." Several other similar processes have been developed, processing a wide variety of wet waste feedstocks (including Turkey offal!) as reported by changing world technologies [65].

A hydrothermal conversion process to produce (blending) fractions *for transport fuels* from wet biomass was developed by Shell

Research BV, a development which was taken over later by Biofuel BV (the HTU® process see Chapter 21). The primary main product aimed for is biocrude, an oily type of product with a considerable lower oxygen content compared to the feedstock biomass and suitable to be refined to different liquid transportation fuels. A more or less similar process but operating with a homogeneous alkaline catalyst in combination with a heterogeneous zirconia catalyst has been developed (mentioned in Chapter 20). A third example operating at slightly higher temperature and pressure was developed by Aalborg University and further developed by Steeper Energy [66]. This hydrofaction® also employs a basic catalyst (K_2CO_3). At the University of Twente, like for supercritical conditions, a high-throughput testing system was developed for rapid investigation of different conditions based on throw-away quartz capillaries which allowed also visual inspection during reaction conditions (see Fig. 1.61).

A rather different approach based on aqueous phase reforming of biomass fragments like sugars and related conversion products was developed by Virent "Bioforming" on the basis of the earlier mentioned pioneering work of Huber et al. [61].

The aim is here to produce intermediates which can be converted in a refinery setup to different liquid fuels like gasoline. Excess hydrogen produced in the aqueous phase reforming steps (or from other sources) is used in different feed preparation and refining/finishing steps. This process would be able after biomass fractionation/pretreatment to accept second generation (lignocelluloses) feedstock to produce a wide range of drop in fuels replacing fossil-derived transportation fuels. While this technology looks extremely interesting (Virent has established partnerships, among others, with Royal Dutch Shell) we still have to wait for the technology to reach commercial stage under economically valid conditions.

1.3 Future of Biomass in Energy Generation

1.3.1 Introduction

It can be expected that in the coming 50 years mankind will grow to 9–10 billion individuals, a somewhat smaller number than followed from earlier estimates but still an enormous population, difficult to be supported in a sustainable manner by our planet. People will still need energy in the form of food; feed for their cattle; and energy for agriculture (mechanization, fertilizers, etc.); processing and transport of food and cooking; heating, cooling, and in-house activities; and communication. They will equally need energy for materials to be produced from raw or recycled materials: metals, cement, polymers, chemicals, paper, glass, etc.

Moreover, in modern and future societies transportation of people and goods/materials will request a considerable fraction of the total energy budget. Groeneveld [23] estimated an average energy consumption for a future mankind of 4–5 kW per person on a continuous basis. This is more than twice the average consumption of today's population but less than half of the average consumption of present-day US inhabitants (12 kW).

For a future lifestyle of high quality for all people, the efficiency of energy generation for satisfying our needs should be considerably increased. Only this would allow the poor people of the world to satisfy their rightful desire for a better life. The UN "Millennium Development Goals" formulated in 2000 as target for 2015 are to be followed up by the post-2015 goals, already formulated by the UN high-level panel on the post-2015 agenda for the 2030 target (see Table 1.10).

Table 1.10   Post-2015 agenda, as formulated by the UN high-level panel

- Leave no one behind: end hunger and ensure that every person achieves a basic standard of well-being.
- Put sustainable development at the core: only by mobilizing social, economic, and environmental action together can we eradicate poverty irreversibly and meet the aspirations of eight billion people.
- Transform economies for jobs and inclusive growth.
- Build peace and effective, open, and accountable institutions for all.
- Forge a new global partnership: a new spirit of solidarity, cooperation, and mutual accountability must underpin the post-2015 agenda.

While these five "large transformations" give certainly worthwhile and challenging targets it is not so clear how these can be met from an energy generation point of view apart from the call to put sustainable development at the core. This should be clearer from a second large global movement, although not directly connected to the UN thrust. On the initiative of the US Academy of Engineering embraced among others by the academies of UK and China, a list of the engineering grand challenges was published [68] which indicate the routes toward a successful response to the challenges faced by mankind. Here the connection with technology, science, engineering, industrial development, and policy is made explicitly. Of the engineering grand challenges a few are addressing the energy field explicitly, like solar energy, fusion, and carbon sequestration.

Solar energy includes nearly all renewable energy sources as indicated in Fig. 1.8 and Fig. 1.9 and it should be given an extra high priority as the successful response to many of the other challenges heavily depend on the availability of inexpensive, renewable and sustainable energy. With some exaggeration one could say that this energy challenge is "the mother of all challenges."

Humans and the rest of the living creatures depend on the energy flow of the sun's radiation which allows their existence and the material cycles involved (containing of the elements C, H, O, N, S, P, K, etc.) of which the carbon cycle deserves most of the attention. In the future, the carbon cycle should be closed for the present biosphere without a large input of fossil carbon.

Carbon has to be recovered by plants mainly from the atmosphere in the form of CO_2 at an extremely low concentration of 0.04% volume and brought to a high concentration in the plant material. This material includes stored hydrocarbons and fats which form the main fuel reserves for plants and all other creatures like animals and human beings who depend on plants for their energy to live.

For the humans, living in such high numbers on earth their own energy was not sufficient and several other energy sources like animal power, wind, burning biomass, etc., were introduced in several forms. As discussed before the worldwide use of fossil fuels caused the increase of CO_2 concentration in the atmosphere with more than 40% compared to preindustrial times as is nowadays generally accepted. This value causes concern about climate change in which the CO_2 concentration plays a dominant role at least with a high probability according to the Intergovernmental Panel on Climate Change (IPCC) panel's report [30]. For many scenarios the model calculated increase in temperature will not remain below

2°C accepted as a maximum allowable increase of the temperature of the atmosphere. The precautionary principle would be guiding us to a rapid phase out of fossil energy carriers unless the CO_2 problem can be effectively solved, for example, by carbon capture and storage (CCS). This CCS, although not contributing to the ideal of true carbon recycling, prevents fossil carbon from accumulating in the atmosphere and buys time for carbon-free energy generation like photovoltaics (PV), wind, etc., and true carbon recycling using biomass or even recover previously emitted carbon from the atmosphere via biomass. CCS development seem to be more difficult than anticipated and activities seem to slow down in recent years. The route away from fossil fuels is a technological challenge without precedent. The problem is not that there will be a shortage of fossil fuels in the coming century. Proven reserves of coal and natural gas are still sufficient. The increasing accumulation of CO_2 in the atmosphere, together with the global distribution of these fossil resources and the insight that all material flows should be closed for a sustainable future, is the main driver.

As explained in Section 1.1 the solar input of renewable energy is extremely large but distributed over a large area and current exploitation (PV, wind, etc.) requires in most cases some form of storage. The relative growth rate of these renewable sources is high but it has been argued [69] that as soon as contribution to the world's energy production would pass a few percent, history has shown that growth slows down for several reasons and it would take several decades for a real large impact. This remains to be seen, however, if this is valid for "drop in's" energy flows like electricity and synfuels which can be considered as drop-ins to be used in existing infrastructures.

1.3.2   The Future Role of Biomass

Biomass at first glance seems to be an excellent replacement for fossil fuels as it is globally based on the same scheme: "combustion" of carbon and hydrogen containing fuels. All forms of fossil fuels can, in principle, be synthesized from biomass. However, the quantities of fossil fuels consumed at present are much higher than can ever be met by biomass without too large interference with the cycle of life on our planet, as will be explained below.

Yearly about 8 Gton carbon of fossil origins is emitted into the atmosphere (see Fig. 1.18), while the biomass on earth is exchanging 60–80 Gton.

To replace the 8 Gton fossil carbon with carbon of biomass we need 1.1–3 times that quantity as the combustion value of fossil fuels per atom carbon is somewhat higher (see Table 1.11) and to compensate for conversion losses if, for example, lignocellulose biomass has to be converted to a liquid hydrocarbon. The carbon efficiencies are relatively low for these processes (in the range of 0.6–0.3). This would mean that at the global recycling flow of carbon of the biomass, up to 20% should become man manipulated. This seems to be dangerously high. With supporting and convincing arguments Kopetz (Chapter 3) from the World Biomass Energy Association proposed for 2035 +/– 150 EJ biomass primary energy consumption (three times the present value), which would correspond to a carbon flow of 4.8 Gton carbon inside the carbon cycle flow of life, corresponding to 6% of this flow, and this seems to be a realistic possibility. So a lion's share of future energy in 2035 should come from renewable energy on a noncarbon basis. Moreover a reduction in use of energy should come from energy savings and lifestyle changes (dematerialize our needs, etc.). Even so, if this 150 EJ of bioenergy is to be reached in 20 years from now it deserves a lot of our attention and efforts.

Table 1.11   Heating values of different fuels

Compound		LHV (MJ/kg)	HHV (MJ/kg)	HHV per mol C (kJ/mol C)
Carbon dioxide	CO_2	0.0	0.0	0.0
Methane	CH_4	50.2	55.7	890.6
Hexane	C_6H_{14}	44.8	48.4	693.9
Butane	C_4H_{10}	45.4	49.2	714.0
Octane	C_8H_{18}	44.5	48.0	683.8
Diesel	$C_{12}H_{24}$	44.0	47.2	660.5
Gasoline	C_8H_{18}	44.5	48.0	683.8
LPG	C_3H_8	46.1	50.1	734.3
Methanol	CH_4O	19.9	22.7	726.0
Ethanol	C_2H_6O	26.8	29.7	683.5
Glucose	$C_6H_{12}O_6$	14.1	15.6	467.1
Wood	$C_6H_{10}O_5$	17–19	20–22	585.9

Source: Derived from *Handbook of Chemistry and Physics*, edited by David R. Lide, 84 Ed., ISBN 0-8493-0484-9.

1.3.3 Will Developments of Biomass Conversion Technologies Be Ready, and What Conversion Technologies of Biomass Will Be Flourishing?

Of course within the framework of this chapter only very general observations can be made. The first point is competition with fossil fuels. While in a lot of places and applications due to special circumstances, biomass technology can survive and even slowly grow, generally on a cost basis competition with coal, gas, and oil is difficult. It should be realized, however, that the harmful and potential very dangerous long-term effects of fossil fuels are not incorporated or at least insufficiently taken into account in the economies. Worldwide stringent and reliable measures like taxation, caps, and other actions will be required to shift the balance. This is the most important factor for the future of biomass and many other renewable and sustainable energy sources. Nevertheless, with the present insufficient action to promote renewable energy and especially biomass the technological developments have already been remarkable in many areas. In Chapter 4 an interesting historical overview on what happened in Europe, supported by the EU, can be found.

Let's first consider biomass in heat applications. These are of great interest and according to Kopetz it will show the largest expansion of biomass energy toward 2035. Several technologies are available to support this. If we want to leave no one behind, the smallest scale of biomass utilization namely cooking in the developing world as practiced by billions of people request a rapid introduction of safe, efficient, and affordable cookers. Van der Sluis in Chapter 6 gives an introduction into this subject and an interesting example. Efficient charcoal production for cooking is extensively discussed by Stassen (Chapter 14). Also methane fermenters can be used for heating and cooking (see Chapter 18). At a larger size central heating systems for individual houses went through great improvements of safety, efficiency, and improved automation and delivery of pellet-sized wood and wood chips.

For larger systems suited for densely populated city areas (half of the world population lives there) several solutions for district heating can be applied, such as retrofitting of gas and oil furnaces. Next to direct cofiring, gasifiers can be installed for cofiring of biomass with fossil fuels. Fast-pyrolysis oil production to allow compact biomass storage and transport over large distances and firing in existing furnaces is an emerging option. A wide variety of solutions which could be faster developed if competition with fossil fuels would be gradually improved by governmental action. In some countries (such as Sweden, Hungary, and Italy) rapid growth of biomass utilization in this way is already realized. For electricity generation almost the same holds.

Biogas producing fermenters are used for electricity production (and of course for replacement of other tasks fulfilled by natural gas) and on the largest scale (firing biomass, like wood, in large [coal] power plants, see Chapter 7) is now a mature technology. Gasification in cofiring may considerably broaden the range of application if economics can be improved and also pyrolysis when developed to fully commercial operation may strongly promote biomass to electricity and fuels. In comparison to most other renewable electricity (based on PV, wind, thermal solar, etc.) the storage capacity of biomass and its conversion products is a great asset and systems based on mixtures of renewable sources will be attractive.

With respect to transportation fuels we can expect that electricity will be important as an energy source, especially in the cities, but for air traffic and long-distance transport, especially for trucks and shipping a liquid fuel has great advantages and is difficult to replace. Biofuels have taken an important place already and because they can be considered to a certain extent as drop-in fuels bioethanol and biodiesel can be applied without too much changes of the infrastructure: the mass produced and surprisingly cheap internal combustion engines, turbines, etc.

If lignocelluloses feedstocks can be the bases for producing these fuels under economic conditions this can increase biofuel production enormously. Apart from production via fermentation of the sugar fraction to alcohol, thermochemical processes may become very important in the future. Pyrolysis and corefining with mineral oil (see Chapter 16), lignin, and sugar conversion to fuel-blending components may contribute to the liquid fuel pool but certainly syngas production via biomass gasification and its conversion to diesel, gasoline, methanol, ethanol, etc., could still become important routes to biofuels from lignocellulose feedstocks. Although it should not be given a high priority according to Kopetz and while syngas production via biomass gasification met in the past with a lot of hurdles, we would plead here for a strong R&D program for developing liquid (and gaseous) fuels from biomass. This could even occur in synergetic lining up with electricity production from other renewable sources (wind, PV, etc.). This will be explained below.

1.3.4 Biomass-Based Fuels for Storage and Transport with Input from Other Renewable Sources

Other forms of renewable solar-based energy sources like PV and wind primarily produce electricity. To cope with daily and seasonal variations one of the options is to use the excess renewable electricity for water electrolysis. This option seems to be attractive as electrolysis of water can be 75%–80% efficient on the basis of the electrical energy input. This efficiency can, in principle, further increase at higher temperatures and also pressurized versions could be developed. Of products via the reaction $2H_2O \xrightarrow{\text{electrical energy}} 2H_2 + O_2$, usually hydrogen formed at one of the two electrodes is considered as fuel for peak shaving at high demands of electricity via combustion in turbines or fuel cells. However, hydrogen is not easy to store and therefore also its conversion with CO_2 to methane or other hydrocarbons is considered and studied for power stations [70].

Further development may also lead to simultaneous electrolysis of CO_2 and H_2O leading to syngas mixtures of CO and H_2 on one electrode and oxygen at the other electrode. This oxygen is, in principle, available for all kind of purposes, among others, for oxygen combustion or gasification of biomass.

Although electrolysis is studied and demonstrated for peak shaving of power stations its cost is presently still high because of capital cost and almost linear upscaling due to parallelization of cells. Hydrogen could be introduced in several ways in processes to produce synfuels for storage or for transportation fuels based on biomass or other carbon sources (see Fig. 1.62). The first route (1) is the usual route in which part of the biomass is used (oxidized) for upgrading. The carbon efficiency (defined as kg C in product over kg C in biomass feed) is low (0.6–0.3) because of the CO_2 emission.

Carbon efficiency is important if green carbon has a high value and fossil CO_2 emission is expensive, for example because of taxation. The higher value of the improved fuel and its value as a drop-in product to replace fossil fuels represent the value of this route. Route 2 shows the introduction of renewable "carbon emission free" hydrogen in upgrading leading to a higher carbon efficiency.

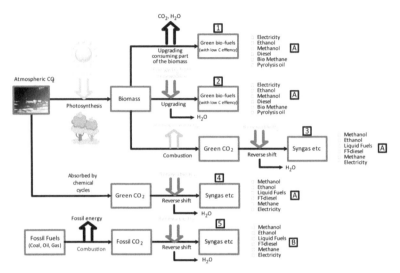

Figure 1.62 Introduction of carbon-free renewable energy via green electricity into biomass-to-fuel routes. Two related routes without biomass are also displayed (4 and 5). Electricity is mentioned in the product slate to indicate that all the fuels can be stored for electricity demanded peak shaving.

In route 3 the biomass is combusted its energy content used and the "green" CO_2 is upgraded to new fuels with hydrogen allowing for a large storage capacity of excess electrical energy.

Route 4 uses directly absorbed CO_2 by chemical means and would have the same advantages at route 3, thus bypassing biomass altogether. It requires the recovery of CO_2 from a large volume of air by some sort of technical solution. Living biomass realizes this with self-assembling systems, creating a very high specific surface area for CO_2 absorption.

Route 5 uses "fossil CO_2" recovered from the flue gas (11% CO_2) and is therefore not green but uses the fossil carbon twice if later emitted to the atmosphere. This route is studied by electricity companies mostly aiming at methane as intermediate storage fuel. Only routes 1–4 should have a license to emit the (short-cycle) CO_2. These routes look attractive but require future developments of the electrolysis systems to low-capital-cost devices and cheap electricity from renewable carbon-free sources (or fusion and "clean" fission). Cheap electricity happens at certain moments and places already, but it remains highly uncertain. It is clear, however, that if "fossil CO_2" is to be made more expensive, short-cycle CO_2 from biomass could be an interesting product for storage of green energy by making fuels out of it with a license to emit the CO_2 in a complete recycle.

1.4 Conclusions

1. Vegetation, plants in different forms, has created the biosphere and the atmospheric composition fit for complex life with a very stable and robust carbon cycle: Vegetation remains our principal ally to keep it that way.

2. Humans in rapidly growing numbers supported by advancing technologies were not satisfied by muscular power, animal power, wind, and combustion of biomass. From 1800 on they fuelled their civilizations by fossil carbon-based sources and developed very efficient heat and power generation via efficient and often surprising cheap and mass-produced equipment, like motors, turbines, transport systems, etc. In a later stage the whole electricity system was built up on this basis via generators, electromotors, lightbulbs, etc.

3. This explosive expansion now has reached its limits as carbon (as CO_2) builds up in our atmosphere due to a fossil input which is >10% of the life's carbon cycle. Mankind has to stop this high input of fossil carbon (in CO_2) as it threatens our lives by drastic climate changes in the coming decades.

4. Future energy generation should be based on fossil carbon-free (or at least lean) energy production. Nearly all energy will come directly or indirectly from the solar radiation which reaches our earth's surface and atmosphere several thousand times' the human needs. However, the most important future renewable sources (PV, wind, etc.) lack sufficient storage capacity and have a fluctuating output (day/night, seasons), etc.

5. Biomass cannot replace all fossil fuel, although after conversion all fuel types (and most chemicals, materials, etc.) can be produced from biomass. Within 20 years its production could reach 150 EJ/year (see Chapter 3), the equivalent of about 0.3 times the present fossil fuel production. It can play, however, an essential role in reaching the all renewable society because of providing long-time storage, decentralized backup, certain forms of transport fuels like for aviation, shipping, long-distance trucks, etc.

6. A wide range of conversion processes have been developed for biomass to improved fuels (solid, gas, liquid) which all have their specific application in very different sectors/markets.

7. Biomass pretreatment and combustion for heat and power is well developed at all different scales. Cofiring and retrofitting to modern equipment have been proven. Gasification to combustion gas plays a well-established role here. For developing countries on the smallest scale (cooking, etc.) much has to be done.

8. Syngas from biomass is promising for synfuels but already for a long time. Economics and dilemmas caused by the typical scale of operations have not been solved yet.

9. The relative young technology of fast pyrolysis is wrestling itself to the market, despite its advantages. Economics are not easy and competition with fossil sources tough. Successful large-scale demos are required.

10. Vegetable oil for biofuel has no technical issues anymore.

11. Fermentation to biogas went through decades of development, found widespread applications, but has still promising development items for all markets, from developing areas to high-tech industrial applications.

12. Fermentation to bioethanol reached widespread application. Brazil made enormous progress in several decades and the United States (corn) is also a large producer. Using lignocellulosic feedstock is under development and facing its moment of truth.

13. Wet thermochemical processes are not yet well developed but remain attractive in future biorefineries and decentralized applications.

14. Biomass conversion to top fuels comparable to the best fossil fuels often involves important conversion losses and results in limited carbon efficiency (say 25%–45%) and energy conversion (say 25%–50%). Introducing cheap (peak) electricity (e.g., from seasonal variations) could result in much higher carbon efficiencies. Also "green" CO_2 could provide storage capacity for renewable electrical energy. Electrolysis of water and water/CO_2 mixtures can provide syngas with high efficiencies and subsequently synfuels with a license to emit the green CO_2 again everywhere.

15. Fossil fuel has to carry the (high) cost of climate change before a fair competition can take place with all types of renewable energy. There is enough fossil fuel in the earth but we should leave most of it there and catch the carbon from our atmosphere and bring it back in a closed cycle. Plants can be very helpful reaching these aims.

Acknowledgments

The contributions to this chapter by Stijn Oudenhoven, Rens Veneman, Sam Odu, Peter Jacobs in the art work, and Yvonne Bruggert in the redaction are gratefully acknowledged. Marijke Buitink is acknowledged for language advice.

References

1. NASA Solar System Exploration, http://solarsystem.nasa.gov/planets/profile.cfm?Object=Sun&Display=Facts.

2. Phillips, K.J.H. (1995). *Guide to the Sun*, 47–53, Cambridge University Press. ISBN 978-0-521-39788-9

3. Niele, F., *Energy Engine of Evolution (Shell Global Solutions)*, Elsevier. ISBN-13:9780444-52154-5

4. Pringle, H. (2012). Quest for fire began earlier than thought, *Science NOW*, Association for the Advancement of Science.

5. Hassan, A.Y., Hill, D.R. (1986). *Islamic Technology: An Illustrated History*, 54, Cambridge University Press. ISBN 0-521-42239-6

6. Research Machines ple. (2004). *The Hutchinson Dictionary of Scientific Biography*.

7. Jenkins, R. (1936). Savery, Newcomen and the early history of the steam engine, in *The Collected Papers of Rhys Jenkins*, 48–93, Cambridge, Newcomen Society.

8. Hulse, D.K. (1999). *The Early Development of the Steam Engine*, TEE, Leamington Spa, UK. ISBN 85761-107-1

9. Reynolds, P. (2003). George Stephenson's 1819 Llansamlet locomotive, in *Early Railways 2: Paper from the Second International Early Railways Conference*, 165–176, Ed. Lewis, M.J.T., Newcomen Society, London.

10. Wise, D.B. (1974). Lenoir: the monitoring pioneer, in *The World of Automobiles*, 1181–1182, Exec. Ed. Ward, I., Orbis, London.

11. Feynman, R. (1964). *The Feynman Lecturer on Physics*, Vol. 1, Addison Wesley, USA. ISBN 0-201-02115-3

12. Smith, J.M., van Ness, H.C., Abott, M. (2005). *Introduction to Chemical Engineering Thermodynamics*, 7th edition, McGraw Hill Higher Education. ISBN-13 978-0071247085

13. Braams, C.M., Stott, P.E. (2010). *Nuclear Fusion: Half a Century of Magnetic Confinement Fusion*, Research Taylor & Francis. ISBN 978-0-7503-0705-5

14. Fleischmann, M., Pons, S., Hawkins, M. (1989). Electrochemically induces nuclear fusion of denterium, *Journal of Electroanalytical Chemistry*, **261**, 2A.

15. International Energy Agency (2012). *Key World Energy Statistics.*

16. NASA, January 14, 2009, *Climate and Earth's Energy Budget.*

17. Fridleifsson, J.B., Bertani, R., Huengens, E., Lund, J.W., Ragnarsson, A., Rybach, L. (2008). The possible role and contribution of geothermal energy to the mitigation of climate change, in *IPCC Scoping Meeting on Renewable Energy Sources*, 59–80, Eds. Hohmeyer, O., Trittin, T., Luebeck, Germany.

18. ASKJA Energy (2014). The independent Icelandic energy portal, http://askjaenergy.org/iceland-introduction/iceland-energy-sector/.

19. Rance Tidal Power Station, http://en.wikipedia.org/wiki/Rance_Tidal_Power_Station

20. *The REN2/Renewable Global Status Report* (2012), see also www.ren21.net.

21. International Energy Agency WE 0-2013 Traditional biomass use database.

22. Palz, W. (2011). *Power for the World*, Pan Stanford, Singapore. ISBN 978-981-430-337-8

23. Groeneveld, M.J. (2008). *The Change from Fossil to Solar and Biofuels Needs Our Energy*, Introductory lecture, University of Twente, October 2008 (printed by the University of Twente).

24. En.wikipedia.org/wiki/artificial photosynthesis.

25. Mahmoudkani, M., Keith, D.W. (2009). Low-energy sodium hydroxide recovery for CO_2 recovery from air, *International Journal of Greenhouse Gas Control*, **3**, 376–384.

26. Maegaard, P., Krenz, A., Palz, W., Eds. (2013). *Wind Power for the World: The Rise of Modern Wind Energy*, Pan Stanford, Singapore. ISBN 978-981-4364-93-5

27. Maegaard, P., Krenz, A., Palz, W., Eds. (2013). *Wind Power for the World: The Rise of Modern Wind Energy, International Reviews and Developments*, Pan Stanford, Singapore. ISBN 978-981-4411-89-9

28. Post, J.W., Veerman, J., Hamelers, H.V.M., Euverink, G.J.W., Metz, S.J., Nijmeijer, K., Buisman, C.J.N. (2007). Salinity-gradient power Evaluation of pressure retarded osmosis and reverse electro dialysis, *Journal of Membrane Science*, **288**(1–2), 218–230.

29. http://www.shell.com/global/future-energy/scenarios/2050.html.

30. IPCC (2014). *Climate Change 2014: Impacts, Adaptation and Vulnerability*, WGII AR5 summary for policymakers, http://ipcc-wg2.gov/AR5.

31. Huang, Y., et al. (2006). Biomass co-firing in a pressurized fluidized bed combustion (PFBC) combined cycle power plant: a techno-environmental assessment based on computational simulations, *Fuel Processing Technology*, **87**(10), 927–934.

32. Knoef, H.A.M., Ed. (2012). *Handbook Biomass Gasification*, 2nd edition. BTG Biomass Technology Group BV, the Netherlands.

33. Higman, C., van der Burgt, M. (2011). *Gasification*, Elsevier Science. ISBN 978-008-0560-90-8

34. Marshall, H.A., Smits, F.C.R.M. (1982). Exxon catalytic coal gasification process and large pilot plant development program, *Proceedings of the 9th Annual International Conference on Coal Gasification, Liquefaction and Conversion to Electricity*, 357–377, Pittsburgh, PA, August 3–5, 1982.

35. Nanou, P. (2013). Biomass gasification for the production of methane, 173, PhD thesis, University of Twente, Enschede.

36. http://www.nuon.com/nieuws/nieuws/2012/toekomst-willem-alexander-centrale-   buggenum-onzeker/ (visited on 01-12-2014).

37. Van de Beld, L. (1996). Cleaning of hot producer gas in a catalytic adiabatic packed bed reactor with periodic flow reversal, in *Development in Thermochemical Biomass Conversion*, 907–920, Ed. Bridgwater, CPL Press.

38. http://www.ecn.nl/docs/library/report/2005/c05009.pdf.

39. Yamada, S., Shimizu, M., Miyoshi, F. (2004). Thermo select waste gasification and reforming process, *JFE Technical Report*, 21–26.

40. Van der Meijden, C.M., Veringa, H.J., Van der Drift, A., Vreugdenhil, B.J. (2008, June). The 800 kWth allothermal biomass gasifier MILENA, in *16th European Biomass Conference and Exhibition*, Valencia, Spain, ECN-M-08-083.

41. Masson, H.A. (1986). The AVSA fluid bed combustor-gasifier project, in *Advanced Gasification*, Eds. Beenackers, A.A.C.M., van Swaaij, W.P.M., D. Reidel. ISBN 90-277-2212-9

42. Henrich, E. (2007). The status of the FZK concept of biomass gasification, *2nd European Summer School on Renewable Motor Fuels*, Warsaw, Poland, August 29–31, 2007.

43. Glaser, B., et al. (2001). The "Terra Preta" phenomenon: a model for sustainable agriculture in the humid tropics, *Naturwissenschaften*, **88**(1), 37–41.

44. Kersten, S., Garcia-Perez, M. (2013). Recent developments in fast pyrolysis of ligno-cellulosic materials, *Current Opinion in Biotechnology*, **24**(3), 414–420.

45. Knezevic, D., et al. (2009). Hydrothermal conversion of biomass. II. Conversion of wood, pyrolysis oil, and glucose in hot compressed water, *Industrial and Engineering Chemistry Research*, **49**(1), 104–112.

46. Scott, D.S., Piskorz, J. (1982). The flash pyrolysis of aspen-poplar wood. *Canadian Journal of Chemical Engineering*, **60**(5), 666–674.

47. Hoekstra, E., et al. (2012). Heterogeneous and homogeneous reactions of pyrolysis vapors from pine wood, *AIChE Journal*, **58**(9), 2830–2842.

48. Brem, G. (2000). *Gasification, Pyrolysis and Liquefaction of Biomass at TNO*, Presented at ECOS 2000.

49. Wagenaar, B.M., et al. (1994). Pyrolysis of biomass in the rotating cone reactor: modelling and experimental justification, *Chemical Engineering Science*, **49**(24, Pt 2), 5109–5126.

50. www.kior.com (visited on 01-12-2014).

51. Pyne newsletter 04, http://www.pyne.co.uk/Resources/user/PYNE%20Newsletters/PyNews%2004.pdf.

52. http://www.nesteoil.com/default.asp?path=1,17765,17766,17780,17781 (visited on 01-12-2014).

53. *Smithfield Foods Fact Sheet: Brief History of Anaerobic Digestion*, http://www.smithfieldfoods.com/media/39085/history-of-anaerobic-digestion.pdf.

54. http://www.afdc.energy.gov.

55. https://nicholas.duke.edu/cgc/HydraulicFracturingWhitepaper2011.pdf.

56. http://www.virent.com/news/virent-receives-epa-registration-for-its-bioform-gasoline/ (visited on 01-12-2014).

57. (2012). *Supercritical Water: A Green Solvent; Properties and Uses*, 218p, Ed. Yizhak, M., John Wiley & Sons, Hoboken. ISBN 978-0470889473

58. Knezevic, D. (2009). *Hydrothermal Conversion of Biomass*, PhD thesis, University Twente.

59. Potic, B., et al. (2004). A high-throughput screening technique for conversion in hot compressed water, *Industrial and Engineering Chemistry Research*, **43**(16), 4580–4584.

60. Cortright, R.D., Davda, R.R., Dumesic, J.A. (2002). Hydrogen from catalytic reforming of biomass-derived hydrocarbons in liquid water, *Nature*, **418**, 964.

61. Huber, G.W., Shabaker, J.W., Dumesic, J.A. (2003). Raney Ni-Sn catalyst for H_2 production from biomass-derived hydrocarbons, *Science*, **300**, 2075.

62. Antal, M.J., Allen, S., Schulman, D., Xu, X., Divilio, R.J. (2000). Biomass gasification in supercritical water, *Industrial and Engineering Chemistry Research*, **39**, 4040.

63. http://www.ikft.kit.edu/english/138.php.

64. http://www.epa.gov/ncer/sbir/success/pdf/enertech_success.pdf.

65. http://peswiki.com/index.php/Directory:Changing_World_Technology_-_Thermal_Depolymerization_Process.

66. http://steeperenergy.com/about-u/what-we-do.

67. http://www.un.org/sg/management/pdf/HLP_P2015_Report.pdf.

68. http://www.engineeringchallenges.org/Object.File/Master/11/574/Grand%20Challenges%20final%20book.pdf.

69. Kramer, G.J., Haigh, M. (2009). No quick switch to low-carbon energy, *Nature*, **462**(3 Dec.).

70. http://www.eon.com/en/media/news/press-releases/2013/8/28/eon-inaugurates-power-to-gas-unit-in-falkenhagen-in-eastern-germany.html.

About the Authors

Wim van Swaaij studied chemical engineering sciences and obtained his PhD at the Technical University in Eindhoven, the Netherlands. From 1965 to 1972 he worked for Shell Research BV, for two years in Nancy (ENSIC) but mainly in Amsterdam, on fluidization, reaction engineering, and process development. His last position was as section leader for gasification processes.

In 1972 he became full professor of chemical engineering sciences and later of biomass at the University of Twente (UT), the Netherlands, where he and his coworkers studied fundamentals of multiphase reactors, developments of novel high-intensity reactors, energy conversion technologies, polymerization reactors, and biomass-to-fuel conversion processes. He has (co-)authored over 500 scientific papers and several handbooks and promoted 60 PhD students.

From 1992 till 2002, Prof. Van Swaaij was scientific director of the Dutch National Graduate School on Process Technology (OSPT) and of the Process Technology Institute Twente (PIT). He is/was a consultant for several multinational firms: AKZO, DSM, Unilever, Shell, Solvay Duphar, TNO, BASF, and Cortis-Dow. He is a member of the Royal Netherlands Academy of Sciences (KNAW) and other learned societies and a honorary member of the Royal National Society of Dutch Engineers. He has received several prizes and rewards for his scientific work, like the "Grand Prix du Génie des Procédés" of the French Academy of Sciences (1996), doctorate honoris causa by the Université de Nancy (1996), the Royal Dutch Shell Prize (2004), and the Golden Tesla medal for energy research (1999). Prof. Van Swaaij is Knight in the Order of the Netherlands Lion.

 **Sascha Kersten** holds a degree in chemical engineering from the University of Amsterdam, the Netherlands. His master's thesis was on modeling and control of complex chemical plants. He began his career at the ECN (the Dutch Energy Foundation), working on modeling of dynamic systems. At the ECN he also started to work on biomass conversion. In 2002, he finished his PhD thesis concerning biomass gasification in circulating fluidized beds (Prof. Van Swaaij).

Since 2003 Sascha is working at UT. Currently he is the chair of the Sustainable Process Technology Group. He is working on energy-efficient separations, CO_2 capture and utilization, water purification, recovery/recycling of scarce elements, and thermochemical conversions of biomass.

Chapter 2

Bioenergy: Global Potentials and Markets

Wolfgang Palz

Retired European Union official, Renewable Energy R&D, European Commission, Brussels, Belgium

w@palz.be

2.1 Introduction

Bioenergy is a major player in the world's energy markets. It is a key component of the recent emergence of the renewable energies (REs).

In 2013, the "modern" forms of biomass contributed 7% to the world's energy consumption. It is the lion's share of the 10% all REs together contributed to meet the global demand for energy [1], not to mention the 9% of traditional biomass for cooking. About 2600 million people in the villages of the less developed countries (LDCs) rely on it for essential needs [2]—in reality the source of major health problems for them and environmental degradation.

In Europe, a kind of benchmark region for the world's energy markets, two-thirds of the 13% RE contribution to energy supplies

Biomass Power for the World: Transformations to Effective Use
Edited by Wim van Swaaij, Sascha Kersten, and Wolfgang Palz
Copyright © 2015 Pan Stanford Publishing Pte. Ltd.
ISBN 978-981-4613-88-0 (Hardcover), 978-981-4669-24-5 (Paperback), 978-981-4613-89-7 (eBook)
www.panstanford.com

comes in the forms of modern bioenergy for heating and cooling, fuel liquids for transport, and electricity.

Since the 1980s the world's reliance on modern bioenergy never stopped increasing. And the trend will no doubt continue. In March 2014, one of the major chiefs of the energy business in France recognized that "there is a new energy world on the horizon; and we are resolved to commit us for this cultural and industrial revolution." Remarkable for a nuclear country's energy representative . . .

Important quarters, in particular in Europe, are aiming at a 100% RE world on the horizon of 2050. By then, various sources expect a bioenergy contribution of up to 30% of global energy consumption. While most of the other REs like solar and wind energy occur only as intermittent sources, bioenergy covers a central role for security of energy supply. Without bioenergy, 100% RE scenarios on commercial grounds are unrealistic.

The basics of biomass are indeed very different from those of more direct conversion of the sun's energy such as photovoltaic (PV) and wind power; they are much more complex. Biomass involves a lot of resources—a variety of waste streams and crops—a bunch of conversion technologies, be they thermal or biological and chemical, and also many logistical aspects, the incidence of trade, and last not least the many utilization schemes.

Among the REs, bioenergy has often become a particular target of attack by some environmentalists. It is true that in practice there have been examples of unsustainable uses of modern biomass, too. In particular, some have tried to associate the terrible deforestation that is ongoing in much of the developing world with biomass production and use. Others dislike the production of fuel liquids on arable land. We will address these problems in a paragraph on environmental ethics here below.

It is a trivial fact of life that everything that grows on the earth will eventually end up as biomass. Even the fossil energies like coal, oil, and gas are examples of it. We are today living from an important "side effect," the oxygen the green matter growing on the earth's surface for 1000 millions of years have left behind in our atmosphere. All that oxygen, 21% of the atmosphere, was produced by water splitting through photosynthesis.

It is often forgotten that we, humans, are an integral part of the biosphere. We better respect it if we do not want to harm ourselves.

We all have in our digestion system more than a pound of bacteria—we could not live without them. A professor at Collège de France in Paris explained to us the other day: "Take a teaspoon filled with simple ground from your garden; it contains a billion of bacteria and other creatures."

2.2   Photosynthesis and Production Yields

Professor Melvin Calvin from Harvard put it like this: "An ideal solar collector has already been designed. Requiring virtually no maintenance, it is economical and non-polluting; it uses an established technology and stores energy. It is called a plant."

Photosynthesis can be described by the following equation [3]:

$$6CO_2 + 6H_2O + \text{Sunlight} = C_6H_{12}O_6 + 6O_2$$

The actual complexity of photosynthesis in a simple leaf is such that even today it is not completely understood. A minimum of eight photons are required simultaneously for the process. The decomposition of water and the resulting liberation of O_2 through photosynthesis were experimentally established by Ruben and Kamen in California only in 1941. If oxygen were not permanently reproduced by the living cover of the earth's plants, in about 3000 years all of it in the atmosphere would have been consumed [4]. Globally photosynthesis produces an estimated 220 bn (1 billion = 1000 million) dry tons of biomass per year, equivalent in energy value to 10 times global energy use.

The maximum photosynthetic efficiency is 6.7% [3]. In comparison to PV this is poor. Today a commercial PV panel has up to 25% light conversion efficiency. A fuel, hydrogen, can be generated by subsequent electrolysis at an overall efficiency of 20%. But hydrogen is not fit for feeding animals and us. For life on the earth nature has done the right choice, carbohydrates produced by photosynthesis!

The maximum efficiency on the ground is achieved with "C4 plants." Their name comes from the first product in the photosynthesis process that is a four-carbon sugar. Examples are maize, sorghum, and sugarcane. They are a lot more productive than

"C3 plants." The latter come as trees, rice, wheat, and soybean and account for 95% of the global plant coverage.

The maximum efficiency of C3 plants as compared to C4 plants is given as 3.3% only. This has to do with extra losses through photorespiration and light saturation at lower light intensities.

In practice maximum biomass production rates in dry tons per hectare per year, taking into account locally available irradiation from the sun are estimated by the authors in [3] as follows: 136 tons/ha·year (tons per hectare·year) in England for C4 plants and half of this for C3 plants. For the US farm belt they calculate respectively 208 tons/ha·year for C4 plants and half of it for C3. Note that these are theoretical maxima efficiencies. In practice as we shall see later we will be talking of 10 to 30 tons/ha·year.

2.3 Biomass Resources

First come to mind the residues of biomass. The energy content of residues from forestry and agricultural activity, including dung, amounted in 1985 globally to 111 EJ/year[1], which corresponded to one-third of all commercial energy use at that time [3]. Residues amounting to approximately 30% to 50% of agricultural and forest production can be estimated globally for over 5000 million tons of that production. In Europe the amount of straw from cereal harvest comes in the range of 100 million tons/year. Part of it is of interest for energy production. Residues from the production of wood from forestry and its use are considerable. They are estimated at more than 60% of harvested raw wood. The Global BioEnergy Partnership (GBEP) in Rome estimates that 65% of the world's RE is of woody origin. It is for most part derived from such residues.

In the 1990s the potential of technically recoverable residues for energy purposes in the European Union (EU) was estimated at almost 10% of energy consumption [5]. The actual production of energy from solid biomass was identified for the EU in 2013 as 82.3 Mtoe [6]. That is in the range predicted in 1990. Most of that solid biomass comes from residues. There is also an EU directive on biowaste or municipal wastes (2008/98/CE). It concerns wastes such as organic wastes from gardens and parks, kitchen waste,

---

[1] EJ (Exa Joule) = 1000 Peta (10 to the 15th) Joule = 23.9 Mtoe = 278 TWh_{th}

etc. The EU has produced 7.7 Mtoe from such wastes in 2009 [7]. Separately one has to take into account wastewater, sludge, etc. They are an important feedstock for biogas of which the EU has produced 8.3 Mtoe in 2009.

In terms of biomass crops, 4100 million ha of forestry existed globally in 1987 compared to an agricultural cropland of 1480 million ha [3]. The forestry on stock was 417 000 million m^3, or approximately 200,000 million tons of living material: an incredibly high amount. The annual increment is of 12,000 million m^3. A closer look shows that **globally the wood cover is shrinking**. This phenomenon is badly affecting the LDCs, whilst it is typical for the Organisation for Economic Co-operation and Development (OECD) countries like the United States, Germany, France, etc., to harvest only less than half of the annual forest increment: in the industrialized countries a lot of CO_2 is currently fixed in the growing forests, thus helping to alleviate climate change. From 1990 to 1995, 56 million ha of forests have disappeared worldwide but in the industrialized countries the forest coverage increased by 9 million ha [8]. Just in 2012, Germany has planted 350 million trees and Poland more than 1000 million: Europeans like their trees! Nevertheless, the conclusion is, **conventional forestry is underexploited in the industrial countries and offers itself as a sizable resource for new energy supply.**

Moreover we got the large potential for energy plantations. Short-rotation forestry (SRF) and plantations of grasses, switch grass, *Miscanthus*, and the like and dedicated crops like sorghum and others are suitable contenders for woody feedstock generation. A lot of experience has been accumulated with them over recent years. A good example is Brazil. The country lacks hard coal reserves and produces since many years charcoal for its industrial needs, for instance, in the steel industry. Brazil produces 6.9 million tons of charcoal per year via SRF plantations employing eucalyptus trees [2]. The growing time for eucalyptus SRF in Brazil is two to three years. In temperate climates one employs growing times for SRF trees of up to seven years.

Energy plantations raise in general terms the question of excessive extraction of nutrients from the ground. This should be avoided at least to some extent by leaving in the field during harvest the leaves and other plant material that is of not much interest for energy use. It will also help to bring the ashes that contain the extracted minerals back to the fields after combustion of the biomass.

With respect to the global energy crop potential on a large scale one has first to address the question of land availability for new plantations. A first opportunity is marginal land. Such land may have been discarded for the production of food or animal feed because of its lower productivity. But for energy plantations the quality criteria aren't as high. Thanks to a new mobilization for energy crops such marginal land will gain in biological quality and economic value.

A more difficult question is the conversion of high-quality agricultural land to the generation of energy instead of foodstuff or animal feed. Again, as in many cases when addressing the global biomass issue, one has to distinguish between the case of the industrialized countries and that of the developing world. They are most different from each other. In the OECD countries we notice—despite immigration of the poor, Latin American (LA) people to the US, Africans to the EU—that the number of inhabitants is by and large stable and food consumption stagnates. On the other hand, productivity of food and feed production on domestic land increased continuously over many years in the past by 1% to 2% per annum. And it keeps increasing. Europe had to deal with "milk lakes," "meat mountains," and other descriptions of the economic nonsense to produce an excess of high-value food stuff for which there is no market. Long ago already, the EU Commission as the main player of the common agricultural policy (CAP) had to address the problem. Policies changed over the years, but always well sticking to the principle that Europe must stay autonomous for its food supply and its farmers must be protected. Almost at any cost. EU directives introduced early on the principle "Set Aside" land, sometimes obligatory, sometimes on a voluntary basis, allowing farmers even to produce alternatives to food and feed such as biomass. In more recent years the trend went toward direct payouts to farmers for not using their land for agricultural crops with additional obligations to maintain its biological value.

Already in 1992 an interservice group of the EU Commission bringing together nine of its directorate generals (DGs) competent on the matter met on my initiative and produced a report on its deliberations [9]. One of the findings was that the Set Aside regime of 15% for all arable crop producers was available for nonfood purposes.

On the horizon of 2025, a quarter of Europe's agricultural land is becoming available for energy plantations with an even higher proportion beyond.

In the United States one-third of the cropland was already kept idle in 1990 [3] for the same reasons described above.

There is a wide variety of suitable species for the generation of woody material on marginal and converted arable land. In Ref. [10] 90 different species are reviewed. Choices depend on the local climate, the final market destination and the expected production cost, transformation, and logistical considerations. Harvesting should not exceed half of total production cost. Hauling over 40 km is considered uneconomical. The energy ratio, that is, the energy input for the establishment, fertilizers, herbicides, harvesting, and hauling as a fraction of harvested energy is highly positive as it lies typically between 10 and 15.

One of the best yields of up to 65 dry tons/ha·year has been achieved with eucalyptus in Brazil. Sorghum can yield in temperate climates 30 dry tons/ha·year; hybrid poplar, and switch grass can yield 15 dry tons/ha·year; and salix up to 12 dry tons/ha·year. Ultimate production costs are dependent on the details of implementation. In general terms, when products come as pellets and chips on the market, they are currently considerably cheaper than the competing electricity and oil and gas from fossil sources.

So far we have addressed energy plantations on land. Water is also an important potential supplier of biomass that was early recognized when interest in modern bioenergy was arising in the 1970s. A plant of choice are the algae. They have a high conversion efficiency of the sun's light and come in a wide variety of types and sizes that grow in fresh or saline waters. Microalgae grow with some of the best efficiencies. Development of this route has, however, not led to commercial results. One of the reasons is that harvesting and drying of the material is cumbersome and eventual costs are prohibitively high. Energy plantations of algae in special ponds turned out to get infected from undesirable pollutants in the open air. Protection under glass cover generates material of too high cost for energy purposes, but of potential interest for the chemical of pharmaceutical industry.

Lastly, on the review of resources one has to address the production of feedstocks for liquid biofuels as they are employed for transport. In 2013, the world produced 115,000 million liters of such biofuels, 3.5% of all transport fuel consumption [11]. The main feedstocks are sugarcane, corn, sugar beet, wheat, and others for ethanol, and palm tree oils and colza for bio-oils. The recurring question is **whether the energy balance of alcohol production is positive at all.** It is indeed lower than that for woody biomass generation, but it is definitely positive when residues from the plantation are employed as energy in the distillation process instead of fuels of fossil origin [9].

The generation cost of bioethanol was for a long time subsidized to make it fit for the market, a usual phenomenon for a newcomer on any market. Nowadays it is, generally speaking, cost competitive with conventional transport fuels [12].

The highest yields are obtained in Brazil with sugarcane. Brazil started the "Pro-Alcohol" program in the 1970s and gains today half of its needs for gasoline from alcohol. To achieve this tremendous result an important step was the introduction in 2003 of the **flexible fuel market**, whereby car engines are tuned to accept by a simple switch either full ethanol or a blend.

The United States has for climatic reasons no other choice than to use corn (maize) that is less productive. It does so in ever increasing quantities since the turn of the century and has arrived at an average of 10% alcohol in its gasoline by 2012. At this level alcohol has run into a "blend wall" that was since two years flattening the increase in the expectation of new federal regulations to increase the use of ethanol even further. In the longer run, the United States is very ambitious for the further introduction of alcohols for transport: within a renewable fuel standard (RFS) there is a mandatory goal to expand the use of ethanol[2] from 9 bgy (billion gallons per year) in 2008 to 36 bgy in 2022.

Europe doesn't play in this area in the same league as Brazil and the United States. A particularity is that it is the subject of harsh attack by European environmentalists who mix up the issue with deforestation in the Amazon and oil palms in Indonesia. In particular the macroeconomic advantages of biofuels for Europe are

---

[2]Ethanol comes 0.64 toe/ton, 27 MJ/kg, 20 MJ/L, 115 bn L = 2.3 EJ = 55.2 Mtoe.

too often left out in the discussion: thanks to the by-products of colza oil generation suitable for animal feed, Europe could stop the import of soybeans for that purpose; likewise does the DDGS by-product of the generation of alcohol from wheat displace the import of protein for feed. **For every ton of sugar beet or cereals used for ethanol production one-third will re-enter the food chain as protein or gluten** [13].

Anyway, as a consequence of the opposition of some environmentalists, an EU directive of 2009 demanding a 10% biofuel blend in its transport fuel by 2020 was downgraded since then.

In 2013 the EU had a consumption of 13.6 Mtoe of biofuels, 4.7% of its consumption of transport fuels. Unlike the world market leaders in the United States, **Europe has the particularity to use four times more biodiesel than bioalcohol in its transport.** European feedstocks are sugar beet or wheat that have respectively a typical productivity yield of 5.3 and 1.9 tons/ha.year of alcohol. The yield of oil generation from colza is at 1.3 tons, even lower than that of alcohol production from wheat. Keeping these figures in mind and the higher costs associated with lower yields for bio-oils, the predominance of biodiesel in the European market is surprising.

One way of interpretation is the fact that in the general European transport market diesel is by and large dominating. It is financially favoured by governmental regulations as excise taxes for diesel are lower than for gasoline. The initial idea was certainly that diesel cars deserve preference as their engines are more efficient—until one had recently to recognize that diesel exhaust emissions are more dangerous for health. As a result of this situation, refineries sit on an excess of gasoline and bioethanol generation would only worsen the situation.

2.4 Biomass Conversion and Application

Firing biomass for cooking and heating, a tradition going back to prehistoric times, is still used by humans today on a large scale.

In the LDCs 2600 million people still rely nowadays on biomass [2]. It is expected that this number will not change very much on the horizon of 2030. Woody material and dried dung is fired in cookstoves and other stoves in a most unsustainable way. Millions of people die each year from the emissions of the primitive stoves used. Collection of fuelwood raises enormous environmental and social problems.

The energy efficiency is a disaster on its own. The efficiency of the stoves is less than 20%. It is even worse for charcoal that is the preferred fuel in the towns and cities of these countries—with an increasing trend as more and more people come to live in cities. The traditional earth kilns that are in use for charcoal production have no better efficiency than 8% to 15%. Used in stoves of a small efficiency of roughly 20% the overall efficiency for use of the fuelwood input is actually no better than 2%! A detailed report of the GBEP/Food and Agriculture Organisation (FAO) points out the many hurdles to change this terrible situation to the better [2]. A "Global Alliance for Clean Cookstoves" was initiated and together with UNEP "The Climate and Clean Air Coalition."

Modern forms of biofuels come as pellets and chips next to the traditional hardwood. They are not employed in the LDCs but in the industrialized countries.

In the industrialized countries the firing of wood in decentralized home installations and the like is currently much expanding.

For instance, in France 7.4 million people used fuelwood for heating in 2012, 30% more than 10 years before.

A most convenient fuel are wood pellets[3]. They can be used as floating fuel similar to oil in automating heating installations. In 2011 Germany used 1.4 million tons of pellets in 155,000 household burners.

In 2011 the world produced 18 million tons of wood pellets. Most of them are consumed in Europe—11.4 million tons, of which some 20% were imported. This equals some 8% of all solid biomass consumption in Europe.

Next to individual heating installation one has to mention the heating networks. They used to be very popular in the communist countries in Eastern Europe—and were supplied by heat from much polluting coal. In modern forms they are promoted on a large scale in Northern Europe with biomass. Even in France 400,000 households are heated today by networks supplied by the heat generated from biomass. It is important to note that the majority

---

[3]Dry wood or pellets come roughly at 5 kWh/kg; a ton of dry wood equals roughly 2 m^3.

of the heating networks in Europe are fed by the heat component of combined heat and power (CHP) systems. In this way the overall efficiency of use of the biomass is very high, generally well over 50%. The power stations employed for CHP are of intermediate size. Were they of excessive size would the heating networks become too large, expensive, and inefficient. For instance, nuclear power plants that use to be very large never work as CHP plants.

In recent years biomass supply has become increasingly popular for large power plants as well, even though they are not of the CHP type. Employing larger shares of biofuel instead of coal provides lower CO_2 emissions and can meet stricter regulations in view of the climate change issues. There exist now already many cofiring power stations of biomass with coal in the range above 100 MW, namely, in Belgium, Poland, or the United Kingdom. **The largest dedicated biomass power plant is since 2012 in operation in Tilbury, UK.** At a power capacity of 750 MW it is supplied with 2.3 million tons of pellets to a fluid-bed combustion system.

Biomass can obviously also be converted to mechanical power only. On small scale, engines are available to produce power in different routes. The Stirling engines employ just the heat of contact to drive a piston. They are little developed. Other engines work like steam engines, employing the heat derived from biomass to produce steam first, just like locomotives. Remember that the first steam engine by James Watt in the late 18th century was powered by biomass. Today they lost all interest: efficiency was too low on that scale. The engines powered by a gasifier have virtually disappeared from the market today. But they had a prestigious past in Europe. Invented by a certain Imbert from the French army in World War I (WWI), passenger cars, buses, and all kind of vehicles equipped with small transportable wood gasifiers shared rallies on European roads and across the Alps in the 1920s and 1930s. In those days the driving force of promotion had become the German J. Linneborn. He had all in all by the end of World War II (WWII) produced 500,000 Imbert car gasifiers in industrial plants in France.

A large industrial sector is the thermal treatment of biomass to generate an intermediate product, instead of direct combustion. To this end the solid biomass is heated at various temperatures, depending on the product sought, under limited access of oxygen/air. Torrefaction of woody material is achieved at temperatures between 200°C and 320°C. The product is exempt of water but keeps much of its cellulose and hemicellulose components. It has an energy density twice as high than the feedstock. The process was in particular developed in view of the use of biomass in power plants.

At higher temperatures in the range of 400°C the biomass can be converted to pure carbon, that is, charcoal. The world produced 47 million tons of charcoal in 2011, a large share of it in the developing countries.

At 500°C it is possible to generate a liquid by flash pyrolysis. The biomass is first decomposed into micro particles and then exposed a very short time to the heat. It is a relatively cheap way to produce a liquid from the solid biomass. The problem are the many tar-like components of the liquid that requires upgrading, for instance with hydrogen. (The hydrogen can be obtained by conventional steam reforming). The process has been the subject for much development in recent years. At last a prototype of 50,000 tons/year was installed in Finland recently. The investment amounted to €30 million. The product is used for heating purposes.

At temperatures above 800°C one achieves gasification. Thereby the biomass is decomposed into H_2 and CO. The process is in commercial use on a large scale.

There have been some attempts in Europe to employ gasification as an intermediate step to produce a liquid as well. The most important was Choren in Germany, at times supported by Shell and Volkswagen. There is a choice of technologies to generate a suitable syngas from biomass, CO, and H_2. Combined in a later step, via methanol a commercial biofuel liquid for transport can be produced (they called it Sunfuel). Technically the process is viable, but the end product turns out to be excessively expensive. Similar experience had been achieved in South Africa after WWII with the liquefaction of hard coal that employed also the Fischer–Tropsch synfuel process. The Choren attempt had to be abandoned.

Next, one has to come to the generation of ethanol and oil liquids from biomass. It was mentioned earlier that they have seen considerable commercial success in recent years.

The conversion of sugar to alcohols that is gained from cane or sugar beat is achieved by yeasts, monocellular fungi. It is the same that it used since millennia for wine making. Producing ethanol from starch, such as corn and wheat, requires a preliminary decomposition of the starch to sugars via hydrolysis. One chooses an enzymatic process. All together it is an economical way to produce ethanol that is widely used today.

Since many years researchers in the United States and elsewhere have tried to develop a process to make cellulosic alcohol. Unfortunately it turns out to be very complex. So far the results have resulted in very uneconomical products. It sounds trivial that the lignocellulosic feedstock has to be broken down to sugars, but it is not. Decomposition is in principle possible via acid or enzymatic hydrolysis. The problem is the cost of the enzymes that have blocked so far any significant access to market implementation. Today bets are open: the US federal Administration targets for 2022 a major share of US alcohol consumption being cellulosic alcohol.

We are closing this round on biomass conversion by addressing the most important conversion agent of them all: bacteria and other microbiological stuff. They take indeed care of degrading and decomposing all living matter ultimately again to CO_2 and water. Without them we would rapidly sink and suffocate under a mountain of accumulating matter.

It turns out that the work of bacteria is highly complex. There is a wide range of different bacteria that rely on assistance of enzymes they may produce themselves or other microorganisms. An important distinction is whether they belong to the aerobic family that need air/oxygen for operation or to the anaerobic family that need operation outside of oxygen. The anaerobic ones were the first living species on the young earth that had no oxygen in its atmosphere yet. All humans have them in their intestines.

Of practical importance is as one alternative aerobic digestion generating compost that is an important fertilizer. Anaerobic digestion on the other hand is employed for biogas production. Biogas consists of methane and CO_2 in equal parts. The feedstocks that go into biogas production are liquid wastes, effluents, etc. Since some years the addition of whole plants of maize as introduced in Germany enlarged the range of such feedstock and increased the energy content. It extended the potential market for biogas in the country, too, as 90% of the residues available for biogas were already in use. **Germany, the European leader for biogas, derived in 2013 4% of its electricity supply from biogas in 7850 installations.**

Recently one has started injection of biogas when stripped of its CO_2 content into conventional gas networks. It could make up 15% "natural biogas" consumption in Europe by 2020. Like natural gas biogas becomes in this way also of interest for transport. For this innovation Sweden is ahead in Europe with 82 ktoe of biogas employed for transport in 2012.

In 2009 Europe generated 8.3 Mtoe of primary energy from biogas. From this amount 25 TWh of electricity were produced, less than a quarter of it in CHP plants. This corresponded to a quarter of all electricity produced from biomass in Europe that year. A major share of the biogas plants are owned by farmers: 4.3 Mtoe were gained from agricultural residues. Landfill gave 3 Mtoe of biogas that year and sewage sludge 1 Mtoe.

Last not least one has to remind the long-time world leadership of China for biogas. Already in the 1980s China had in its rural areas over 5 million production units running.

2.5 Ethical Environment of Bioenergy

Together with the overall criteria of cost competitiveness for bio energy in the mainstream of markets, biomass must come as an environmentally friendly energy or we better leave it aside in any strategy for the future. At a conference in Brussels in 1989 [14], Jacques Delors, the EU Commission president, pointed rightly out that "we had to look for harmony with nature and give up on a millennia old tradition in Europe and the West associated with Judeo-Christianity that man had to dominate nature, or as Francis Bacon had formulated it, that man is the centre of the world. No, man has responsibility toward nature!"

As a matter of fact, many environmentalists don't see bioenergy as part of such environmental ethics. As mentioned before a particular angle of attack concerns the large-scale generation of biofuels for transport from agricultural crops such as cane, corn, or oil palms. One is indeed tempted to mix up the fact that the industrialized countries—but also Brazil and countries in Africa—direct feedstocks for food and animal feed toward the energy markets whilst millions, maybe tens of millions of people are starving of hunger. Putting forward such arguments overlook the reality in the LDCs where disorganization, corruption, civil wars, and the like are among the elements that are responsible for the extreme poverty of large fringes of the population. I was myself employed for some time at the EU Commission Service in charge of helping the ACP countries with essential food aid for survival needs. A lot is done in the direction north–south. But nobody responsible can really want that—for instance, Europe supplies all of Africa with all its needs of food and feed. It would be the ultimate kill of the domestic African farmers. The UN FAO in Rome has particular competence on the subject. In a report on poverty and food security [15] it points to the small-scale bioenergy projects that can improve food security and help boost rural communities. Following the report **underinvestment for years in agriculture handicaps seriously food production and this coupled with rural poverty is a key driver of world hunger.**

Europe derived actually important benefits from its shy attempt to promote the bio fuels for transport. European car drivers will have noticed that leaded gasoline has disappeared from all filling stations. The elimination of lead was a consequence of the introduction of bioethanol in gasoline. This is not a minor achievement. Imagine that for decades European cars in the hundreds of million were polluting massively all the land with lead from the exhaust gases of their cars—because Europeans unlike the Americans, for instance—like those powerful engines that need high-octane gasoline. And for that one used the lead. By now all gasoline, not only E10, is blended with ETBE, an ether of bioalcohol. For low concentration it must not be displayed on the filling outlet so that the opponents of biofuel don't even realize that they got it in their tank. Biofuels have a high octane number and aren't polluting like lead.

An important reproach to bioenergy and all REs is that it is too costly and society can't afford it. It is currently an argument in Germany, where the government takes it to stop in 2014 most support for the renewables. It is true that the kilowatt-hour price to consumers in the country increased from 2000 to 2011 by €14 cents, but contrary to the claims of the opponents in industry and administration only €2.5 cents of it was due to the public support to the renewables.

In the United States, the conservatives predicted *economic apocalypse* when in early 2014 the Environmental Protection Agency came up with the proposal to cut 30% of the emissions of the existing power plants by 2030. But the argument that cutting emissions would imply immense costs was proven to be wrong. To cut emissions and achieve a positive effect building clean infrastructures would imply no more than 1% of cumulative growth over the next 20 years.

Conservatives all over the world claim that subsidies for RE development are excessive. In reality it is the old polluting fossil energies that are maintained in life with globally over $500,000 million a year. In a paper published in 2013 it was made public that in the EU €35,000 million went in 2011 as subsidies to atomic energy and €26,000 million to conventional fossil energies and a further €40,000 million to cover the consequences on health. That was a lot more than the €30,000 million budget in support of clean renewables.

Last not least there is the argument of a better quality of life with the REs. Nobody can deny that it is good that the farming community benefits from the market development of bioenergy. There is biogas and other income from market introduction of residues, new alternative crops. The rural world in the industrialized countries deserves such new support to maintain the population there in place and preserve an essential tissue of biological well-being and diversity.

Decentralized supply by bioenergy and the other renewables is more appropriate than the conventional highly centralized supply structures to meet the decentralized energy demand and create local jobs, industry, and income. It is the image of a better world, in sharp contrast to what is demonstrated presently, for instance, in Belgium and France, where the highly centralized nuclear electricity supply industry is putting forward the thread of possible power cuts.

Local communities have started organising the dissemination of RE installations on the ground. That may also give a new opportunity to involve the poor. Alone in Germany, Europe's leading industry nation, 320,000 people are currently cut from the power supply networks as they can't pay the bills.

2.6 Outlook: The Clean Energy Revolution

Bioenergy together with all the other forms of modern REs has benefited from a tremendous market growth worldwide. Since some

30 years an unstoppable dynamic has developed that is no doubt going to continue and even accelerate into the future. Energy has become a matter of the people. Individuals and communities alike were getting involved in widespread deployment. Business and wealth were coming about. About 38% of the 6.5 million new jobs that were created worldwide concerned bioenergy. All polls suggest that there is wide support of the people to furthering RE.

Together with an ever-increasing overall energy efficiency is the widespread proliferation of RE indeed the only credible route to limit climate change to an increase of the mean temperature on the earth to 2°C, whilst CO_2 concentration in the atmosphere is capped at 450 ppm—it stood at 400 ppm in 2013. Next we are going to illustrate with some key figures the likely growth for bioenergy and RE at large. Starting from 2013, the last year for which reliable energy statistics are available, we are going to summarize what some of the most relevant global operators in energy have found out on the horizons 2020, 2030, and 2050.

2.6.1 2020

The International Energy Agency (IEA), an arm of the OECD in Paris, anticipates for 2020 [11] the following global market increases:

- Biofuels for transport 140,000 million liters from 115,000 million liters in 2013.
- Biopower capacity 130 GW, an increase of 50% over the 88 GW in 2013.
- All RE power capacity, including hydropower 2555 GW; also an increase of 50% over the 1690 GW in 2013. For the industrialized countries it is expected that 80% of all new power installations will be RE.

The EU has decided in 2009 an "Energy and Climate" directive which has the following targets for 2020 which are legally binding. The trend in 2014 was that these targets would actually be met.

- In Europe as a whole, or on average in all EU member states, the share of the RE in final energy consumption will double to 20%.
- Within these 20% the lion's share will go to bioenergy: 13% of consumption in 2020 will come from bioenergy (175 Mtoe).

- The bioenergy contribution will roughly go for 60% to the heating market, 20% for transport, and 20% for electricity.
- 35% of all electricity will be of RE origin. 6.3% of all electricity supply will be from biomass.

For Germany, a leading country for RE, the federal government foresees for 2020 [16]:

- A share of 35% for RE to electricity supply. In 2005 it was just 10%, which increased to 28% in 2014.
- In 2013 biomass provided half of the 13% of RE supply—38% as solids and 12% as biogas.
- In 1990 bioelectricity provided a little 0.2 TWh; in 2012 it had exploded to 36 TWh.

2.6.2   2030

"Sustainable Energy for All" by the UN initiated by its secretary general in 2012 aims roughly on the horizon of 2030 at the following:

- Doubling the RE share in the global energy mix
- Doubling the average daily deployment rate of RE power capacity to 500 MW up to 2030
- Increasing and reorienting energy investments from $1300 billion a year to $1800 or $2400 billion a year

Bloomberg New Energy Finance in London is a rich source of analytical figures [17]:

- Modern REs (excluding large hydropower) will by 2030 account for 20% of global electricity generation from 5% now.
- Average capital investment for power generation will be on average $362 billion per annum for all the coming years. (This figure is not inconsistent with the UN figures here above, as those concerned all energy, not only power.)
- Power demand will only grow by 16% in Europe and 13% in the United States. Chinese demand will double. This will be little compared to the sixfold electricity consumption there between 1990 and 2011.
- China would be slowly switching away from coal. But by 2030 the country will still derive 46% of its electricity from coal.

On the other hand the further introduction of RE to 20% of electricity consumption will beat that of new gas-fired generation to 10%.

- Emissions would fall in Europe by 56% and in the United States by 13% but increase in China by 43%.
- A factor they expect to help RE and hinder the construction of new fossil-fuel power capacity will be the cost of capital.
- The world's energy supply will not be enough decarbonized without new policy intervention in developing countries. "We cannot simply rely on the market to deliver the outcome."

2.6.3 2050

The European Renewable Energy Council (EREC), which grouped (before its recent financial troubles) most of the EU's RE associations, came up with a 100% RE supply scenario for Europe:

- Europe's final energy consumption of 1200 Mtoe would fall by 15% toward 2050.
- Bioenergy would contribute some 30% of supply, doubling its share of 2020.

For Germany a whole variety of 100% RE supply scenarios have been published. We refer here to that of Fachagentur Nachwachsende Rohstoffe (FNR), a German federal agency for biomass [18]:

- Germany would have a total energy demand of 7000 PJ, all covered by RE.
- 32.4% would be met by biomass:
    - o 14.4% from energy crops
    - o 7.2% from forestry
    - o 6% from agricultural residues
    - o 4.8% from other residues

2.7 Conclusion

Until the Industrial Revolution in England some 200 years ago mankind relied exclusively on biomass for its energy needs. Animals provided the needed power. Electricity did not yet exist for exploitation by humans.

It followed the coal age, followed—as some of us will still remember—the oil age and the nuclear age. Now, after the turn of the century we have embarked on a new one, the Solar Age. We are not coming back to the Stone Age in this way. Our modern bioenergy is different from that of our ancestors. Typically it is also utilized to generate power. And bioenergy is complemented by a large range of newcomers, solar heat and solar PVs, wind power, and modern hydropower.

With all the bunch of clean, inexhaustible natural energies which we have brought to industrial and commercial maturity, unbeatable in cost on the markets, the way is open for a better world.

Renewable Power for the World, Power for the People, Power for Peace, Power for the Poor.

References

1. REN21, Paris, Renewables 2014, *Global Status Report.*

2. GBEP/FAO, Rome 2014, *Towards Sustainable Modern Wood Energy Development.*

3. Hall, D.O., Casillo-Calle, F., Williams, R., Woods, J. (1992). Biomass for energy: supply prospects, in *Renewable Energy Sources for Fuels and Electricity*, 593, Eds. Johansson, T.B., et al., prepared for the UN Rio Conference in 1992, Island Press, Washington DC. ISBN 1-55963-138-4.

4. Hall, D.O., Rao, K.K. (1981). *Photosynthesis*, Edward Arnold, London. ISBN 07131-28275.

5. Eurosolar (1994). *The Potential of Biomass in Europe, an Assessment*, Bonn, for the EU Commission, unpublished.

6. ObservER (2014). Various statistics on biomass in Europe, the latest from 2014, *Systèmes Solaires, Le Journal des Energies Renouvelables*, Paris, ISSN 0295-5873.

7. ObservER (2014). Waste barometer, *Systèmes Solaires, Le Journal des Energies Renouvelables*, Paris, ISSN 0295-5873.

8. AGRA Europe No. 2605, Brussels 1996.

9. EU Commission (1992). *Biomass, a New Future?*, Brussels, April 1992, Internal EU report, unpublished.

10. Little, Jr., E.L. (2014). *Common Fuelwood Crops, a Handbook for Their Identification*, Communi-Tech Associates, Morgantown, WV.

11. IEA (2014). *Medium Term Market Report*, Paris 2014.

12. EU Commission COM(2001)457, Draft directive on the promotion of use of biofuels.

13. ePURE, the European Renewable Ethanol Association, and COPA-COGECA, the Agricultural Associations, Brussels, May 2011.

14. EU Commission (1989). *Environmental Ethics*, Brussels 1989, ISBN 92-826-13894.

15. FAO (2011). *Bioenergy and Food Security, Analytical Analysis*, Rome 2011.

16. Bdew, The German Federal Association for Energy and Water, 2014.

17. Bloomberg New Energy Finance, VIP comment, July 2014.

18. FNR, German Federal Agency for Renewable Feedstocks, 2013.

About the Author

Wolfgang Palz (WP) holds a PhD in physics (Dr.rer.nat.) from the University of Karlsruhe, Germany, from 1965. In 1965–1970 he was a professor of semiconductor physics in Nancy, France. In 1970–1976 he was in charge of power systems development at the French National Space Agency, CNES, in Paris. In 1973 he was co-organizer of the UNESCO Congress "The Sun in the Service of Mankind" in Paris. He was also a keynote speaker at the 40th anniversary event of that Congress at UNESCO in Paris in 2013. In 1976/78 UNESCO published his book *Solar Electricity* in seven languages.

In 1977–2002 WP was an official of the EU Commission in Brussels, the executive body of the European Union. In 1977–1997 he managed the development program of renewable energies; it included policy development and contracting to the European industry and academia of the EU Commission's budget (almost $1 billion over that period). The R&D program comprised the sectors of solar architecture, solar energy, wind energy, biomass, and ocean energy.

In 1997 WP became an EU Commission counsel for renewable energy deployment in Africa and, besides, advised the EU commissioner for energy on the EU white paper on renewable energy (RE) issued that year.

From 2000 to 2002 WP was a member of the energy committee (Enquête Commission) of the German Parliament, the Bundestag, in Berlin, to establish an energy strategy for Germany on the time horizon 2050. In 2005/2010 he was a consultant of the EU Commission for photovoltaic (PV) programs in Latin America.

WP was a co-organizer of the "Great Wall World Renewable Energy Forum," which took place in Beijing, China, from October 24 to 26, 2006.

In 2010 WP edited the book *Power for the World: The Emergence of Electricity from the Sun*, published in September by Pan Stanford in Singapore (ISBN 978-981-430-337-8). In 2013 WP edited the book *Solar Power for the World* and two books on wind power with Danish Folkecenter, all with the same publisher in Singapore. The same year a new book, *Biomass Power for the World: Transformations to Effective Use*, to be published by 2015, was started with the University of Twente in the Netherlands.

WP is bearer of an Order of Merit of Germany (Bundesverdienstkreuz), has been recognized as a wind energy pioneer in Britain, and has received the European Prizes for biomass, wind energy, and solar PV. In September 2010 he received the Swiss Solar Prize, together with the pioneers B. Piccard and A. Borschberg, the pilots of the airplane *Solar Impulse*. In August 2011 WP received the ISES Advancing Solar Energy Policy Award.

By 2014, WP was:

- Member, the French National Aid Commission, Investments for the Future, Paris
- Member, the Board of REN21, Paris
- Advisory Board Member, American Council on RE, ACORE, Washington DC
- Member of the Global Bio-Energy Partnership GBEP, FAO Rome
- Board Member, Eurosolar Russia, Moscow
- Chairman for Europe, World Council RE (WCRE)
- Honorary Board Member of ISES, Freiburg, Germany

Chapter 3

The Future Role of Bioenergy in the Global Energy System

Heinz Kopetz
Holländargatan 17, 111 60 Stockholm, Sweden
hg.kopetz@netway.at

The global energy system is based predominantly on fossil fuels. Over the last decades the use of fossil fuels grew continuously until now, 2013. But more and more people start to realize that a continuous further growth in the use of fossil fuels will endanger the well-being of future generations, because the combustion of fossil fuels is the main cause of climate change. In 2011 according to the working paper 1 [1] of the Intergovernmental Panel on Climate Change (IPCC) 90% of the global CO_2 emissions were caused by the use of fossil fuels and CO_2 came up for 75% of all greenhouse gas emissions.

In the meanwhile the concern about climate change is growing in many countries. The gathering of more than 100 heads of states or governments at the climate summit in New York on September 23, 2014, was a strong demonstration of the growing awareness of the climate problem. But the core message how to mitigate

Biomass Power for the World: Transformations to Effective Use
Edited by Wim van Swaaij, Sascha Kersten, and Wolfgang Palz
Copyright © 2015 Pan Stanford Publishing Pte. Ltd.
ISBN 978-981-4613-88-0 (Hardcover), 978-981-4669-24-5 (Paperback), 978-981-4613-89-7 (eBook)
www.panstanford.com

climate change successfully did not reach the public audience. It will not be enough to provide several billion dollars to countries in development, it will not be enough to push for more e-mobility and better thermal insulation of buildings! If the industrialized countries have to reduce the greenhouse gas emissions by 90% within 30 years in order to achieve the 2°C target—and this is reality—they have to leave the fossil fuels in the earth's crust and replace them by renewable energies. And this decision should be taken now to give the business community and financing institution clear signals on how to invest.

It is not surprising: important and powerful parts of the business world and the public dislike this truth. They promote new technologies like carbon capture and storage (CCS) or bio-CCS to reduce CO_2 from the atmosphere, because these technologies would allow us to go on with the combustion of fossil fuels as usual hoping that the next generation will collect the carbon from the atmosphere and put it back into the earth crust (bio-CCS). But these concepts are not at all convincing. They are similar to the discussion of nuclear fusion that was seen as the golden key to all future energy problems fifty years ago. I remember the saying of a representative of Electricité de France (EDF) during a meeting in Brussels with respect to nuclear fusion: "C'est une energie de l'avenir, et la restará." The same will happen with CCS because nobody can explain convincingly why we should first dig out the carbon and then dig it back using 70%–80% of the energy in the fossil carrier. This is that inefficient and expensive as compared to the accelerated deployment of renewable energies that every decision-making body following rational arguments and not the pressure of the fossil industry will favor renewable energy instead of CCS.

Certainly improved energy efficiency in all parts of the energy system is an important part of the new energy strategy. But it will not be sufficient. In many parts of the world the energy demand will grow in the next years, because the consumption is still very low and the population is growing.

The key strategy to mitigate climate change is the transformation of the energy system from fossil sources to renewable energies. Go 100% renewable should be the message of the century. But this is the long run goal that might divert the attention of the urgent steps needed in the next 20 years until 2035. Calculations of the World

Bioenergy Association (WBA) on the basis of IPCC reports illustrate that without relying on CCS the use of fossil fuels should be halved until 2035 in order to attain the 2°C target. This can be achieved if the majority of the countries on the globe starts a consequent policy in favor of all renewables such as wind, solar power, hydropower, geothermal energy, and biomass and curb all subsidies for fossil fuels. Which role will biomass have to play in an energy system using only 50% of the fossil fuels of today given an increased demand for energy in some parts of the world and a reduced demand in other parts?

3.1 Differences between Renewable Energy Sources

It is important to analyze the basic properties of different renewable energies and fossil sources.

(a) Renewables without feedstock cost: Wind, solar power, solar-thermal power, hydropower, and geothermal power have no feedstock cost. They use the solar radiation in direct form (solar power, solar-thermal) or in indirect form (wind, hydropower). Geothermal is an exception using energy from the earth and not direct from the sun. All these energies have high capital cost but low running cost. As soon as the installations are constructed they deliver energy at low marginal cost. It is more economical to let them run, and shut down other generating units if there is an overproduction. These energies deliver mainly electricity. Their fast development explains partly why the transition of the energy system is frequently reduced to a transition of the electricity sector. The problem of these energies is the intermittence of the supply and the difficulty to store the excess electricity in a cheap way. Another aspect is the difference in the quantity of generated electricity from solar- and hydro-installations between the winter and summer seasons. Some publications promote the use of electricity for heating overlooking the fact that the peak demand for heat appears during the winter period, whereas the peak generation of electricity from hydropower and solar power occurs during the summer period. Only windmills offer electricity all around the year in a more or less stable way. These renewables without feedstock cost cannot replace fossil fuels in the heating sector in the next 20 years.

(b) Biomass (the renewable with feedstock cost): Biomass is chemical stored solar energy. The plants absorb carbon from the air in the form of CO_2 and build up carbon hydrates via the photosynthesis. Solar energy powers this process and is stored in the carbon hydrates (biomass). This stored CO_2 and energy is released either by decay of the biomass as it happens in nature or by using it for food, feed, energy, or material purposes. Hence sustainably produced biomass is a carbon neutral energy source. Its use does not increase the carbon content of the atmosphere, as long as the production of biomass is equal or bigger than its use—this is one of the important criteria of sustainable biomass. The production of biomass as feedstock for energy is related to different cost elements such as the cost of land, production, collecting, transporting, storing, and processing biomass.

Biomass can be converted to all forms of final energy such as heat, transport fuels, and electricity on the basis of a broad variety of conversion technologies. Bioelectricity is more expensive than electricity from hydropower or wind. In most countries of the Northern hemisphere bioheat is cheaper than heat from fossil fuels with the exception of those countries with subsidies for or a very low taxation of fossil fuels. Biofuels are at present the only available alternative to fossil transportation fuels for heavy trucks, big ocean ships, planes, and tractors used in agriculture and long-distance traffic in individual cars. In any case the cost of final energy from biomass are always determined by the cost of the feedstock and the cost of the conversion—finance, depreciation, operation, and maintenance costs. These facts have to be considered when developing a bioenergy strategy.

(c) Fossil energy carriers such as carbon, oil, and gas have also feedstock cost and conversion cost. As opposed to biomass the combustion of fossil fuels adds additional carbon from the earth's crust to the atmosphere, blows up the natural carbon cycle, and causes climate change. In addition in many parts of the world the production of fossil fuels is declining, making more and more countries dependent on the small group of producer countries. Climate change and the upcoming scarcity in many parts of the world are the main reasons why our societies have to leave fossil energies as fast as possible.

3.2 Development Goals for Biomass to Energy

The underlying assumption of this text can be summarized as follows:

> *The global society—the citizens, the governments, the business community—is willing to achieve the 2°C target in this century and consequently to transform the energy system in a first step until 2035 accordingly, without relying on CCS. This requires us to reduce the use of fossil fuels by 50% until 2035.*

Even if electricity from wind, solar power, and hydropower grows rapidly until 2035, for generating more renewable electricity in 2035 than the total generation of electricity today biomass will also have to grow fast and deliver 140–150 EJ primary energy by 2035 [2]. The main target markets of biomass will be heat, transport fuels, and electricity generated in cogeneration units or in installations, especially biogas, that help to stabilize the electricity grid.

At present globally 50% of the final energy is used in the form of heat. Theoretically this demand could be reduced strongly by new energy-saving buildings and by better insulation of existing buildings. Yet the realistic potential to reduce the demand for heat until 2035 is limited. The cities are already built, and not much more than 20%–25% of the houses will be rebuilt until 2035. Even in cities with ambitious policies, for example, the city of Graz, the annual rate of thermal insulation of existing buildings was not more than 1.2% in the last years, in many European cities it is lower. As experience proves the heat demand after insulation is in average 30% lower than before. As a result under *ceteris paribus* conditions in the city of Graz, the heat demand by thermal insulation was reduced by 0.4% per year; this would mean −8% in 20 years if the program of insulation is continued. On the other hand, in North America, Europe, and Japan more than 80% of the buildings are heated with fossil fuels or electricity mainly derived from fossil or nuclear fuels. A switch from fossil to renewable heat such as biomass is without alternatives in order to reduce the use of fossil fuels fast enough until 2035. Thermal insulation and the construction of new low-energy buildings are very important, but it takes decades to reduce the heat demand significantly. It might be possible to reduce the heat demand by 30%–40% until 2050 but certainly not until 2035.

Due to economic and technological reasons it is more difficult to replace fossil fuels in the transport sector than in the heating sector. Yet also the transport sector has to be changed to renewable fuels among which until 2035 biofuels will remain the most important ones—first-generation and advanced biofuels will have to play their role.

Putting this consideration together it can be estimated that the contribution of biomass to the final energy consumption in 2035 should be in the size of 125–130 EJ, of which the biggest share should be heat, about 10% biofuels, and a small part—about 4%—electricity.

3.3 Current Contribution of Biomass

How do these figures relate the present contribution of biomass? The total primary energy supply of biomass rose from 43.2 EJ in 2000 to 54.9 EJ in 2011, which is a growth of 1 EJ/year. In 2011, Asia was leading in the use of biomass followed by Africa—in these two continents 70% of the total biomass was used. A big share of the biomass is traditional biomass for cooking in millions of households in Africa and Asia [3].

Europe achieved the fastest growth of biomass—82% in the last 11 years—reaching 6.1 EJ. The Americas were leading in the production of biofuels, especially the United States, with corn ethanol, Brazil with sugarcane ethanol, and Argentine with soy-based biodiesel. In 2014 in the United States the first commercial plant producing advanced biofuels from cellulosic materials like corn stalks started operation. The total production of biofuels was growing in the last years reaching in total 115 billion liters. This growth occurred in the Americas, Africa, and Asia—Europe lost its leading position in the biofuels industry due to unreliable political frame work conditions.

In 2011 the electricity production from biomass reached 422 TWh (422 TWh electricity equals 1.5 EJ final energy; assuming an average effciency of 25% in the conversion of biomass to electricity, a quantity of 6 EJ primary energy from biomass is needed to generate 422 TWh electricity) and the leading continents were Europe and North America.

Biogas contributed 1.1 EJ in 2011, leading continents being Asia and Europe. Electricity from biogas in Europe reached almost 40 TWh, this is 10% of global electricity from biomass.

In 2011, the contribution of biomass to final energy reached 46.5 EJ out of this 92% was used as heat, 5% for transport, and 3% as bioelectricity.

These data show that at present the bulk of biomass also goes to the heat sector in many places, especially in Africa, in an inefficient and not sustainable way.

3.4 Developing the Potential of Biomass

Within the described strategy toward the 2°C target sustainable biomass should develop three times faster worldwide until 2035 than in the last decade. How can this happen? This question leads to the issues of biomass potential, conversion technologies, strategies for market penetration, and appropriate support policies.

In 2011 the primary biomass supply was derived as follows: 89% solid biomass, 5% biofuels, 4% wastes, and 2% biogas.

In a more systematic way we distinguish three sources of biomass: forestry, agriculture, and waste. These sources include main products, by-products, and different waste streams as explained in Table 3.1. Currently more than 80% of the biomass comes from forestry.

Table 3.1   Different sources of biomass

	Forestry	Agriculture	Waste streams
By-products (Examples)	Felling residues, bark, branches, sawdust, wood chips, residues of the wood industry	Manure, straw, bagasse, kernels, empty fruit bunches from oil palms, other residues	Municipal solid waste (MSW) Organic waste from industry
Main products (Examples)	Stem wood, firewood	B2 energy crops: cereals, corn, oilseeds, energy grass, short rotation coppices, etc.	Sewage sludge

In regions without a developed bioenergy industry most of agricultural and forestry by-products and residues such as bark, straw, sawdust, and others are not used but burned on the fields, dumped somewhere, left in the forests or brought to landfills, thus releasing heat, CO_2, or CH_4. This is also the case of huge amounts of organic waste being produced in the urban agglomerations of the world. All these losses of by-products from agriculture and forestry and organic waste sum up to at least 60 EJ of energy.

Also existing forests offer additional feedstock for bioenergy. Only one-third of the global forests (4 billion ha) are in utilization [4]. Some countries like Sweden or Austria demonstrate that sustainable forest management leads to a higher wood production per hectare, thus allowing an increase of the carbon stock in the forests and simultaneously harvesting more wood. Improving forest practices in many parts of the world as it is done in some European countries, and increasing the forest area by additional 200 million ha could deliver an additional 25 EJ energy.

Improved land use is a further source for more biomass. At present 1.5 billion ha land is used for crop production—12% of the total land area. More than 600 Mio ha land in addition can be used for rain-fed agriculture and planting of new forests [5]. If the land needed for the growing world population is abstracted, the residual land could deliver more than 25 EJ.

Hence a better use of waste and by-products, of existing forests, and of the available land offer sufficient potential to deliver additional 95 EJ biomass summing up to 150 EJ primary energy from biomass. Yet various steps are needed to mobilize this potential:

- First priority: better use of by-products and waste for energy by technology transfer, better training, and reliable economic framework conditions.
- The improvement of the agricultural and forest management methods worldwide
- The plantation of new forests and a further reduction of deforestation with the aim to increase in total the global forest area by 200 Mio ha until 2035.
- The regaining of degraded land for agriculture and for fighting against desertification by a combination of afforestation and agriforestry. The policies to maintain the fertility of the land

and to protect fertile land against misuse and degradation have to be strongly improved.

- A close cooperation between governments, international organizations, the private business sector, and farmers and forest owners.

3.5 Guidelines and Critical Voices

Biomass offers many possibilities in terms of conversion technologies and final energy output. A set of rules for the further development of biomass is helpful to avoid wrong decisions.

3.5.1 Principles of Biomass Development

- Regional use: If biomass is used in the region where it grows the cost of transport can be minimized and the expenditures of the consumer for energy can be kept in the region thus creating regional economic growth. Therefore local and regional production and utilization of biomass should be given a first priority. This principle finds its limitation in areas where the production of biomass is bigger than the demand for bioenergy or vice versa. Under these circumstances the national or international trade of biomass makes sense.
- Efficient conversion of biomass to bioenergy: Although biomass is renewable it is not unlimited available. Therefore as much as possible of the primary energy stored in biomass should be converted to final energy that is used as heat, electricity, or transport fuel. The efficiency in the conversion of biomass to bioenergy should reach 75% or more. Electricity alone plants using biomass are therefore not complying with this principle.
- Sustainable production of biomass: The supply of biomass has to follow the rules of sustainability. Not more biomass for energy must be harvested than is produced in a region, and simultaneously fertility of the soil, water supply, and biodiversity have to be safeguarded. This principle of sustainability not only refers to the biomass production, but it should be applied to food and feed production in general.

- Stability of support measures: As the experience in many countries shows a stop-and-go policy in the support measures leads to negative consequences for the development of renewables; it undermines the confidence of investors in the policy. Therefore support measures shall be well considered, and as soon as they are put in operation they should remain stable for a given period of time. This is not only valid for bioenergy but for all renewable energy sources.
- Integration with other renewable energy sources: Biomass is an important renewable energy source but also other renewables such as wind, photovoltaics, solar heat, hydropower, and geothermal energy have to play their role. Biomass as stored solar energy can supplement other renewables. Biomass is particularly well suited to supply energy during winter time when the demand is high and other renewables like hydropower, solar heat, and photovoltaics have a rather low performance. Therefore the future role of bioenergy should always be seen in relation to other renewables.

3.5.2 The Carbon Neutrality of Biomass

It is sometimes argued that biomass, especially forest biomass, is not carbon neutral. This objection is not true. The use of forest biomass is carbon neutral, because the carbon contained in wood originates from the atmosphere and it is released to the atmosphere by wood decay or by combustion. Before a tree can be burned it has to grow by absorbing carbon from the atmosphere. Theories on carbon debt and payback time of biomass are not credible, because they are based on the unrealistic assumption that trees are first burned and then grown! It is a fundamental requirement of sustainable forestry that the carbon stock in forests remain stable or increases over time. Deforestation and unsustainable forest management lead to a decline of the carbon stock in the forests—this has to be avoided. The forests are part of the global carbon pool atmosphere—biosphere within which the carbon moves as part of the natural carbon cycle.

The carbon released by burning fossil fuels is not part of the natural carbon cycle. It rapidly increases the CO_2 content of the atmosphere. Therefore, the use of fossil fuels creates a carbon debt that will be a huge burden for future generations.

3.5.3   Land Use, Food Production, and Biofuels

Critical voices argue that biomass for energy competes with the food supply of a growing population. The biggest part of biomass to energy comes from forests, from by-products of agriculture, and from waste streams. These bioenergy sources do not compete at all with the food supply. These critics concern the 5% biomass going to the production of transport fuels, yet they often overlook the complexity of the issue.

Conventional biofuel production not only delivers ethanol and biodiesel but also protein feed, with the quantities of these both produced on a similar scale. In 2010 protein production related to conventional biofuels based on corn, cereals, canola, and soybean delivered 79 million tons of protein feed corresponding to the protein production of 29 million ha soybean that is more than a quarter of the global demand for soybean cake. Hence conventional biofuel production chains are a vital part of both global fuel and protein supplies [6].

In addition as examples show the development of bioenergy goes hand in hand with improved investment and higher productivity in agriculture. Making agriculture and forestry more efficient on a sustainable basis, improving the education and agricultural practices in many parts of the world, and improving the land policy in the sense that fertile land is better protected against degradation, desertification, and other uses create the basis for a boost in food production and an increased biofuel production.

The more fossil fuels are left in the earth in order to protect the climate the more some parts of the land will be needed to produce renewable raw materials for material use and energy. A concept that denies this fact supposes that the use of fossil fuels can be continued forever.

In the years 2008/2009 food prices increased on the world market and also the biofuel production was growing to about 80 billion liters. At this time biofuels were blamed as cause of increasing food prices although scientific evidence for this assertion was missing. Now we have the year 2014, biofuel production increased to 115 billion liters, the world population is more than 300 million people bigger than five years ago and prices for important agricultural commodities are extremely low, reducing the income of farmers in many parts of the world. The main reason for this phenomenon lies primarily in the big variability of global harvests depending on positive or negative weather conditions in the main growing regions, such as sufficient rain at the right time, adequate temperatures, and no floods or droughts, and secondly in the withdrawal of instruments to stabilize agricultural commodity markets as the consequence of a liberal market philosophy.

3.6 Bioenergy: The Markets

The final energy demand comprises the demand for heat, electricity, and transport. Electricity is used for electricity-specific purposes such as light, stationary work; communication and information; and other specific services and partly for heat and mobility. In the following table (Table 3.2) the electricity consumption in total is presented, including the share of electricity used for heat and transport. In the future a bigger share of electricity will be used for heat and mobility.

Table 3.2   Global demand for final energy, 2010

	EJ	%
Heat	200	55
Electricity	64	18
Transportation fuel	99	27
Total	363	100

Source: Ref. [5].

In any case Table 3.2 shows that heat is dominating the final demand for energy.

3.7 Biomass to Heat

Paradoxically the heating sector is often overlooked in the discussion about energy strategies, although such a big share of final energy is used as heat: heat for cooking, heat for warm water and space heating, and high-temperature heat for industrial processes. This strong demand explains why the biggest share of biomass, used for energy, goes to the heat sector.

Half of the world population uses biomass as firewood or charcoal for cooking and heating, in some regions on the basis of nonsustainable forest management and declining forest area [6]. A specific problem is the low efficiency of these systems: if the charcoal is produced using traditional production methods below 30% efficiency, as this is often the case and the efficiency of the cooking stove is also under 30%, then less than 10% of the wood energy is transformed to heat. The improvement of these systems—highly efficient equipment, better forest management—in combination with a better electricity supply should be a pivotal point in the development and energy policy of these countries.

On the other side, in the Northern Hemisphere the forest area is growing and the heat market is supplied predominantly by oil, natural gas, and electricity, although biomass is cheaper than oil: Europe, summer 2012, cost per unit primary energy:

- Wood chips, delivered at plant   €15–20/MWh
- Pellets price CIF Rotterdam   €28/MWh
- Crude oil, Brent, September 9, 2012   €52/MWh

Hence, the heat market offers huge possibilities for biomass. In densely populated downtown areas, district heating is the best concept. It offers the chance to combine the heat supply coming from biomass with efficient electricity production. Outside the central areas wood pellet stoves and boilers are the choice—bigger consumers prefer wood chips. Industrial heat users transform dried biomass to wood powder and inject the wood powder into adapted oil burners to escape the economic burden of growing oil prices.

A few countries demonstrate how biomass to heat can be pushed.

- Sweden: Thirty years ago environmental taxes on fossil fuels were implemented, making heat from biomass cheaper than using oil or coal. Now, less than 5% of the heat supply comes from fossil fuels. The change of big district heating systems in cities from fossil fuels to biomass is a logistic and technological challenge. Many cities demonstrate how this can be achieved. In Stockholm, for instance, in 2014 a new biomass boiler with a capacity of 300 MW in combination with an innovative new supply concept is just in construction.

- Pecs, Hungary: A 150 MW wood boiler and a 150 MW straw boiler recently completely replaced coal and gas. It took several years to implement an efficient logistic chain for the supply of 200,000 tons of straw annually.
- Italy: Another approach is used to promote renewable heat. Due to the lower heat demand district heating systems in this country are barely in use. Here, a combination of tax policies and financial incentives encouraged many private consumers to switch from fossil fuels to pellet stoves. Italy has the fastest-growing market for pellets—more than 2 million pellets stoves have been installed over the last 15 years.

In view of the year 2035 and in compliance with a climate mitigation strategy a strong push for renewable heat, especially based on biomass, is needed. But also the other sources of renewable heat such as geothermal and solar energy should be developed (Table 3.3).

Table 3.3 Proposed development of renewable heat to 2035

Energy sources	2010, IEA (Mtoe)	2035, WBA (Mtoe)
Fossil sources	139	66
Bioenergy	47	112
Other renewable energy sources	2	21
Total renewable heat	49	134
Total heat	188	200
Hereof renewable	26%	67%

Source: Ref. [2].

This rapid growth of use of biomass for heat requires specific strategies for different countries. In countries north of 45° latitude biomass in the form of wood chips, other wood residues, and agricultural by-products such as straw should be increasingly used in district heating plants, plus the use of heat from biomass cogeneration plants. In the eastern part of Europe, district grids using natural gas as energy source are installed in many cities. This fact offers big opportunities for biomass such as straw or wood. The lack of capital for the replacement of gas boilers by more expensive biomass boilers hinders this transformation. Specific European funds for renewable heating systems could accelerate this transformation to renewables. But also the service and manufacturing sector will become a market for this kind of biomass-sourced heat.

In addition a rapidly growing pellets industry will supply pellets for the residential, manufacturing, and service sectors and for electricity generation. Pellets have a high energy density and can therefore be traded worldwide at low cost as long as the main consumer is along the coastline. In developing countries with a high share of traditional biomass and charcoal the main focus will have to be to improve efficiency and the technologies used and to drive the move to sustainable production.

Reliable, stable incentives by government policies to the heat consumer—be it high taxes on fossil fuels or investment grants—are proven measures to transform the heat market. Hence, the main responsibility for the transformation of this sector lies on governments by creating attractive framework conditions.

3.8   Biofuels for Transport

In the year 2010 the share of renewables in the transport sector, including road, train, ship, and air transport, was low. Out of the total consumption of 2376 Mtoe (99.5 EJ) only 24 Mtoe (1 EJ) was covered by electricity and 60 Mtoe (2.5 EJ) by biofuels, mainly ethanol and biodiesel. As 80% of electricity was not renewable in 2010 the total amount of renewables in the transport was only 64 Mtoe, corresponding to 2.7%.

Biofuels for transport are any liquid or gaseous fuels derived from organic material. Sugarcane and sugar beet, grain such as corn and wheat, cassava, and oilseeds are biomass sources for conventional biofuel production—they are often called first-generation biofuels. Cellulosic biomass, organic waste, and algae have the potential to become an important basis for advanced biofuels—they are also called second-generation biofuels. Also biogas transformed to biomethane, and pure vegetable oils are part of the biofuel portfolio. In this text the terms "conventional" and "advanced" are used for the classification of biofuels instead of the terms "first-" and "second-generation" fuels.

Biofuel production is a rather small part of the total contribution of biomass to energy. The total production of biomass reached 55.4 EJ, whereas biofuel came up for 3 EJ in 2013, 5% of total biomass.

The global biofuel production was growing at a rate of 19% annually in the first decade of this century. In the period between 2010 and 2012 this growth slowed down, probably as a consequence of the high commodity prices in the years 2008 and 2009. Meanwhile, by 2013 production increased reaching 115 billion liters (see Table 3.4).

Table 3.4 Global production of liquid biofuels

Year	Biofuel production	
	Billion liters	PJ
2000	17.8	430
2005	34.9	860
2009	91.0	2340
2010	104	2640
2011	107	2760
2012	106	2740
2013	115	2960

Source: Refs. [3, 7].

Ethanol and biodiesel play an uneven role. The ethanol production, mainly based on corn and sugarcane, is much more important than the biodiesel production; yet biodiesel production, mainly based on soya oil and canola oil, is growing continuously. The next table (Table 3.5) offers data until 2012.

Table 3.5 Global production of ethanol and biodiesel

Year	Bioethanol production		Biodiesel production	
	Billion liters	PJ	Billion liters	PJ
2000	17.0	398	0.80	28.2
2005	31.1	728	3.80	134
2009	73.2	1713	17.8	627
2010	85.0	1989	18.5	651
2011	84.2	1970	22.4	788
2012	83.1	1945	22.5	792

Source: Ref. [3].

The global biofuel production is dominated by the Americas. In 2011 about 70% of the global production came from North and South America, followed by Europe and Asia. In Africa the biofuel production at the industrial scale just started recently.

In 2012 the leading producer of ethanol was the United States with 50.4 billion liters, followed by Brazil (21.6 billion liters) and Europe (4.2 billion liters). Thailand had a production of 0.7 billion liters.

Concerning biodiesel the situation was different. Europe was the leading producer in 2012 with 9.1 billion liters, followed by the United States (3.4 billion liters) and Argentina (2.98 billion liters). Argentina increased its biodiesel production over the last years rapidly using soybean as feedstock. The production in Thailand was 0.9 billion liters in 2012.

The use of upgraded biogas as transport fuel is not of importance so far. Europe used in 2011 about 0.4 billion m^3 biomethane.

The production of commercially produced advanced biofuels is just starting.

Many activities are going on concerning the use of biofuels for the aviation sector. Biofuels are the only available alternative to fossil fuels for the aviation and shipping sectors.

3.9 Conventional Biofuels and Protein Production

The rapid growth of conventional biofuels such as ethanol and biodiesel initiated many discussions as they are closely related to land use and food production. Especially in Europe, in the last years an intensive debate on the future use of conventional biofuels took place, with the result of political uncertainty and the reduction of the 10% blending target to probably 7%. In the United States recent policy reviews have brought uncertainty to the market, too. Also in Brazil the economic conditions are not favorable for the ethanol sector due to internal reasons. In other parts of the world, especially in Asia, the situation is better.

One reason for this debate is the misunderstanding of the interrelation between biofuel production and the feed market.

The production of conventional biofuels is in many cases connected with the production of protein. In this context we can distinguish between two types of crops for the biofuel production:

Crops A: deliver feedstock for *biofuels and protein feed* for the market

Crops B: deliver feedstock for *biofuels without protein production* for the market such as sugarcane and oil palm (tropical crops)

Corn, cereals, rapeseed, and soybean used as a feedstock for biofuels also produce protein feed. These protein feeds like Dried Distillers Grains with Solubles (DDGS) or rapeseed cake contain different quantities of protein per ton. To make the yields better comparable they are expressed in Table 3.6 for the fuel in energy units (toe) and for the protein in tons of soy cake equivalent with 44% protein.

As can be seen in Table 3.6 the highest yields in terms of biofuels can be achieved with sugarcane for ethanol and oil palms for biodiesel. On the other hand corn and soybean, and to a lesser extent wheat and canola, also deliver important quantities of protein feed. Table 3.6 gives an overview based on average yields in the main growing regions of the mentioned crops. In terms of total output—biofuels and protein feed—oil palm, corn, and sugarcane are the highest-yielding crops.

Table 3.6 Comparable fuel and protein production for different crops per hectare (unit toe for fuel and tons for soybean equivalent for protein feed)

Crop	Fuel (toe/ha)	Protein feed soybean cake equiv. (t/ha)
Ethanol		
Corn	1.9	1.9
Wheat	1.3	1.4
Sugarcane	3.1	–
Biodiesel		
Soybean	0.4	2.1
Rapeseed	0.9	1.4
Oil palm	4.3	–

Source: Ref. [6].

On the basis of these data it is possible to calculate the total protein production related with the conventional biofuel production.

In 2010 the production of biofuels based on corn, cereals, rapeseed, and soybean reached 57.9 Mt biofuels and 79.1 Mt protein feed corresponding to the protein production of 29 million ha soybean, which is more than a quarter of the global demand for soybean cake. Hence conventional biofuel production chains are a vital part of both global fuel and protein supplies.

To summarize, conventional biofuels achieved a rapid growth in the first decade of this century. This growth is slowing down now, especially in Europe and Brazil, whereas in Asia and Africa biofuel production is increasing. Also in the coming years most of the biofuels produced will be conventional biofuels; the first advanced biofuels plants is just starting operation. Conventional biofuels have a big importance for the protein production, especially biofuels based on corn, canola, and soybean.

The fast growth of the biofuel production, experienced in the first decade of this century, will not continue. The International Energy Agency (IEA) published recently an outlook expecting 140 billion liters biofuels in 2020, mainly as conventional biofuels. This equals an annual growth of 3%–4%.

But also in the year 2035 the WBA does not expect a dominating role of renewables in the transport sector. The WBA proposes an increase of first-generation fuels, of advanced biofuels, and of biomethane for transport to more than 300 Mtoe 12–14 EJ in total.

3.10   Biomass to Electricity

In 2011 almost 72 GW biomass power capacity was in place worldwide. Bioelectricity is mainly produced using solid biomass or biogas. Many utilities are investing in electricity from biomass—be it new plants or cofiring to existing plants—to avoid CO_2 emissions. This is an important development that helps to build up efficient global supply chains for solid biomass.

In principal electricity should be produced in combined heat and power (CHP) plants; but this is today not always possible. Exceptions might also be justified in areas with an oversupply of agricultural residues and no nearby heat demand. China delivers a positive example: In the last year many straw-fired dedicated electricity plants have been built using straw, that was burned before on the fields. On the other hand, in Australia millions of tons of straw are just burned after the wheat harvest without any use.

The rapid growth of wind and solar electricity will change the markets. In future electricity from biomass should partly compensate for the fluctuation of wind and solar electricity. This is a particularly interesting option for electricity from biogas.

As the costs of electricity from biomass are higher than the market price, reliable long term government policies are needed. A system of feed in tariffs for electricity as developed in Germany is the most successful solution. But also government support for the construction of district heating grid is required, to create a basis for efficient CHP solutions. Growth expectations for bioelectricity until 2035 range between 16 and 26 EJ primary energy from biomass.

3.11 Outlook

In the next 20 years many innovations in the energy field will again take place—just as many innovations were observed to happen over the last 20 years. These innovations will facilitate the transformation to a 100% renewable energy system.

The main challenge until 2035 should be the transformation of the electricity and heating sector to renewables; renewable transport should not be the main concern as long as big quantities of fossil fuels are used for heat or electricity generation. The transformation of the transport sector should follow later, depending on new experiences to be gained over the next two decades. It also can be expected that the demand for heating/cooling will go down in the longer run due to the better energy efficiencies of buildings; then a larger share of biomass would be available to be used for the generation of transport fuels after 2035.

The WBA is convinced that the best answer to the new IPCC report on climate change is an accelerated transition to an energy system without reliance on fossil and nuclear energy resources. Government policies and investment decisions should be directed toward this goal. A delay in the transformation of the energy system deepens the problem. It leads to additional global warming and would require even bigger adjustment efforts in the future. 100% renewable energy is possible within less than 40 years! The faster we take this road, the better for the future!

References

1. Working Group I Contribution of the IPCC Fifth Assessment Report (2013). *Climate Change 2013. The Physical Science Basis, Summary to Policymakers.*

2. World Bioenergy Association (2013). *Bioenergy Magazine 5,* Stockholm.

3. World Bioenergy Association (2014). *Global Bioenergy Statistic Report,* Stockholm.

4. Kopetz, H. (2013). Renewable resources: Build a biomass energy market, *Nature,* **494**(7435):29–31.

5. IEA (2012). *World Energy Outlook,* Paris.

6. World Bioenergy Association (2013). *Fact Sheet Biofuels,* Stockholm.

7. IEA midterm review, 2014.

About the Author

 **Heinz Kopetz** is president of the World Bioenergy Association, Stockholm, since 2012. In 2013 he also served as chairman of RENAlliance, a network of the global industry associations of renewable technologies such as wind energy, solar energy, hydropower, geothermal energy, and bioenergy.

Dr. Kopetz graduated from the University of Life Sciences in Vienna and the Iowa State University, USA, in economics in 1966. After his studies he served as managing director of an agricultural agency in Austria. In the 1980s he developed the concept of the *farmer as the energy producer* and implemented financial and educational programs in this field. During this time he also founded the Austrian Biomass Association, was engaged in the deployment of district heating systems using biomass, and started the first programs of short rotation forestry in Austria.

During his professional career Dr. Kopetz was engaged in different positions related to energy, such as member of the supervisory board of a public utility for electricity and district heating; and was chairman of the European Biomass Association (AEBIOM), Brussels; and member of the Advisory Committee for Energy of the European Commission, Brussels.

Dr. Kopetz works with issues such as potential sustainability, carbon neutrality, and efficient utilization of biomass and has written several books on these subjects.

Chapter 4

EU Biomass R&D: The Beginnings
(Laying the Ground for Biomass Becoming Mainstream in Sustainable Energy Supply in Europe)

Wolfgang Palz[a] and Henri Zibetta[b]

[a] Retired European Union official, Renewable Energy R&D,
European Commission, Brussels
[b] FAC Consult, Renewable Energy Sources and Technologies, Brussels
w@palz.be

4.1 Introduction

The European Commission in Brussels, the executive arm of the European Union, the EU, previously called the "European Communities, or EC," started its R&D programs in the early 1970s. Originally, the EU had six founding nations. Later, in the early days of the R&D programs when the United Kingdom, Ireland, Denmark, and Greece had joined, the EU had 10 member countries.

The European Commission started the new R&D programs with one on the environment. It was followed in 1975—by a decision

Biomass Power for the World: Transformations to Effective Use
Edited by Wim van Swaaij, Sascha Kersten, and Wolfgang Palz
Copyright © 2015 Pan Stanford Publishing Pte. Ltd.
ISBN 978-981-4613-88-0 (Hardcover), 978-981-4669-24-5 (Paperback), 978-981-4613-89-7 (eBook)
www.panstanford.com

of the EU Council of Ministers to represent the member countries and by the European Parliament—by a program on solar and other renewable energies (REs). The European Commission had early on set up a special directorate general to manage the programs centrally from Brussels. The German Dr. Günter Schuster (×) became its first DG and organized it very much the way it still works today. He structured the programs later on in the framework programs (FRPs), which simplified the decision process by the EU institutions and up to now gave guidance to all the programs' implementation.

In the 1970s the EU R&D programs had a budget of a few million euros only—really little in comparison to the EU structural funds and those going to the common agricultural policy. Nowadays (2014), the number of member countries having proudly reached 28, the EU FRP No. 8, running from 2014 to 2020, also called "Horizon 2020," reached a budget of 80 billion (80,000 million) euros, of which 6 billion euros are earmarked for the energy sector, including 2.4 billion euros for Euratom (fusion and atomic energy).

The first RE R&D program was running from 1975 to 1979, followed by a second four-year program (1979–1983) and others later on. The first years served primarily to get organized. From 1979 on the program had reached full speed. Program implementation took place with full involvement of the EU member countries, whose representatives were sitting in consultative committees on all levels: assessments and program strategic outlines, selection of contractors, and evaluation of results. All contractors were selected through proposals in reply to calls for proposals in the *Official Journal of the European Union*. They had to be in line with the various strategies defined previously. The calls for proposals were specific and led to homogeneous programs with well-defined objectives, whereby all contractors were investigating different aspects of the same overall problem. An important tool for implementation of our RE programs were the contractors' coordination meetings. They were held throughout Europe also with the aim to get contractors to know each other—in the early days it was typical that researchers knew well their colleagues in the United States but had no idea of those working a few kilometres across the next national border. We also encouraged the organization of specialized conferences and the participation of our contractors to make sure that our programs were not hiding in protected circles but were accessible to the interested public. The series of conferences on biomass we started in 1980 is still ongoing. It was from the beginning the biggest of its kind worldwide.

It was considered a privilege to be an EU contractor, and competition was strong. Contracts were grouping transnationally universities, public entities, and private entities. The European Commission took the decisions awarding the contracts, but financially cofunding had to come from the private parties and the member countries. Through the latter, our program was a leading and integrating part of Europe's R&D in the specific fields, but it did not come in addition to national programs.

As said before the RE programs got really on-stream with the one starting in 1979 over a four-year period. It comprised 420 contracts, which were centrally managed from Brussels. The overall budget of all these contracts amounted to 110 million euros, to which the EU contributed 46 million euros from its own budget. (In the first four-year RE R&D program, 1975–1979, it had been of 12 million euros only.)

The biomass R&D program, as integral part of the RE program, consisted of 87 contracts with an EU budget contribution of 8.8 million euros for 4 years.

A particularity of the program was its openness and transparency. Contractors were encouraged to communicate their findings and to expose them to the worldwide community of experts in the field. Thousands of communications were actually given by the contractors in association with the program in specialized networks or public meetings. Well over 25,000 pages of text were published in proceedings, books, and scientific journals. The present paper can only give a taste of all the activities in not more than "homeopathic quantities."

Under a multilayer hierarchy at the European Commission in Brussels W. Palz served from 1977 up to 1997 as the division head for the RE R&D program. Directly responsible for the R&D program on biomass was for many years Dr. Giuliano Grassi, assisted by H. Zibetta. The program could also benefit from the involvement of many leading European experts from the member countries. In the biomass field, key experts for the European Commission were Philippe Chartier from Paris and David Hall from London.

From left to right, top row: Pietro Moncada, Manuel Reis, Theo Steemers, Wolfgang Palz, Giancarlo Carratti, Henri Zibetta, Arthouros Zervos
From left to right, bottom row: Giuliano Grassi, Anne Baubin, Giovanna Ceraulo, Diana Tavianucci
Below: Wolfgang Palz, Head of Division E-3 (Renewable Energies)
(Photo: C. Weiss)

Figure 4.1 The European Commission's Renewable Energy R&D Division in 1991. By courtesy of H. Zibetta.

4.2   Early Strategies

Already in the midseventies, biomass was considered as having the most important potential amongst all RE sources and being economically competitive even in the short term. Three main biomass energy sources were considered as immediately available: cereal straw, livestock waste, and timber residues. The principle of "energy crops" was at its very early conceptual stage.

A landmark of early biomass strategies in Europe was the 230-page book *Energy from Biomass in Europe*, published in 1980 in London [1], with later translations into German, French, and Italian. The book goes back to a study contract of GTS Ltd. in England with the European Commission. The main authors were L. White and J. Low, assisted by a European expert group of G. Hansen (Denmark), A. Hurand (France), D. Kearney (Ireland), G. King (United Kingdom), G. Lettinga (the Netherlands), R. Morandini (Italy), H. Naveau (Belgium), and K. Wagener (Germany). The book was edited by Palz and Chartier. The book's assessment provided a great amount of data and scientific references. It reviewed all promising routes for bioenergy from wastes and surpluses, the possibilities for energy crops, and the techniques for utilization. In his preface to the publication G. Schuster, the DG of the EU R&D programs underlined the study's result that—against all controversy on its utilization— biomass could meet within a few years up to 5% of the EC's 10 member countries' energy needs from residues and another 5% from energy crops.

In the R&D program for the first time "energy farming" was considered as a promising branch for agriculture in the future, taking into account the large amount of hectares which had to be set aside of food production by the common agricultural policy (9 million ha in 1986, 12 million ha in 1990, and 15 million ha later).

On the basis of the knowledge available in the midseventies, the European R&D strategy started on the following axes:

• Improvement of the production yields per hectare for dedicated energy crops
• Research on identifying and selecting annual or perennial plants with high-energy potential
• Research on biomass energy from algae
• Improvement of the harvesting techniques and machinery

- Development of the forestry biomass collection techniques
- Development of agricultural and industrial biomass waste collection and recovery schemes
- Research and development on transport and storage technologies and processes
- Research and development on dedicated, specific energy conversion technologies

The "1st E.C. Conference on Energy from Biomass," held in Brighton, United Kingdom, in 1980, was another milestone in the development of a European biomass R&D strategy. The public congress with hundreds of delegates was organized not only with the usual presentation of papers but also with specialized round tables on each subject. The conclusions of all discussions were recorded and went into the proceedings, which were published shortly after in a 1000-page report [2]. It was actually a world congress with many experts from the United States, Brazil, and other nations. A summary of recommendations came to the following conclusions:

- Conversion technologies cannot be separated from local considerations about resources, infrastructure, and end-use applications.
- Thermochemical routes for conversion are generally more efficient from an energetical point of view than the biological ones when the biomass is rather dry.
- The biological routes are generally more efficient in terms of nutrient and organic matter recycling. Anaerobic digestion offers the possibility of coupling depollution of liquid effluents with energy production.

There was also general consensus among experts that liquid biofuels should be a priority of action. Methanol offered better prospects than ethanol, it was claimed—against the opinion of Sergio Trindade from Brazil, who presented in a keynote the Proalcool program of Brazil which had just started. Ethanol from lignocellulosic matter was identified as an option for research. And one kept also in mind direct liquefaction of biomass by thermochemical means.

The preference to methanol given in these assessments was justified by the consideration that the productive yields of lignocellulosic matter is indeed higher than those achieved for sugar and starch crops, the feedstocks for bioethanol. But seen from today in 2014, experts in those early days got it all wrong. "The devil is in the details." It was ethanol which made it since then massively into the mainstream of the world's transport fuel markets. Methanol and ethanol from lignocellulose, too, did not yet really make it out of the R&D field. One had to learn it the hard way: pilot plants were successfully built but the economics and logistical considerations stood in the way of commercial success.

One should underline that the development of biomass energy sources and technologies in developing countries was also mentioned for the first time at the conference in Brighton as part of the European R&D strategy on renewable energies.

From the nineties the CO_2 impact of global warming increased the interest for European R&D on biomass. From this moment, the low CO_2 emission rate of biomass exploitation was considered as one of the three-key factor for further European R&D on the subject: energy, environment, and social development. The objectives of the European biomass R&D program became more specifically targeted to:

- security of long-term energy supply in Europe;
- contribution to the development of industrial new markets;
- improvement of the environment by utilizing residues and wastes;
- decreasing the greenhouse gases emission rates;
- diversifying the agricultural production and the management on marginal lands; and
- providing opportunities for less developed regions in Europe.

Decentralized electricity production and local cogeneration was also part of this strategy.

4.3 Fundamentals and Assessments

4.3.1 Photosynthesis

The maximum theoretical efficiency of plants for converting the sun's radiation into living matter is generally assumed as 6.7%. This figure goes actually back to a study contract of the EU biomass R&D

program with Dr. Jim Coombs, an independent expert from Reading in England, formerly employed by Tate and Lyle Ltd. The results of this extensive study were published in 1983 by Reidel/Kluwer, the Netherlands, for the European Commission in a 200-page book [3]. The book, *Plants as Solar Collectors, Optimizing Productivity for Energy*, had next to Coombs David Hall and Philippe Chartier, who had supervised the study as coauthors.

The study demonstrates that the thermodynamic efficiency of the photochemical conversion for monochromatic light involving 8 quanta has a theoretical maximum at 34%. This efficiency is further reduced, taking into account the photosynthetically available energy due to the particular solar radiation spectrum, light scattering and reflection by the leaf surface, photorespiration, and others.

A final conversion efficiency of 6.7% is then calculated on the following assumptions:

- 50% loss, as only half of the solar spectrum, the one between 400 and 700 nm, is photosynthetically usable;
- 20% loss due to reflection and transmission by leafs;
- 72% loss representing quantum efficiency requirements for CO_2 fixation in 680 nm light (assuming 8 quanta/CO_2);
- 40% loss due to respiration.

If 10 quanta were assumed for CO_2 fixation the efficiency would come out at 5.5%.

In conclusion the study provides a good estimate of the maximum conversion efficiency but it is not scientifically rigorous to the "last decimals after the comma."

The study goes also into details of the photosythetic particularities of C4 and C3 plants.

In terms of energy crop cultivation the study recommends that crops be able to grow well on poorer soils and have a high capacity to take up nutrients. The plants should have good water use economy. They should have a high carbohydrate (cellulose) content, resulting in a higher efficiency of use of fixed carbon since less carbon will be lost in respiration than would be the case for plants which accumulate other metabolites such as proteins and oils. A low protein content will also reduce nitrogen requirements. Low rates of dark maintenance respiration and photorespiration would also be an advantage.

The RE R&D program had next to biomass an extra program on water splitting to produce hydrogen and oxygen via photochemistry, photoelectrochemistry, or photobiology. The program was carried out with dozens of contractors for many years and gave an opportunity for the publication of a whole series of books on its results by Reidel/Kluwer, the Netherlands, for the European Commission [4]. The aim of the R&D was to better understand the photoconversion mechanisms, aiming at the maximization of hydrogen production, among others, by living cells (bacteria, algae), or through photosynthetic membranes. Eventually this program was given up considering the fact that solar photovoltaics (PV) can be used to produce hydrogen at much higher efficiencies.

4.3.2 Residues and Crops

Biomass R&D was at its early stage in the 1980s. The first European studies on vegetal energy crops and agricultural residues assessed a potential of 1 to 7 tons of dry material per hectare and year (straw, bagasse, etc.), a potential of 10 to 15 tons/ha·year of dry material for annual plants, and 5 to 15 tons/ha·year for short-rotation forestry (SRF). The difficulty for exploiting the existing natural vegetation was raised at this time because of its low density of supply and because of the problems of accessibility of the resource. At this point, the mechanization and the transport constraints were identified as some of the main obstacles to the development of biomass energy.

Studies on microalgae showed an important potential of productivity (in the natural environment, in controlled logoon conditions, or in photoreactors). European R&D recommendations focused on the improvement of the productivity and collection rates, on the control of pathogens, on the development of specific harvesting technologies, and on the development of adapted conversion technologies. Many technological and economical barriers still had to be overcome [5].

4.3.3 Conversion Technologies

At the beginning of the second European biomass energy R&D program (1979–1983), two main routes for biomass conversion were clearly identified for further investigation:

1. Biological conversion processes for wet materials (from animal and vegetal sources):
   - Anaerobic digestion (or biomethanization): Biogas is a methane-rich fuel produced in a digester (biological reactor). The substrates are converted inside the digester into methane by specific microorganisms.
   - Acid or enzymatic hydrolysis of lignocellulosic waste from agriculture and industry for the obtention of fermentable sugars.
   - Alcoholic fermentation of sugars—from food and nonfood crops or from hydrolyzed lignocellulosic materials—for the production of ethanol. It can be used as a fuel or as an additive in conventional fossil fuels (the addition of 10% to 50% ethanol was expected at medium term in the 1980s, without modification of the engines). Today, the reality is not so far from these expectations. But engine developments had to be implemented by the car industry. At this time, the use of ethanol in transport fuels was strongly stimulated by new EU directives, limiting or even eliminating the use of lead-derived additives. It was then common practice to use toxic lead-derived additives for improving engine performance. Ethanol plays now partially this role of an octane booster in modern combustion engines, together with other less toxic molecules.

2. Thermal conversion technologies for dry materials (with less than 10% water content):
   - Direct combustion: Advanced direct biomass combustion technologies were developed from the early nineties. It was targeted mainly to the European electricity industry, through the improvement of furnace performances, process automatization, control of the emission of particles, and fluidized-bed combustion for high-scale plants. The objective was the reduction of the CO_2 emission rate in conventional thermal power plants.
   - Charcoal production: In the 1970s less than 25% of the total roundwood harvested worldwide was converted to charcoal. In Africa on average the figure is rather 30% to 40%. Africa consumed then approximately 370 million m^3

annually of fuelwood, denuding large areas of trees. Only 33% of Africa's original tropical forests were left in the 1970s [6].

- Pyrolysis: Derived from charcoal production processes, pyrolysis for fuels also consists of the partial combustion under controlled conditions but maximization of the production of a liquid fuel. R&D on the fast-pyrolysis process started in the nineties, improving the production of the liquid phase. The advantages of pyrolysis were related to a high energy content, to the long-term storage potential, and to easier transport conditions. Additionally, it was possible to produce light transport fuel after hydrogenation of this bio–crude oil. Nevertheless, the economical conditions for the industrial implementation of such a technology were still to be investigated in the mideighties.

- Gasification: The production of a producer gas from wood is a well-known process for hundreds of years. This gas can be directly used as a fuel or under special conditions as a syngas for methanol production. For the first time, methanol was produced from gasification synthesis gas by Lurgi (Frankfurt) in 1983 with the support of European R&D funds.

From this moment started the discussion about the most appropriated composition of synthesis gas for methanol production and about the breakeven point of plant size for economical performance.

In parallel, evaluation reports assessed the various possibilities for using low-heating-value and medium-heating-value producer gas for electricity production with gas turbines, diesel engines, and fuel cells.

4.3.4 Logistics

Although mechanization, transport, and other logistic aspects were considered as a main barrier to the development of bioenergy, less R&D on these topics was addressed in the 1980s.

It is through pilot and demonstration projects that the logistic aspects were really taken into consideration. European R&D focused on biomass energy logistics from the mideighties on the following: harvesting methods (felling, concentration, predrying, chipping), transport, storage, and drying.

Specific studies were carried out acccording to the species (annual plants, perennial plants) and according to the conversion technologies (biological or thermochemical conversion processes).

The construction of the first dedicated harvesting machines for sweet sorghum started in the early nineties [7].

4.4 Development of New Feedstocks

4.4.1 Short-Rotation Forestry

The European R&D on this topic started in the eighties with the selection of species adapted to marginal lands (excluded from food agriculture); their genetic improvement, including hybridization (alders, poplars); the understanding of the physiology of such trees; and their reaction to stress conditions (high density, lack of water, diseases) [7]. Research concerned also cultivation techniques (spacing, irrigation, harvesting, transport of the raw material).

SRF trials (coppice and single-stem) were conducted in France, Italy, Portugal, and the United Kingdom with a wide variety of species amongst which were poplars, willows, and eucalyptus (this one being unfortunately unable to resist to frost). By the end of the eighties, SRF was ready to be integrated in the common agricultural policy.

Available evidence suggests that biomass forestry would constitute the best land use on between 1 and 5 million ha, that is, between 7% and 30% of the 15 million ha, if agricultural surpluses and the associated subsidies were to be brought under control (F. C. Hummel in Ref. [8], Vol. 1). In 1988 the only large-scale modern forest biomass plantations existed in Spain and Portugal. They were of eucalyptus grown for pulp on 10- to 14-year rotation. There were some 800,000 ha of these plantations, and they were among the most profitable investments. In Northern Europe poplars and willows grow equally fast but make greater demand on soil fertility and moisture.

Within the biomass R&D program promissing species were the following (E. Teissier du Cros in Ref. [8] Vol. 3): among conifers hybrid larch and western hemlock in the United Kingdom and hybrid larch, Sitka spruce, and sugi in France. Among broadleafs mainly willows were promising in northwestern Europe, poplars when enough water was available, alder on marginal land, and eucalyptus in the Mediterranean zone. Cost of production for a farmer-operated system achieving 12 dt/ha·year with coppice amounted in 1988 to 50 euros/dt (C. P. Mitchell in Ref. [8], Vol. 3). Poplars offer the best prospects as the planting materials are well known, and are available; they are easy to establish; productivities are high; and the risk of disease is less than with willow.

4.4.2 Grasses

At the beginning of the nineties, the European R&D program on biomass gave priority to C4 annual crops for agricultural land. The high-yielding plants with C4 photosythesis are generally tropical and subtropical species and are intolerant to relatively low temperatures during much of the growing season in northern Europe. But a small number of C4 species grow naturally in temperate climates. Within the program in Britain and Ireland three perennial C4 plants, two cordgrasses and galingale, were successfully tried (S. P. Long et al. in Ref. [8], Vol. 3). Up to 19 dt/ha·year have been demonstrated in trial plots. Switchgrass, another C4 grass, was extensively tested across Europe, with best yield observed in Greece [9].

The yield seems to be higher than that for ryegrass, reed canary grass, and giant reed, the C3 plants which were investigated, too.

One should also mention the excellent results obtained with *Cynara cardunculus* (cardoon). Spain coordinated an EU network, the "Cynara Network" with laboratories in six member countries involved; 20 dt/ha·year have been demonstrated [10].

4.4.3 Agricultural Energy Crops

Sorghum, an annual C4 crop, was the subject of extensive investigation. It is already an important grain crop cultivated on

45 million ha worldwide for food and animal feed. Within our new research on sorghum for nonfood application, yields for fiber-sorghum up to 30 dt/ha·year have been demonstrated in France (G. Gosse in Ref. [11]). The variety sweet sorghum is of interest as it accumulates sugar in its stems (D. Delianis in Ref. [11]). The sugar juice must be separated from the bagasse. Because of the low purity of juice, a syrup, the crystallization of sorghum sugar juice is difficult. Hence the sugar is best used by valorization to alcohol. The best varieties tested were Keller, Wray, Cowley, and others. Yields of up to 45 dt/ha·year have been obtained in southern Europe. Sugar percentages ranged from 9% to 12% on fresh stem matter: sugar yields of up to 12 tons/ha·year were estimated. From this amount 10,000 L/ha·year of alcohol production are possible.

Another interesting energy crop, actually a kind of giant grass, is *Miscanthus*, a C3 plant. The European Commission's "DG Agriculture" has compiled a handbook on *Miscanthus* in 1997 as part of its Food Agro Industry Research (FAIR) program [12]. Third-year yields of up to 24 dt/ha·year were reported for irrigated sites in southern Europe and up to 18 dt/ha·year on nonirrigated sites in Germany and Britain. It was observed that the biggest obstacles to crop establishment are the ability to survive the first winter. *Miscanthus* is a more environmentally acceptable crop than others in terms of soil erosion minimization, biodiversity, use of resources, and nutrient leaching.

Finally the Jerusalem artichoke should also be mentioned [10]. Around 14% to 16% sugar content has been achieved. And the sunflowers and oilrape should conclude this round.

4.4.4 Algae

Algal biomass was considered in the EU R&D program from the beginning of the 1980s, but the effort was limited in time. An important aspect was the study of the algae growth in the lagoon of Venice, where algae grow in their natural environment, the sea. Then we had some effort on basic research of microalgae for hydrocarbon production and biological research on closed tubular systems (Wagener in Ref. [7]).

4.5 Improvement of Biological Conversion Technologies

4.5.1 Biogas

Hundreds of biogas plants already existed in Europe in the eighties. On the basis of the existing situation, the European R&D focused on process and technology development, on the potential contribution to the European energy mix, on the economical feasibility and payback time of the biogas plants, and on the operating problems and reliability.

Pilot projects were implemented for the treatment of animal manure, agroindustrial waste, and algae, in mesophilic or thermophilic conditions, and in one-step or in two-step processes, producing a low-heating-value gas (biogas) for combustion purposes (heat/electricity).

One of the main contributions of the European biomass R&D program was the establishment and updating of a detailed databank of the existing biomass plants in Europe [13]. In this report published by the European Commission in 1990 on the basis of a study report by Nyns and Pauss from the UCL in Belgium, 743 biogas plants in Europe are identified. There are also 223 biogas plant builders listed. The report has in particular noticed that biogas plants have a tendency to proliferate, in the neighborhood of specialized research centers across Europe which facilitate promotion and servicing.

In a dedicated 330-page practical handbook, *Biogas Plants in Europe*, from 1984, all possible aspects of the subjects had already been reviewed previously in all details [14].

4.5.2 Other Fermentation Processes

European R&D was important in this sector since the beginning of the eighties: fermentation after hemicellulose and cellulose hydrolysis (fungi, actinomycetes, clostridia), acid hydrolysis, improvement of yeast metabolism, improved bacterial metabolism (*Zymomonas*), modified yeasts, improved processes and fermenters, and new and improved processes for alcohol separation.

Since then, ethanol for fuel was a main sector of the European R&D.

4.6 Improvement of Thermal Conversion Technologies

4.6.1 Charcoal Making

Carbonization of dry lignocellulosic materials takes place when temperature rises under controlled air intake. The reaction above 270°C is exothermic and energetically self-sufficient. The temperature inside the reactor climbs and stabilizes around 400°C–450°C. This process is empirically well known since ancient times [15], with very low yields: the open-pit technology and later the earthmound kiln method are still used in some developing countries.

After World War II (WWII), the rapid-pyrolysis process was developed in the United States, making the production of charcoal much more efficient and profitable.

In 1985 the European Commission published *Handbook of Charcoal Making: The Traditional and Industrial Methods*, which goes back to a study contract of the biomass R&D program with Walter Emrich. It reviewed on 270 pages all new methods with due regard to the by-products of charcoal making, that is, pyrolytic oils and tars [6].

In 1635, analyzing the process of the charcoal industry, chemist Rudolf Glauber pointed out the existence of potentially useful by-products in the condensates of the earthmound kiln method.

During WWII, the charcoal industry developed on a significant scale. The liquid by-products (today called pyrolysis oils) were used as raw material, providing many derivates: methanol, aldehydes, organic acids, acetone, tars, etc.

From the midseventies, the European biomass R&D program focused on this technology. R&D addressed material supply (transport, storage, drying, preprocessing), equipment improvement, technology and process improvements, safety and environmental improvement, and end-use markets (briquetting, home fuel, fuel for the industry, activated carbon, lump charcoal, etc.).

4.6.2 Pyrolysis

In fast pyrolysis, biomass is decomposed into vapors, aerosols, and some charcoal. After condensation a dark brown liquid is formed. Its heating value is about half that of conventional fuel. The process requires a carefully controlled reaction temperature of approximately 500°C. It is endothermic and requires a heat input. The residence time of the finely ground biomass is limited to only two seconds. Hence the process is also called *flash pyrolysis*. Up to 75% of the solid biomass is eventually obtained as bio-oil. Different types of reactors have been investigated: the bubbling fluid-bed reactor, circulating and transported beds, ablative pyrolysis, and others. The reactor represents, following our experts' view, only 10% to 15% of the total system cost, the rest being associated with logistics: biomass reception, storage and handling, drying and grinding, product collection and storage, and last not least upgrading. The latter can be achieved through the addition of methanol as a solvent, hydrocracking, or vapor cracking (A. V. Bridgwater in Ref. [9]).

It is important to note that pyrolysis was a key component of the European Commission's R&D program. As mentioned earlier, the main aim was to generate a cost-competitive, environmentally friendly liquid fuel. Logistical considerations were also to be taken into account, as biomass collection and supply may impact the acceptable size of the installation. With respect to the competing routes of liquid biofuel production from lignocellulosic material— such as the one via gasification to a synfuel described here below— pyrolysis looked more promising. Expressed in simple terms, it must be cheaper to keep the biological chains in the biomaterial to some extend intact, as is the case with pyrolysis, than to decompose it to elementary molecules which have afterward to be recomposed to a bioliquid.

Bridgwater from Aston University in Birmingham was a key expert on pyrolysis who advised the European Commission's program on this subject. Pyrolysis experiments were carried forward, for instance, in the Netherlands and Italy. Finland was until today the most committed to the advancement of pyrolysis (K. Sipilä et al. in Ref. [16]). At VTT in Finland production was tested in various reactors with different raw materials. Boiler tests, diesel engine tests, and gas turbine tests were carried out.

Nevertheless, progress to achieve a commercial bio-oil product was slow. One of the reasons is that the bio–crude oil is a complex mixture of components which lacks stability and is difficult to store and which does not mix with petroleum products.

4.6.3 Gasification

Having evaluated the large experience gained in Europe with small individual gasifiers for automobile transport, it was decided to stay away in the biomass R&D program from any further development of this route. Interest for gasification of solid biomass was then aiming at larger systems, in particular in view of woody biomass utilization in larger power plants. A good example is the IGCC biomass plant in Värnamo, built by Sydkraft AB in Sweden with Foster Wheeler. It was the world's first IGCC plant utilizing exclusively wood as fuel. The plant has a pressurized circulating fluid-bed gasifier and generates 6 MW_e and 9 MW_{th} for district heating. By June 1998 it had already performed 5400 hours of gasification runs (K. Stahl and M. Neergaard in Ref. [16]).

Having defined a particular strategic interest for the conversion of solid biomass into liquid fuels, the EU program managers took early on the initiative to start a pilot program to generate biomethanol from solid biomass via its gasification to syngas. Syngas has the particularity with respect to simple gasification with air that nitrogen which makes up close to 80% of our air has to be excempt from the syngas. Syngas is composed exclusively of CO and H_2.

The EU biomass R&D pilot program lasted for three years, from 1982 to 1984. After a call for proposals four projects were selected and built: one at Clamecy with Creusot Loire in France; one at Lurgi in Frankfurt, Germany; one at Sulmona by AGIP/Italenergy in Italy; and one with John Brown Wellman in England. The full-size pilot projects had a throughput of 1 ton of woody biomass per hour. All four projects proved to be technically viable, although two of them did not just use pure oxygen for gasification but applied new routes. Syngas was produced in all four plants, and some of it was converted by Lurgi into biomethanol.

The program was a good example of European cooperation of large industrial companies, as the four contractors included some of Europe's biggest companies of mechanical engineering.

On behalf of the European Commission the results of the program were evaluated following joint criteria by Ton Beenackers and Wim van Swaaij of University of Twente in the Netherlands. The program and its results were published by the European Commission in 1985 in an extensive 230-page book [17].

Although the program was successful from a technical point of view it also had the value to demonstrate the economic weakness of this syngas route. The plant at Clamecy could benefit from a follow-up phase by financial support of the French government, but ultimately there was consensus among all experts that this route was for excessive cost not really promising for liquid biofuel production. There was no EU follow-up of this program.

4.6.4 Direct Combustion

Co-combustion of biomass with coal has become since the turn of the century an important market in Europe. There is also a European directive on co-incineration. Large power plants up to 100 MW and above are concerned by this route. Wood firing is associated with particular problems like fouling. On the other hand it can improve the combustion of coal and capture SO_2, for instance. Torrefaction of pellets and chips has also emerged as an important issue on the cofiring market.

4.7 Integration and Pilot Projects

The early European R&D studies aimed at the control of mechanisms and at the stabilization of the production and conversion processes. Then, investigations started about the feasibility of large-scale integrated pilot projects. This evolution was the direct consequence of the oil shocks which started in the 1970s.

The so-called integrated projects consisted not only of the integration of technologies from top to bottom but also of the parallel integration of the processes, for example, biomass for energy and pulp for paper.

The large European biomass for energy network (LEBEN) targeted the development of energy-autonomous villages in Europe

[18]. These projects initiated and financially supported by European R&D funds aimed at the local develompent of energy from biomass in remote villages. Initiatives were started in the Abruzzo region in Italy and spread later to other regions in southern Europe. The objective was to integrate all existing technologies for this purpose: mapping of the various biomass resources, transport, infrastructure and related facilities, potential conversion technologies, analysis of the end users' demand, development of appropriated distribution routes, and analysis of the social and economical performance (integrating all environmental impacts).

The LEBEN projects were associated with the EU regional development funds.

Finally one should mention Action de Préparation, d'Accompagnement et de Support (APAS), a special budget of 25 million euros decided in 1994 by the European Parliament in favor of our RE activities [18]. It served the development of initiatives and concepts for large-scale integration of the REs in regions, cities, and utilities. The program started with over 70 signed contracts in 1994. In the field of biomass, a series of networks was set up for promoting electricity generation from biomass.

4.8 Promotion and Dissemination

4.8.1 Biomass Conferences

The "1st E.C. Conference on Energy from Biomass" was held at Brighton in 1980. More than a contractors' meeting, it was the first time that all European and non-European scientists were able to meet on the subject of bioenergy. Although most of them were at their early stages, all the R&D subjects were presented and discussed. The resulting proceedings were considered as a world reference.

From this time, some 17 conferences were organized on the same basis until 2014.

It was an opportunity for European scientists from all the sectors concerned by the biomass energy industry, not only on technical subjects, but also on European strategic orientations in a world context, to gather.

All these conferences focused on the improvement of the management of biomass resources, on the promotion of the competitiveness of biomass for energy and industry, and on the social and environmental impacts of this RE source (including its potentially negative impacts).

In accordance with the European agricultural, energy, environmental, and social policies, the overall objective of these conferences was the establishment of an effective political commitment in order to develop a European bioenergy industry.

4.8.2 Books

Many books have been published by the European Commission in cooperation with commercial publishers in the frame of the biomass R&D program. In a series with Reidel/Kluwer in the Netherlands eight books on biomass were published between 1981 and 1986. Others were published in cooperation with Applied Science Publishers, Elsevier Applied Science, Springer, TÜV Rheinland, CPL Press, and others.

4.8.3 Newsletters

Between 1991 and 1995, nine newsletters on the EU RE R&D programs, called SOLAR EUROPE, were published in cooperation with the magazine *Systèmes Solaires* in Paris.

4.8.4 Renewable Energy Fora

With the aim to improve coherence within the European Commission's R&D program on REs and strengthen the contact of the REs' communities of experts and contractors in the biomass field with those on solar energy and building integration, PV, or wind energy, broad fora have been organized. An important one found place in Saarbrücken, Germany, in 1988, with the participation of Karl-Heinz Narjes, vice president of the European Commission, and the minister Oskar Lafontaine.

We organized another major forum of this kind in 1993 at the UNESCO House in Paris.

4.9 Conclusions

Here are our conclusions:

- Very ancient technologies can be boosted through R&D, reaching high-level technical efficiency.
- New concepts and processes usually start on a nonviable economical basis.
- Potentially promising technologies never develop without R&D support.
- Only R&D, demonstration, and information/dissemination enable us to reach economical efficiency.
- Technological evolution applied in society follows a long-term, step-by-step process, even if the fundamentals are already well known decades in advance.
- Progress was sometimes facilitated through the various existing EU policies (agriculture, energy, industry, environment).
- Local projects, taking into account all the specific socioeconomical, environmental, and industrial aspects in so-called integrated projects, led to successful results.

Since the time when the EU R&D programs were started in the 1970, the market implementation of the REs in their modern forms achieved now by 2014 tremendous strength. A main driving force was R&D, but it was not the only one. Once political initiatives in the EU member countries and in the European Parliament had led to higher attention in public support budgets, industry got organized—mostly new innovative ones, leaving the conventional energy operators waiting on the sidelines. As the markets started to develop the laws of mass production came into play. This was particularly visible in the PV market were kilowatt-hour costs were divided by a factor of 20 in the last 20 years. Also in biomass the market costs came down through mass production. There are ups and downs in market opportunities—the barrel price of oil may go down or certain governments knock down the support for biogas— but most importantly, politics and climate experts start at last to understand that a limit to the threatening climate change is entirely dependent on the further dissemination of the REs in the world's energy markets.

And bioenergy has and always had a leading role in the markets for the REs. Should the world approach indeed a 100% RE supply scenario on the horizon of 2050, as many experts believe, bioenergy's share could, following this author's view, achieve a share of up to 60% [19–22].

References

1. White, L., Low, J., Hansen, G., Hurand, A., Kearney, D., King, G., Lettinga, G., Morandini, R., Naveau, H., Wagener, K. (1980). *Energy from Biomass in Europe*, Eds. Palz, W., Chartier, P., Applied Science, London, EUR 6809 EN; *Energia da Biomassa in Europa*, CSARE, Venezia.

2. Palz, W., Chartier, P., Hall, D. (Eds.) (1981). *Energy from Biomass, 1st European E.C. Conference*, Applied Science, London, EUR 7091.

3. Coombs, J., Hall, D., Chartier, P. (1983). *Plants as Solar Collectors, Optimizing Productivity for Energy*, Reidel/Kluwer, Dordrecht, EUR 8575; (1985). *Biotechnologie zur Energieerzeugung, Verbesserungsmöglichkeiten der Produktivität von Energiepflanzen*, Verlag TÜV, Rheinland, Cologne, EUR 8575 DE.

4. Hall, D., Palz, W. (Eds.) (1981). *Photochemical, Photoelectrochemical, and Photobiological Processes*, proceedings of the EC contractors' meeting, Cadarache, October 1981, Reidel/Kluwer, Dordrecht, EUR 7666.

5. Palz, W., Grassi, G. (Eds.) (1981). *Energy from Biomass*, Vol. 2, Proceedings of the Workshop on Biomass Pilot Projects on Methanol Production and Algae, Brussels, October 1981, Reidel/Kluwer, Dordrecht, EUR 7667.

6. Emrich, W., EU Commission (1985). *Handbook of Charcoal Making: The Traditional and Industrial Methods* (*Energy from Biomass*, Vol. 7), preface by Palz, W., Reidel/Kluwer, Dordrecht, EUR 9590.

7. Grassi, G., Palz, W. (Eds.) (1982). *Energy from Biomass*, Vol. 3, Proceedings of the EC Contractors' Meeting, Brussels, May 1982, Reidel/Kluwer, Dordrecht, EUR 8044.

8. (1988). *EUROFORUM New Energies*, Proceedings of the International Congress, Saarbrücken, Germany, October 1988, Palz, W., chairman of the organizing committee, H.S. Stephens & Associates, for the EU Commission 1988, EUR 11884.

9. Palz, W., et al. (Eds.) (2002). *Twelfth European Biomass Conference*, proceedings of the conference, Amsterdam, Netherlands, June 2002, ETA Florence and WIP Munich, ISBN 3-936338-10-8.

10. Fernandez, J., et al. (1996). *European Energy Crops*. Country report for Spain.

11. (1996). *First European Seminar on Sorghum for Energy and Industry*, Proceedings of the Meeting, Toulouse, France, April 1996, INRA and Ademe for the EU Commission.

12. *Miscanthus Handbook*, Hyperion, Ireland, for the EU Commission DG VI, December 1997.

13. Pauss, A., Nyns, E.-J. (1990). *Biogas Plants in Europe: An Updated Databank*, UCL, Belgium, for the EU Commission, EUR 12896.

14. Demuynck, M., Nyns, E.-J., Palz, W. (1984). *Biogas Plants in Europe: A Practical Handbook* (*Energy from Biomass*, Vol. 6), Reidel/Kluwer, Dordrecht, EUR 9096.

15. *Plinius the Ancient, Historia Naturalis*.

16. Sipilä, K., Korhonen, M. (Eds.) (1998). *Power Production from Biomass III, Gasification and Pyrolysis*, Proceedings of a Symposium, Espoo, Finland, September 1998, VTT Energy.

17. Beenackers A., van Swaaij, W. (Eds.) (1986). Advanced gasification, methanol production from wood, *Energy from Biomass*, Vol. 8, Results of the EEC Pilot Programme, preface by Palz, W., Reidel/Kluwer, Dordrecht, EUR 10407.

18. Newsletter "Solar Europe" for the EU Commission's RE R&D program, Systèmes Solaires, Paris.

19. Palz, W. (1995). *Future Options for Biomass in Europe*, at Workshop on Energy from Biomass and Wastes, Dublin Castle, Ireland.

20. Palz, W. (1996). *Biomass in Europe: A Global View*, at the Global Climate Change Forum, September 1996, Columbus, OH.

21. Palz, W. (1997). *European Renewable Energy Perspective*, at the First World Sustainable Energy Trade Fair Conference, May 1997, Amsterdam.

22. Palz, W. (1998). *Die Zukunft der Biomasse in Europa*, at Der Holzkongress, May 1998, Schwäbisch Hall, Germany.

About the Authors

Wolfgang Palz (WP) holds a PhD in physics (Dr.rer.nat.) from the University of Karlsruhe, Germany, from 1965.  In 1965–1970 he was a professor of semiconductor physics in Nancy, France. In 1970–1976 he was in charge of power systems development at the French National Space Agency, CNES, in Paris. In 1973 he was co-organizer of the UNESCO Congress "The Sun in the Service of Mankind" in Paris. He was also a keynote speaker at the 40th anniversary event of that Congress at UNESCO in Paris in 2013. In 1976/78 UNESCO published his book *Solar Electricity* in seven languages.

In 1977–2002 WP was an official of the EU Commission in Brussels, the executive body of the European Union. In 1977–1997 he managed the development program of renewable energies; it included policy development and contracting to the European industry and academia of the EU Commission's budget (almost $1 billion over that period). The R&D program comprised the sectors of solar architecture, solar energy, wind energy, biomass, and ocean energy.

In 1997 WP became an EU Commission counsel for renewable energy deployment in Africa and, besides, advised the EU commissioner for energy on the EU white paper on renewable energy (RE) issued that year.

From 2000 to 2002 WP was a member of the energy committee (Enquête Commission) of the German Parliament, the Bundestag, in Berlin, to establish an energy strategy for Germany on the time horizon 2050. In 2005/2010 he was a consultant of the EU Commission for photovoltaic (PV) programs in Latin America.

WP was a co-organizer of the "Great Wall World Renewable Energy Forum," which took place in Beijing, China, from October 24 to 26, 2006.

In 2010 WP edited the book *Power for the World: The Emergence of Electricity from the Sun*, published in September by Pan Stanford in Singapore (ISBN 978-981-430-337-8). In 2013 WP edited the book *Solar Power for the World* and two books on wind power with Danish Folkecenter, all with the same publisher in Singapore. The same year a new book, *Biomass Power for the World: Transformations to Effective Use*, to be published by 2015, was started with the University of Twente in the Netherlands.

WP is bearer of an Order of Merit of Germany (Bundesverdienstkreuz), has been recognized as a wind energy pioneer in Britain, and has received the European Prizes for biomass, wind energy, and solar PV. In September 2010 he received the Swiss Solar Prize, together with the pioneers B. Piccard and A. Borschberg, the pilots of the airplane *Solar Impulse*. In August 2011 WP received the ISES Advancing Solar Energy Policy Award.

By 2014, WP was:

- Member, the French National Aid Commission, Investments for the Future, Paris
- Member, the Board of REN21, Paris
- Advisory Board Member, American Council on RE, ACORE, Washington DC
- Member of the Global Bio-Energy Partnership GBEP, FAO Rome
- Board Member, Eurosolar Russia, Moscow
- Chairman for Europe, World Council RE (WCRE)
- Honorary Board Member of ISES, Freiburg, Germany

Henri Zibetta (M.Sc. Eng.), of Swiss and Belgian nationalities, was born in Bellinzona in 1960. He is living in Brussels, Belgium, since 1968.

From 1980 to 1985, Mr. Zibetta studied applied sciences and civil engineering at the Free University of Brussels. In 1985, he graduated in master in science of engineering from the Free University of Brussels. From 1985 to 1987, he was a research engineer and member of the scientific staff at the Free University of Brussels.

Mr. Zibetta was at FABI-URTB as editor of *Belgian Engineers' Information Bulletin* and has also worked as a freelance expert for the European Commission, R&D Directorate, Renewable Energy Unit. He was member of the staff, head of the energy unit, and project coordinator at Mens & Ruimte. He is founder of FAC Consult and was its managing director from 1989 to 2014. He is involved in consulting, studies and publications on energy, and the environment and natural resources.

Chapter 5

Large European Bioenergy Network (LEBEN): A Strategy for Commercial Exploitation of Biomass at the Regional Level

Giuliano Grassi

European Biomass Industry Association, Rond-Point Schuman 6, B-1040, Brussels, Belgium

giuliano.grassi3@gmail.com

"Large-scale exploitation of biomass in the EC through the introduction of large integrated bio-energy schemes" (LEBEN projects) may offer in the long term an interesting contribution in alleviating the present problems of the common agricultural policy (CAP) with a maximum potential of generating 500 to 600 million toe per annum of clean energy.

This idea of LEBEN was initially proposed by Grassi in 1984. After 30 years it can still be considered a valid concept of an instrument to **promote rural development** through the expansion of agriculture and forestry, **creating jobs** and revitalizing communities in marginal areas, offering help in solving the problem of agricultural surpluses

Biomass Power for the World: Transformations to Effective Use
Edited by Wim van Swaaij, Sascha Kersten, and Wolfgang Palz
Copyright © 2015 Pan Stanford Publishing Pte. Ltd.
ISBN 978-981-4613-88-0 (Hardcover), 978-981-4669-24-5 (Paperback), 978-981-4613-89-7 (eBook)
www.panstanford.com

by **developing alternative land uses**, and providing **substantial benefits as far as the environment is concerned**, both locally and globally.

It has been applied partially in the frame of the present important EC Biomass Action Plan (diversification of agriculture activity, implementation of INTERREG projects, structural funds, social funds activities, etc.).

In the following pages detailed information is presented concerning the history, concept, development, scope, object, and content of about 16 regional LEBEN assessment study projects which were implemented during the years 1985–1992, with the coordination and strong support of the R&D program DG XII.

5.1 Introduction

Following the first oil shock of the 1970s the Commission of European Community (CEC) had initiated (under the leadership of Dr. W. Palz) a wide R&D Dem program on renewable energy. In May 1978 I moved from EURATOM association—Commisariat à l'énergie atomique (EURATOM—CEA/EDF) nuclear fast breeder program (Super-Phoenix projects) to a position at the European Commission (EC) DG XII Renewable Energy Program, managing the photovoltaic bioenergy wind energy subprograms.

In the year 1984, following the presentation at the European Parliament of the project "IDEA" by Prof. Umberto Colombo (Club of Rome), proposing the full exploitation of the biomass potential in the EC and conscious (from the preliminary but wide EC R&D Dem. results) of the large possible energy-socioeconomic-environmental benefits deriving from sustainable exploitation of endogen biomass of any type, I got the idea to sensitize regional authorities (as legal organizations responsible of the use of land and of rural development) about the possibility to exploit their biomass in the context of the common agricultural policy (CAP) on the European level. I took the initiative to promote the large European bioenergy network (LEBEN) (a network of large regional diversified bioenergy projects) in most of the EC 12 countries of that time.

LEBEN was not an official activity of the CEC. However, many of the process technologies and activities proposed and being considered at that time for the participating regions were derived from R&D Dem. work carried out by the CEC (feasibility study, pilot plants, demonstration plants, concerted action, etc.), covering relevant disciplines and technological processes in the areas of biomass production, harvesting, thermochemical conversion, biological conversion, and its end use. Products of major interest include heat and power generation, biofuels (biogas: biomethanal, bioethanol; bio-syngas), pulp for paper, fertilizers, composites, etc.

Thus during the period 1985–1992 about 16 regional LEBEN projects were assessed, most with some financial support, coordinated by my unit (DG XII).

5.2 Growth of the LEBEN Concept in the Context of the European Dimension

The beginning and promotion of this network of regional biomass projects was based on my personal conviction (supported also by numerous and highly professional experts assisting the EC) manifested strongly in many of my speeches of that time as follows:

> We are approaching the time when very large agro-energy industrial projects could be introduced throughout the entire community. Our research and development programmers are actively supporting relevant technologies valid also for large-scale production, conversion and utilization of fuels and materials from biomass at a regional scale.

This prediction wanted to emphasize the ever-increasing economic, social, and environmental importance of sustainable renewable agricultural and forest resources, including agricultural and solid wastes, as potential valuable commercial (commodity) sources of energy and materials in the context of the European dimension. This taking into account that all national and regional governments are continuously confronted with a wide variety of challenging policy issues concerning the use of energy, environmental health protection, and socioeconomic development.

Evolution and technological advances in the processing of biomass resources allowed more and more agriculture and forestry feedstock to yield an increasing range of products, not just for the use of biomass as an energy source which was not new, because in the past most of the energy needs were indeed met by the direct combustion of biomass.

Furthermore the rising cost of oil processing, the public concern about the environment and safety of nuclear power, and the increasing ability of the bioenergy sector to compete with conventional fuels had created the potential and promising perspectives for the exploitation of biomass on a massive scale. This is now confirmed (September 2014) by a recent study by the agency International Renewable Energy Agency (IRENA), quantifying that the potential contribution of bioenergy could reach 60% of all renewable energy and 20% of total energy needs.

LEBEN was imagined to play a significative role in addressing these new emerging areas of public policy of particular interest also on a regional level, wherein lies the decentralized public governance authority.

Ensuring that this potential would be fully exploited in Europe was the challenge faced by LEBEN, represented by a group of interested scientific experts and representatives of governments and industry brought together on an ad hoc basis.

It was intended to serve as a model for public sector technological transfer activities and to stimulate the establishment of decentralized regional-oriented programs and of networks of stakeholders, providing support for the commercialization and deployment of innovative biomass technologies. The overall goal of LEBEN was to encourage the use of all biomass energy technologies which are technically feasible and cost effective, with its major focus on the transfer of current and reliable information to potential bioenergy users, planning activities best suited for each region, including preliminary technoeconomic information.

Although the idea of LEBEN was not taken seriously at first, with the exception of the Abruzzo region (Italy), which was the first to consider it and ready to propose strong financial support for LEBEN Abruzzo. Successively the concept of LEBEN gradually gathered more and more strength, attracting interest from an ever-growing list of regions within the member states of the European community as well from Eastern Europe, China, Australia, and Latin America.

In 1988 LEBEN involved 16 EC regional programs coordinated by the EC DG XII. The knowledge gained provided the first results and information to optimize the work on a regional level to the administrations of these regions and confirmed the important role and contribution of biomass to their energy requirements with several associated benefits.

However, the LEBEN concept produced some expected confusion, considering its wide scope:

- The wide range of raw materials from agriculture or forest activities
- The wide variety of conversion and utilization technologies
- The many forms in which biomass can be introduced into industry
- The geographical considerations and situation in each region

Furthermore, at the beginning of the biomass programs most of the projects were conceived as individual and site-specific installations (biogas plants, bioethanol plants, power generators, gassifiers, heating plants, etc.), while by the adoption of large integrated schemes as proposed in LEBEN a much wider range of markets could be served simultaneously (electricity, heat, biofuels, paper, fertilizes, solid fuel, industrial rare materials, fertileness, desalinated wastes, composites for building, etc.) requiring integrated processing configuration.

Hence, whilst the marketplace dictates the wanted products and their prices, a combination of climate, geology, land use patterns, and social conditions of each regional LEBEN will determine the limits, nature, and amounts of final biomass-derived products. However, it is the acquired scientific and engineering know-how (present and future) which will determine the ability to match biomass raw materials to the market's needs.

The economics and time frames of the technological developments are thus crucial for an optimized and sound establishment of the LEBEN concept and for its deployment in the European Union (EU) regions. However, the old concept of LEBEN can still be considered (after 30 years from its origin) a valid instrument for diversification of energy supply, diversification of agriculture, and forestry activity with considerable benefits, especially now having a much greater range of commercial processing technologies at our disposal.

As a concluding remark, the development of the LEBEN concept can also offer an interesting contribution in alleviating the main problems of CAP (overproduction of food) by exploiting the estimated maximum EU biomass potential of 500–600 Mtoe/year.

5.3 LEBEN Projects and the European Financial Mechanisms

The agricultural communities in all member states share common problems. However, at the same time there are marked differences in agriculture, forest practices, land use, climate, geology, populations, and rural incomes on a regional basis. Depending on the crop there is overproduction in some areas with underproduction in others. As a whole the community is increasingly concerned about the costs of subsidies under CAP; pollution of land, air, and water; waste treatment; rural poverty; migration of rural populations; and reliance on imported fuels, timber, and wood products, and some animal feedstuffs. The variation in regional economies is recognized in the financial mechanisms administered by the EC and the European Investment Bank, as illustrated in Fig. 5.1, in relation to eligibility for assistance under the terms of the community's structural funds. These help measures relate to economic development of regions, promotion of employment, environmental protection, energy conversation, new sources of energy, provision of infrastructure, and adaptation of activity sectors such as agriculture and forestry. It is thus perhaps no coincidence that a number of the first LEBEN projects were established in regions with the greatest problems but also availability of significant financial supports measures, as indicated here below.

5.4 Development of an Integrated Approach

There was a strong feeling that the rise of biomass for energy and industrial purposes could only be competitive on a large scale, as envisaged in LEBEN, but also if the coproduction of industrial commodities was enclosed with the adoption of an integrated and coherent system for the selected production techniques and market products and especially if all the selected production techniques and markets were assembled into an integrated and coherent production system.

Identification of viable markets, however, would also require wider links to societies with which the LEBEN projects should coexist compatibly with other ongoing socioeconomic activities.

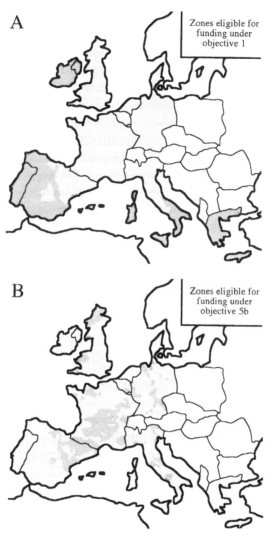

Figure 5.1   (A) Areas qualifying for assistance under objective 1 from various instruments of the CEC (ERDF, ESF, etc.) aimed at promoting the development and adjustment of the regions where development is lagging (i.e., where per capita GDP is less than, or close to, 75% of the community average). (B) Areas qualifying for assistance in view of the reform of CAP under objective 5 from the EAGGF, ERDF, ESF, etc. Objective 5a concerns adapting production, processing, and marketing structures in agriculture and forestry, whilst objective 5b is aimed at promoting the development of rural areas.

Therefore the integration of these supplementary bioenergy-bioindustrial activities into the regional general context was considered of major importance for good success, facilitating the social, economic, and cultural implications. Whilst the search for new uses of biomass continues to date, in the past years a number of significant market possibilities were already established. These included electricity generation and production of compost and organic fertilizers, heat and chemicals, liquid transport fuels, pulp for paper, and novel polymers. Biomass could provide the raw material, whilst reliable commercial technology was available to meet the market needs. Economics remained the key issue. However, although the time was difficult for investors due to the risk of novelty, only by investing the needs of the future may be realized.

5.5 European Contest: Coping with the Changing Needs

Also at that time (1980s–1990s) among others, six problems were of major concerns to the EC for which the LEBEN projects were considered as being able to provide a positive contribution:

- Security and diversification of long-term energy supply (renewable)
- General environmental problems and waste accumulation
- Use and management of surplus and marginal lands
- Protection of the natural environment and rural society
- Employment in rural districts
- Cooperation with Third World countries

5.6 Land Use, Employment, and the Environment

5.6.1 How LEBEN Could Impact Europe's Future

At a summit meeting held in Houston in July 1990 leaders of the world's seven most industrialized nations met to discuss the future of the planet. Two of the three issues highest on the agenda at this meeting were of direct relevance to LEBEN:

1. The setting of ambitious targets for limiting greenhouse gases which cause global warming
2. The phasing out of agricultural subsidies for farmers

However, reducing atmospheric carbon dioxide levels by introducing punitive taxes, as well as solving agricultural overproduction by setting land aside to become derelict, generates other problems.

On the contrary the implementation of large biomass LEBEN programs was considered too drastic and able to have dramatic affects on both the major problems identified above. However, the greenhouse effect is just on the range of environmental problems which LEBEN aims to address. Land use, reductions in subsidies, and set-aside payments would hit hardest at smaller farms, especially those on marginal land, if no alternative to conventional agriculture is identified. It seems that about 20 million ha of agricultural land will be abandoned, and in these regions the EU implementation of LEBEN projects can reverse declining fortunes, offering an alternative of nonfood use of biomass production.

In several areas the decline of local agriculture has led already to the destruction of rural communities. When conventional agricultural activities become uneconomic, farms close down, the subsequent unemployment and loss of income encourage the population to go to the city in search of wealth, and all which remains is empty villages.

As an instrument of rural development LEBEN aims to reverse this trend. Because of its integrated approach combining energy, industry, and agriculture, the LEBEN concept provides economic alternatives to conventional agriculture. In fact because 50%–60% of the final production cost in the processing of biomass is manpower—*the processing of about 500 tons of dry matter of biomass justifies the creation of a new job; thus a large number of diversified jobs can be generated.* If the full potential EU biomass could be exploited by LEBEN activities about 2 million jobs could be created. These would not only replace the jobs lost due to the decline in conventional agricultural activity and the small number of jobs lost for the reduction of conventional fuel import but also add many more, especially in less developed rural areas.

As far as concerns of the environment, its coverage by scientists, politicians, legislators, and the media has brought matters of the environment to everyone's attention. Every day, media headlines reveal a new environmental story: the greenhouse effect, toxic emissions, water pollution, acid rain, urban pollution, etc.

Although the raising of public awareness now means that we all know the main problems the planet faces, simple changes are not enough. There are no short-cut solutions to environmental problems. What is required is wide-ranging actions which attack the roots of environmental damage at all levels—local, national, and global. It is exactly this type of approach which is offered by LEBEN. For example, on the local impact, traditional cultivation on hilly areas has stripped thousands of hectares of natural forestry. This deforestation has damaged the fragile fertility of these areas, allowing soil erosion and nutrient loss from steep slopes. In the last decades the abandoning of agricultural activities in marginal areas has accelerated the process. This kind of waste of land use is what LEBEN aims to reduce by replanting abandoned land with forestry. The trees will stabilize the soil, provide a barrier to soil erosion by wind and surface water run-off, and recycle nutrients.

Rotation systems proposed by LEBEN will stabilize the ecology of these areas, whilst serving as a source of raw materials, reducing the risk of fire. The use of LEBEN technology can result in such wood residues being collected for commercial use and converted to produce energy and industrial products.

At the national level large use of biomass can reduce emissions of chlorine, sulfur, and oxides of nitrogen, contributing to a reduction in acid rain and air pollution.

5.7 Cooperation among LEBEN Projects

Why should the EU unite in the development and deployment of large biomass regional projects like LEBEN?

The removal of trade barriers within the EU, community harmonization, the deep reform of CAP, and the possibility to supply a large amount of clean energy and biomaterials provide a strong basis for general economic and in particular much desired new rural prosperity. The transition from local and national to European markets requires strong cooperation among the EU regions:

- to demonstrate the integration of regional system performance for biomaterial and energy production;
- to acquire operating experience of these large-scale systems;
- to optimize their processes and utilization technologies;
- to guarantee substantial investments and support for innovation;
- to develop markets for the bioproducts;
- to ensure continuity of supply;
- to mitigate the costs of biomass;
- to harmonize the quality standards of products; and
- to eliminate legislative barriers to LEBEN activities.

When I first proposed LEBEN in 1985 it was for economic reasons. The large scale and the concept of integration were established as a means of pooling the know-how which would have been gained in different technologies, so lowering the costs for everyone. An operational **database** was initiated to which different LEBEN projects could be linked to absorb the generated information, disseminating it to other projects.

The exchange and **training** of personnel were also envisaged to improve operational efficiency. We also financially supported the development of an **expert system** as an instrument for decision makers to enable them to assess how best to fit their own particular needs—in practical terms which biomass resources would be most appropriate in specific regional situations.

5.8 Biomass Resources for LEBEN Projects

Typical regional biomass types are the following:

- Wood from conventional forests and from coppice and short-rotation forestry using a wide range of species, best suited to local conditions (i.e., poplars, willows, pines, eucalyptus, alders, *Robinia*, etc.) (Fig. 5.2)
- Agricultural residues from the cultivation of food crops (i.e., straw, corn stock, bagasse, etc.)
- Wastes from food processing, paper pulp production, organic fractions of nonhazardous industrial wastes, etc.
- Energy crops

Figure 5.2   Short rotation forestry activity.

In particular three promising high-productivity crops under intensive scrutiny were considered for LEBEN:

- Sweet sorghum, of great interest for the south of Europe (Fig. 5.3)
- *Miscanthus*, of great interest for North-Central Europe (Fig. 5.4)
- *Cynara*, of great interest for semiarid soils (Mediterranean regions) (Fig. 5.5)

Figure 5.3   Sweet sorghum.

Figure 5.4   *Miscanthus.*

Figure 5.5   *Cynara.*

In Fig. 5.6 are shown the sites and dimension of two energy crops (sweet sorghum and *Miscanthus*) supported and coordinated by DG XII. Under the advice of Prof. Stander (Polytechnic Institute of Munich) we established the first European network on *Miscanthus*.

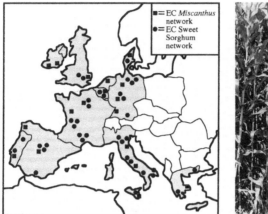

*Studies on the highly productive C4 grasses (Sweet Sorghum and **Miscanthus**) are being coordinated through several networks covering production trials, physiology and propagation as well as end use. The photograph illustrates Sweet Sorghum growing at around 35 tonnes (dry matter) per hectare in Greece.*

Figure 5.6   Sites and dimension of sweet sorghum and *Miscanthus*.

At the end of the 1980s, to reduce the huge agriculture surpluses and considerable CAP subsidies in the EC-12 it was estimated that an area of about 20 million ha of agriculture land had to be taken out of production, offering an opportunity for large-scale forestation and energy crop production schemes. A set-aside program was thus introduced, covering about an 8 million ha area of agricultural land. At that time also many million hectares of marginal land had been already abandoned for economic (low-productivity) reasons.

5.9   Processing to Provide Feedstock from Regional Forests and Farms

Intensive activity was envisaged for the harvesting, drying, processing, storage, and transport (logistics) of the regional LEBEN biomass.

The solid biomass resources which are dispersed, available on a seasonal basis, have low physical and energy density and contain quite a lot of water; thus they require to be converted into suitable feedstock for cheaper transportation and subsequent processing and end use. Therefore these must be dried, and their energy density must be increased (pelletization) or converted into liquid-pyrolysis bio–crude oil gaseous biofuels (biogas-biosyngas). At the end of the 1980s, for the exploitation of the large amount of diversified biomass regional resources, liquefaction of solid biomass (by flash/rapid pyrolysis) into bio–crude oil was considered the most appropriate method for economic reasons, taking into account investment and processing costs, for the following reasons:

- The difficulty of pelletization of different agroforestry and waste resources
- The considerable activity on pyrolysis (pilot demonstration) by the EC having captured all the more advanced technological progress in Canada and the United States (four technologies) and on the development of the four new EC technologies
- The considerable activity and positive results on the stabilization of bio–crude oil (six months) and upgrading (hydrogenation), although at that time the competitiveness of biofuels with oil price at 15 $/bbl was impossible without taking into account indirect and social costs

5.10   Conversions and Market Products

The conversion technologies of major interest envisaged at that time (1985–1990) included **biological conversion** and **thermo-chemical conversion**.

5.10.1   Biological Conversion

5.10.1.1   Anaerobic digestion

The microbial decomposition of wet organic matter in the absence of oxygen is a spontaneous reaction which occurs widely in nature. Traditionally this type of microbiology has been applied to the treatment of sewage sludge and widely adapted as a means of product energy on farms from manure. Increasing numbers of systems treating industrial effluents are now being built in response to environmental legislation, and large amounts of gas are being recovered from waste disposal landfill sites.

5.10.1.2 Fermentation

The production of bioethanaol by classic yeast-based fermentation, followed by distillation as a means of product recovery, remains the basis of the largest biomass energy schemes, typified by those using sugarcane in Brazil or maize in the United States (see Fig. 5.7). The high cost of agricultural raw materials as a result of CAP makes bioethanol production by this route less favorable from an economic point of view if sugar beet or wheat is used as feedstock. The use of alternative feedstock such as sweet sorghum may change this. As indicated above such a crop figure prominently in proposed LEBEN activities.

Figure 5.7 Bioethanol micro distillery (A. Nardi).

5.10.1.3 Composting

The conversion of fibrous wastes, wood chips or bark to compost is also brought about by microbial action, but unlike anaerobic digestion or fermentation the organisms require oxygen. Large-scale plants have been built, a number designed to use wood wastes or the separated organic fraction of municipal solid waste (MSW). In such a plant the reaction is monitored, controlled, and enhanced by forced aeration, taken at different stages of stabilization, which can be used as soil conditioners, contribute valuable plant nutrients, and improve soil water-holding capacity and structure. Again, the EC at the time was supporting a small network aimed at consolidating information to LEBEN activities as a means of disposing of residues from biomass processing as well as providing a valuable soil conditioner.

5.10.1.4 Combustion

Direct combustion of low-moisture biomass remains the major route of utilization at present. Considerable amounts of wood are burned for domestic heating and cooking, whilst many industries use wastes generated in-house as an energy source. On a larger scale biomass fuels can be used to generate steam for power production (see section Gasification). The problems with such large-scale power production relate to the need to ensure a regular supply of fuel on a year-round basis. Large storage facilities are required and transport costs are high, reflecting the low energy density. Storage, in turn, can cause potential problems of fire risk. Direct combustion has the disadvantage that the heat output is difficult to control rapidly over a wide range. The solution of many of these problems is to upgrade the biomass to a fuel with a higher heating value.

Figure 5.8   Combustion plant.

5.10.2   Thermochemical Conversion

There are three basic methods of converting biomass to higher-grade fuels using heat, as indicated next.

5.10.2.1   Pyrolysis

Current thermochemical activities related to LEBEN projects are concentrated on pyrolysis. Pyrolysis is one of the most attractive thermochemical routes by which biomass can be upgraded to fuels and chemicals. The process consists of heating solid biomass, such as woodchips with a low moisture content, at a high temperature in the absence of air (oxygen) to produce solid (charcoal), liquid (pyrolysis oil), or gaseous fuels.

The conversion of biomass to solid or liquid gives it a much greater commercial value. It becomes easier to handle, store, and transport. The energy content on a weight basis is increased.

The pyrolysis liquid can be further upgraded to produce gasoline substitutes compatible with internal-combustion engines. On a larger scale the addition of pyrolysis oil to a petroleum oil refinery will lower the average sulfur content of the fuel fraction, with obvious environmental benefits.

A number of processes were analyzed to optimize an efficient, pollution-free system. These include the construction of a 1 ton/hour slow-pyrolysis fluid-bed pilot plant in Raiano (Italy) by Consorzio Alten. The recovered main products from this plant were charcoal and oil. Studies on the feasibility of using the pulverized material as carbon-oil-methanol slurries led to the development of a stable, low-viscosity product with a carbon content of 52%. Since the oil has handling, storage, transport, and utilization properties similar to bunker oil and a heating value of around 20 GJ/ton, it can be used in existing boiler systems. The risk of polymerization of pyrolysis oil was reduced drastically from few days to some months.

5.10.2.2 Liquefaction

This is a relatively low-temperature, high-pressure catalytic process, often carried out in a reducing atmosphere (hydrogen or carbon monoxide) or using a hydrogen donor system. Since liquefaction is carried out in the liquid phase, heat transfer is improved, whilst the reducing atmosphere results in a product with low oxygen content, improving the quality and calorific value. The high cost of pressure reactors, the need for feed preparation (drying and size reduction), and unresolved problems of feeding slurries at high pressure, as well as separation of product from solvent, have reduced interest in this technology as far as LEBEN is concerned.

5.10.2.3 Gasification

Gasification is a high-temperature process in which a solid fuel is reacted with a limited amount of oxygen to convert the feed to a mixture of gases including hydrogen and carbon monoxide. In small gasifiers, as illustrated in Fig. 5.9, the oxygen may be derived from air. In this case the fuel gas will be diluted with residual nitrogen from the air and hence a low-heating-value product will result. Typically, an energy efficiency of around 75% can be achieved with moderately wet wood in small air-blown systems. Large oxygen gasifiers can produce a higher-value gas, which may be catalytically upgraded to hydrocarbons or methanol. Such systems were under no commercial development and so not considered for LEBEN at that time.

Figure 5.9   Gasification in a small gasifier in which the oxygen may be derived from air (Tommaso Guicciadini). See also Color Insert.

5.10.2.4   Pollution-free paper pulp

Agricultural residues, such as fiber from sweet sorghum, are suitable for paper pulp production of reasonable quality. However, conventional straw-based factories have been closed due to problems of pollution associated with the large amounts of water and chemicals used and the high costs, making the product noncompetitive. The alkaline sulfite anthraquinone methanol (ASAM) process offered, thus, an alternative route to paper, which could compete with existing pulping systems and would not cause environmental damage (Fig. 5.10). The production costs and energy needs of this process were investigated as part of LEBEN activities, with expected costs 20%–30% lower than those of conventional paper-producing methods such as the Kraft or sulfite processes. The capital investment costs were similar to those of a Kraft mill. Since all bleaching effluents are recycled together with the black cooking liquor, pollution is reduced to zero. The process gives high yields of strong pulp, as indicated in Table 5.1. Production trials (5 tons of paper) were carried out using sweet sorghum bagasse as feedstock, by which 1,000 copies of an EC book (summarizing the history of the development on sweet sorghum at the EC-DG XII) were printed (see Fig. 5.11).

Figure 5.10 The ASAM pilot plant of Kraftanlagen, Heidelberg.

Table 5.1 Results from the ASAM process applied to low-quality wood and sorghum bagasse

Property	Small diameter coppice wood	Sweet sorghum bagasse
Tensile strength (m)	22.80	20.10
Specific gravity (g/smc)	0.76	0.65
Opacity (%)	97.30	
Liquor: wood ratio	4:1	4:1
Brightness (%ISO)	47.10	33.80
DPW	2.95	
Kappa	10.00	6.90
Tear (CN)	80.00	53.00
Yield (%)	50.00	53.00

COMMISSION OF THE
EUROPEAN COMMUNITIES

PROMISING INDUSTRIAL
ENERGY CROP

SWEET SORGHUM

RECENT DEVELOPMENTS
IN EUROPE

Giuliano Grassi
Pietro Moncada P.C.
Henri Zibetta

Photographs:
Tommaso Guicciardini

Published
September 1992

Figure 5.11 EC book summarizing the history of the development on sweet sorghum at the EC-DG XII. This publication had been printed on paper produced by bagasse of sweet sorghum (ASAM process).

5.11   List of LEBEN Regional Projects

See Table 5.2 for a list of the first 16 LEBEN projects.

Table 5.2 List of the first 16 LEBEN projects

Country	Site
Germany	Saar
Greece	Evrytania
Ireland	Cork
Italy	Abruzzo
	Compania
	Metaponto
	Sardinia
	Sicily
	Tuscany
	Umbria
Netherlands	Groningen
Portugal	Coimbra
Spain	Catalunya
	Castilla y Leon
	Galicia
United Kingdom	Scotland

Although in 1992 the total number of LEBEN projects coordinated was 16 (as listed in Table 5.2) the total number and expressions of interest reached the considerable number of 31 regions, as shown in Fig. 5.12.

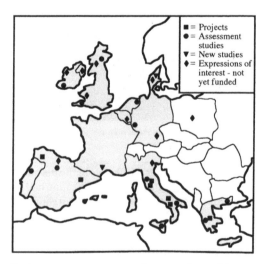

Figure 5.12 Projects, assessments, and expressions of interest for LEBEN in 31 regions.

5.12   Preferred Market Products of LEBEN

A number of (pre)feasibility studies have indicated the type and range of markets which can be served by agroforestry biomass from commercial processing technologies (Fig. 5.13).

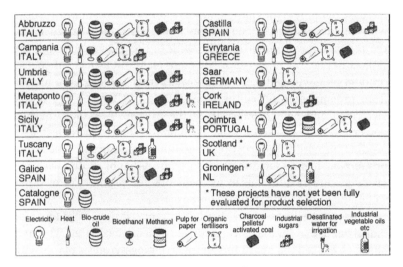

Figure 5.13   Alternative products obtainable from biomass.

5.13   Activities Carried Out (1992) in the First 16 Regional LEBEN Projects

Next a few of the 16 regional LEBEN projects are described.

5.13.1   LEBEN Abruzzo

5.13.1.1   The birth of LEBEN

The mountainous Abruzzo region on the Adriatic Coast of southern Italy is the site of the first LEBEN project. It was leading the way for the European LEBEN program. Covering over 1 million ha, of which some 50% is agricultural land and 20% is forested, it has at present over 100,000 ha of unexploited coppice woodland.

It was anticipated that if nothing was done to reverse the decline in rural activity, over 20% of the region (200,000 ha) would have become disused by the turn of the century.

These problems were being addressed by a regional development program which aimed to improve exploitation of the potential productivity, restore a balanced socioeconomic situation, and increase employment, paralleling the objective of LEBEN. These were the driving forces behind the ambitious project approved by the regional government in 1985 with the objective of having the project in place around 1995. The main technical aspects of energy production envisaged were as follows: pyrolytic bio-oil production from forestry and agricultural residues as well as energy crops, electricity and heat production in the existing Avezzano thermal power station (27 MWe) using biomass fuel, hydroelectricity production within the existing irrigation complex, and bioethanol production from specific sweet sorghum ethanol crops, with by-products and residues used as feedstock for production of other value-added products (biogas compost).

The possibilities and feasibilities or various options for industrial use of biomass were currently being evaluated. The first phase envisaged the possibility of constructing pyrolysis plants with a total capacity of 180,000 tons/year to provide bio-oil to feed the thermoelectric power station Avezzano. It was expected to generate 200 million kWh/year. In addition an ethanol factory with a capacity of 500 to 1000 tpd was envisaged to use molasses, surplus starch products, and sorghum syrup.

The second phase envisaged the completion of a network of pyrolytic units for conversion of 320,000 tons of biomass per year as well as the development of an agroindustrial complex for value-added products.

The total investment, in excess of 227 million ECU, could be available with the community's financial support (L64/86) and regional financial support (LR31/82 and LR16/86) totaling about 70% of this.

To ensure the best results from all the integrated activities planned, a consortium was formed involving local authorities, industry, and private contractors.

On June 6, 1985, I was invited (see dissemination activity) by the European Parliament in Brussels (Alliance Libre Europeenne) to make a presentation of "LEBEN—PROJEC ABRUZZO." The assessment study evaluated that with a total 227 M€ investment (120 M€ for bioenergy, 24 M€ for hydroelectricity, 83 M€ for the agroindustrial sector) such investment could have been recovered in about four years (but with the vital support of the EC structural and social funds), thus able to produce a considerable socioeconomic impact due to the creation of 3670 jobs: 1500 for bioenergy, 20 for biopower, 50 for hydroelectricity, and 2100 for agroindustry. However, also if there was strong support from the regional authorities and the EC the implementation of LEBEN Abruzzo did not materialize due to the negative advice of the major Italian energy organizations.

5.13.2 Umbria

5.13.2.1 Plantations for power in the green heart of Italy

The LEBEN project in Umbria aimed to help solve the problem of agricultural surplus by expanding existing forest plantations to pro-vide feedstock for a new power station. Umbria is a temperate re-gion 8456 km^2 in central Italy with a verdant countryside, which has earned it the name "the green heart of Italy." Agriculture thrived in the region until in the 1980s, when farming was largely abandoned in hilly and mountainous areas. The high productivity of the fertile plains has been continually exploited and there is now an increas-ing problem of crop surplus. There is thus a need in the region for alternative crops and suitable conversion systems to provide raw materials for local energy and industry markets. In collaboration with Ente Nazionale per 1'Energia Elettrica (ENEL) the aim was to use fast-pyrolysis technology to convert local biomass resources to biofuel which could be used to fuel an electricity-generating power station. There was an opportunity for this project to be developed rapidly as some land has already been planted with short-rotation forestry for the production of wood chips. At the time these were used in tobacco drying but could be exploited on a larger scale as py-rolysis feedstock. To study the feasibility and impact of this project a regional consortium was set up involving farmers, private opera-tors, Ente di Sviluppo Agricolo Umbria (ESAU), and Ente Nazionale Energie Alternative (ENEA) (the Italian public research institute for renewable energies), as well as local authorities, businesses, and for-estry cooperatives. The consortium is investigating socioeconomic implications of the project, availability of marginal land, and local environmental sensitivities. Information from these sources is being used to evaluate the possibility of generating valuable by-products to determine the best areas for energy plantations, maximize the production of electricity, and create new industries based on biomass products, such as paper manufacturing.

5.13.3 Grand Designs for Tuscany

In Tuscany, home to the art treasures and architecture of Florence and Siena, the LEBEN project has "creative designs" for 2400 ha of marginal land. A master plan was drawn up for the exploitation of the area's biomass via a processing plant. Tuscany has large areas of very marginal land for which there appears to be no other income-generating use. A wide range of possible activities was investigated covering growth of nonfood plant species as well as aspects of harvesting and cultivation, with special attention given to the possibilities of reducing the use of fertilizers and pesticides.

To ensure optimization of the project and to maintain progress a consortium and promoting committee was established involving local authorities, industry, private contractors, and the University of Siena. The emphasis was on economically viable projects. The initial outline incorporates a very flexible industrial scheme: inputs can be easily changed and outputs can be diversified into a number of market opportunities. Initial cost–benefit analysis has indicated an 11.5% rate of return on investment.

Elsewhere in Italy LEBEN projects are also being planned for Metaponto, Sicily, and Sardinia.

5.13.4 SPAIN: Galicia

5.13.4.1 Generating energy and wealth

The Spanish electricity-generating company, Union Electrica Fenosa, was behind a project to replace conventional energy sources in Galicia with residual forest biomass. Galicia, in the northwest of Spain, covers an area of almost 3 million ha; 67% of this is forested.

Union Fenosa planned to use these massive natural resources to produce bio–crude oil which could be used in the production of electricity in an existing thermal plant as well as in the boilers and furnaces of local industries. Activities are centered on the Ferrol region, an area which requires urgent re-industrialization. Once the success of the scheme has been proven, activities will be expanded throughout the region. The first stage of the project, a feasibility study to establish the potential of the region's biomass and the suitability of conversion processes, showed promising results. Work is underway to investigate practical aspects of harvesting, transport, and pyrolysis, and a global techno-economic viability study was carried out. The project, coordinated by Union Fenosa, includes participation by the University of Madrid, engineering firms INTECSA and INFYE, pyrolysis experts WIP, and many other organizations. The first phase was supported by the CEC and by Union Fenosa on a 50:50 basis. As the project entered the second phase it was anticipated that further expenditure of over 200 million pesetas by Union Fenosa would be supplemented by a similar contribution from regional sources.

5.13.5 Castilla y Leon

The Castillian region in the northwest of Spain covers an area of 80,000 km^2 rich in forestry and agriculture, which offers a potential yield of 2 million tons of biofuels. A LEBEN project was established to investigate how the potential of the region's resources could be used in the future. Information obtained from small rural industries in the region suggested that in many cases they could effectively satisfy most of their energy needs with biomass.

5.13.6 Catalunya

In the hinterland of Catalunya a LEBEN project aimed at supplementing the local 160 MWe power station's low-grade coal fuel with derivatives of the abundant biomass in the region was established. The project links national and local objectives to reduce pollution, create jobs, and protect the forestry and hydrological ecology of the region, which was suffering from the decline of its two main local industries, mining and textiles. The LEBEN project offered an opportunity to help revitalize the economy of the region by exploiting its agricultural and forest potential. The region could generate a total of almost a quarter of a million tons per year of biomass. Assessment studies have shown that collecting accessible brushwood residues would yield 80,000 tons/year to be used as biofuel feedstock, whilst planting rapid-growing species on abandoned farmland would yield another 80,000 tons/year. Intense poplar production on wet wasteland would add a further 20,000 tons/year. Planting sweet sorghum and Jerusalem artichoke on marginal land would yield 35,000 tons/year for conversion to bioethanol. Pyrolysis again offered the best opportunity to utilize much of this potential resource. The price of pyrolytic derivatives would be similar to that of the lignite powder currently used to fuel the local power station, offering socioeconomic and environmental advantages.

5.13.7 Ireland

5.13.7.1 Smiles on biomass

In Ireland, West Cork has been identified as the region which offers the greatest opportunity for integrated agroenergy projects as sustainable enterprises. The southwest coast of Ireland, warmed by the Gulf Stream, has an exceptional climate, boasting Italian gardens in which herbs and flowers normally found on the Adriatic and Mediterranean coasts thrive. Besides the attractive climate it also has a local university able to provide a suitably educated workforce and an abundance of marginal agricutural land. However, much of the potential it offers is going to waste—20% of the workforce is unemployed and much of the land was unused. Hyperion Systems, which is the local LEBEN participant, believed that biomass energy could dramatically alter this situation. Following an extensive feasibility study, the project swung into action: planting of sweet sorghum and *Miscanthus* has started and a 50 kW pilot plant and a biogas conversion unit have been opened. In addition, the imminent selling off of state forests planted 30 years ago would have created a surplus of timber residues, which could also be exploited. Sean McCarthy of Hyperion Systems was optimistic about the future of LEBEN projects in Ireland, indicating that "there is an immediate market for paper pulp and organic fertilizers and there is considerable potential for biomass to replace imported power."

5.13.8   Poplar Biomass in the United Kingdom

In central Scotland the Forestry Commission (FC) was considering a scheme to spruce up a region currently scarred by derelict industry by planting it with 20,000 ha of forestry, including up to 5000 ha of alders and poplars. Besides improving the environment the FC was considering the industrial use of forest residues and coppicing systems for energy production, which could be used by paper and pulp mills in the area. The scheme included a research program to identify the most suitable tree species as well as to train and demonstrate to farmers how they can maximize forestry potential on their land. An initial feasibility study and economic assessment was carried out.

5.13.9   Greece

5.13.9.1   Rural renaissance in Evrytania

Evrytania, an area of Greece, low in wealth but rich in natural resources, has a significant potential for the large-scale production of bioenergy and the generation of hydroelectric power (Fig. 5.14).

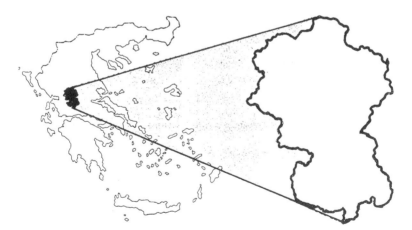

Figure 5.14   Evrytania, Greece: 187,000 ha of forest and pastures with a population which has dwindled from 40,000 to around 26,000 over three decades.

At that time the region was suffering from migration, with rural workers moving to Athens and neighboring regions in search of work. However, those who remained were looking for wealth closer to home. Half of the region was covered in forests offering 120,000 tons of timber per annum, of which only 58,000 tons were currently being used. Timber plants were only active for one-third of the year and used only 50% of the biomass available, the timber left behind representing a recurring fire risk. By exploiting this wasted potential in the timber industry, planting on abandoned agricultural land, and using new technologies, such as oil pyrolysis for energy production, LEBEN Evrytania could aim to provide permanent jobs, stimulate rural growth, and reduce the threat of environmental damage. Project leader Prof. Spiros Kyritsis emphasized the need to produce energy in the region as it was expensive to transport energy from outside.

He recently stated that "we have to invest in plantations to ensure continuity of supply so that production of biomass can be matched to final use. Because the raw materials are dispersed the possibility of developing a mobile pyrolytic system should be considered. Our main priority is to find appropriate production. We are looking forward very much to producing energy products economically."

5.13.10 Portugal

Without any oil or coal of its own Portugal's most important resource was biomass. In the future the country aimed to exploit this renewable resource to provide up to 10% of its energy requirements. The Portuguese Centre for Biomass Energy was created and universities throughout the country were working on projects to increase the intensity of forest management as well as to promote and develop new equipment and technology to produce steam and electricity for industry from plants using biomass and urban waste. Among these projects one aimed to utilize the waste materials from the country's huge cork industry. Goncalvos dos Santos, at a LEBEN meeting, indicated that "scientists are trying to define new technologies so that we can be more efficient in industrial areas. The policy of the secretary of state is very much behind us—the sector can grow in the next two years."

5.13.11 Germany

LEBEN's first project in Germany is in Saar, where it was hoped that 5% of the region's heat requirements, currently met by imported fossil fuels, could be replaced through exploitation of biomass. The region has a low population density and hence a relatively low overall energy demand. The possibility of partially replacing fossil energy with integrated biomass systems using gasification, combustion, and pyrolysis technology for decentralized energy generation was currently being evaluated. A diverse range of raw materials was available in the region for this project, including grass, straw, and woodchips, as well as sewage sludge and manure. By converting effluent and agricultural wastes to fuels the LEBEN Saar project aimed to improve the local environment, whilst establishing innovative sustainable energy systems.

See Figs. 5.15–5.17 for LEBEN dissemination actions.

Dr G Grassi (CEC) addresses a group of biomass experts at a LEBEN workshop in the offices of CIEMAT (Spain)

Figure 5.15   LEBEN dissemination action.

5.14   Concluding Remarks

The concept of LEBEN projects based on a massive exploitation of biomass of any type at the regional scale (after 30 years from its origin) can be still considered a valuable instrument for accelerating the contribution of bioenergy to the total energy mix of regions, countries, and the EU. Substantial general and local benefits will be

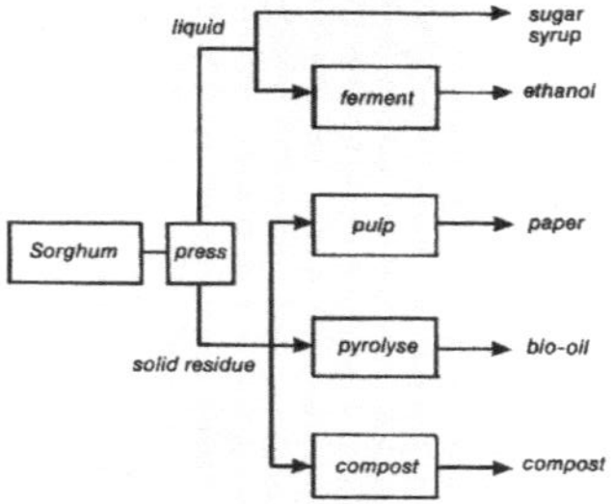

Large European Biomass Energy Network

Introducing LEBEN news

October 1989

Pyrolysis pilot plant in Raiano, Italy.
Capacity is about 500 kg/h dry wood
Output is 25% wt char and 20% wt bio-oil

What contribution will be made to the regions

The anticipated benefits to the regions involved in LEBEN include socio-economic benefits of increased employment; environmental benefits of lower pollution levels including recycling of carbon dioxide and lower sulphur emissions; industrial benefits of new industries and new products.

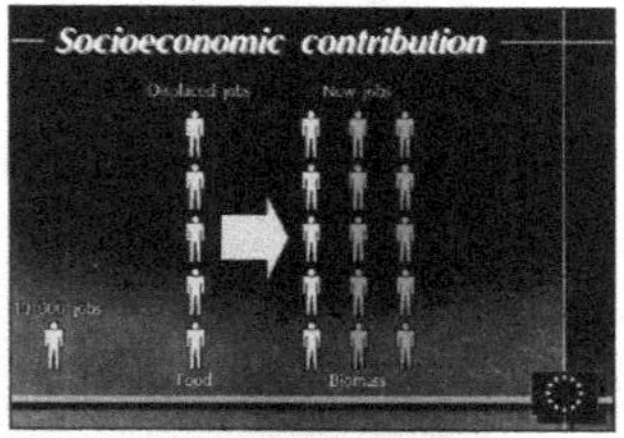

Socioeconomic contribution

Integrated use of sorghum

The way in which a new crop could be exploited to produce a full range of energy and other added value products is illustrated by the following flow sheet:

Sorghum → press →
- liquid → ferment → sugar syrup / ethanol
- solid residue → pulp → paper
- pyrolyse → bio-oil
- compost → compost

R, D & D activities of common interest for the LEBEN-projects

BIOMASS PRODUCTION

- Short Rotation Forestry (poplar, willow, eucalyptus, Robinia, Miscanthus)
- Sugar Crops (sweet sorghum, Jerusalem artichoke)
- Soil Fertility Conservation and Improvements
- Land Recovery from Biomass Production to Food Production

HARVESTING, PRETREATMENT OF BIOMASS

- Development of sweet sorghum harvester
- Development of short rotation forestry harvester
- Technology development for sugar-juice extraction, concentration, sterilisation, storage
- Biomass collection in remote areas
- Pneumatic transportation of biomass

BIOLOGICAL CONVERSION

- Acid hydrolysis of lignocellulosic material
- Enzymatic hydrolysis of lignocellulosic material
- Pentose and lignin valorisation

THERMO-CONVERSION

- Development of modular standard pyrolysis units
- Up-grading of pyrolytic fuels
- Testing of pyrolytic fuels in boilers (oil and slurries)
- Testing of pyrolytic oil in gas turbines
- Testing of pyrolytic oil in oil refineries
- Standards definition of pyrolytic products for different markets
- Standards definition for wastes disposal, combustion of pyrolytic fuels etc
- By-products utilisation

DEVELOPMENT OF INTEGRATED ACTIVITIES (INTEGRATED PILOT-PROJECTS)

- Pulp for paper from biomass
- Organic fertilisers
- Biomass test centres

COORDINATION AND COOPERATION

- Development of an expert system for LEBEN-Projects
- Studies and analysis (socio-economic, environmental etc)
- Formation, exchange of personnel
- Exchange of information and experience

Figure 5.16 LEBEN dissemination action.

generated in terms of employment and of the environment, assuming that the criteria of sustainability are adopted.

Figure 5.17 LEBEN dissemination action.

However, the implementation and effective development of LEBEN projects in the EU will require a complex level of analysis of modern integrated energy planning. Variables such as regional, social, and economic costs, viability of innovative technologies and systems (i.e., biorefineries), and environmental impacts must be analyzed and quantified. Specific software packages must be developed, that is, (a) programs which review supply and demand of biomass and agricultural-economic-demographic scenario and costs, (b) development of expertise systems to assist planners and decision makers; (c) establishment of regional biomass systems, and (d) risk evaluation systems. The implementation of a significant LEBEN program could avoid the abandonment of more marginal agricultural lands and the destruction of rural communities, producing also socially and economically attractive solutions to the problem of CAP. The creation of a large number of diversified jobs, especially in rural areas, should require a deep extensive analysis of possible social credits.

Acknowledgments

Dr. G. Grassi gratefully acknowledges the assistance and considerable documents and information provided by former EC biomass expert Dr. James Coombs, CPL, U.K. Additional material has been taken from reports, proceedings of LEBEN workshops, and LEBEN News (October 1990–April 1992), as well as publications from the European Community R&D programs.

About the Author

Giuliano Grassi has a degree in mechanical engineering from the University of Pisa (1956) and a postgraduate degree in nuclear metallurgy and technology from Aston University.

Dr. Grassi has a huge experience in the nuclear and renewable energy sectors, for a total of 56 years—21 years in the nuclear reactor field, mostly in the development and construction of advanced Pu-, U-, and Na-cooled fast breeder projects (Rapsodie, Phoenix, Superphoenix 1300MWe) in France (EURATOM CEA). He was active abroad for several years, of which 4 years were spent in the United Kingdom, 7 in France, and 35 in Brussels, Belgium.

He was manager of the renewable energy (photovoltaic, wind, bioenergy) R&D programs of the European Commission (EC) for eight years and was responsible, for seven years, for the bioenergy R&D program of the EC in Brussels. Since 1994, he is the secretary general of the European Biomass Industry Association (EUBIA, Ron-Point Schuman 6, B-1040 Brussels).

Dr. Grassi owns numerous patents in the nuclear and renewable energy fields. He has won several awards, including Linnerborn Prize, Vienna, 1994 (EC Conference); Bioenergy Pioneer, Florence, 2000 (World RE Conference); EUROSOLAR Prize, Berlin, 2006 (Bioenergy and Solar Architecture Conference); EUROSOLAR, Italy, Rome, 2006 (RE Meeting); Best Renewable Energy Partnership with Developing Countries, awarded by the European Commission, Brussels, 2004; Bioclimatic Architecture with PV, Tenerife, 2010; Solar Prize for PV, Italy, Zurich, 2012.

His present activities include several EC bioenergy projects; definition and promotion of commercial biorefineries; industrialization of large-scale pelletization and torrefaction plants; industrialization of mobile pelletization units; advanced small-scale cogeneration units; logistics for large-scale supply of refined biomasses; smart urban districts (100 RE); biochemicals from biomass; and biofertilizers. He has published numerous books and scientific reports. His recent contributions to publications include *Power for the World: The Emergence of Electricity from the Sun* (2012) and *Biomass Power for the World: Transformations to Effective Use* (2015).

Chapter 6

Energy from Biomass: Cooking without Fossil Fuels (a Journey into the World of Cooking in Developing Countries)

Paul van der Sluis

Philips Research, High Tech Campus 34, 5656 AE Eindhoven, the Netherlands
paul.van.der.sluis@philips.com

6.1   Introduction

Cooking of food is more important than most people realize. In fact, there is evidence that the large human brain could not have developed without cooking of food [1, 2]. Given a maximum limit of time per day for food collection and eating, the current human body and large brain cannot be fed on raw food alone. Cooking food softens tough fibers and speeds up the process of chewing and digestion. It also allows more calories to be extracted from the same food. So the cooking made it possible to grow larger brains and even leave time to use it intelligently.

That suggests that without cooking there would be no modern humans. Cooking can thus be regarded as the most important

Biomass Power for the World: Transformations to Effective Use
Edited by Wim van Swaaij, Sascha Kersten, and Wolfgang Palz
Copyright © 2015 Pan Stanford Publishing Pte. Ltd.
ISBN 978-981-4613-88-0 (Hardcover), 978-981-4669-24-5 (Paperback), 978-981-4613-89-7 (eBook)
www.panstanford.com

invention. There is scientific evidence that this happened around 1.9 million year ago [3], although more conclusive evidence is of more recent date.

Therefore, it is amazing to realize that an invention so crucial for the development of modern humans has not changed for more than 2 billion people today. In developing countries the three-stone fires which are currently used have hardly changed over the last 1.9 million years. There are very good reasons for this: the "stove" is portable, the stove can cook quickly, the temperature can be increased or decreased by adding or withdrawing wood pieces, they take all sizes and shapes of pot and can be used with quite a few types of organic fuel. But most importantly: the stoves are free and the fuel is free or cheap.

These advantages do come at a price: Nearly two million people a year die prematurely due to illnesses attributable to indoor air pollution from cooking [4]. And because the smoke is unpleasant, many people transform the wood into charcoal, which is then subsequently burned in inefficient stoves. This practice is better from a health perspective, but is disastrous in terms of efficiency: up to 90% of the energy content of the wood is lost in the nonindustrial manufacture of charcoal. This practice contributes to unnecessary deforestation and high greenhouse gas emissions.

But also the wood burning stoves are inefficient. That means time and money is spent on fuel which could be spent more usefully. That also contributes to unnecessary deforestation, greenhouse emissions and black carbon. And because so many of these stoves are in use the effects are noticeable worldwide. The picture in Fig. 6.1 shows the stove emissions of just one small village in Nepal (Ghusang, just north of the Anapourna mountain range). The picture also clearly demonstrates why simple chimneys in practice give no improvements in the exposure of people to the toxic emissions.

However, when enough money is available, people prefer more sophisticated stoves which do not give smoke in the air and soot on the cooking vessels, which do give immediate response and do not require tending. These are the stoves which are used in the developed countries. However, all these devices use fossil fuel in one way or another.

Figure 6.1   Stove emissions of one small village in Nepal (Ghusang, just north of the Anapourna mountain range).

6.2   Initial Stove Development

Fifteen years ago, I was transforming my household into one which does not require fossil fuels to run. For all functions affordable solutions were available, except for cooking. There were no commercial devices available that could deliver this. My search for these devices brought me in contact with the cooking problems of the two billion people living in developing countries.

These problems have not passed unnoticed. Between 1982 and 1992, the Chinese National Improved Stove Program reported the installation of 129 million improved stoves in rural households. This was both in terms of size and in terms of coverage (50%) the largest program to date. Similar programs were rolled out in other Asian and African countries [5]. Still the adaptation is low enough that 2 million people die each year.

Apart from these programs there is a community of stove designers, research centers (such as Aprovecho [6], or the clean cookstove research program of the United States Environmental Protection Agency [7]) and disseminators. The latter has culminated into the Global Alliance for Clean Cookstoves. The alliance seeks to mobilize high-level national and donor commitments toward the goal of universal adoption of clean cookstoves and fuels. Its ambitious goal is to foster the adoption of clean cookstoves and fuels in 100 million households by 2020. It brings together private, public, and nonprofit stakeholders and is currently the prime source for information related to clean cooking solutions and cooperation partners.

However, this was not available when I was looking for a solution for my cooking question: Easy and clean cooking without using fossil fuels. Most so-called improved stoves at that time (and still) do had lower emissions and higher efficiencies but very moderately. Being used to cook with natural gas or electricity my standards were very high. Only biogas digester–stove combinations have properties which come close. But living outside the tropics and having no cow available (I live in a city) a biogas digester is not an option. I considered charcoal stoves as inappropriate because most of the energy content of the fuel is lost making charcoal from wood.

So the stove had to be wood fueled, needed to burn without smoke and soot, had to have an easy way to regulate fire power, and had to be quick to start up. No stove design using wood fuel which I could find came close. Making use of the Internet and the stove community which was active at that time, it became clear that real advances in stove performance could best be gasifier stoves or fan stoves.

The gasifier stoves were pioneered by Dr. Thomas B. Reed. Several different types were constructed. In these stoves clever airflow gasifies the wood first and then burns the resultant gasses. The emissions of such a stove can be very low, but disadvantages are that fuel size and quality is rather critical, stove power regulation is not easy, and the stove is batch fueled. The latter means that fuel is loaded into a cold stove and lit. The stove stays on until the fuel is gone. I personally wanted an easier stove, but I have studied the gasification well, since it can give such low emissions. That is also why there are still avid proponents of this technology such as Paul S. Anderson [8] and why several gasifier stoves are in the market. The most successful gasifier stove is the fan-driven Oorja stove, initially brought on the market by BP but nowadays marketed by Firstenergy [9]. The fuel sensitivity of the stove is circumvented by using standardized pellets and the power control is improved by using speed control of a fan.

Fan stoves were available for the camping market. Some of them are available even to date such as the Sierra stove (ZZ Manufacturing, Inc.) [10] or the wood gas stove [11]. The fan stoves burn twigs and wood pieces and fire power is regulated by regulating the fan speed. The fan also makes the ignition procedure very fast, which adds to convenience.

These stoves are very useful for their purpose but as such were not useful for me because they lack durability, are too low power, and still have too high emissions. The latter is also evident from the blackening of the cooking vessels. These stoves also need external batteries, which are not very practical for a stove which needs to be used every day.

So I tried to build a stove which combines the clean combustion of the gasifier stoves with the practicalities of the fan stoves. The stove would get a voltage-controlled fan which forces combustion air into the combustion chamber. The combustion air is divided into two streams (see Fig. 6.2). Near the bottom so-called primary air is blown in and near the top secondary air is blown in. The primary air is mainly meant for gasifying the wood. In principle wood can be gasified by heat alone, but burning a bit of the wood near the bottom helps to keep the combustion processes going. And once the volatile gasses are gone, the primary air is needed to volatile the charcoal. The resultant gasses are then burnt by the secondary air.

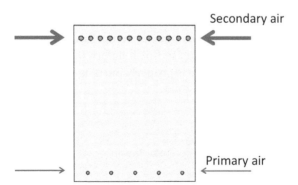

Figure 6.2   Airflow into the combustion chamber.

The first parameter to set was the thermal power of the stove. This is mainly governed by the size of the combustion chamber. The aim was a power of 2 kW, which is close to the maximum power of a standard cooking appliance in the developed world. That initial prototype had a combustion chamber of just 10 cm diameter and a depth of 11 cm. Over the course of several months I optimized the amount of primary and secondary air which is blown into the combustion chamber. The aim was to minimize the blackening of the cooking vessel and to burn all charcoal which is created during the combustion. It was also noted that combustion improved by having the air blown in at high speed. This is presumably because in this way mixing of combustible gasses with the air is better. That does mean that the fan should generate a high pressure and force the air in through small holes. In the end, through careful optimization it was found that about one-fourth of the air is primary air and three-fourth is secondary air.

One essential innovation which reduces the emissions further was the fact that we force the secondary air downward. A normal flame has a tendency to rise because it is hot. When the secondary air hits the stream from the opposite side, the resultant airflow and flame will move upward. See Fig. 6.3 for horizontal forcing. However, just above the combustion chamber (yellow) there will be a cooking vessel at maximum 100°C. That is incredibly cool for a flame. So the flame cools, combustion is quenched, and the results are products of incomplete combustion. That means soot, smoke, and carbon monoxide—in other words high emissions. If the secondary air is forced downward, the collision with the stream from the opposite side will result in a flow downward. So the flame stays a lot longer inside the combustion chamber at temperatures beyond 1000°C. This allows the combustion to finish before the flame hits the cool cooking vessel. If the stove is properly burning the flame looks like a rolling torus.

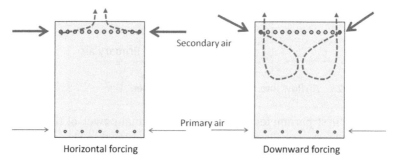

Secondary air

Primary air

Horizontal forcing

Downward forcing

Figure 6.3   Horizontal and downward forcing.

Below is a picture of the flame which clearly shows the airflow downward into the middle and then exiting along the edge after being in the combustion chamber for a long time (Fig. 6.4).

Figure 6.4   Image of a flame showing the airflow downward into the middle and then exiting along the edge after being in the combustion chamber for a long time.

Yet there is another feature which contributes to low emissions. The stove is operated by adding blocks of fuel to the fire from time to time. Especially directly after adding fuel the rising gasses can be too cool to ignite. That is the reason that the secondary air is preheated. This makes sure that the combustible gasses which rise up in the stove are ignited properly all the time. The air is preheated by letting it flow along the outside of the combustion chamber but on the inside of the insulation, which keeps the outside of the stove cool to the touch. This airflow therefore also cools the outside of the stove. In the current version of the stove this insulation is performed by metal reflectors.

To have a symmetric airflow into the combustion chamber (to get the rolling flame) the fan is mounted below the combustion chamber. The fan is a brushless DC fan for reasons of longevity. Thanks to the computer industry such fans are cheap, efficient and have incredible lifetimes.

The fan which is responsible for the very clean combustion [12] does introduce a problem: a reliable electrical power source is now required. Cooking is a very important daily activity and the stove has to be close to 100% reliable. That means that a fan-operated stove needs a very reliable fan and a very reliable power source. From an extensive analysis I derived that a thermoelectric power source would be a logical choice: no moving parts should guarantee long service-free operation and recharging is not required. The fire acts as the heat source and the fan cools the Peltier on the cold side. A small rechargeable battery would start the fan during the time the stove is still cold. When the stove heats up more and more power for the fan would come from the thermoelectric power source. At full power, the thermoelectric power source runs the fan and recharges the start-up battery. The start-up battery is then ready for use next time. The energy flows are handles with a small electronics board. The mechanical construction is schematically depicted below (Fig. 6.5).

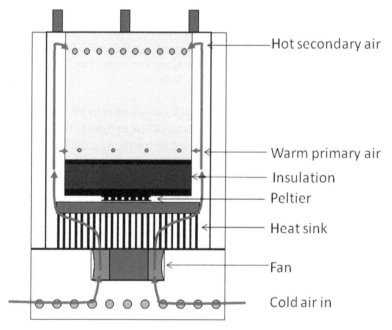

Figure 6.5   Schematic depiction of the mechanical construction of a Peltier powered fan stove.

The fan sucks in cold air from the environment. It blows the cold air directly on the heat sink, which keeps the cold side of the thermoelectric generator cool. The air warms up to about 100°C and part is used as primary combustion air. The remainder flows along the combustion chamber, heats up to about 400°C and is used as secondary air. Between the combustion chamber and the thermoelectric generator insulation is installed to limit the maximum temperature the thermoelectric generator is subjected to. The bottom of the combustion chamber gets hotter than a thermoelectric generator can handle. This stove was used in my household for several years.

6.3 Philips Stove Program

The whole stove episode could easily have ended with just one stove in my garden. However, at that moment there was a Philips wide business challenge to come up with proposals for profitable and sustainable solutions for emerging and developing markets. Among the selected themes were health, well-being, clean energy, and food processing. The stove fitted perfectly. The lower emissions are good for the cook's health. The absence of smoke improves the well-being of the cook. Cooking is food processing. Moreover, the product could be used by the 2.5 billion people who still rely on solid fuels to cook on in the emerging and developing markets. Such a large number of products means the market has the potential to become a profitable business.

So I got permission to carry out exploratory work. This gave me the opportunity to send the stove to the Aprovecho Research Center. This center is dedicated to researching, developing and disseminating clean cookstove technologies for meeting the basic needs of refugees, impoverished people, and communities in the developing world. They are active for decades and have the capability to test the performance of stoves and compare them. This stove was optimized semi-empirical using only the soot deposition on the bottom of a cooking vessel during a fixed cooking task as a guide. I was thrilled to learn that this stove was the cleanest wood burning stove they had tested to date. Stoves are tested with a standardized laboratory test, called the water boiling test. This test brings 5L water to the boil and then simmers it for a certain period of time. See the graph in Fig. 6.6 for the results.

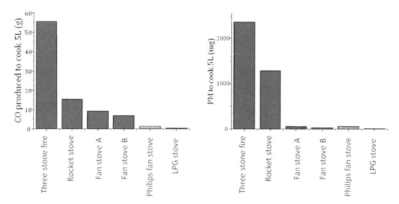

Figure 6.6   Comparison graph of the performance of different stoves.

In the graph two important metrics relating to health damaging emissions are shown: Carbon monoxide (CO) and particulate matter (PM). The latter is more commonly known as smoke. Carbon monoxide is considered a proxy for all products of incomplete combustion. The three-stone fire is the reference fire and is representative for current daily practice. The rocket stove is an improved stove design without a fan. Fan stove A and B are stoves which incorporate a fan, but emit about five times more CO. The liquefied petroleum gas (LPG) stove is considered the cleanest possible stove when the stove is based on combustion. From the graph above it is clear that the Philips stove is the cleanest-burning stove, almost as clean as an LPG stove.

At the moment we were just very happy about this. Later, however, we learned the significance of this fact. From the exposure–response relationship of child pneumonia as a function of PM of less than 2.5 μm, it is clear that improved stoves which reduce the emissions by a factor of 2 or even 4 give hardly a health benefit. Only stoves which come close to the emissions of an LPG stove give a significant health benefit [13]. This is the real importance of our low emissions.

The record-low emissions of the stove also mean that global warming consequences of this stove will be low. The reason is that the carbon in products of incomplete combustion has in general a higher global warming potential than the same amount of carbon in pure CO_2. So if part of the carbon is not burned completely to CO_2 but remains CO or becomes soot, the net effect is an increased global warming. Smoke is more complicated because it consists of soot which is black and has a warming effect and so-called white smoke (the most visible part of smoke), which has a cooling effect. The Philips stove also scores very good in this respect. The end result is that the Philips stove has record low greenhouse gas emissions. See the graph below for the greenhouse effect of cooking in the case of renewably harvested wood (Fig. 6.7) [12]. In the case of renewably harvested wood the net effect on global warming of CO_2 is zero. Do note that although the 250 g CO_2e/l does not seem much, it has to be multiplied with the total number of stoves on the planet and the amount of food cooked per stove per day. Assuming 500 million stoves and 10 l cooked food/day the total is approximately 500 million tons of CO_2. That is comparable to all the CO_2 emissions from the consumption of fossil fuels of Canada in 2011 or 1.5% of the world's emissions from the consumption of fossil fuels in 2011 [14].

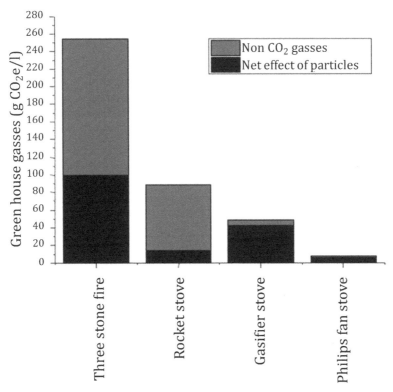

Figure 6.7   Greenhouse effect of cooking in the case of renewably harvested wood.

I also looked at the efficiency of the stove. The efficiency is very important. Stove efficiency directly translates to fuel use and thus to cost of usage. That cost can be either monetary or in time. But both ways, savings will make a stove attractive. Using a modified water boiling test the Philips stove was compared to a three-stone fire [12]. The Philips stove needed just 45% of the wood, or a 55% fuel saving. That is very significant. But the absolute efficiency of the Philips stove is still just below 50%. To understand why this is not better we have to look at the energy transfer from the fire to the cooking vessel. In principle there are two ways of heat transfer (Fig. 6.8). Direct radiation from the fire to the cooking vessel (depicted by the pink dashed arrows) and by convection of the hot exhaust gasses from the fire. Both can be increased by reducing the gap between cooking vessel and opening of the combustion chamber. However, during cooking the cook wants to add wood pieces through this gap without removing the cooking vessel. We therefore had to make the gap larger than is wanted from an efficiency perspective. An optimal gap would increase the efficiency to 60%. Another method is the so-called skirt. A metal sheet is fitted around the cooking vessel with a 5 mm gap and is connected to the combustion chamber. This forces all gases closely along the cooking vessel instead of allowing them to wander off to the side and to just heat the environment. I have measured an efficiency of over 80% with a skirt. However, a skirt makes additions of fuel impossible, and needs to be fitted to the cooking vessel quite precisely. This is a procedure not liked by cooks. The alternative is to permanently connect the skirt to the cooking vessel, but that makes every cooking vessel more expensive and the narrow gap between the cooking vessel and skirt is difficult to clean from o.a. soot. And if a thick layer accumulates, the effect is negated. So we have chosen not to use a skirt.

Another option would be to add a heat exchanger structure to the bottom of the cooking vessel. That also allows efficiencies of 80%. But these structures are difficult to make and thus expensive. They are also very difficult to clean and when they are dirty the effect diminishes. Since the cooking vessel is not part of the stove and these vessels are hardly used, we are not reporting data with such cooking vessels.

The last option to reduce losses from hot air wandering away from the cooking vessel would be to lower the total amount of air blown into the fire. From measurements of airflow and combustion rate it can easily be figured out that the Philips stove uses a huge amount of air which is not per se required for stochiometric combustion. However, measurements also show that if we reduce the total amount of air, the emissions of the stove rise. This is a balance we have to choose. Since these low emissions are really required to make a significant health impact we have chosen for surplus air at the expense of somewhat higher fuel use. Reducing the emissions by just a factor of 2 has hardly a health impact [13]. In fact reducing our emissions further will still have an additional health benefit and is therefore one of our current research directions.

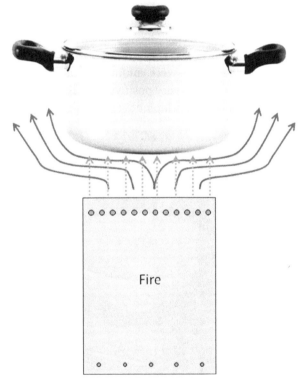

Figure 6.8　Two ways of heat transfer: direct radiation from the fire to the cooking vessel (dashed arrows) and by convection of hot exhaust gasses from the fire.

Next, I tried to get consumer feedback. I was happy with the stove as a cooking appliance, but would cooks from developing countries be happy? So I contacted a colleague who worked in our research laboratory in Bangalore, India, whether he could help me with preliminary consumer tests. He was very willing to help. So I had a few copies of the stove made by Philips. My Indian colleague brought the stoves to his parent's village in the Indian countryside. Although people really liked the stove, it was considered way too small. People looking at the stove believed it was too small to cook even an egg. Clearly not true at 2 kW thermal power, but we have to take (potential) customers comments very seriously.

The next step was to make seven larger stoves. Because the thermoelectric generator was still in its infancy, the stoves were driven with an external battery pack. Extensive field tests were carried out with this stove in India (see Fig. 6.9). The comments were very positive. People especially liked the controllable fire, the high heat output (4 kW now), the very quick start-up time (also thanks to the fan), the ashes which were confined inside the stove, and the fuel savings. Because of the very positive consumer feedback, it was decided to start product development for the woodstove.

Figure 6.9   A stove driven with an external battery pack. See also Color Insert.

6.4 Philips Stove Products

From now on the stove was destined to become a regular Philips product. That means a team of people would develop a product from the research prototype including lifetime tests and documentation and set up production of that stove (in India).

The first stove type was a thermoelectric-powered stove with 4 kW of power. It got a wide base for stable operation. See Fig. 6.10 below. This stove was produced in small numbers and was used extensively in the field in India.

Figure 6.10   A thermoelectric-powered stove with 4 kW of power.

Two different problems were found in the product. First of all, we saw extreme corrosion in some of the stoves, especially stoves which were used in dry areas. This was in the end traced back to elevated chloride levels in the wood of the tropical acacia tree growing in soil with elevated salt levels. The chloride quickly corroded even the best-quality stainless steel. To solve this problem a large program was setup to find the right material for the combustion chamber. This includes lifetime testing, which is a lot of time-consuming manual labor. The stove was designed to last for 5000 hours of operation. See Fig. 6.11 below. In the end a stainless steel combustion chamber lined with heat-resistant tiles was found to be adequate.

Figure 6.11   Stoves designed to last for 5000 hours of operation.

The second problem was reliability of the thermoelectric generator. Well known for its reliability because there are no moving parts, we saw generator failures with less than a few thousand hours of usage. This was in the end traced back to thermal cycling. The thermoelectric device is just 4 mm thick. At operation there is a temperature difference of 200°C over the thickness of the generator where it is zero when not in operation. This generates cyclic forces on the generator. Designed to live for five years and to be used twice a day, the generator should survive this more than 3500 times. But it did not. A program was started to develop a generator which could survive especially the cycling together with the supplier.

To get a product to market quickly, it was decided to start producing a stove without a thermoelectric generator. It has a large rechargeable battery pack. Once charged it will operate for about a week. This stove was made originally in India. However, we encountered an unforeseen coincidence of market conditions and distribution problems. The stove project was put on hold.

In 2010, Philips, the United Nations Industrial Development Organization (UNIDO), the Industrial Development Corporation (IDC), and the African Clean Energy (ACE) company started exploring possibilities to introduce the Woodstove to the African market. In this public–private partnership, UNIDO shares its network and experience, Philips provides its technology, knowledge, and intellectual property to ACE to manufacture the stoves, and IDC will share experience in distribution networks. Currently the stoves are produced by ACE [15] and distributed in sub-Saharan Africa (see Fig. 6.12). Everything is in place now to make sure my stove will not remain one stove in my garden.

Figure 6.12   The latest stove design as of 2014.

References

1. Karina Fonseca-Azevedo and Suzana Herculano-Houzel (2012), *Metabolic constraint imposes tradeoff between body size and number of brain neurons in human evolution,* PNAS 2012 109 (45) 18571–18576.

2. *Catching Fire: How Cooking Made Us Human* by Richard Wrangham, Basic Books, New York, 2009.

3. Organ C, Nunn CL, Machanda Z, Wrangham RW (2011) *Phylogenetic rate shifts in feeding time during the evolution of Homo.* Proc Natl Acad Sci USA **108**:14555–14559.

4. GLOBAL HEALTH RISKS, Mortality and burden of disease attributable to selected major risks, World Health Organization 2009.

5. Kirk R. Smith, Gu Shuhua, Huang Kun, Qiu Daxiong, *One hundred million Improved Cookstoves in China: How Was It Done?* World Development 1993 21 (6) 941–961.

6. http://www.aprovecho.org/lab/home

7. http://www.epa.gov/airscience/air-cleancookstove.htm

8. http://www.drtlud.com/

9. http://www.firstenergy.in/

10. http://www.zzstove.com/

11. http://biomassauthority.com/portable-biomass-camping-stove/

12. Nordica MacCarty, Damon Ogle, Dean Still, Dr. Tami Bond, Christoph Roden, Dr. Bryan Willson, Laboratory Comparison of the Global-Warming Potential of Six Categories of Biomass Cooking Stoves, 2007, Aprovecho Research Center, also available online: http://www.scscertified.com/lcs/docs/Global_warming_full_9-6-07.pdf

13. http://ehs.sph.berkeley.edu/krsmith/Presentations/2012/2012_HAP_CRT_Nepal.pdf

14. U.S. Energy Information Administration. http://www.eia.gov/

15. http://www.ace.co.ls/

About the Author

Paul van der Sluis (1962) studied chemistry in Utrecht, the Netherlands. His PhD in crystallography and crystal growth was also carried out in Utrecht. After a short postdoc on crystal growth, he started to work at the Philips Research Laboratories in Eindhoven, the Netherlands. In the more than 25 years that he has been active there, he has worked on very diverse subjects, ranging from X-ray diffraction, switchable mirrors, semiconductor memories, and magnetic resonance imaging (MRI) to cook stoves. These cook stoves are currently internationally recognized as being the cleanest wood burning stoves in production to date.

Dr. van der Sluis has (co-)authored over 70 papers and has more than 15 US patents to his credit. At Philips Research he received the Gillis Holst award.

Chapter 7

Biomass Cofiring

Sjaak van Loo[a] and Jaap Koppejan[a,b]

[a]Procede Biomass BV, PO Box 328, 7500 AH Enschede, the Netherlands
[b]Biomass Combustion and Co-firing, International Energy Agency,
9, rue de la Fédération, 75739 Paris Cedex 15, France
sjaakvanloo@procede.n, jaapkoppejan@procede.nl

7.1 Introduction

The term 'cofiring', in general, refers to co-combustion of two different fuels. Cofiring of biomass with coal in conventional coal-fired boilers represents a combination of renewable and fossil energy utilization that derives the greatest benefits from both fuel types.

It capitalizes on the large investment and infrastructure associated with the existing fossil-fuel-based power systems, while requiring only a relatively modest investment to include a fraction of biomass in the fuel mix.

When proper choices of biomass, coal, boiler design, and boiler operation are made, gaseous pollutants (SO_x, NO_x, etc.) and net greenhouse gas emissions (primarily CO_2) decrease in a reliable and cost-effective manner.

Cofiring can be realized in a relatively short time frame with relatively low technical risk. However, improper choices of fuels,

Biomass Power for the World: Transformations to Effective Use
Edited by Wim van Swaaij, Sascha Kersten, and Wolfgang Palz
Copyright © 2015 Pan Stanford Publishing Pte. Ltd.
ISBN 978-981-4613-88-0 (Hardcover), 978-981-4669-24-5 (Paperback), 978-981-4613-89-7 (eBook)
www.panstanford.com

boiler design, or operating conditions could minimize or even negate many of the advantages of burning biomass with coal and may, in some cases, lead to significant damage to equipment.

In many countries, cofiring is one of the most economic technologies available to achieve CO_2 reduction on a significant scale.

Replacing an existing fossil fuel in a large power plant is often preferred over building a new but somewhat smaller dedicated biomass plant, because economies-of-scale effects make that larger plants can accommodate additional unit operations (e.g., superheaters, economizers, reheaters, etc.) to improve on thermal and electrical conversion efficiencies. Therefore, cofiring in large thermal power stations can lead to an overall saving of fuels in comparison to independent fossil and biomass plants. Further, better use of existing flue gas–cleaning facilities present in such large-scale installations can be made.

7.2   Cofiring Concepts

The share of biomass that is cofired as part of a fuel mix can be defined either on a mass or on a thermal basis. The cofiring ratio on a mass basis refers to the weight of biomass divided by the total weight of the fuel mix (i.e., coal plus biomass). On a thermal basis, however, the cofiring ratio refers to the ratio of the heating value of the biomass content of the fuel mix to that of the whole mixture.

Biomass can be integrated in an existing pulverized coal-fired power plant in different ways, as schematically shown in Fig. 7.1.

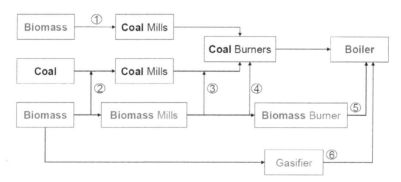

Figure 7.1   Cofiring concepts. Courtesy of W. Livingston, Doosan, UK.

The technology options relevant to pulverized coal-fired power plants can be categorized as follows:

- Direct cofiring:
  - o Milling of (pelletized) biomass through modified coal mills and firing in existing coal burners
  - o Mixing biomass with coal upstream of the coal mixing and firing in existing coal burners
  - o Mixing premilled biomass with milled coal and firing of the mixed fuel in existing coal burners
  - o Firing of premilled biomass in existing coal burners or directly into the boiler
  - o Firing of premilled biomass in dedicated biomass burners
- Indirect cofiring: Gasification of biomass and combustion of the product gas in the boiler
- Parallel cofiring: Combustion of biomass in a separate boiler and utilization of the steam produced within the power plant

7.2.1 Direct Cofiring

The direct cofiring approach can be implemented through five options (see Figs. 7.2–7.5).

Figure 7.2 Dumping wood chips on a coal conveyor before the mills. Courtesy of Delta Electricity, Australia.

Figure 7.3   Vertical spindle ball and ring mills at Drax Power station, UK, converted to 100% biomass. Courtesy of W. Livingston, Doosan, UK.

The first option is milling of pelletized biomass in existing coal mills. The coal mill tends to break the pellets back to the original dust-size distribution. Generally the maximum heat input from the mill group is significantly derated, commonly to around 70% of that of coal.

The second option is the mixing of biomass with the coal upstream of the coal feeders. The mixed fuel is sent through the coal mills and distributed across all of the coal burners' supplies by the mill at the required cofiring rate.

This has been the preferred approach for coal-fired power stations embarking on biomass cofiring for the first time. It is, in principle, the simplest option and would involve the lowest capital cost. The expenditure is principally on the biomass reception, storage, and handling facilities. This approach is particularly attractive when there are concerns about the security of supply of the biomass materials and about the long-term security of the financial support through subsidies for cofiring. It would, however, involve the highest risk of interference with the coal-firing capability of the boiler unit, and this approach may only be applicable to the cofiring of a limited range of biofuel types and at very low biofuel-to-coal cofiring ratios.

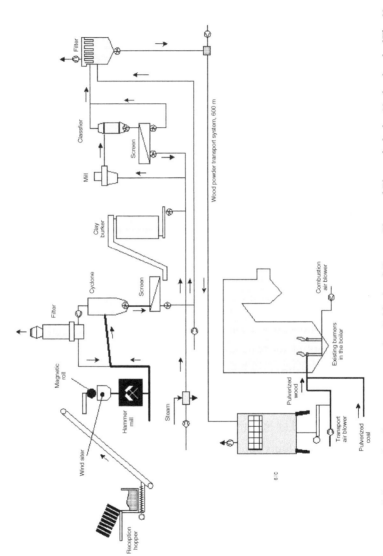

Figure 7.4 Diagram of the wood cofiring system at the Gelderland Power Plant of Electrabel, the Netherlands (*The Handbook of Biomass Combustion & Co-firing*, Eds Sjaak van Loo & Jaap Koppejan, ISBN: 978-1-84407-249-1).

Figure 7.5   Biomass and coal fuel storage at the Dolna Odra cofiring plant in Poland. Courtesy of I. Obernberger, Bios-Bioenergy, Austria.

As biomass materials have a high content of volatile matter, released at temperatures above about 180°C, the mill inlet temperature must be controlled to prevent release of volatile matter that can lead to explosions and mill fires. The key issue in mill safety is avoiding hot primary air coming into direct contact with dry fuel, particularly during certain mill operations such as:

- planned and emergency shutdowns;
- restarts after emergency shutdowns; and
- loss of coal or intermittent coal feed incidents, etc.

The third and fourth options involve separate handling, metering, and comminution of the biofuel and injection into the pulverized fuel pipework upstream of the burners or at the burners. These options require the installation of a number of biofuel transport pipes across the boiler front, which may already be congested. It may also prove to be more difficult to control and to maintain the burner's operating characteristics over the normal boiler load curve.

When the milled biomass is fired through the existing pulverised coal burners, there are a number of significant technical issues. Because the particle size distribution of the milled biomass is much coarser than that of pulverized coal, the particle drying and heating times tend to be higher, resulting in delayed combustion. In most cases it is necessary to modify the burners and, potentially, the flame detectors to provide a robust solution for both cofiring and 100% biomass firing applications.

The fifth option involves the separate handling and comminution of the biomass. The biomass can be premilled either off-site or on-site, providing flexibility in fuel supply.

The dedicated biomass burners are normally based on conventional pulverized coal burners or cyclone burners.

If the existing coal-firing capability is to be maintained, additional burners may be required for auxiliary biomass firing. Appropriate locations for the biomass burners are not easy to find, particularly as retrofit, and additional furnace penetrations and burner support structures are required. Fuel and air supply systems for the biomass burners have to be installed, and flame-monitoring equipment for the biomass flames are required.

Overall, the installation of dedicated biomass burners is regarded as being expensive and a relatively high-risk approach to biomass cofiring. This approach represents the highest capital cost option but involves the least risk to normal boiler operation.

7.2.2 Indirect Cofiring

The indirect cofiring approach is essentially based on the gasification of chipped wood fuels and the combustion of the product fuel gas in a coal-fired furnace. The main product of the gasification process is a low-calorific-value fuel gas, with the calorific value depending principally on the moisture content of the fuel. Other major products are:

- biomass ash materials, including alkali metals and trace metals;
- tars and other condensable organic species, and;
- Cl, N, and S species.

In terms of the nature and cost of the installed equipment, indirect cofiring is equivalent to the replacement of the comminution equipment by a gasifier.

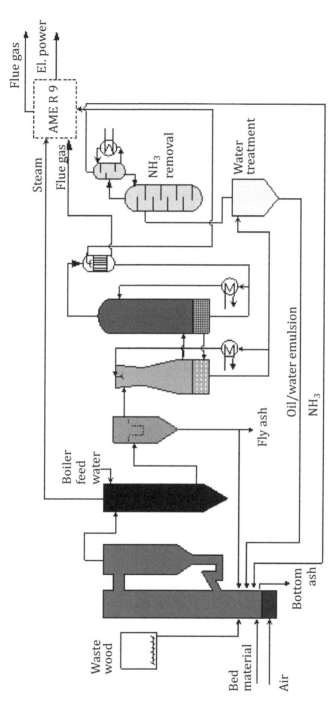

Figure 7.6 Original diagram of the AMERGAS biomass gasification plant at the Amer Power Plant in Geertruidenberg, the Netherlands. (Willeboer, W. (2000). AMERGAS biomass gasifier starting operation, in *The Use of Coal in Mixture with Wastes and Residues II. Proc. EU Sem.*, Cottbus.)

One of the key issues with the indirect cofiring approach is the degree of the fuel gas cleaning before it can be directed into the furnace for co-combustion. The low-capital-cost option involves no cooling or cleaning of the gas, with all of the products of the gasification process being carried forward to the furnace through hot-gas ductwork at 800°C–900°C. There are consequently some risks to the operation and availability of the boiler. Operational experience indicates that these risks are small, albeit at a very low biomass cofiring ratio, and with a relatively clean wood fuel. These risks will obviously be more appreciable at higher cofiring ratios and with demolition and other waste wood materials.

The alternative is to cool and clean the fuel gas prior to combustion in the boiler furnace. This is obviously a more expensive option but can provide a greater degree of fuel flexibility, and the risks to the operation and availability of the coal-fired boiler are largely avoided.

7.2.3 Parallel Cofiring

The parallel firing concept basically involves a fully separate combustion installation for biomass/waste to produce steam, which, in turn, is used in the coal-fired power plant where it is upgraded to higher conditions, resulting in higher conversion efficiencies. Though the investment in indirect cofiring and parallel firing installations is significantly higher than in direct co-combustion installations, advantages such as the possibility to use relatively difficult fuels with high alkali and chlorine contents and the separation of the ashes are reasons why these concepts can be justifiable.

7.3 Fuel Properties

Biomass fuels used for cofiring may range from woody (ligneous) to grassy and straw-derived (herbaceous) materials and include both residues and energy crops. Woody residues are generally the fuels of choice for coal-fired boilers, while energy crops and herbaceous residues represent future fuel resources and opportunity fuels, respectively. Biomass fuel properties differ significantly from those of coal and also show significantly greater variation as a class of fuels than coal does. For example, ash contents vary from less than 1% to over 20%, and fuel nitrogen varies from around 0.1% to over 1% (Fig. 7.7). Other notable properties of biomass relative to coal are a generally high moisture content (usually more than 25% and sometimes more than 50% as-fired, although there are exceptions), potentially high chlorine content (ranging from near 0 to 2.5%), relatively low heating value (typically about half that of bituminous coal), and low bulk density (as low as one-tenth that of coal per unit heating value). Each of these properties affects the design, operation, and performance of cofiring systems.

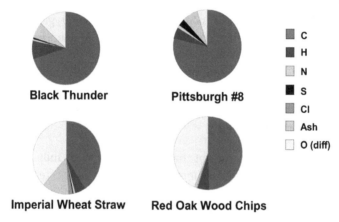

Figure 7.7   Typical ultimate analyses of biomass and coal. All of the components are presented clockwise, starting from the top with C. Courtesy of Larry baxter, USA. See also Color Insert.

The influence of the type of biomass fuel on boiler design is given in the overview in Fig. 7.8.

When proper choices of biomass, coal, boiler design, and boiler operation are made, gaseous pollutants (SO_x, NO_x, etc.) and net greenhouse gas emissions (primarily CO_2) decrease.

7.3.1   Fuel Pretreatment

The biomass feedstock that arrives at a power plant may need to undergo a preliminary stage of size reduction to bring down the particle top size to about 4 mm. This has proved to be problematic, particularly for wood-based materials, as the

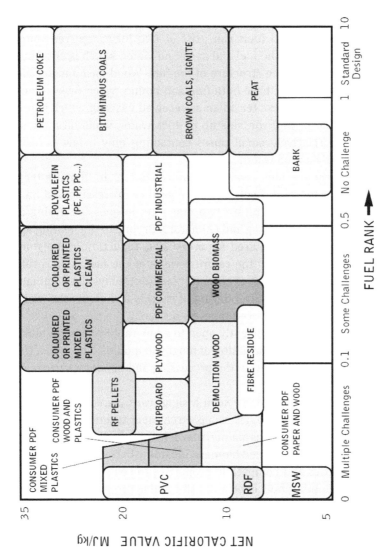

Figure 7.8 Influence of fuel type on boiler design. (Veijonen, K., Vainikka, P., Järvinen, T., and Alakangas, E. (2003). Biomass co-firing: an efficient way to reduce greenhouse emission. EUBIONET.)

conventional agricultural wood-milling equipment (e.g., hammer mills) has generally inadequate capacity. Significant development, however, has been made over the recent years with the introduction of purpose-built equipment for preliminary size reduction of wood-based materials.

Moisture in solid fuels can exist in two forms, as free water within the pores of the fuel and as bound water, which is absorbed to the interior surface structure of the fuel. Wood-based materials, being rather porous, have both free and bound water. Green wood contains about 45% water on an as-received basis, of which half is bound water. Lignite contains up to 40% water, while bituminous coals have relatively small pores containing only a few percent water, mostly bound water.

Obviously, the high level of moisture content in biomass fuels produces technical difficulties with regard to processing and handling of the fuels. There are two major techniques for drying of biomass fuels, (1) heating and (2) natural drying. In the heating approach, the energy required for every 20% reduction in moisture content is about 1% of the calorific value of the fuel. As a result, the heating approach is not economic, and nowadays the dominant drying technique is natural drying. With this technique, the moisture content can be reduced to levels below 20% if the structure of the raw biomass is sufficiently open to allow rapid release of moisture and mitigate the formation of fungi and moulds. The rest of the drying should occur within the processing and handling system, as well as in the boiler.

The last five years have seen a significant increase of interest in torrefaction technologies as a pretreatment technology for solid biomass. This interest has mainly been driven by the characteristics of the torrefied and densified biomass, including better transportation characteristics and compatible properties to coal, such as heating value, grindability, bulk energy density, and hydrophobicity.

In the torrefaction process, biomass is heated to a temperature of approx. 250°C–320°C at low oxygen concentrations, so all moisture is removed as well as a fraction of the volatile matter of the dry biomass. Ideally, the energy contained in the released volatiles is equal to the heating requirements of the process so that a thermal efficiency exceeding approx. 90%–95% can be achieved.

Due to the substantial weight loss and a relatively smaller loss of calorific content, the heating value of processed biomass per mass unit increases significantly. Through subsequent densification, the volumetric energy density can be increased up to a factor of three compared to the original biomass fuel.

Through the torrefaction process and depending on its severity, fibrous, tenacious, and hydrophyllic properties of biomass can be altered so that the product becomes brittle (therefore easy to grind) and hydrophobic. These behavioural changes can have significant advantages in the supply chain, since both the logistics and the combustion aspects can be made simpler, more cost effective, and compatible with coal.

Currently at least 40–50 torrefaction initiatives can be identified, about equally divided between Europe and North America. These installations intend to demonstrate the technical and economic feasibility of torrefaction as a viable pretreatment option and of the torrefied product for cofiring in existing pulverised coal-fired power plants. Several of these installations in both Europe and North America have a name tag capacity up to several hundred thousand tonnes. As of yet, however, only a handful are actually producing, and the greatest challenge is therefore related to successful technical and economical demonstration of the individual technologies.

7.4 Operational Experience

There has been rapid progress over the past years in the development of the coutilization of biomass materials in coal-fired boiler plants. Cofiring technology is presently being routinely practised at commercial scales in Canada, the Netherlands, Belgium, Finland, Denmark, and many other countries. In many other countries around the world, cofiring trials have been executed successfully. An inventory by IEA Bioenergy Combustion and Co-Firing of the application of cofiring worldwide has indicated that about 250 coal-fired power plants have experience with cofiring biomass or waste, at least on a trial basis (Table 7.1). Typical power stations where cofiring has been implemented are in the range from approx. 50 MW_e to 700 MW_e. Cofiring ratios vary from as little as 1% on a heat basis to full conversion (100%).

Table 7.1 Overview of operational experience with cofiring of biomass

Country	BFB	CFB	CFB, BFB	Grate	PF	Unknown	Total
Australia					8		8
Austria		3		1	1		5
Belgium					1		1
Canada					7		7
Denmark		1		4	7		12
Finland	42	13	6	4	10	6	81
Germany				1	4	22	27
Indonesia	2						2
Italy					6	1	7
Netherlands					6		6
Norway		1					1
Spain		1				1	2
Sweden	3	7		2	3		15
Taiwan		1					1
Thailand		1					1
UK		2			16		18
USA	1	5		5	29		40
Total	48	35	6	17	98	30	234

Source: Courtesy of IEA Biomass Combustion and Cofiring

The majority are pulverized coal boilers, including tangentially fired, wall-fired, and cyclone-fired units. Bubbling and circulating fluidized bed boilers and stoker boilers have also been used and can be considered very fuel flexible. The cofiring activities have involved all of the commercially significant solid fossil fuels, including lignites, subbituminous coals, bituminous coals, anthracites, and petroleum coke. These fuels have been cofired with a very wide range of biomass material, including herbaceous and woody materials, wet and dry agricultural residues, and energy crops.

Initially the majority of biomass cofiring was by premixing of biomass, in the form of granules, pellets, and dust; with coal on the coal conveyor; and milling and firing of the mixed fuel in existing coal burners. In this way a range of biomass materials was cofired at modest cofiring ratios, generally less than 10% on a heat input basis, resulting in only modest impacts on the boiler.

As a general guideline for measures that have to be taken for the burner system:

0–10 wt%   In general no or minimal measures with respect to burner settings

10–50 wt%   Possible modifications of the preprocessing necessary; adaptation of the burner settings, for example, with respect to combustion air; possible limitations on the types of biomass that can be cofired using the original burners

50–100 wt% Adaptation of the burner settings or complete replacement of the existing burners by dedicated biomass burners

To achieve higher cofiring ratios and conversion of coal-fired stations to 100% biomass firing, the following approaches are being adopted:

- Mixing premilled biomass with milled coal and firing of the mixed fuel in existing coal burners
- Milling of pelletized biomass through modified coal mills and firing in existing coal burners

The most successful conversions have used the existing equipment, such as mills, pipework, and burners, as much as possible.

Examples of successful conversions are:

- Vasthamnverket, Helsingborg, Seweden
  Originally a 200 MW pulverized coal boiler was converted in the late 1990s to firing 100% wood pellets, processing the pellets through the roller mills.
- Hasselby Heat and Power Plant, Sweden
  The pore and heat plant was successfully converted from coal firing to 100% wood pellet firing in 1993 and is still in operation.
- Drax Power Station, UK
  In 2010 two large vertical spindle ball and ring coal mills were converted to 100% wood pellets and are still in commercial operation.

Experience has indicated that the heat input from the converted mills is significantly derated to around 50%–70% of that with coal, but this can be supplemented by additional mills.

Clearly, any biomass cofiring scheme would be very carefully examined to identify the technical risks and the costs associated with any interference to normal power plant operations. The potential costs are likely to be high compared to any additional revenues from cofiring. These risks are of two types:

- Reductions in plant availability and flexibility in operation
- Increases in maintenance and replacement costs associated with the biomass handling and firing equipment and with the boiler plant

The above-mentioned risks are underpinned by a series of technical issues, which can be classified into four main groups, namely:

- fuel preparation, processing, and handling issues;
- combustion-related issues (e.g., flame stability, burnout, etc.) affecting plant operation and control;
- ash-related issues (slagging, fouling, corrosion, etc.); and
- emissions.

The selection of the cofiring option would commonly be made on the basis of minimum interference with normal plant operations, and this may favour the installation of dedicated biomass handling, processing, and firing equipment if the ratio of biomass in the fuel mix is greater than certain limits. These dedicated unit operations inevitably lead to higher capital cost options. Firing biomass involves risks of increased plant outages and possible interference with the operation of the burners, the furnace, the boiler convective section, and the environmental control equipment. The risks of increased plant maintenance costs would be associated principally with increased ash deposition and accumulation within the boiler, the potential for increased corrosion rates of high-temperature boiler components, and interference with the operation of NO_x, SO_x, and particulate emission reduction equipment (Fig. 7.9). Since 2000, significant progress has been made in improving the fundamental understanding of the underlying mechanisms and obtaining hands-on experience in order to demonstrate that cofiring schemes can be implemented at acceptable risks under a wide variety of circumstances.

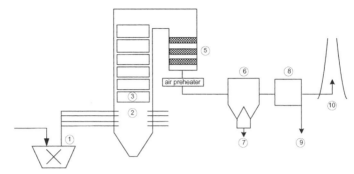

Figure 7.9   Effects of biomass cofiring on a coal-fired power station. (1) Grinding equipment: reduced capacity and lifetime; (2) combustion chamber: slagging; (3) superheater: high-temperature corrosion; (4) heat exchanger: depositions and erosion; (5) De-NO$_x$ installation: capacity, poisoning; (6) electric precipitator: capacity; (7) ash: utilization; (8) De-SO$_x$ installation: capacity; (9) utilization of residues from desulphurisation; (10) fuel gas: emissions (*The Handbook of Biomass Combustion and Co-firing*, Eds Sjaak van Loo and Jaap Koppejan, ISBN: 978-1-84407-249-1, 2008).

7.5   Summary

When considering the broad range of experiences with cofiring biomass in coal-fired power plants of various configurations worldwide, one can conclude that:

- cofiring has been demonstrated successfully in about 250 installations worldwide for most combinations of fuels and boiler types;
- cofiring offers among the highest electrical conversion efficiencies of any biomass power option;
- cofiring biomass residues in existing coal-fired boilers is among the lowest-cost biomass power production options;
- well-managed cofiring projects involve low technical risk; and
- in addition to mitigation of CO$_2$ emission, cofiring biomass in existing coal-fired boilers usually also leads to reduced emissions of NO$_x$, SO$_x$, and other fuel gas components.

It can therefore be concluded that cofiring of biomass in existing coal-fired boilers provides an attractive approach to nearly every aspect of bioenergy project development.

There are, however, a number of barriers to a wider use of cofiring technology, the majority of which are nontechnical.

Coal-fired power plants are under public discussion due to their high emissions relative to natural gas at current low gas prices. Furthermore nondispatchable renewables result in:

- low power prices;
- lower merit order position for coal-fired units; and
- lower number of full-load-equivalent hours.

The costs and risks associated with the still poorly developed supply infrastructure for biomass materials in most countries are obviously also key barriers to the further development of biomass cofiring. Improvement in the current situation can only come with the development of further biomass coutilization projects.

About the Authors

Sjaak van Loo has a background in chemical engineering (MSc Chemical Engineering, University of Twente, the Netherlands) and started his career as a researcher at TNO (Netherlands Organisation for Applied Scientific Research). He specialized in bioenergy in the early 1990s and was involved in several biomass combustion–related R&D projects, including the founding and management of the Dutch Bio-Energy Information Centre, the management of IEA Bioenergy Task 19 (biomass combustion), and Task 32 (biomass combustion and cofiring).

Since 2003, Sjaak is director and co-owner of Procede Holding BV, an R&D and innovation-oriented SME specialized in sustainable process technology development (www.procede.nl).

Jaap Koppejan graduated from the University of Twente (1993, MSc applied physics) and has specialized in bioenergy technology development and implementation through various assignments at UN-FAO, TNO, and other organizations. He has been involved in several biomass combustion–related R&D projects, including management of IEA Bioenergy Task 32 (biomass combustion and co-firing) and the combustion and cofiring section within the European ThermalNet expert network. Other projects are related to resource assessments, feasibility studies, policy studies, and project implementation.

Since 2007, Jaap is director of Procede Biomass BV, a subsidiary R&D company of Procede Holding BV, involved in the development and commercialization of bioenergy technologies. He is cofounder of Bio Forte BV, an energy service–providing company using innovative biomass combustion technology concepts.

Chapter 8

Biomass in Combined Heat and Power Production

Sara Kärki

Fortum Power and Heat Oy, New Energy Concepts, Keilaniementie 1, Espoo, POB 100, FI-00048 Fortum, Finland
sara.karki@fortum.com

8.1 Introduction

In combined heat and power (CHP) production or cogeneration plants both electricity and useful heat are generated in one power plant. The heat can be utilized for industrial processes or for district heating. A typical district heating system has two pipelines going to the end user: The first one delivers hot water to a heat exchanger located in the user's property. In the heat exchanger thermal energy is transferred from the district heating water to a heat transfer medium of the central heating system. The second pipeline returns from the customer to the CHP plant, bringing back the cooled water for reheating. In this way, the district heating water is not consumed in the process and only thermal energy is conveyed from a central location to customers distributed around the district heating network

Biomass Power for the World: Transformations to Effective Use
Edited by Wim van Swaaij, Sascha Kersten, and Wolfgang Palz
Copyright © 2015 Pan Stanford Publishing Pte. Ltd.
ISBN 978-981-4613-88-0 (Hardcover), 978-981-4669-24-5 (Paperback), 978-981-4613-89-7 (eBook)
www.panstanford.com

in a community or an industrial park. This enables the utilization of the high efficiency of CHP production, but as a drawback, as district heating pipelines are typically constructed underground, district heating requires significant capital investments to the district heating network.

Biomass has its advantages as a renewable fuel having the capability to rebind and use CO_2 produced and released to the environment during combustion and thus it is seen as a sustainable source for energy production. However, biomass combustion is not completely straightforward and several technoeconomic issues have to be considered:

- Origin, availability, and price of the biomass dependent on the location of the plant
- Relatively high costs of harvesting, transportation, and storage caused mainly by low-energy density of most biomass
- Technical challenges related to the use at the CHP plant, including conveying of biomass and combustion-related challenges caused by biomass impurities and ash

In addition to the technical issues related to biomass combustion in a CHP plant, there are other more practical challenges that need to be taken care of. For instance, CHP plants are usually located near communities to avoid too long pipelines from the production site to the end users. This means that all biomass has to be transported to the site that can be even in a center of a city. When compared to fossil fuels, such as coal, the traffic required to bring the same energy in form of biomass to the site is significantly higher. For instance, when the volume of coal and biomass chips are compared, approximately seven times more space in storage and transportation capacity is needed [1]. The less energy-dense the fuel is, the more resources are required for the logistics. This also correlates with the economical distance for supply and transportation of the fuel, as well as unwanted side effects like increased traffic followed by noise, pollution, and increased safety risks due to higher amount of heavy traffic among other traffic.

Despite the issues related to the use of biomass in energy production, using it in a CHP plant is the most efficient way to produce both heat and electricity from biomass. There are also several combustion technologies that have been developed and optimized for the use of biomass. CHP production using biomass as fuel is an established technology and an existing, large-scale business, especially in the Nordic countries.

In this chapter the following topics will be discussed in more detail:

- Fortum's CHP production
- District heating and technologies enabling CHP production using biomass
- Energy and resource efficiency
- Future of CHP: New ways to further improve resource efficiency

8.2 Fortum's Heat and Power Production

8.2.1 Fortum's Presence and Production

Fortum is a power and heat company operating in the Nordic countries, Russia, Poland, and the Baltic countries. Fortum's purpose is to create energy that improves life for present and future generations. The company generates, distributes, and sells electricity, heat, and related expert services. Fortum has over 1.3 million electricity sales customers in the Nordic countries [2].

Fortum is within the 20 largest electricity-producing companies in Europe and Russia with electricity production of 68.7 TWh in 2013. However, regarding heat generation, Fortum is a large heat producer worldwide with heat production of 42.8 TWh heat in 2013. Thus, Fortum's core competences are in efficient heat, cooling, power, and CHP production. Figure 8.1 illustrates Fortum's geographical presence, production, and sales of power and heat in 2013 [2].

Fortum produces heat and power both separately in power production plants and heat boilers, as well as in CHP plants. Fortum's CHP plants are located in the Nordic countries, Baltics, Poland, and Russia. Most of our CHP plants are located in communities and are connected to district heating networks and/or industrial facilities using heat for their processes [2].

We produce electricity and heat in CHP plants from various fuels, such as different types of biomass, including wood, wood chips, and forest residues; bio-oil; waste; coal; natural gas; as well as peat. The installed electricity production capacity of all Fortum's CHP plants in Europe and Russia in 2013 was 4943 MW$_e$. In addition to the CHP production plants, we have also heat-only boilers (HOBs) taking care of peak loads in heat consumption and running the district heat network in a flexible way. The total heat capacity produced both with CHP plants as well as HOBs was 21 659 MW$_h$ [2].

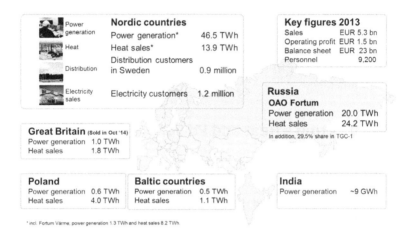

Figure 8.1   Fortum's geographical presence and key figures in 2013. See also Color Insert.

8.2.1   Transition from Fossil to Solar Economy

Sustainability is at the core of Fortum's strategy and we take economic, social, and environmental responsibility. We believe that the growing energy demand will be fulfilled in the future with electricity production. Therefore Fortum focuses on competence in hydropower, nuclear, and CHP production, as well as reduction of environmental impacts. Fortum also develops new sustainable energy systems, forms, and products [2].

Although, we still use fossil fuels in power and heat production, we believe that the world is in transition from fossil fuels to renewable energy originating from solar resources. The future solar economy is based on high-efficiency systems using bioenergy, geothermal energy, hydropower, ocean, wind, and sun as fuels for energy production. As CHP production is highly efficient and when based on bio- or waste fuels also minor in CO_2 emissions, it is one of the main focus areas in Fortum [2].

8.3 District Heating and Technologies Enabling CHP Production Using Biomass

8.3.1 District Heating and Heat Consumption

8.3.1.1 District heating network

District heating is available in Western and Eastern Europe, especially in the Nordic countries, Northern America, Japan, Korea, China, and Mongolia. The principle of district heating is that centrally and locally produced heat is transported to the user via piping to avoid distributed heat production and to benefit from high efficiency of CHP generation. At the CHP plant fuels are combusted to produce thermal energy that is transferred to the heat transfer medium, which is most commonly water. The thermal energy generated in combustion and transferred to the water is then fed to a district heating network, where the hot water flows in pipes to the users. The district heating water temperature varies between +65°C and 115°C, depending on the outside temperature and the momentary heat demand. [3].

At the user's end a central heating system typically conveys the heat from the district heating network to individual houses via heat exchangers. The district heating is also used for heating the hot water for households. After the district heating water has transferred the heat to the heat distribution system of the property, the cooled-down water is returned via another pipe to the CHP plant. Most European district heating systems use water as the heat transportation medium due to its high heat capacity and environmental safety.

A district heating network is a large investment that is usually built gradually similar to other infrastructure investments. Typical district heating pipes are well-isolated metal pipes that are constructed underground to avoid heat losses and decrease the risk of freezing of smaller pipes during cold winter times. In many densely inhabited locations space is limited and constructing several pipelines above ground level would not be practical or even possible.

8.3.1.2 Heat consumption

Heat demand and consumption in communities are dictated by the normal consumption of hot water in households as well as the need for heating the houses, which is dependent on the outside temperature. From these the first mentioned varies on an hourly basis and less on a seasonal basis, creating peak hours in the morning and evenings throughout the year when people use hot water, for instance, for showering. The second reason causing fluctuation in the heat consumption is dictated by diurnal variation when the outside temperature is usually lower during night than day time, as well as the seasonal variation when the outside temperature is lower during winter than in summer. The variation of the outside temperature thus also causes variation in the heat demand for heating the houses. In many industries heat is used for processes. In industry the heat consumption is, however, more stable as industrial facilities typically want to maximize the running time of their processes. As an example the heat consumption of Espoo City shows clearly the seasonal variation of heat consumption, as illustrated in Fig. 8.2.

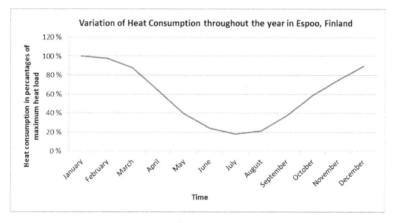

Figure 8.2   Seasonal variation of heat consumption in Espoo, shown as a percentage (%) of maximum load based on monthly averages.

Since the heat consumption in communities fluctuates, the heat production capacity has to fluctuate. Although district heating is usually based on centralized heat production, and often also on centralized CHP production, the district heating network has technical and physical limitations, requiring additional local heat production at the far ends of the network. Building a central heat production plant equivalent to the maximum heat demand is not economical since those maximum peak hours of the year are very few. As the capacity of the CHP plant increases, the capital investment to the plant similarly increases. When the seasonal heat demand is converted to the utilization period of maximum load, shown in Fig. 8.3, the technoeconomical inefficiency of sizing a CHP plant based on the maximum load becomes evident. Thus, it is more economical to invest into smaller and cheaper HOBs, with lower unit capital costs but higher operation costs, producing the heat for those few peak capacity hours.

In analogy, it is not rational to oversize the district heating network either and therefore it is built to manage normal base loads. Then, during morning and evening peak hours, heat is produced also in more local heating plants, that is, HOBs, typically using more expensive fuels or by utilizing heat storages. Therefore, as the demand for heat increases during peak hours, also the heat producer and district heat network operator will start optimizing the system and starting up smaller heating plants.

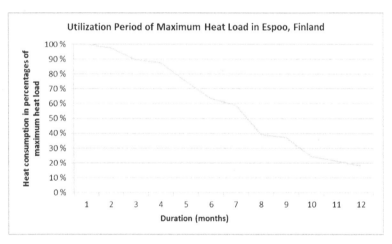

Figure 8.3 Seasonal heat consumption converted to utilization period of maximum heat load in Espoo, shown as a percentage of maximum load.

Due to the techno-economical dimensioning of the district heating network, there are also technical limitations related to the capacity and the capability to transfer heat around the district heating network. For instance, district heating water has to be pumped from a central production unit to the end users who can be even dozens of kilometers away from the central heat production unit. In addition, the heat transfer capability of the district heating network is limited by the physical dimensions of the pipelines, as well as the installed pumping capacity. In addition, because the networks can be built gradually within decades some of the major gridlines may have grown too small when the community has expanded and thus there might be bottlenecks in the system, preventing effective heat transfer from one location to another. Therefore, during times when the heat consumption is highest, the heat produced in a central location cannot be physically transferred to the furthest points of the network. Thus, also due to this reason some small heat production plants are usually needed. As these plants run only for limited hours during the morning and evening peak hours and during the coldest times of the year, they have to be flexible, reliable, and quick in startup and shutdown. The technologies used for this type of small heat only plants are very well-known burner technologies usually based on more expensive fossil fuels such as heating oils or natural gas.

8.3.2 CHP Plants

There are many types of CHP plants but the principle in all of them is still the same: First, heat produced in combustion is converted to kinetic energy in a turbine and to electricity in a generator, and the remaining thermal energy is transferred to a heat transfer medium used, for instance, for district heating. The main difference between a CHP plant and a condensing power plant is that the remaining thermal energy might not be enough to be utilized in a useful way, and therefore, a CHP plant is designed so that some extra thermal energy is used for priming the heat to useful temperature with the expense of electricity production.

There are many configurations and technologies for the collection of the heat produced during power generation in a useful form. One way is to use a steam power plant, where fuel is combusted in a boiler/furnace from which the heat is transported through heat exchangers, such as superheaters, to a water cycle. In the heat exchangers high-pressure steam is produced. That steam runs through a steam turbine, which transforms the kinetic energy to electricity in a generator. The heat for district heating or industrial processes can be recovered in various ways, for example, by running the turbine in back-pressure mode so that the steam exists the turbine in a temperature still high enough to produce district heat (with a temperature up to 115°C). Another way is to take a bled steam stream and prime the temperature of the remaining water so that it can be used for district heating. Other technologies suitable for cogeneration are combustion or gas turbines, combined cycle processes, and engines fueled with gas or diesel [3].

Although the above-mentioned technologies can cogenerate heat and electricity, many of them use either gaseous or liquid fuels. When concentrating on CHP production from biomass, it has to be recognized that gaseous or liquid biofuels are relatively rare. For instance, different waste bio-oil or even methane from digestion of wastes exists but only in limited locations. Liquid and gaseous biofuels have very limited availability and are usually more expensive. These fuels are of course technically as appropriate as or even better fuels than solid biomass, but due to price and availability these are considered as marginal fuels for CHP production. In fact, most renewable fuels that can be thought of as the resource for primary energy, that is, fuels for electricity and heat production, are solid biomass. Therefore, a steam power plant is the main CHP technology used and several variations for combustion of solid biofuels have been developed. Thus, following sections will focus on the special characteristics of solid biofuels and technologies designed for production of CHP from solid biomass.

8.3.3 Challenges of Biomass in Energy Production

Use of solid biofuels is much more challenging than liquid or gaseous fuels and requires more equipment and attention since conveying solid fuels is more difficult than liquids or gases. One of the main challenges is the physical form of biomass as the biomass typically has both low material density and low energy density. The low density of biomass makes transportation relatively expensive and often not that practical, as the transportation is not usually limited by weight but the volume of biomass. Also, separate fuel-receiving areas are needed with equipment designed to move the low-density fuel. While fuel handling also requires attention for fossil solid fuels, such as coal, the lower energy density of biomass requires more space and makes well-functioning logistics a necessity. Thus, harvesting, logistics, and conveying of the fuel have to be well organized all the way from the forest or the field to the boiler at the power plant site.

As the density and form of biomass are also different from and more challenging than fossil solid fuels, conveying of biomass at the power plant site has to be well planned and technologies well selected to avoid blocking of equipment and feeding lines. Solid biomass is challenging in other ways too: the energy density is typically quite low compared to, for example, coal or other fossil fuels and the moisture content is high, even up to 50–60 wt.% [1]. In Nordic conditions during winter time, the water in the biomass can be frozen. Because the majority of the mass of biomass can be water or ice, that part is not useful as energy and a lot of energy is consumed for evaporation of water. As the energy density of the fuel is relatively low, also the combustion requires some special attention in order to still maintain high efficiency. One way to manage with the low-energy content and high water content is to co-combust biomass with some other fuel such as peat, coal, or waste or use higher-moisture-content biomass in a mix with better-quality biomass.

As saleable products, electricity and heat, can only be produced from the organic part of biomass, utility owners usually pay only for that part of the biomass. Thus, it is typical that biomass trade is based on the lower heating value of biomass as it is received at the site and the price is agreed on the basis of the actual energy content of the fuel. On the other hand, for biomass sales and logistic companies, the operation costs are accumulating from harvesting, transporting, and storing the total mass. For both parties it is clear that the quality of biomass also varies and depends on many things that are discussed below. Thus, there might be a conflict of interest between the seller and the buyer of biomass. Therefore, the quality of biomass is very important and, for instance, online measurement methods have been under development for measuring water and/or ice content of biomass.

8.3.3.1 Variation of fuel quality

Since biomass is an organic fuel, the quality varies depending on harvesting location and type of biomass. For instance, woody biomass is completely different from agricultural biomass in both hemicellulose, cellulose, and lignin ratios and ash content. The ash content for agricultural straw can be around 5 wt.% for dry biomass or even higher when it is around 0.5–2 wt.% for woody biomass The variation of the ash content of woody biomass is dependent on the wood species as well as the composition of the biomass, while needles, leaves, bark, and branches contain more ash components than the stem of the wood. In addition, soil composition and, for instance, fertilizers have an effect on the inorganic components accumulating to the biomass [1].

The time between cutting and using the biomass has an effect on the quality as, for instance, some organic components evaporate (volatile organic components [VOCs] evaporate), resulting in lower energy content of the fuel, together with a proportionally higher ash content. On the other hand, during storage, water is also evaporated, which decreases the amount of energy consumed for evaporation of water during combustion. There are several reasons why the ash content of biomass is important. The inorganic components, such as alkali metals and chloride, affect the combustion and can create possible problems such as high-temperature corrosion of metals, sintering of the fluidized-bed material, and slagging and fouling of the flue gas ducts, among other issues. Since fresh biomass contains more green leaves and needles with a higher chloride and alkali metal content than dried biomass, using fresh, green biomass increases the risk for corrosion. Thus the age of cut biomass has both positive and negative effects on the quality of biomass. Several research projects have been conducted to find out the optimum time for using biomass after cutting [1].

Sintering of the fluidized bed is relevant in case fluidized-bed combustion technology is used (more discussed below). Sintering of the fluidized-bed material is caused by the melting of alkali metal salts present in biomass. As these metal salts melt they start to build larger particle formations, making the fluidization of the bed material difficult and resulting in unstable combustion caused by an uncontrolled temperature gradient in the bed and freeboard. In a worst-case scenario, this may result in unwanted interruptions in the operation, spoilage of the bed material, and extra maintenance work. As sintering only occurs in high temperatures, sintering can be managed with good control of combustion temperature and fuel quality.

High-temperature corrosion of metals is caused by the inorganic components of the biomass ash such as K, Na, Ca, and Cl, which go through various reaction chains in the combustion, resulting in chemical compounds reacting relatively easily with metals, such as the superheater materials or boiler walls. Especially with high concentrations of Cl in biomass the risk of severe corrosion increases, leading to reduced life time of boiler components and even to breakdowns during operation. This causes unscheduled downtime and lost profits as well as difficulties to provide heat for customers who may be dependent on heat delivery. The problem with high-temperature corrosion is that currently there are no proven and reliable online methods to monitor the progress of corrosion. Now, the condition of crucial components such as superheaters can only be inspected during the yearly outage periods. High-temperature corrosion can be managed with controlling biomass fuel quality, maintaining material temperatures low enough, and adding additives or fuels containing some sulfur to the fuel mix. As sulfur keenly reacts with alkali metals, chloride is free to form HCl, which in low concentrations is not that corrosive to metals. Since peat contains sulfur and it is commonly available in Finland it has been used for controlling the combustion environment without the need for adding commercial additives.

In some cases biomass contains a lot of inorganic components that form thick hard layers on the heat exchanger surfaces and in the flue gas ducts, preventing effective heat transfer from the boiler, for example, to the steam cycle, as well as even limit the flue gas flow. This is called *slagging and fouling*. The formation of these layers can be problematic since they may be difficult to remove with, for example, soot blowing during operation and once they become too thick and start to cause problems in the steam cycle or the flue gas flow. This may also lead to downtime due to extra maintenance work.

8.3.4 Technologies Enabling CHP Production Using Biomass: Fluidized-Bed Boilers and Grate-Fired Boilers

For cogeneration of heat and power from solid biomass, steam power plants are most commonly used. A typical CHP plant with a steam cycle consists of:

- a boiler, combusting solid fuels, such as biomass or wastes, and transforming the thermal energy generated in combustion via heat exchangers to water;
- a steam cycle with a steam turbine;
- a generator to produce electricity; and finally also
- other heat exchangers to transform the low-temperature heat, for example, to a district heating network or an industrial process using heat.

Figure 8.4 illustrates a simplified scheme of a steam cycle power plant cogenerating power and heat for district heating.

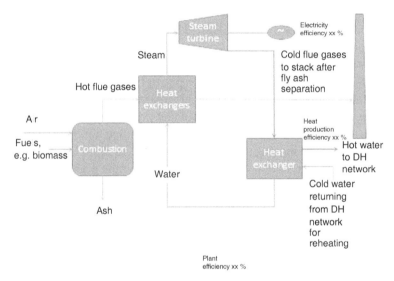

Figure 8.4   Simplified scheme of a steam power plant cogenerating power and heat for district heating.

There are two well-suiting technologies that can overcome the technical challenges related to biomass combustion, fluidized-bed boilers and grate-fired boilers. Grate-fired boilers have been used for long times for combustion of solid fuels and the grate can be designed for various different biomass types. In grate-fired boilers, biomass is fed on a fixed or mechanical grate that can have a sloping structure to facilitate the moving of biomass during combustion. In a grate-fired boiler the biomass is fed on the grate and air is fed through and on top of the grate. Biomass ash is removed from the end or the bottom of the grate. Fluidized-bed technology on the other hand has been used since the 1970s in boilers. In a fluidized bed a solid heat transfer medium, such as sand, is fluidized with air from the bottom of the boiler, creating a fluidized bed together with the biomass particles and the flue gases formed in the combustion [3].

The advantages of fluidized-bed technology are good mixing of the bed and the fuel enabling flexible fuel mixing and stable combustion with low emissions, even when running in partial loads. The heat transfer from the bed to the steam cycle is effective enabling high thermal efficiency. Due to good heat transfer and high heat capacity of the bed also biomass with very high moisture contents can be burned effectively. Fluidized-bed technology also tolerates well variations in the fuel mix, which is important for combustion of biomass since quality variation is typical for biomass. The operation of a fluidized-bed boiler is also simple due to the quick response time and easy control. Also only limited pretreatment of the fuel is needed mainly focusing on metal removal and possibly some size reduction. Fluidized-bed boilers also require little maintenance since they have only a few moving parts. The disadvantages are relatively high electricity consumption caused by maintaining the pressure difference over the fluidized bed created with air blowing through the bed and possible erosion of heat transfer pipes located in the bed. There are, however, designs, such as masonry around the bed superheaters, that prevent erosion. One practical advantage of fluidized-bed technology is that the boilers are relatively compact and are thus suitable in smaller plant sites, for example, in communities [3].

The advantages of grate-fired technology are that the moisture content and particle size of the biomass can be high and that the internal electricity consumption is low. The disadvantages of grate-fired boilers are poorer and slower control compared to fluidized-bed boilers. Although the grate technology can be designed and selected for various types of fuels, as technology is defined, it is more sensitive to the fuel quality variations than the fluidized-bed technology. The amount of combustion air is also high and still the amount of residual carbon from combustion can be high also, resulting in larger emissions. In case the grate is mechanical with moving parts, the need for maintenance work can also be high [3].

8.4 Energy and Resource Efficiency: Future CHP Boosts Efficiency with Integrated Processes

8.4.1 System Efficiency

There are different technologies that maximize electricity generation or heat generation. With condensing power, where solid fuels are combusted and the energy is converted to electricity via a generator, relatively high electricity efficiencies can be obtained. Typical electricity efficiency of a condensing steam power plant is around 38%–42%, but it can be even up to 45%, depending on steam temperatures and how well the configuration is optimized. In case there is a need for heat in addition to electricity comparing the efficiencies is more complicated since there are two products with different sale values. The value of electricity can be easily defined, since each producer can sell the electricity to the grid and get the price dictated by the electricity market. Heat on the other hand is a more local product and the value is also dependent on temperature of the heat produced. Thus comparing the overall efficiency of cogeneration and separate generation of heat is more challenging. One way to look at this is, however, to take a step back and evaluate system efficiency.

A simple example will clarify this. In case one would need 1 MW of heat and 0.5 MW of electricity, there are three possible ways to produce the required heat:

(A) Heat in an HOB using gaseous, solid, or liquid fuels;

(B) Direct electrical heating; or

(C) Cogeneration of heat and power.

In case A, illustrated in Fig. 8.5, electricity could be produced with condensing power with a typical efficiency of 38% from fuel to electricity. To produce the required 0.5 MW_e electricity, one would thus need 1.35 MW_f of fuels. In addition to this, another 1.2 MW_f of fuels would be needed for the production of 1.1 MW heat with a typical HOB efficiency of 90% and heat losses of 10% in the district heating grid during heat transfer. The total amount of fuels required to produce 1 MW of heat and 0.5 MW electricity is thus 2.6 MW.

Case A: Separate electricity and heat production

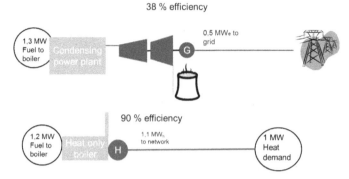

Figure 8.5   Separate electricity and heat production in two plants.

In case B, illustrated in Fig. 8.6, electricity can be again produced with a typical efficiency of 38% from fuel to electricity, consuming 1.3 MW fuels. Direct electrical heating is very efficient since practically all energy of the electricity can be converted to heat, but a few percent can be lost in the electricity grid. Thus to produce the same 1 MW of heat 2.7 MW fuels is required. In case B, the total amount of fuels needed to produce 1 MW of heat and 0.5 MW of electricity is thus 4.0 MW.

Case B: Heat production from electricity

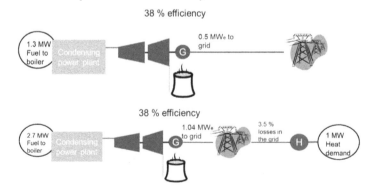

Figure 8.6   Electricity production and heat production using electricity.

In case C, illustrated in Fig. 8.7, heat and power are cogenerated in a CHP plant. In this case, the plant would use 1.8 MW of fuels to produce 0.5 MW of electricity. This would result in an efficiency of 28% from fuel to electricity. This way of defining the efficiency, however, neglects the fact that some of the fuel energy is used on purpose for heat generation with the expense of the efficiency of electricity production by running the power plant in back-pressure mode. With this plant the same energy can be used to produce 1.1 MW_{th} thermal energy that can be fed to the district heating network. This results to approximately 1 MW of heat at the end user if 10% of the heat is lost as heat losses in the district heating network. Thus in case C, the total amount of fuels needed to produce 1 MW of heat and 0.5 of MW electricity is only 1.8 MW, which is significantly lower than in cases A and B.

Case C: Combined heat and power generation

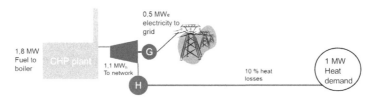

Figure 8.7   Cogeneration of heat and power.

This simple calculation shows that the resources are used most efficiently when they are used in CHP production as for case A 1.3 times more fuels is needed and for case B 2.2 times more fuels is needed when compared to case C to produce the same amount of heat and electricity. In addition to the benefit of high energy and resource efficiency, CHP generation takes advantage of lower emissions and use of resources such as the need for transportation of fuels, that is, traffic that accumulates throughout the whole supply chain.

8.4.2   Future of CHP: Boosted Resource Efficiency via Integrated Processes

CHP production is a highly resource- and energy-efficient way to generate heat and electricity. However, there are ways to improve the efficiency even more by integrating different processes to the CHP environment. For instance, there are many new technologies aiming to produce liquid biofuels for the transportation sector starting from lignocellulosic raw materials. Many of these processes that start from biomass have one fundamental challenge: Fossil gasoline and diesel, to be replaced with the bio-option, are pure hydrocarbons with practically no oxygen. In lignocellulosic biomass roughly half of the mass is oxygen, which makes its removal the most important target of the whole upgrading process. But the challenge with oxygen removal is that oxygen cannot be removed as O_2 but it usually is removed either as H_2O or with lighter oxygen-containing components or as CO or CO_2. This means that the yield of the end product is inevitably going to decrease, and the more oxygen is removed usually the lower yield, especially with thermochemical processes. While the yield of the end product decreases, similarly the amount of byproducts usually increases. This makes the efficient use of by-products important.

For example, although ethanol, still containing some oxygen and not being a pure hydrocarbon, can to some extent be blended with fossil gasoline, most production processes result in large amounts of other by-products, such as lignin. In fact, the aimed end product might not actually be the main product at all and utilization of the by-products for something useful becomes a necessity for the economics of the process. The same applies also to other processes such as pyrolysis followed by upgrading as well as gasification, torrefaction, and various sugar conversion routes producing char, gases, low-energy-content liquid side streams. Typically, either these processes also produce surplus heat that is not consumed in the process or require process heat, for instance, for drying of the feedstock. All processes also require some electricity. Thus, the CHP environment is a very potential parent site for integration of new processes with both surplus and deficit mass and energy flows.

Technically CHP plants are very logical places to utilize these side streams for CHP production, since biomass-fired fluidized-bed boilers are fairly robust to different types of biomass and fuel mixes. There are various synergies supporting the integration of thermochemical processes to the CHP production.

For the thermochemical process the benefits are that:

- an existing site with human resources, services, as well as existing biomass logistics can be utilized;
- surplus heat produced in the process can be used for district heat production and low-temperature heat streams can be integrated with the CHP and used in an efficient way;

- the process design and sizing can be more flexible, since the process doesn't necessarily have to be self-sustaining and heat and energy can be produced at the CHP also for the process; and
- by-products can be combusted effectively with existing equipment with only relatively limited modifications enabling the reduction of investments in additional boilers, flue gas treatment, and other auxiliary systems.

For the CHP process the benefits are:

- utilization of low-temperature heat streams for, for example, drying of biomass before the thermochemical process;
- benefiting of a longer yearly running time in case the process uses heat; and
- preheating of different process streams with low-temperature heat, for example, from returning district heating water.

Also, in case the process is endothermic and the CHP would have too high capacity, for instance, if an industrial heat customer has closed down, the CHP could benefit from a new heat customer.

8.5 Conclusions

8.5.1 Key Points from the Utility Point of View

A power plant operator is considering both maximizing its profits but also managing the risks related to the operation and the significant investments made. Since in most cases more than 50% of the operational costs of a CHP plant are related to fuel, the price of fuel is of key importance for profitable operation. While the quality of biomass or more generally fuels decrease, the operational and investment risks increase due to possible problems in logistics, fuel availability, and price–quality ratio of the fuel. The operator, especially in CHP production, is responsible for production of a local product that can be essential for its customers. For instance, if the district heat is used for heating the whole community, unscheduled stops in operation may lead to serious consequences especially in cold regions if heating of houses is interrupted. Therefore, no major changes risking the production are tolerated. There are only limited possibilities to optimize fuel quality without reliable methods to measure online the changes in fuel mixture. Thus, since online measurement methods for monitoring the quality of biomass are still under development, it is important to maintain constant, known fuel quality with regard to the energy content, ash content, and particle size distribution.

Also, the availability and price of biomass has to be managed as power plants have to operate reliably throughout decades. While building district heat networks is quite capital intensive, the usefulness of district heating is proven in densely inhabited areas. On the other hand building a power plant in the middle of a city is not completely unproblematic due to issues related to land use limitations, emissions, and logistics of solid fuels, amongst others. Although there are many different technologies to be used for CHP, some of them are more lucrative than others. The winning technologies are those that are flexible for the fuel quality and fuel mixes alternating with time as biomass availability, regulations, and political decisions change the operation environment.

Although CHP production is highly efficient way to produce heat and electricity, a drawback is that the electricity generation is driven by the heat load available, that is, the heat consumption. In the Nordic regions, this is not a major disadvantage as heat and electricity demands are typically coincident but in warmer regions this is an issue due to the shorter heating period. Additional CHP-integrated processes can improve this situation by providing a heat customer to the CHP plant with a steady year-round heat demand, also enabling longer-term electricity production and making combined heat and power production profitable in warmer regions as well.

In the future, when more and more of both primary energy and even fuels for transportation sector are produced with bioresources, the CHP environment has a true potential in improving the resource efficiency and even turning some otherwise unprofitable processes to profitable.

8.5.2 Fortum's Approach to CHP Production from Biomass

Fortum believes that future energy production is based on solar- and CO_2-lean resources instead of fossil fuels. Thus, biomass, together with energy and resource efficient production, is in Fortum's focus, in addition to other CO_2-lean technologies such as solar power and heat production, geothermal energy, hydropower, and wave energy. We strive to increase the share of biomass in our power plants. We test different biomass types; for instance, olive stones amongst other biofuels have been combusted successfully in the Värtan CHP plant originally fueled with coal located in Stockholm, Sweden. To overcome the technoeconomic challenges related to biomass use, flexible and efficient technologies and new innovations are required. We invest in biomass fueled and multifuel cogeneration plants and in January 2013 the construction of a 410 MW bio-CHP plant started at Värtan, Stockholm. The plant is expected to be in operation in 2016 and will be one of the world's largest biofuelled CHP plants [2].

In addition to utilizing existing flexible bio-CHP technologies, Fortum also actively develops the cogeneration concepts and technologies, together with various research and commercial partners. As an example of a process boosting resource efficiency Fortum has invested to an integrated pyrolysis production plant built and closely integrated to Fortum's CHP plant in Joensuu, Finland. By producing a new product Fortum Otso™ bio-oil, fossil heating oils can be replaced and used for district heating in HOBs or industrial applications.

References

1. Alakangas, E. (2000). *Suomessa Käytettävien Polttoaineiden Ominaisuuksia / Properties of Fuels Used in Finland* (in Finnish). VTT Valtion teknillinen tutkimuskeskus.

2. Fortum Oyj (2013). *Fortum Annual Report 2013*.

3. Koskelainen, L., Saarela, R., Sipilä, K. (2006). *Kaukolämmön Käsikirja/ District Heating Handbook* (in Finnish), Energiateollisuus ry. ISBN 952-5615-08-1.

About the Author

 **Sara Kärki** is a project manager in Fortum Corp., Finland, and works on different biomass research and development projects.

Sara received her master of science (technology) degree from Tampere University of Technology, Finland, in 2010, and she majored in power plants technology. She has worked for Neste Oil, Finland, on optimization of natural gas delivery. Since 2005 she has worked in Fortum on district heating, biosphere modeling in nuclear waste safety research, and renewable energy research and development.

Currently, Sara coordinates research and development projects that focus on biomass combustion, high-temperature corrosion, and biomass quality issues in CHP production. She also develops other biomass-based processes such as second-generation pyrolysis technologies that can be integrated to a CHP plant. In addition, she works on quality, environmental, health, and safety issues of pyrolysis bio-oils.

Sara's expertise is in biomass combustion, CHP production, and research and development of new biomass-based processes and products.

Chapter 9

Solvent-Based Biorefinery of Lignocellulosic Biomass

Paulus Johannes de Wild and Wouter Johannes Joseph Huijgen
Energy Research Centre of the Netherlands (ECN),
Unit Biomass and Energy Efficiency, Westerduinweg 3,
1755 LE Petten, the Netherlands
dewild@ecn.nl, huijgen@ecn.nl

9.1 Introduction

9.1.1 Biomass as a Renewable Resource for Chemicals, Fuels, and Other Products

Our modern society is unsustainable because it very much depends on fossil resources for chemicals, fuels, materials, and energy. The unlimited exploitation of oil, natural gas, and coal by mankind increasingly causes global problems such as climate change, environmental pollution, depletion of natural resources, and geopolitical instability. While renewable resources such as sun, wind, geothermal energy, hydropower, and biomass all contribute to

Biomass Power for the World: Transformations to Effective Use
Edited by Wim van Swaaij, Sascha Kersten, and Wolfgang Palz
Copyright © 2015 Pan Stanford Publishing Pte. Ltd.
ISBN 978-981-4613-88-0 (Hardcover), 978-981-4669-24-5 (Paperback), 978-981-4613-89-7 (eBook)
www.panstanford.com

the transition to a sustainable energy system, biomass is the only source of renewable carbon for chemicals, fuels, and materials [1].

Most terrestrial biomass consists of three major biopolymers: cellulose, hemicellulose, and lignin. Depending on the type of biomass, inorganic constituents (ash) and organic extractable material may be present as well, albeit generally in smaller quantities. Cellulose is a primary sugar source for bioconversion processes and consists of tightly linked glucose monomers. Hemicellulose is a secondary sugar source and consists of shorter polymers of various carbohydrates. Lignin consists of phenolic moieties, polymerized in a complex three dimensional network structure. Inorganic constituents are predominantly salts of alkali and alkaline earth metals such as potassium, sodium, magnesium, and calcium. Also silicon compounds such as silicates can be present in significant quantities. Finally, lignocellulosic biomass contains a variety of extractable organic material (extractives) such as fatty acids, proteins, terpenes, macromolecular phenolics, and oligomeric sugar species.

Lignocellulosic biomass can be used as such for several applications such as construction timber, rubber, textiles, and heat and power. In addition, the chemical structure of lignocellulosic biomass merits its exploitation (as feedstock) for chemicals, fuels, and performance materials. However, this requires physicochemical treatments, analogous to the refining of fossil oil. Unfortunately, the oil refinery infrastructure in its present form cannot be deployed for biomass because of fundamental differences in chemical structure and characteristics between biomass and fossil oil. The intrinsic heterogeneity of lignocellulosic biomass requires the development of (a combination of) new transformation technologies in order to enable the production of chemicals, fuels, and performance materials.

9.1.2 Biomass Fractionation

Typical thermochemical biomass conversion technologies such as combustion, gasification, and pyrolysis are mostly targeted on energy applications, either directly (combustion) or via the production of secondary energy carriers (syngas from gasification, pyrolysis oil/gas). From an economic perspective, this is not very efficient because the energy applications of biomass are generally low value and the added value of the generated heat and power likely does not make up for feedstock and processing costs. Although biomass locally can be deployed as a source for heat and power, large-scale energy applications seem better achieved from other renewable sources such as sun and wind for electricity. However, biomass is also a renewable source of carbon and its conversion into value-added materials, fuels, and chemicals can be economically much more attractive than its use for energy applications.

At present very few cost-effective processes exist for efficiently converting lignocellulosic biomass directly to components that are suited for producing chemicals, fuels, and other products. In general this is due to the fact that each of the major constituents (lignin, cellulose, and hemicellulose) demands distinct processing conditions such as temperature, pressure, catalysts, reaction environment, etc., in order to effectively break apart its polymer structure and selectively convert it into the desired product. In general, the highly intertwined character of lignocellulosic biomass, featuring a plethora of physicochemical linkages and (partly) overlapping thermochemical stabilities between the main biomass constituents, makes a selective and effective fractionation into its major constituents challenging. In addition, any deconstruction treatment inevitably leads to reactive species that often counteract further fractionation because of repolymerization and recondensation reactions.

A key issue for a successful biorefinery is a cost-effective and efficient fractionation of lignocellulosic biomass into its main constituents (hemicellulose, cellulose, and lignin), thus enabling dedicated further processing of each of the main fractions separately. This has been acknowledged for quite some time and already in the seventies and eighties of the last century (1st and 2nd oil crises) extensive R&D efforts took place to explore the (liquid-phase) fractionation of lignocellulosic biomass for chemicals, fuels, and other products [1]. Often, the starting point was the proven technology of biomass pulping in water at elevated temperatures and in the presence of certain "cooking" chemicals as practiced by the pulp and paper industry to liberate the cellulose fibres for further upgrading to paper. In general, lignin, hemicellulose, extractives, ash, and cooking chemicals end up in the so-called black liquor that is partially dewatered and combusted to meet the energy requirements of the pulp mill and to recover the cooking chemicals. Although some companies (Metso, Domtar) recover lignin, post-treatment of the black liquor for higher-value materials is generally not pursued.

Within a biorefinery, the so-called organosolv process can be used to solubilize lignin and hemicellulose from lignocellulosic biomass in order to efficiently split the biopolymers cellulose, lignin, and hemicellulose, for example, for further hydrolytic processing of the carbohydrate polymers to liberate sugars for subsequent (bio)chemical processing. Organosolv involves high-temperature treatment (typically around 200°C) of the biomass with a water-miscible organic solvent such as ethanol or acetone and optionally an acidic catalyst such as sulphuric acid. During organosolv, the lignocellulosic biomass is fractionated into a cellulose-enriched solid product stream (pulp) and a liquid product stream (liquor), comprising lignin and hemicellulose derivatives.

9.1.3 Staged Solvolysis: Step-Wise Unraveling Lignocellulosic Biomass

Looking at the physicochemical differences between the main constituents of lignocellulosic biomass (extractives, inorganics, hemicellulose, cellulose, and lignin), it seems logical to explore the potential of a sequential multistep or staged fractionation approach to effectively unravel the complex biomass structure into its main fractions. Although few examples show the potential of a step-wise approach to unravel the major constituents within the complex lignocellulosic biomass (see Section 9.2.3), neglecting constituents such as inorganic matter (ash minerals) and extractives may lead to suboptimal fractionation results due to various and unknown interferences between inorganics, organic extractives, and the major fractions hemicellulose, cellulose, and lignin.

To improve the biorefinery of lignocellulosic biomass, a governing principle could be that each fraction (including inorganics and extractives) is targeted according to its specific physicochemical characteristics, starting with the least severe treatment conditions. This option offers some clear advantages compared to direct transformation of the whole biomass. Firstly, it enables the valorization of all important biomass constituents, including the nonstructural fractions. Secondly, it offers the possibility to obtain the main biomass fractions (hemi)cellulose and lignin in superior quality because of the effective removal of the other (interfering) compounds and the prevention of undesired side reactions such as the formation of "pseudolignin."

The following sequence can be envisioned:

1. (Soluble) ash removal by washing with cold acidified water
2. Removal of extractives at elevated temperature with water and organic solvents
3. Hemicellulose auto- or catalytic hydrolysis
4. Lignin and cellulose separation by organosolv delignification

In an integrated biorefinery the sequence above could be seen as the solvent-based primary biorefinery in which lignocellulosic biomass is pretreated and fractionated into products that are further processed in a secondary biorefinery. This secondary biorefinery uses dedicated technologies for each of the fractions from the primary biorefinery, as visualized in Fig. 9.1. Basically, each step results in a product that can either be fractionated further or processed via a different route such as pyrolysis, gasification, chemical conversion, or biotechnological ways (e.g., anaerobic digestion and fermentation).

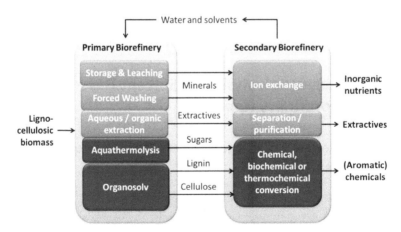

Figure 9.1   Two-step biorefinery concept featuring staged solvolysis in the primary biorefinery and dedicated processing in the secondary biorefinery. See also Color Insert.

This chapter describes solvent-based processes for the biorefinery of lignocellulosic biomass, both directly in a single step as well as in a staged approach. The focus will be on the latter as a fractionation concept that is targeted at sequential solvolytic processes to recover minerals, extractives, hemicellulose-derived sugars, cellulose, and lignin as primary fractions for further processing using dedicated technologies.

Figure 9.2 gives a schematic overview of the envisaged staged solvolysis concept. It should be noted that, depending on the location, size, and type of biomass, the first three stages may be bypassed.

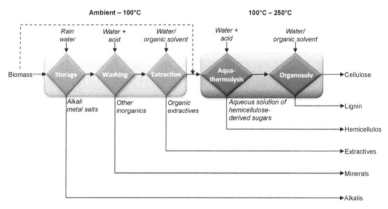

Figure 9.2   Staged solvolysis of lignocellulosic biomass for primary fractions.

9.2 Staged Solvolysis as a Solvent-Based Biorefinery Approach

9.2.1 Removal of Inorganic Matter (Ash) by Aqueous Leaching

The major and minor elements in biomass, in decreasing order of abundance, are generally C, O, H, N, Ca, K, Si, Mg, Al, S, Fe, P, Cl, Na, Mn, and Ti. Biomass is a complex heterogeneous mixture of organic matter and, to a lesser extent, inorganic matter. Table 9.1 presents a general overview of the composition of (lignocellulosic) biomass [2].

Table 9.1   Phase composition of biomass

Matter	State and type of constituents	Phases and components
Solid organic matter	Noncrystalline	Structural ingredients (hemicellulose, lignin), extractives, others
	Crystalline	Structural ingredients (cellulose), organic minerals such as Ca-Mg-K-Na oxalates, others
Solid inorganic matter	Crystalline	Mineral species from different mineral classes (silicates, oxyhydroxides, sulphates, phosphates, carbonates, chlorides, nitrates, others)
	Semicrystalline	Poorly crystallized mineraloids of some silicates, phosphates, hydroxides, chlorides, nitrates, others
	Amorphous	Amorphous phases such as various glasses, silicates, others
Fluid matter (mostly inorganic)	Liquid, gas	Moisture, gas, and gas–liquid inclusions associated with both organic and inorganic matter

Source: Taken from Ref. [2].

Alkali metals play a crucial role in thermal degradation of biomass. Especially, potassium catalyzes both volatile combustion and char burnout reactions, at least under low-temperature combustion [3]. In addition, experimental evidence suggests that alkaline additives act as homogeneous catalysts in various hydrothermal degradation reactions such as the conversion of glucose into formate salts [4].

Potassium is a key plant nutrient and present in varying amounts, depending upon the type of biomass, its growing conditions, and time of harvest. Of all the metals present in biomass, it is potassium that has the greatest influence on its thermal conversion properties. In general, the metal species in biomass are important in thermochemical conversion processes because of their impact on secondary degradation reactions, slagging, fouling, and other ash-related problems [3].

Alkali metals can be removed from the biomass, for example to prevent their catalytic effects on the fractionation of biomass. Natural leaching by rainfall during the period between biomass harvest and collection can already substantially reduce the content of alkali metal salts, especially chlorides. Alkali metal salts are soluble in (acidified) water, and just leaving the biomass on the field and exposing it to rain during the period between harvesting and collecting can lead to a removal of 30%–45% of potassium and chloride, respectively [5]. So, on-field leaching can offer an additional low-cost, on-farm strategy option to improve biomass quality for further biorefinery purposes.

Industrially, leaching of minerals from biomass commonly takes place in the sugar refining industry, where leaching is applied to remove the sucrose from the solid bagasse residue. During this process, inorganic material is substantially removed as well. Design variables related to leaching are the ratio water to biomass (and the presence of acids), the temperature of the water, the leaching time, and the application of dedicated mechanical means to dewater the biomass. The addition of acid to the leaching water increases the efficiency of the alkali metal leaching, although most of the alkali metals can be removed by water only. The increased potassium content in the leachate might render this stream suitable as feed for the fertilizer industry provided that there are no heavy metals present [6].

9.2.2 Solvent Extraction of (Organic) Extractable Material

Lignocellulosic biomass such as wood and straw contains extractives, including lipids, phenolic compounds, terpenoids, fatty acids, resin acids, steryl esters, sterol, waxes, and proteins. Generally, the extractive content varies between 2% and 5% for woods and 5% and 15% for straws but can be as high as 30% for feedstocks like olive tree prunings [7]. Although extractives generally contribute little to the overall wood composition, they can exert a significant influence on properties such as mechanical strength, color, and thermochemical stability. For biorefinery it might be desirable to remove the extractable compounds either to valorize them or to avoid interference of the extractives during the fractionation and/or conversion of the main constituents—hemicellulose, cellulose, and lignin.

Shebani et al. studied the effect of water-soluble and organic solvent–soluble extractives on the thermochemical stability of four wood species via an elaborate thermogravimetric analysis (TGA) study [8]. In general, the results suggest that a higher cellulose and lignin content leads to a better thermal stability of wood in different temperature regimes. It was also found that hot-water extractives decompose at lower temperatures than hemicelluloses and appear to protect the hemicelluloses. Ohtani et al. assumed that hot-water extractives suppress the decomposition and loss of hemicelluloses by acting as a protector for hemicelluloses during the alkaline cooking [9]. Removal of these extractives would render the hemicellulose more susceptible for thermochemical conversion such as acid-catalyzed hydrolysis.

Extractives can be removed from the biomass by solvent extraction, where the solvent can be water or a suitable and recoverable organic substance. For instance, Speaks et al. (1997) describe a process for extracting volatile organic compounds and higher-molecular-weight compounds ("pitch") from wood particulates. Thus, the emission of volatile organic compounds into the atmosphere during the processing of wood particulates into commercially useful products, such as oriented strand board, particle board, chipboard veneers, and various pulp and paper products is virtually eliminated [10]. In the extraction process, a solvent or mixture of solvents is used to remove wood extractives, including volatile organic compounds and pitch, from the wood particulates. The best results were obtained with acetone.

The example above merely illustrates the removal of extractives as undesired compounds for the further production of wood pulp. Within the framework of lignocellulosic biorefinery, an efficient removal of extractives is a desirable asset, too, because, in general, it enables the production of purer cellulose and lignin fractions, and it may increase the enzymatic digestibility (hydrolysis) of the cellulose when compared to the case in which extractives have not been removed.

Recently, Smit et al. (2014) described a treatment process for lignocellulosic biomass via an improved organosolv procedure that involves removal of extractives (world patent WO 2014/126471) [11]. A preferred procedure consists of an aqueous pre-extraction at a temperature in between 20°C and 60°C followed by an extraction with an organic solvent such as ethanol or acetone at a temperature in between 30°C and 80°C. During the two-stage extraction procedure, nonstructural components such as salts, proteins, fatty acids, triglycerides, waxes, terpenes, and resin acids are removed, leading to an improved purity of the cellulose and lignin fractions, better lignin properties (less pseudolignin formation), and increased enzymatic cellulose digestibility. Hydrophilic components (e.g., salts and water-soluble proteins) are predominantly washed away during the aqueous pre-extraction. Lipophilic components (lipophilic proteins, fatty acids, triglycerides, waxes, terpenes, resin acids, etc.) are predominantly extracted during the treatment with an organic solvent. An example of an application of extractives within a biorefinery itself is given in patent WO2014/098589. In this patent, Smit and Huijgen present a method in which the protein containing aqueous extract from biomass types such as grass and straw is used later on to improve the enzymatic digestibility of cellulose [12]. The authors hypothesize that the extracted proteins adsorb on residual lignin present in the cellulosic substrate resulting after pretreatment. This presumed adsorption prevents inactivation of hydrolytic enzymes by irreversible binding to lignin.

It should be noted that the organosolv procedures, as described above, are particularly suitable for the production of high-purity cellulose and lignin. The hemicelluloses have relatively limited thermochemical stability and tend to break down as a result of the relatively high temperatures employed during organosolv. In general, hemicelluloses are hydrolyzed into C_5 and/or C_6 sugar monomers, which may be subsequently dehydrated to furans such as furfural and hydroxymethylfurfural (HMF) under the influence of acid. These furans are reactive and tend to be converted to other compounds via recondensation and repolymerization reactions among themselves or with lignin-derived species. These secondary products are undesired and in general less valuable than products such as monomeric hemicellulose sugars or their primary furanic conversion compounds. In addition, these degradation products can pollute the cellulose and lignin fractions, rendering them less suitable for direct application or further processing.

After the removal of inorganic and organic extractives from the lignocellulosic biomass by suitable aqueous and organic extraction schemes, a next step in a solvolysis cascade can be the selective hydrolysis and optional depolymerization of the hemicellulose constituent, leaving a purified solid lignocellulose complex for further processing.

9.2.3 Removal of Hemicellulose by (Auto-) Catalytic Hydrolysis

In 2009 De Wild et al. published research on a hybrid biorefinery concept. The concept encompassed the selective removal of hemicellulose by a so-called aquathermolysis approach (treatment with hot pressurized water) as a first process step, followed by a second pyrolysis step with the remaining lignin–cellulose complex [13]. The fact that hemicellulose can be quantitatively removed from biomass by treatment with hot pressurized water has been known for some time [14–20]. The treatment results in water-soluble oligomeric and monomeric sugars and their degradation products such as furfural and HMF, both of which are valuable chemicals [21, 22].

The hydrolysis and further dehydration of the hemicelluloses is catalyzed by acetic acid, originating from the acetyl groups of hemicellulose. Simultaneously, water-soluble components like alkali metal ions and water-soluble extractives are leached out from the solid biomass. Cellulose and lignin are relatively unaffected by the hot-water treatment, although parts of the lignin may be solubilized as well, especially when the product liquor is continuously removed from the solid biomass [23]. The amount of dissolved lignin depends on temperature and reaction time. Longer reaction times and/or higher temperatures render the lignin insoluble due to recondensation reactions [15, 24]. The recondensed structures could be thermally more stable than the parent material and less prone to thermal degradation under conditions where cellulose degrades. The removal of the hemicellulose and the apparent increase in thermochemical stability of the (recondensed) lignin implies that in a subsequent pyrolysis step a more selective depolymerization of cellulose is possible, for example, resulting in enhanced yields of levoglucosan and less lignin-derived degradation fragments when compared to direct pyrolysis of the original feedstock. Supported by an extensive nuclear magnetic resonance (NMR) study, the work clearly showed that aquathermolysis results in removal of hemicellulose and that lignin ether bonds are broken. Apart from that, lignin seems hardly affected by the aquathermolysis [25]. Cellulose is also retained, although it seems to become more crystalline, probably due to a higher ordering of amorphous cellulose when the samples are cooled down after aquathermolysis. Figure 9.3 presents a schematic overview of a possible hybrid biorefinery in which aquathermolysis and pyrolysis are combined.

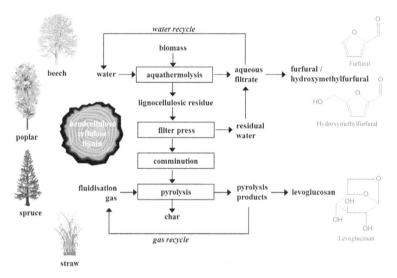

Figure 9.3   Hybrid biorefinery concept [1].

Alternatively, after the removal of the hemicellulose, the resulting (wet) cellulose–lignin complex can be further fractionated by the action of an organic solvent. This organosolv process uses a mixture of water and an organic solvent to extract and thereby separate the lignin from the cellulose.

In general, the aqueous ethanol organosolv process is conducted at conditions that lead to extensive degradation of the hemicellulose sugars. To prevent degradation and to eventually recover the sugars from the hemicellulose fraction, Huijgen et al. applied prehydrolysis, targeted at the hydrolysis of the hemicellulose polymer from wheat straw into monomeric xylose [27]. Although the treatment resulted in a significant amount of xylose released, the lignin yield in the subsequent organosolv delignification step decreased due to the formation of so-called pseudolignin and lignin recondensation during the prehydrolysis step.

9.2.4 Separation of Lignin and Cellulose by Organosolv Fractionation

9.2.4.1 Organosolv fractionation

Organosolv is based on the treatment of biomass with an (aqueous) organic solvent at elevated temperatures [28–30].

Commonly used solvents are ethanol, methanol, acetone, and organic acids like acetic acid and formic acid or combinations thereof. Organosolv processes primarily delignify lignocellulose, with the organic solvent functioning as lignin extractant. Simultaneously, the hemicellulose fraction is depolymerized through acid-catalyzed hydrolysis. In general, organosolv processes aim to fractionate the lignocellulosic biomass as much as possible into its individual major fractions in contrast to other pretreatment technologies such as steam explosion and dilute acid hydrolysis.

While other pretreatment technologies merely make the cellulose fraction suitable for further processing without recovery of a purified lignin fraction, organosolv coproduces lignin with high purity (limited amounts of residual carbohydrates and minerals). In addition, organosolv lignins have a relatively low molecular weight with a narrow distribution and very low sulphur content. Consequently, their application spectrum is broader compared to the more impure lignin-containing residues derived from conventional pretreatments, which are targeted primarily toward the production of cellulose for paper or second-generation bioethanol. These lignin-containing residues are a complex mixture of unconverted carbohydrates, lignin, minerals, and process chemicals or microbial residues. Hardly any applications for such complex by-products have been identified other than combustion for combined heat and power (CHP). A block scheme of the aqueous ethanol organosolv process as developed and practised at the Energy research Centre of the Netherlands (ECN) is presented in Fig. 9.4.

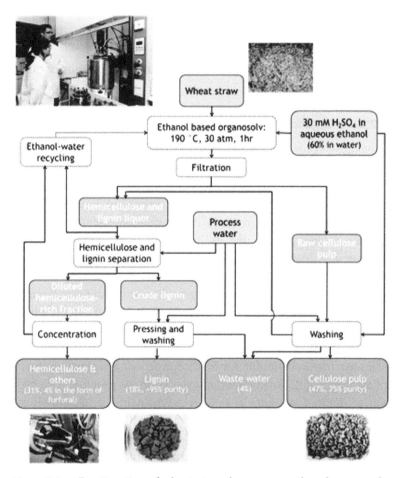

Figure 9.4 Fractionation of wheat straw by aqueous ethanol organosolv [31].

Another example of organosolv fractionation technology is the acetic acid/formic acid process, exploited by the French Compagnie Industrielle de la Matière Végétale (CIMV).

In Pomacle, France, the CIMV operates a pilot plant that takes in 100 kg/hr wheat straw that is treated with a mixture of acetic acid and formic acid [29, 32]. The CIMV pilot plant differs from most other lignocellulosic biorefineries because it produces a high-quality lignin as a separate marketable product instead of a solid fuel. Figure 9.5 presents a schematic diagram of the CIMV biorefinery concept and a photograph of the pilot plant in France.

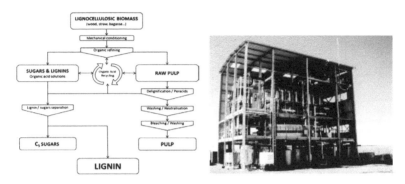

Figure 9.5 Process scheme and photograph of the CIMV organic acid–based organosolv process [33].

Figure 9.6 gives a photographic impression of the CIMV biorefinery products from wheat straw.

Figure 9.6 Typical biorefinery products from CIMV organosolv fractionation [33].

Both the ethanol and the organic acid organosolv processes are examples of modern fractionation technologies for lignocellulosic biomass that are specifically aimed at the production of pure biomass fractions for further processing to value-added chemicals and other materials to enhance the profitability of the biorefinery. A generic view of future organosolv-based biorefineries is presented in Fig. 9.7.

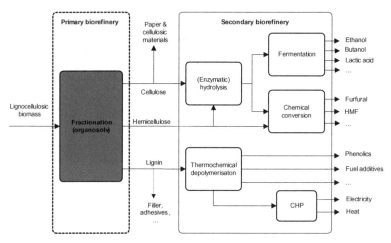

Figure 9.7   General layout of an organosolv-based biorefinery [30].

Finally, Fig. 9.8 is an example of a more detailed concept of an organosolv-based biorefinery, targeted at the fractionation of biomass into hemicellulose, cellulose, and lignin and the integrated further conversion of the hemicellulose pentoses into furfural.

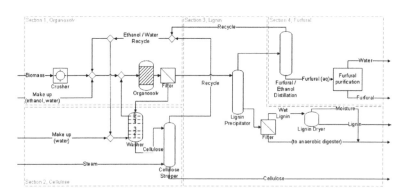

Figure 9.8   Ethanol-based organosolv biorefinery [34].

9.2.4.2 Prehydrolysis–organosolv approach

Various experimental studies on hydrolysis of hemicellulose prior to organosolv treatment have been published during the last years, such as with steam [35], dilute acid presoaking [36], and hydrothermal pretreatment [37–39]. In an approach to improve the ethanol-based organosolv fractionation of lignocellulose, Liu et al. also attempted a prehydrolysis step, aimed at the removal of the hemicelluloses prior to the separation of the lignin and the cellulose [40].

Recently, Huijgen et al. reported for wheat straw a staged approach targeted at the removal of hemicellulose, prior to the subsequent lignin–cellulose fractionation and enzymatic cellulose hydrolysis [27]. Wheat straw was fractionated using a three-step biorefining approach: (1) aqueous pretreatment for hemicellulose prehydrolysis into sugars, (2) organosolv delignification, and (3) enzymatic cellulose hydrolysis into glucose. Prehydrolysis was applied to avoid degradation of hemicellulose sugars during organosolv delignification.

The prehydrolysis–organosolv approach was compared to a direct one-step organosolv treatment (i.e., without prehydrolysis) to elucidate the influence of the prehydrolysis step on the delignification during subsequent organosolv treatment and the resulting enzymatic digestibility of the cellulose fraction. The experimental approach is illustrated in Fig. 9.9.

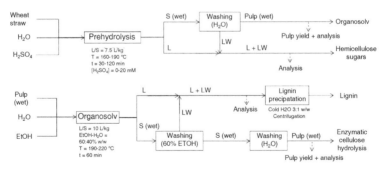

Figure 9.9   Experimental approach of direct organosolv vs. prehydrolysis—organosolv of wheat straw. Reprinted from Ref. [27], Copyright 2012, with permission from Elsevier.

The one-step organosolv treatment using a single arbitrary set of process conditions resulted in 67% lignin removal and 51%

xylan removal from the wheat straw. At the same time, 95% of the glucan remains in the solid fraction. In spite of the 51% xylan removal, the xylose yield is only 4% based on the xylan content of the feedstock. This xylose is mainly present in the form of oligomers. In total, residual xylan and identified xylan derivatives (including a furfural yield of 2%) account for only 56% of the original xylan in the feedstock. The remaining 44% has reacted into unidentified products, including possibly condensates with lignin [41]. The enzymatic digestibility of the produced cellulose-enriched pulp is 51% compared to 15% for untreated raw wheat straw.

Subsequently, prehydrolysis was performed prior to the organosolv delignification. Organosolv was performed at identical conditions as used for the organosolv experiment without prehydrolysis. The influence of prehydrolysis temperature (160°C, 175°C, 190°C), reaction time (30, 60, 120 min), and addition of H_2SO_4 (0, 10, 20 mM) as a catalyst was studied around the initial experiment. The use of a prehydrolysis step prior to organosolv delignification of wheat straw was found to improve the yield of xylose and the enzymatic cellulose digestibility. The maximum xylose yield obtained was 67% compared to 4% in the case of organosolv without prehydrolysis. However, prehydrolysis was found to reduce the lignin yield by organosolv delignification due to the formation of pseudolignin and lignin recondensation during prehydrolysis. This reduction could partly be compensated by increasing the temperature of the organosolv delignification step to 220°C. In addition, the application of a prehydrolysis step was found to substantially improve the enzymatic cellulose digestibility despite the recondensation of lignin and/or pseudolignin on the cellulose. Increasing the organosolv delignification temperature to 220°C resulted in a maximum enzymatic glucose yield of 93% or 0.36 kg/kg straw.

9.2.5 Conclusion: Staged Solvolysis

It is obvious that staged solvolysis can lead to relatively pure, separate fractions that can be further processed by dedicated technologies. However, the added value of the produced fractions should be counterbalanced by the costs related to the use of different solvents and processing technologies. Depending on the type of biomass and the desired products, direct solvent-based liquefaction may be a viable alternative to valorize the biomass. The next section will briefly address some current developments in the field of direct solvolytic processing of lignocellulosic biomass.

9.3 Direct Solvolysis of Lignocellulosic Biomass for Fuels and Chemicals

9.3.1 Introduction

Direct solvolysis is a thermochemical liquefaction process using water and/or organic solvents to convert the solid lignocellulosic biomass into a liquid product. The objective of a total dissolution or liquefaction of biomass is to facilitate transportation, refining, and conversion. Liquefaction may enable the use of biomass as feedstock in conventional crude oil refineries with the use of conventional or modified processing/separation technologies. Currently, (fast) pyrolysis is intensively explored as liquefaction pretreatment for biomass processing in refineries. Despite the fact that modern state-of-the-art fast pyrolysis easily achieves liquid yields up to 80 wt.% (based on the dry biomass weight), the resulting fast pyrolysis oil is still unsuitable for direct processing in conventional crude oil refineries because of its acidity, water content, instability, solids content, etc. [42]. Direct solvolysis-based liquefaction may offer advantages compared to (fast) pyrolysis due to milder and more controlled process conditions.

In the case of water only, the solvolysis process is generally called hydrothermal liquefaction (HTL) which is performed at subcritical water process conditions (10–25 MPa pressure, 280°C–370°C) with or without catalysts. HTL is best suited for wet biomass and allows for the conversion of a wide range of feedstocks in a so-called liquid biocrude product. A detailed description of HTL can be found elsewhere [43, 44].

Wang and Wang recently reported a comprehensive review of the vast area of solvolysis [44]. Solvolysis has been studied with a variety of organic solvents such as (aromatic) alcohols, polyols, acetone, 1,4 dioxane, tetralin and other hydrogen donor solvents. In their review the authors clearly point out that the type of solvent greatly influences yield and composition of the resulting bio-oil, although governing principles that determine the specific effects of solvents during liquefaction still require further understanding [44]. It was proposed that differences in density and polarity of the solvent might be a major reason for their different behaviour during liquefaction [45].

9.3.2 Alcoholysis

An example of the use of methanol and ethanol in liquefaction of biomass is given by Grisel et al., who investigated the acid-catalyzed alcoholysis of wheat straw in 95% methanol and in 94% ethanol (w/w) in the presence of various amounts of sulphuric acid and compared it to the alcoholysis of wheat straw-derived organosolv pulp and commercially available celluloses [46].

It was found that lignocellulosic biomass can be converted in a single step via acid-catalyzed alcoholysis using lower linear alcohols. Main products were furfural and alkyl glucosides. The major by-products were found to be levulinic acid and its alkyl ester. The amount of insoluble humins that are formed was limited. The depolymerization and alkylation of carbohydrates are mainly Brønsted acid-catalyzed and the presence of proton-consuming compounds, such as acid-neutralizing native minerals or chloride, hamper the liquefaction efficiency and the alkyl glucosides yields. The optimum acid dose needs to be adjusted to the acid neutralization capacity of the biomass. The amount of dimethyl ether formed from methanol under these conditions was limited but became readily more prominent at higher temperatures.

Solvent loss due to ether formation from ethanol condensation was much smaller. Delignification prior to alcoholysis did not influence the results much but allowed higher product concentrations and enables a separate lignin valorization. The glucosides can be separated, for example, by means of chromatography, and may be converted into furan building blocks, for example, for the production of plastic precursors, such as 2,5-furandicarboxylic acid.

Eerhart et al. compared direct alcoholysis of wheat straw for the production of furans with organosolv fractionation of wheat straw followed by (1) alcoholysis of the resulting cellulose pulp or (2) enzymatic hydrolysis of the cellulose [47]. It was concluded that all three routes studied technically could be a viable option to produce biofuels and biobased plastics via the furan pathway at efficiencies generally higher than other biorefinery processes such as production of second-generation bioethanol.

9.3.3   Direct Solvolysis in the Presence of Hydrogen Donor Solvents

In the quest for so-called drop-in fuels from biomass, a major challenge is an effective removal of oxygen from the crude bio-oil product. In fast pyrolysis this is addressed by using specific in situ or ex situ catalytic approaches, often in the presence of hydrogen. Also the catalytic hydrodeoxygenation of fast pyrolysis oil is pursued to some extent [48]. Analogous to approaches in the fast pyrolysis field, in direct solvolysis processes the use of hydrogen is investigated, too, for oxygen removal from the solvolysis products. Hydrogen donor solvents such as tetralin or formic acid can be deployed to generate hydrogen in situ during the solvolysis.

Kleinert and Barth have reported on the conversion of lignin to liquids with the aim of obtaining liquid fuels. The high oxygen content and the dominance of aromatic structures in lignins require a high degree of chemical transformation and incorporation of additional hydrogen to give stable, nonpolar, petroleum-soluble liquid products [49]. The authors deployed formic acid/alcohol mixtures as the reaction medium for converting lignin to liquid "oils" in a solvolysis process. During the conversion, both depolymerization and removal of oxygen by formation of water occur in a single step. Formic acid serves as both the hydrogen donor and reaction medium in the solvolysis process. Using an alcohol as cosolvent can improve the liquid yields and H/C ratios. Very little coke (5%) was produced. The liquids produced comprise two phases that can be easily separated, where the lighter organic phase consists mainly of low-molecular-weight alkylphenols and C_8–C_{10} aliphatics. The process is developed to be combined with, for example, bioethanol production from lignocellulosic carbohydrates in a biorefinery concept aimed at converting all fractions of the wood into renewable liquid fuels. The solvolysis process uses formic acid as the reagent for both depolymerization and oxygen removal. Although formic acid is a costly chemical, it may be generated from biomass in an integrated oxidative process.

9.3.4 Direct Liquefaction with Biomass-Derived Phenolic Solvents

Preferred organic solvents for direct solvolysis can be derived from biomass itself and provide a good performance during liquefaction, including a good recoverability. For instance, Azadi et al. reported on an approach by which the hemicellulose and cellulose fractions of biomass are converted through catalytic processes into platform chemicals and transportation fuels using an organic solvent obtained by depolymerization of the lignin fraction [50].

In the paper, the authors aim at the production of second-generation biofuels from lignocellulosic biomass through the intermediate production of oxygenated platform molecules, such as furan intermediates (furfural [FuAl], furfuryl alcohol [FuOH], and HMF), levulinic acid (LA), and γ-valerolactone (GVL). The organic solvent is prepared by the catalytic depolymerization of the lignin part of poplar and consists of alkyl-substituted phenolics. Figure 9.10 shows the roadmap proposed by the authors for the conversion of lignocellulosic biomass to fuels and chemicals.

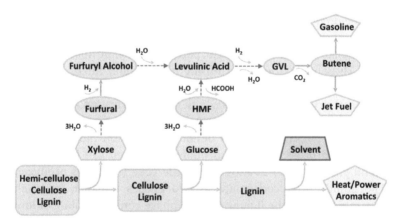

Figure 9.10 Roadmap for conversion of lignocellulosic biomass (rectangles) to chemicals (ovals) and fuels (pentagons), passing through the intermediate formation of C_5 and C_6 sugars (hexagons). Dashed arrows indicate processes that can be carried out using a lignin-derived organic solvent for the production of furfural, HMF, levulinic acid, and GVL [50].

Another example of the use of a lignin-derived solvent for biomass liquefaction is given by Kumar et al., who used guaiacol (2-methoxyphenol) as a solvent for direct liquefaction of lignocellulosic feedstock [51]. In an extensive parameter study, the authors showed that pine wood could be converted to bio-oil in guaiacol with an oil yield of >90% (based on C) at 300°C–350°C without catalysts or reactive atmospheres. Process conditions such as temperature, heating rate, reaction time, and water concentration were optimized to achieve the best compromise between maximizing the bio-oil yield and minimizing the fraction of heavy ends with a molecular weight >1000 Da. Best results were obtained at a temperature in the range of 320°C–350°C, a reaction time >200 seconds, and the addition of up to 10 wt.% of water. A remarkable mechanistic result was that the formation of undesired vacuum residue (heavy ends) apparently is not due to secondary condensation of bio-oil but rather to incomplete primary depolymerization of the reacting biomass with simultaneous dehydration of fragments. Water acts as a catalyst and accelerates liquefaction, but it seems not to affect the final product yields. Final conclusion of the work was that the 30% final vacuum residue is still too high to allow effective recycling of the bio-oil as a liquefaction medium. Alternative approaches are needed such as the use of catalysts or the removal and/or cracking the heavy ends prior to recycling.

These approaches are interesting and merit further research because the possibility to use bio-oil or fractions thereof as effective liquefaction solvents offer several advantages, both from a process and a fundamental chemical point of view. Biomass-derived fractions for solvolysis may be integrated in the whole process, thereby minimizing the need for an external solvent supply. Chemically, the principle "like dissolves like" offers interesting liquefaction possibilities. For instance, in the Noguchi process, phenol was used to liquefy lignin prior to hydrogenation to monophenolic compounds [52]. Kakemoto et al. patented a thermochemical method to manufacture phenols from lignin that was extracted in a solvolysis pulping process by a solvent containing phenols [53]. The subsequent thermochemical deconstruction of the lignin took place in a mixture with a double-ring aromatic solvent such as naphthalene. More recently, Okuda et al. selectively converted lignin at 400°C into single-chemical species using a water–p-cresol mixture [54].

9.3.5 Direct Liquefaction with Ionic Liquids

Alternatively to organic solvents as described in previous paragraphs, also the use of ionic liquids (ILs) and deep eutectic solvents (DESs) has been explored for the pretreatment, fractionation, and/or liquefaction of lignocellulosic biomass. Vancov et al. recently reviewed a pretreatment method for lignocellulosic materials using ILs [55]. The authors focus on the use of ILs for dissolution of cellulose for further enzymatic hydrolysis and fermentation. ILs are nonvolatile solvents that exhibit unique solvating properties. In their review the authors describe the dissolution of cellulose and lignocellulose in various ILs, including key properties such as high hydrogen-bonding basicity, which increases the ability of the IL to dissolve cellulose. As a pretreatment in biofuel production, the review details aspects such as the regeneration of cellulose from ILs, structural changes that arise in the regenerated cellulose and their effect on enzymatic hydrolysis, the potential for IL recycling, and, finally, exploitation of ILs in an integrated bioprocess. ILs could hold the key to unlocking new and exciting processes for the production of biofuels from lignocellulosic materials. ILs have many advantages over traditional solvents, such as their low volatility, which arises from them having no vapor pressure, nonflammability, odorlessness, and thermal stability. ILs can be specifically designed and the liquid's ability to dissolve a variety of polar and nonpolar materials makes them useful for dissolving the complex structure of lignocellulose. Current literature on the use of ILs as a pretreatment in the production of bioethanol showed that the process seems promising because lignocellulose dissolved in ILs can be easily regenerated by the simple addition of an antisolvent. Regenerated cellulose is amorphous and porous and much more responsive to enzymatic saccharification. However, the use of ILs for biorefinery has major shortcomings, especially in process development (including difficulties in recycling of the IL due to build-up of contaminants) and current costs of the solvents.

Brandt et al. have critically reviewed the application of ILs for the deconstruction and fractionation of lignocellulosic biomass [56]. In their extensive review the authors address the solubility of lignocellulosic biomass (and the individual biopolymers within it) in ILs, the deconstruction effects brought about by the use of ILs as a solvent and practical considerations regarding the design of IL-based deconstruction processes.

In conclusion the authors state that ILs have as major advantage that they are able to decrystallize cellulose with a simultaneous disruption of the lignin–hemicellulose network. The possibility of removing lignin with the IL and recovering a separate, possibly more valuable lignin fraction is also an attractive feature. However, deconstruction with ILs will only be viable if its advantages outweigh the major drawback of ILs, their cost in relation to the cost of the feedstock that has to be treated and the revenues for the final products. Other important hurdles/research challenges are the relatively large amount of ILs needed to treat biomass, the recycling of the spent ILs, the end of use recovery of ILs, toxicity of residual ILs on downstream processing, and the health and environmental impacts of ILs in general.

Recently, also the use of DESs, also called low-transition-temperature mixtures (LTTMs), for biomass processing has been proposed [57, 58]. DESs are eutectic mixtures of solids that due to strong hydrogen bonding interactions show a very low transition temperature and are therefore liquid at low temperatures. Examples of DESs include mixtures of hydrogen bond donors, such as urea, oxalic acid, and lactic acid, and hydrogen bond acceptors, such as choline chloride, alanine, and proline. DESs are potentially much cheaper solvents than ILs and are, in principle, biorenewable. Fransisco et al. reported the ability of lactic acid–choline chloride mixtures to dissolve lignin [58]. Since the same DES showed no cellulose solubility, it might be used for lignin–cellulose separation purposes.

Both ILs and DESs have interesting characteristics, but it should be realized that research on the use of these solvents in lignocellulose pretreatment or as part of an integrated bioprocess is obviously in its infancy. At the current stage, it is too early to draw any conclusions here whether, in the end, these solvents could be applied economically in industrial biorefinery processes.

9.3.6 Conclusion: Direct Solvolysis

Direct solvolysis is a thermochemical liquefaction process using water and/or organic solvents to convert the solid lignocellulosic biomass into a liquid product that enables easier transportation, refining, and conversion in conventional crude oil refineries with the use of conventional or modified process technologies. Currently, (fast) pyrolysis is intensively explored as liquefaction pretreatment for biomass processing in refineries. Despite the fact that modern state-of-the-art fast pyrolysis easily achieves liquid yields up to 80 wt.% (based on the dry biomass weight), the resulting fast pyrolysis oil is still unsuitable for direct processing in conventional crude oil refineries because of its acidity, water content, instability, solids content, etc. Due to milder and more controlled process conditions direct solvolysis-based liquefaction may offer clear advantages compared to (fast) pyrolysis. However, at the moment fast pyrolysis seems to be a more mature technology when compared to solvolysis. For example, fast pyrolysis is routinely performed in continuous-reactor systems, while solvolysis processes are currently generally still operated at a small scale and in batch mode. Which technology will prevail at the end of the day is difficult to predict and will depend on the balance between processing flexibility, costs, and quality of final products.

9.4 Conclusions

This chapter presents a perspective on the solvent-based fractionation and liquefaction of lignocellulosic biomass via a brief overview of staged and direct solvolysis processes. Staged approaches offer the clear benefit of separation of all major biomass fractions, enabling the deployment of dedicated upgrading technologies for each of the separate fractions. Of course, the revenues from the final products should provide a sufficient economic margin for capital and operational expenses that come with the different process steps in the staged approach and the further upgrading. Therefore, staged approaches seem to be especially preferable when specific high-value chemicals are the desired products. On the other hand, when the target product is a drop-in biofuel, direct solvolysis may be the technology of choice, analogous to fast pyrolysis. Overall, significant progress has been made in the field of solvent-based biorefinery of lignocellulosic biomass in recent years and solvent-based processes might make a substantial future contribution to the biobased economy.

References

1. De Wild, P.J. (2011). *Biomass Pyrolysis for Chemicals*, PhD thesis, Groningen University, the Netherlands.

2. Vassilev, S.V., Baxter, D., Andersen, L.K., Vassileva, C.G. (2010). An overview of the chemical composition of biomass, *Fuel*, **89**, 913–933.

3. Nowakowski, D.J., Jones, J.M., Brydson, R.M.D., Ross, A.B. (2007). Potassium catalysis in the pyrolysis behaviour of short rotation willow coppice, *Fuel*, **86**, 2389–2402.

4. Jin, F., Wang, Y., Zeng, X., Shen, Z., Yao, G. (2014). Water under high temperature and pressure conditions and its applications to develop green technologies for biomass conversion, in *Applications of Hydrothermal Reactions to Biomass Conversion, Part I*, Springer, Berlin, Heidelberg, *Series Green Chemistry and Sustainable Technology*. ISBN 978-3-642-54457-6

5. Tonn, B., Thumm, U., Lewandowski, I., Claupein, W. (2012). Leaching of biomass from semi-natural grasslands: effects on chemical composition and ash high-temperature behaviour, *Biomass and Bioenergy*, **36**, 390–403.

6. Nachenius, R.W., Ronsse, F., Venderbosch, R.H., Prins, W. (2013). Biomass pyrolysis, in *Advances in Chemical Engineering*, Vol. 42, 75–139, Elsevier. ISSN 0065-2377

7. Cara, C., Ruiz, E., Oliva, J.M., Sáez, F., Castro, E. (2008). Conversion of olive tree biomass into fermentable sugars by dilute acid pretreatment and enzymatic saccharification, *Bioresource Technology*, **99**, 1869–1876.

8. Shebani, A.N., Van Rheenen, A.J., Meincken, M. (2008). The effect of wood extractives on the thermal stability of different wood species, *Thermochimica Acta*, **471**, 43–50.

9. Ohtani, Y., Mazumder, B.B., Sameshima, K. (2001). Influence of the chemical composition of kenaf bast and core on the alkaline pulping response, *Journal of Wood Science*, **47**, 30–35.

10. Speaks, J.R., Campbell, R.O., Veal, M.A. (1997). *Pretreatment of Wood Particulates for Removal of Wood Extractives*. US patent 5,698,667.

11. Smit, A.T., Huijgen, W.J.J., Grisel, R.J.H (2014). *Process for the Treatment of Lignocellulosic Biomass*. Patent application WO2014/126471.

12. Smit, A.T., Huijgen, W.J.J. (2014). *Process for Enzymatic Hydrolysis of Cellulose*. Patent application WO2014/098589.

13. De Wild, P.J., Den Uil, H., Reith, J.H., Lunshof, A., Hendriks, C., Van Eck, E.R.H., Heeres, H.J. (2009). Bioenergy II: biomass valorisation by a hybrid thermochemical fractionation approach, *International Journal of Chemical Reactor Engineering*, **7**, A51.

14. Richter, G.A. (1956). Some aspects of prehydrolysis pulping, *TAPPI Journal*, **39**(4), 193–210.

15. Lora, J.H., Wayman, M. (1978). Delignification of hardwoods by autohydrolysis and extraction, *TAPPI Journal*, **61**, 47–50.

16. Bonn, G., Concin, R., Bobleter, O. (1983). Hydrothermolysis: a new process for the utilization of biomass, *Wood Science and Technology*, **17**, 195–202.

17. Garrote, G., Domínguez, H., Parajó, J.C. (1999). Hydrothermal processing of lignocellulosic materials, *Holz als Roh und Werkstoff*, **57**, 191–202.

18. Mosier, N., Wyman, C.E., Dale, B.E., Elander, R.T., Lee, Y.Y., Holtzapple, M., Ladisch, M.R. (2005). Features of promising technologies for pretreatment of lignocellulosic biomass. *Bioresource Technology*, **96**, 673–686.

19. Liu, S. (2008). A kinetic model on autocatalytic reactions in woody biomass hydrolysis, *Journal of Biobased Materials and Bioenergy*, **2**, 135–147.

20. Yu, Y., Lou, X., Wu, H. (2008). Some recent advances in hydrolysis of biomass in hot-compressed water and its comparisons with other hydrolysis methods, *Energy Fuels*, **22**, 44–60.

21. Lehnen, R., Saake, B., Nimz, H.H. (2001). Furfural and hydroxymethylfurfural as by-products of Formacell pulping, *Holzforschung*, **55**, 199–204.

22. Zeitsch, K.J. (2000). *The Chemistry and Technology of Furfural and Its Many By-Products*, Sugar Series 13, Elsevier Science B.V., the Netherlands. ISBN 0-444-50351-X

23. Bobleter, O., Binder, H., Concin, R., Burtscher, E., Palz, W. (1981). *The Conversion of Biomass to Fuel Raw Material by Hydrothermal Treatment, Energy from Biomass (1st E.C. Conference)*, 554–562, Eds. Cartier, P., Hall, P.O., Applied Science.

24. Biermann, C.J., Schultz, T.P., McGinnis, G.D. (1984). Rapid steam hydrolysis / extraction of mixed hardwoods as a biomass pretreatment, *Journal of Wood Chemistry and Technology*, **4**, 1, 111–128.

25. Habets, S., De Wild, P.J., Huijgen, W.J.J., Van Eck, E.R.H. (2013). The influence of thermochemical treatments on the lignocellulosic structure of wheat straw as studied by natural abundance 13C NMR, *Bioresource Technology*, **146**, 585–590.

26. Luo, J., Xu, Y., Zhao, L., Dong, L., Tong, D., Zhu, L., Hu, C. (2010). Two-step hydrothermal conversion of Pubescens to obtain furans and phenol compounds separately, *Bioresource Technology*, **101**, 8873–8880.

27. Huijgen, W.J.J., Smit, A.T., De Wild, P.J., Den Uil, H. (2012). Fractionation of wheat straw by prehydrolysis, organosolv delignification and enzymatic hydrolysis for production of sugars and lignin. *Bioresource Technology*, 114, 389–398.

28. Zhao, X.B., Cheng, K.K., Liu, D.H. (2009). Organosolv pretreatment of lignocellulosic biomass for enzymatic hydrolysis, *Applied Microbiology and Biotechnology*, **82**(5), 815.

29. Benjelloun-Mlayah, B., Delmas, M. (2011). *Process for the Separation of Lignins and Sugars from an Extraction Liquor*. Patent WO2011/154293.

30. Wildschut, J., Smit, A.T., Reith, J.H., Huijgen, W.J.J. (2013). Ethanol-based organosolv fractionation of wheat straw for the production of lignin and enzymatically digestible cellulose, *Bioresource Technology*, **135**, 58–66.

31. Snelders, J., Dornez, E., Benjelloun-Mlayah, B., Huijgen, W., de Wild P., and Courtin C.M. (2014). Comparison CIMV and ECN organosolv processes, Poster presentation at conference Tomorrow's Biorefineries in Europe, February 11-12, 2014, Brussels, Belgium.

32. Snelders, J., Dornez, E., Benjelloun-Mlayah, B., Huijgen, W.J.J., De Wild, P.J., Gosselink, R.J.A, Gerritsma, J., Courtin, C.M. (2014). Biorefining of wheat straw using an acetic and formic acid based organosolv fractionation process, *Bioresource Technology*, **156**, 275–282.

33. O'Donohue, M.J. (2014). *Insight into BIOCORE results.* Presentation available through https://colloque6.inra.fr/eubiorefineryprojectsfinalconf/

34. Van der Linden, R., Huijgen, W.J.J., Reith, J.H. (2012). *Ethanol-Based Organosolv Biorefineries: Feedstock-Flexibility, Economic Evaluation.* Presented at Nordic Wood Biorefinery Conference, Helsinki, Finland.

35. Chen, H., Liu, L. (2007). Unpolluted fractionation of wheat straw by steam explosion and ethanol extraction, *Bioresource Technology*, **98**, 666–676.

36. Brosse, N., El Hage, R., Sannigrahi, P., Ragauskas, A. (2009). Dilute sulphuric acid and ethanol organosolv pretreatment of Miscanthus x Giganteus, *Cellulose Chemistry and Technology*, **44**, 71–78.

37. El Hage, R., Chrusciel, L., Desharnais, L., Brosse, N. (2010). Effect of autohydrolysis of Miscanthusx giganteus on lignin structure and organosolv delignification, *Bioresource Technology*, **101**, 9321–9329.

38. Romaní, A., Garrote, G., López, F., Parajó, J.C. (2011). Eucalyptus globulus wood fractionation by autohydrolysis and organosolv delignification, *Bioresource Technology*, **102**, 5896–5904.

39. Ruiz, H.A., Ruzene, D.S., Silva, D.P., Macieira da Silva, F.F., Vicente, A.A., Teixeira, J.A. (2011). Development and characterization of an environmentally friendly process sequence (autohydrolysis and organosolv) for wheat straw delignification, *Applied Biochemistry and Biotechnology*, **164**, 629–641.

40. Liu, Z., Fatehi, P., Jahan, M.S., Ni, Y. (2011). Separation of lignocellulosic materials by combined processes of pre-hydrolysis and ethanol extraction, *Bioresource Technology*, **102**, 1264–1269.

41. Huijgen, W.J.J., Reith, J.H., Den Uil, H. (2010). Pretreatment and fractionation of wheat straw by an acetone-based organosolv process, *Industrial and Engineering Chemistry Research*, **49**(20), 10132–10140.

42. Mohan, D., Pittman, Jr., C.U., Steele, P.H. (2006). Pyrolysis of wood / biomass for bio-oil: a critical review, *Energy Fuels*, **20**, 848–889.

43. Toor, S.S., Rosendahl, L., Rudolf, A. (2011). Hydrothermal liquefaction of biomass: a review of subcritical water technologies, *Energy*, **36**, 2328.

44. Wang, H., Wang, Y. (2013). Biomass to bio-oil via liquefaction, in *Biomass Processing, Conversion and Biorefinery*, 153–166.

45. Yuan, X., Wang, J., Zeng, G., Huang, H., Pei, X., Liu, Z. (2011). Comparative studies of thermochemical liquefaction characteristics of microalgae using different organic solvents, *Energy*, **36**, 6406.

46. Grisel, R.J.H, Van der Waal, J.K., De Jong, E., Huijgen, W.J.J. (2014). Acid catalysed alcoholysis of lignocellulosic biomass: towards second generation furan-derivatives, *Catalysis Today*, **223**, 3–10.

47. Eerhart, A.J.J.E., Huijgen, W.J.J., Grisel, R.J.H, Van der Waal, J.C., De Jong, E., De Sousa Dias, A., Faaij, A.P.C., Patel, M.K. (2014). Fuels and plastics from lignocellulosic biomass via the furan pathway; a technical analysis, *RSC Advances*, **4**(7), 3536–3549.

48. De Wild, P.J., Van der Laan, R.R., Kloekhorst, A., Heeres, H.J. (2008). Lignin valorisation for chemicals and (transportation) fuels via (catalytic) pyrolysis and hydrodeoxygenation, *Environmental Progress and Sustainable Energy*, **28**(3), 461–469.

49. Kleinert, M., Barth, T. (2008). Towards a lignocellulosic biorefinery: Direct one-step conversion of lignin to hydrogen-enriched bio-fuel, *Energy Fuels*, **22**, 1371–1379.

50. Azadi, P., Carrasquillo-Flores, R., Pagán-Torres, Y.J., Gürbüz, E.I., Farnood, R., Dumesic, J.A. (2012). Catalytic conversion of biomass using solvents derived from lignin, *Green Chemistry*, **14**, 1573.

51. Kumar, S., Lange, J.-P., Van Rossum, G., Kersten, S.R.A. (2014). Liquefaction of lignocellulose: Process parameter study to minimize heavy ends, *Industrial and Engineering Chemistry Research*, **53**, 11668–11676.

52. Goheen, D.W. (1966). Hydrogenation of lignin by the Noguchi process, *Advances in Chemistry*, **59**, Chap. 14, 205–225.

53. Kakemoto, G., Sagara, H., Suzuki, N., Kachi, S. (1990). *Method of Manufacturing Phenols from Lignin*. Patent application US4900873.

54. Okuda, K., Man, X., Umetsu, M., Takami, S., Adschiri, T. (2004). Efficient conversion of lignin into single chemical species by solvothermal reaction in water-p-cresol solvent, *Journal of Physics: Condensed Matter*, **16**, 1325–1330.

55. Vancov, T., Alston, A.-S., Brown, T., McIntosh, S. (2012). Use of ionic liquids in converting lignocellulosic material to biofuels, *Renewable Energy*, **45**, 1–6.

56. Brandt, A., Gravik, J, Hallett, J.P., Welton, T. (2013). Deconstruction of lignocellulosic biomass with ionic liquids, *Green Chemistry*, **15**, 550–583.

57. Choi, Y.H., Van Spronsen, J., Dai, Y., Verberne, M., Hollmann, F., Arends, I.W.C.E., Witkamp, G.J., Verpoorte, R. (2011). Are natural deep eutectic solvents the missing link in understanding cellular metabolism and physiology?, *Plant Physiology*, **156**(4), 1701–1705.

58. Francisco, M., Van den Bruinhorst, A., Kroon, M.C. (2012). New natural and renewable low transition temperature mixtures (LTTMs): screening as solvents for lignocellulosic biomass processing, *Green Chemistry*, **14**, 2153–2157.

About the Authors

Paul de Wild works as a research scientist at the Energy  Research Centre (ECN), the Netherlands, on thermochemical conversion of biomass within the biorefinery approach. Paul obtained a PhD on the topic of biomass pyrolysis for chemicals from the University of Groningen, the Netherlands, in 2011.

Wouter Huijgen works as a research scientist and project manager  within the biomass and energy efficiency unit at the ECN. His main current research interest is the development of fractionation technologies for the biorefining of both lignocellulosic biomass and seaweeds. In addition, he works on thermochemical conversion of cellulose and lignin as well as on analytics for lignocellulose and seaweeds.

Wouter graduated (Msc) in chemical engineering (specializations: environmental engineering and process technology) from the University of Twente (2001). After his studies, he worked as a PhD student at the ECN. In 2007, Wouter obtained a PhD in environmental sciences from Wageningen University on "CO_2 Sequestration by Mineral Carbonation" in collaboration with Delft University of Technology.

Chapter 10

Synfuels via Biomass Gasification

Nicolaus Dahmen and Jörg Sauer

Institute for Catalysis Research and Technology (IKFT),
Karlsruhe Institute of Technology (KIT), Hermann-von-Helmholtz-Platz 1,
76344 Eggenstein-Leopoldshafen, Germany
nicolaus.dahmen@kit.edu

10.1 Introduction

Synthetic fuels, in short synfuels, stand for chemical energy carriers produced from synthesis gas (syngas) as a mixture of carbon monoxide (CO) and hydrogen (H_2). The so-called biomass-to-liquid (BtL) technologies allow for the use of almost every type of biomass, organic residues, or wastes. The technologies are similar to the already established gas-to-liquid (GtL) or coal-to-liquid (CtL) processes and proceed via gasification of suitable biomass-derived materials to syngas. In Chapter 2 the main types of biomass gasification technologies and reactors are described. Gasification in this context means the decomposition of organic material to the smallest building blocks synthetic chemistry may think of: CO and H_2, which can be reacted to an amazing variety of products and fuels, for example, methane, hydrogen, methanol, ethanol, dimethylether (DME), olefins, gasoline, and diesel. Several catalytic processes

Biomass Power for the World: Transformations to Effective Use
Edited by Wim van Swaaij, Sascha Kersten, and Wolfgang Palz
Copyright © 2015 Pan Stanford Publishing Pte. Ltd.
ISBN 978-981-4613-88-0 (Hardcover), 978-981-4669-24-5 (Paperback), 978-981-4613-89-7 (eBook)
www.panstanford.com

are involved, which are compiled in Chapter 3 in more detail, for example, the direct synthesis of methanol or ethanol from syngas, the methanol-based processes methanol to olefins (MtO), and methanol to gasoline (MtG), the production of fuels by oligomerization of olefins (conversion of olefins to distillate, COD), and the synthesis of a wide range of hydrocarbons by Fischer–Tropsch (FT) processes.

The first experiments on the hydration of CO with H_2 have been carried out by Paul Sabatier already in 1902 mainly yielding methane. While BASF in the 1910th favored methanol production, Franz Fischer and Hans Tropsch invented "their" hydrocarbon synthesis in 1925 as a better alternative to the direct hydrogenation of coal. The first trade name Kogasin nicely describe the product sequence from char (**Ko**ks) converted to syn**gas** and further to gasoline (Benz**in**). According to Fig. 10.1 this is the principle route still today but with different feed materials and a much broader product range.

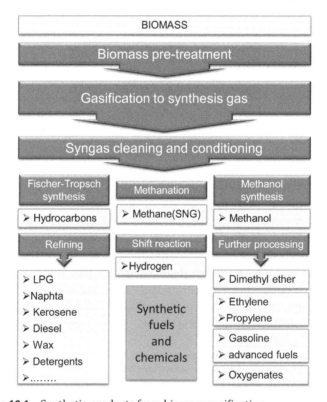

Figure 10.1   Synthetic products from biomass gasification.

10.2   Syngas Generation

As a result of the increasing interest in syngas based products, many gasification methods, reactors, and plant configurations are available or under development. While the chemical syntheses are well established and commercially applied using fossil feedstock, the feedstock-specific front-end processes for conversion of biomass into synthesis gas are still a major point of development. This in particular, if not wood but other types of lignocellulosic biomass with its higher ash content is used. Hydrogen, carbon monoxide, carbon dioxide, and methane are the main products of gasification in the syngas mixture, containing up to more than 70% of the energy of the original feedstock. Gasification temperatures are typically above 700°C, at which the feedstock is converted usually by means of a gasification agent like steam and oxygen or air. Apart from gasification of solid fuels, the conversion of liquid and gaseous fuel is usually referred to as reforming. In general, solid feed materials can be coal, lignite, peat, biomass, and organic waste; liquid fuels are fractions or residues from crude oil refining, pyrolysis tars and oils, gaseous fuels are natural gas or gaseous by-products with a high enough calorific value. That of biomass varies between ca. 15 MJ/kg for air dry wood to ca. 18 MJ/kg for wood pellets.

The elemental composition of the abundant lignocellulosic biomass feedstocks like wood or straw vary between 35–53 wt.% for carbon, 5–6 wt.% for hydrogen, 32–43 wt.% for oxygen, 0.1–1.4 wt.% for nitrogen, 0.01–1.00 wt.% for chlorine, and 0.02–0.5 wt.% for sulfur plus some salts and minerals (the so-called ash content) and moisture. Typically, an H_2/CO ratio in the order of 1:1 is obtained in the syngas which cannot directly be used for all syntheses. By the water gas shift reaction as a separate process step the ratio has to be adjusted to the requirements of the synthesis applied. Major sources of biomass are agricultural and forestry products and residues, including purpose grown bioenergy crops, residues from land cultivation, industrial waste, and to a certain extent municipal solid waste. Regarding this multitude of potential materials, feed-tolerant and flexible pretreatment technologies should be applied to achieve high throughput capacities improving economic viability of the plants.

However, the syngas produced by gasification contains impurities, depending on the type of feedstock and conversion technology applied. Typical are organic tar components and BTX (benzene, toluene, and xylenes), inorganic nitrogen, sulfur, and chlorine containing impurities such as NH_3, HCN, H_2S, COS, and HCl, as well as volatile metals (regarding biomass in particular Na and K), dust, char, and soot. Larger hydrocarbons summarized as tars being produced at low temperature gasification processes, reduce the primary syngas yield, may cause fouling of downstream equipment, coat surfaces, and plug pores in filters and sorbents. Other contaminants are corrosive or poisons to the catalysts in the subsequently following synthesis stages. For raw syngas cleaning, conventional technology is available. Tar constituents and BTX may be removed by either thermal or catalytic cracking, or by scrubbing with an oil-based medium followed by gasifier recycle. The other above mentioned impurities are removed by standard wet gas cleaning technologies. Using absorbing liquids like refrigerated methanol or amines, for example, by the Rectisol or Selexol process, CO_2 and sulfur compounds are removed in separate fractions in a multistage process, resulting in a pure CO_2 product and an H_2S/ COS-enriched Claus gas fraction suitable for sulfur production. In advanced "dry" hot gas cleaning, the residual contaminants are removed by chemical sorbent materials at elevated temperatures up to 800°C. In the case, that gasification occurs already at those pressures required in the subsequently following synthesis, dry pressurized gas cleaning is expected to enable significant energetic benefits.

For biomass gasification, a variety of technologies exist. In view of the large-scale production of high-quality syngas, as required for synfuel production, mainly the types of gasifiers shown in Fig. 10.2 have to be considered here. Direct or indirect gasification using fluid-bed and entrained-flow gasifiers is possible with potential plant capacities of up to several hundred megawatt of thermal fuel input capacity. In indirect gasification the energy required is provided by an external heat source, for example, by hot flue gas, while in direct or autothermal gasification thermal energy is provided by partial internal combustion under addition of an oxidizing agent to the fuel feed. Main differences between the two approaches are compiled in Table 10.1.

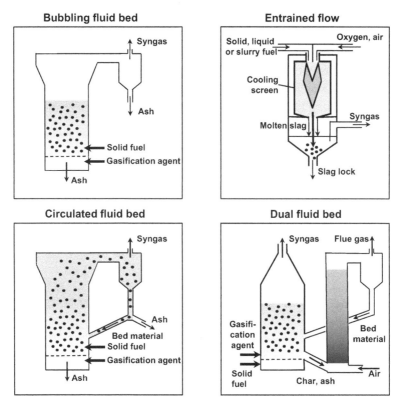

Figure 10.2 Main types of gasification reactors for large-scale biomass conversion.

In fluidized-bed gasifiers, biomass particles are rapidly mixed and heated by hot fluidized-bed materials. Usually air or steam is used as a fluidization and gasification agent. Due to the intense mixing, the gasification reactions occur throughout the whole bed, leading to a uniform temperature distribution. The degree of fluidization can be small (bubbling fluidized bed, BFB) or high (circulating fluidized bed, CFB). BFB gasifiers have a well-defined interface between the reaction zone of the fluidized bed and the above freeboard. They are well known and commonly used because of their robust properties but show relatively high tar formation of the order of 1–2 wt.%. In a CFB gasifier, there is no distinct interface between the fluidized sand bed and the freeboard; the entrained sand and char are separated by a cyclone and are recycled back to the gasifier. The carbon conversion is considerably better than in BFB gasifiers. In addition, CFB gasifiers can be operated at elevated pressures, being advantageous and more economic in regard to hydrogen, liquid fuel, and chemical production. Carbon conversion usually is found to be between 85%–95%. The feed material size is in the order of 10^{-3} m, while in a BFB gasifier pellets and small wood chips may be used.

Table 10.1 Comparison of direct (autothermal) and indirect (allothermal) gasification

	Indirect gasification	Direct gasification
Scheme		
Gasification agent	Steam, CO_2	Air, oxygen (air separation needed)
Reaction temperature	800°C–1,000°C	1,000°C–1,600°C
Features	Longer residence times Larger feed particle size Ash management Higher methane concentration in syngas	Short residence times of seconds Small feed particles Liquid slag management
Issues	Tar formation, decomposition required	CO_2 formation, removal needed in the case of air as oxidation agent: high N_2 content, syngas dilution

To not poison the sensitive synthesis catalysts, the syngas must be of high purity and free of dust and tar. Unlike most fixed-bed and fluidized-bed gasifiers, entrained-flow gasifiers are able to generate a gas practically free of tar with only little methane at the high gasification temperatures above of 1200°C. At residence times of a few seconds carbon conversion of 95%–99% and even above can be achieved. An entrained flow gasifier can be slurry- or particle-fed, usually blown by oxygen or steam as the gasification agent. Any feed material which can be pumped and sprayed pneumatically is suitable. Biomass powders with a particle size in the order of 10^{-4} m can be fed to the gasifier by a so-called high-density flow feed system. However, this requires periodic operation of a sophisticated system of locks, which becomes more complicated as the pressure rises.

To use also biomass and biogenic waste with higher ash and alkali contents, the gasifier may be operated at slagging conditions: depending on the alkali content ash softening or melting already occur below 1000°C. The liquid slag can be handled, for example, by a reactor equipped with a cooling screen or special ceramic refractory lining, allowing the formation of a liquid slag drained out of the reaction volume.

10.2.1 Pressurized vs. Atmospheric Gasification

Soon after the first application of gasifiers to coal conversion operated at atmospheric pressure, large-scale operation of pressurized gasifiers at pressures between 25 to 40 bar became state of the art, some gasifiers have been designed for pressures up to 100 bar. The advantages motivating the use of high pressure were:

- Increase in the reaction rate;
- High specific throughput, high-throughput capacities;
- Increased methane yield at low-temperature operation (for synthetic natural gas [SNG] production);
- Reduction of the gas volume to be treated, for example, in gas cleaning;
- Saving of the work of compression for the subsequent use of the gas produced.

Fuel supply under elevated pressures is effected for solids by lock hopper systems or pneumatic devices and for liquids and slurries by pumps and screw feeders.

Due to economy-of-scale reasons syngas production and its further processing to synfuels demand for large-scale facilities. For comparison, a crude oil refinery has an input capacity in the order of 10 Mio·t/a. On the other hand, biomass usually exhibits low volumetric energy densities and has to be harvested and collected from wide areas from agriculture or forestry and does not constitute a uniform feedstock. Chemical composition and the content of minerals and other inorganic material, usually denoted as the ash content, may vary significantly. Therefore, pretreatment to provide a more homogeneous feed material of appropriate energy density may be useful. Such a pretreatment may be carried out in regionally distributed plants the products of which are transported to larger conversion facilities for syngas and synfuel production. Currently, process combinations with fast pyrolysis and with torrefaction as preparation steps are under development, the characteristic features of which are compiled in Table 10.2.

Table 10.2   Fast pyrolysis and torrefaction for biomass pretreatment

	Fast pyrolysis	Torrefaction
Scheme		
Reactor type examples	Screw reactor, rotating cone, fluid bed, etc.	Rotating drum, screw reactor, Herreshoff oven, etc.
Reaction temperature	450°C–550°C	250°C–350°C
Features	Short residence times of seconds, provided by contact with a heat carrier medium, usually sand Fine feed particles (<5 mm) Typical product yield in wt.% for wood: 20% gas, 20% char, 60% liquid condensates, including ca. 15% of reaction water Gas in optimum case sufficient to provide the process heat	Residence times of 30 minutes Lager feed particles, pellets, chips Typical product yield in for wood: 20 wt.% of gas (9 e%*), 80% of torrefied material (90 e%) with a heating value of ca. 22 MJ/kg Gas in optimum case sufficient to provide the process heat
Issues	More complex technology	Simple technology, limited scale-up for some reactor types

*Percentage of energy related to the energy initially contained in the dry feedstock.

10.3 Syngas-Based Products

10.3.1 Methanol

Methanol is produced from syngas at temperatures between 220°C and 275°C at pressures ranging from 50 to 100 bar using $Cu/ZnO/Al_2O_3$ catalyst systems. The underlying chemistry is summarized in a simplified way in equation (Eq. 10.1) in Table 10.3. Some major methanol technology suppliers are ICI, Lurgi, and Mitsubishi. Blends with petrol containing up to 20 vol.% of methanol can be used to power combustion engines without much modification. In the European community, only 3 vol.% are allowed for gasoline blends mainly for its toxicity. Another option is the use of methanol in fuel cells, either directly in direct methanol fuel cells or indirectly as hydrogen source after reforming.

10.3.2 Ethanol and Higher Alcohols from Syngas

Direct synthesis of ethanol from syngas is intensely investigated. Another approach in this context is the homologization of methanol, that is, the reaction of methanol with syngas to yield ethanol. Higher alcohols can also be formed. The reactions are summarized in Eq. 10.2 and Eq. 10.3. However, ethanol selectivities are still moderate, and there is lack of highly efficient catalysts for these reactions.

10.3.3 Dimethylether

DME can be obtained from direct synthesis according to Eq. 10.4 or via dehydration of methanol (Eq. 10.5). From a mechanistic point of view direct synthesis proceeds also via methanol formation and subsequent release of water but without intermediate isolation of methanol. The process can also be designed to yield both methanol and DME. Established methanol catalysts are employed for methanol formation and typical dehydration catalysts are solid-acid catalysts, among others; alumina; silica-, phosphorus-, or boron-modified alumina; or HZSM-5 zeolithes. Since DME properties are similar to those of liquefied petroleum gas (LPG), it can be used in typical LPG applications, for example, power generation, domestic fuels, or automotive fuels. If DME is employed as an admixture, LPG properties are not significantly affected up to a DME content of around 20%. Compared to LPG, the cetane number is much higher, and DME is, in principle, a suitable, clean fuel for diesel engines. Due to missing carbon–carbon bonds, the combustion of DME leads to practically soot free product gases.

10.3.4 Gasoline via the MtG Process

Conversion of methanol to gasoline in the so-called MtG process is accomplished by zeolitic ZSM-5 catalysts via either fluidized- or fixed-bed technologies. The former was demonstrated by Mobil, Union Rheinische Braunkohlen Kraftstoff AG, and Uhde in a pilot plant at UK Wesseling, and the latter was operated successfully by Methanex in New Zealand licensed by ExxonMobil. The Mobil process yields around 38 wt.% of gasoline, 4 wt.% of LPG, approximately 58 wt.% of water, and a small amount of fuel gas. The underlying chemistry is complex and the multistage process is initiated by the formation of DME through dehydration of methanol (Eq. 10.5). The following chain growth and cyclization reactions proceed via further release of water and can be described, in a very simplified manner, by Eqs. 10.6 and 10.7. These reactions involve not only reactions of DME with itself Eq. 10.7 but also reactions of DME with methanol Eq. 10.6. The resulting hydrocarbon product mixture is free of sulfur as well as nitrogen and exhibits low benzene content. Further advantages of the process are minimal CO_2 emission and high energy efficiency. However, direct production of longer chain alkanes for diesel and jet fuel using typical MtG catalysts is not possible today.

The MtG process can be combined with MtO technology as realized in the so-called Mobil olefin to gasoline/distillate (MOGD) process. Employing this process, olefins are synthesized in the first step followed by olefin oligomerization to gasoline or diesel using a ZSM-5 catalyst. Hence, this route affords, unlike the MtG process, also the production of synthetic diesel or jet fuel. A similar process is the Topsøe integrated gasoline synthesis (TIGAS) process developed by Haldor Topsøe A/S and demonstrated at a pilot plant in Houston. It is, in principal, an improved MtG process combining methanol, DME, and gasoline production in a single synthesis loop, thus circumventing intermediate production and storage of methanol. Different syngas compositions can be employed since methanol/ DME synthesis is flexible in terms of syngas specifications.

10.3.5 Fischer–Tropsch Synthesis

By the FT process (Eq. 10.8), syngas is directly converted into hydrocarbons. It is well established and currently operated, for example, by Sasol (Sasol Advanced Synthol, SAS process in Secunda, South Africa) and Shell (middle distillate synthesis, MDS process in Bintulu, Malaysia, and in the PEARL project in Qatar) by using coal and natural gas, respectively. FT syntheses can be conducted in slurry phase, fixed-, and fluidized-bed reactors, mainly depending on the catalysts and reaction temperatures. With respect to reaction temperatures one can distinguish between low- and high-temperature FT technologies (LT-FT and HT-FT). The former operate at temperatures around 220°C and yield primarily long-chain hydrocarbons such as paraffins and waxes, whereas the latter, operating at temperatures around 340°C, produce mainly naphtha and olefins. Long-chain products obtained via LT-FT processes are hydrocracked in the next step to yield diesel in very high quality. Today, LT-FT syntheses combined with slurry-phase reactors are, at present, the preferred option.

Chain growth proceeds at the surface of cobalt- and/or iron-based catalyst systems, for example, Fe-, Fe/Co-, Fe/Co-Spinel-, Co/Mn-Spinel-, or Cu-doped Co catalysts. Iron catalysts are cheaper than cobalt catalysts and more flexible as well as resistant with respect to syngas composition and quality. But Cobalt catalysts show best performances at a H_2:CO ratio of 2:1 and feature longer lifetimes as well as higher selectivities than iron catalysts.

10.3.6 Hydrogen

Hydrogen can be produced via gasification of biomass, provided that the gasification procedure is optimized to yield a high hydrogen content in the resulting gas mixture. A downstream water gas shift reaction (Eq. 10.9) reduces carbon monoxide contents and thus increases hydrogen yield. Further options for hydrogen production from biomass are supercritical water gasification and dark fermentation. Crude biohydrogen has to be cleaned, compressed or liquefied, or stored in suitable storage media and can be used as fuel for specialized combustion engines or fuel cells.

10.3.7 Methane

Methane can also be obtained by biomass gasification and allows for the use of a wider and complementary variety of biomass types compared to biomass anaerobic digestion. The so-called bio-SNG (substitute natural gas) can be used after cleaning and compression or liquefaction, as fuel for modified spark ignition engines and exhibits high octane numbers. Recently, SNG production is discussed and developed in the context with energy storage in fluctuating renewable energy production systems, for example, in the power-to-gas (PtG) process. From surplus electrical power hydrogen is generated from water by electrolysis and subsequently used for CO_2 hydrogenation to produce methane according to reverse Eq. 10.9 and Eq. 10.10.

Table 10.3 Syngas reactions to synfuels

Reaction to . . .	Formal reaction equation	
Methanol	$CO + 2H_2 \Leftrightarrow CH_3OH$	(10.1)
Ethanol	$nCO + 2nH_2 \Rightarrow C_2H_5OH + H_2O$	(10.2)
	$CH_3OH + nCO + 2nH_2 \Leftrightarrow C_{n+1}H_{2n+3}OH + nH_2O$	(10.3)
Dimethylether	$3CO + 3H_2 \Leftrightarrow CH_3OCH_3 + CO_2$	(10.4)
Gasoline	$2CH_3OH \Leftrightarrow CH_3OCH_3 + H_2O$	(10.5)
	$nCH_3OCH_3 + nCH_3OH \Leftrightarrow "(CH_2)_{3n}" + 2nH_2O$	(10.6)
	$nCH_3OCH_3 \Leftrightarrow "(CH_2)_{2n}" + nH_2O$	(10.7)
Hydrocarbons (Fischer–Tropsch)	$nCO + (2n+1) H_2 \Leftrightarrow (CH_2)_n + nH_2O$	(10.8)
Hydrogen	$CO + H_2O \Leftrightarrow CO_2 + H_2$	(10.9)
Methane	$CO + 3H_2 \Leftrightarrow CH_4 + H_2O$	(10.10)

10.4 Recent Process Developments

The longest tradition in the production of synfuels and chemicals can be found in coal conversion technologies. However, coal was not cost competitive for long times compared to crude oil refining. Recently, new production facilities have been erected, for example, in China with large production capacities to produce methanol, propene, and SNG from coal. Also, natural gas is utilized for Fischer–Tropsch synthesis, for example, in oil- and gas-producing countries like Qatar, Nigeria, and Malaysia. Increasing interest is also devoted to biomass utilization in the context with renewable energy and climate change, say, CO_2 emission reduction targets. However, the value of a biofuel has to be scored in terms of its CO_2 reduction potential. In the EU, the greenhouse gas emission saving from the use of biofuels and bioliquids shall be at least 50% and 60% in 2018 for those installations in which production started on or after January 1, 2017.

Biomass is the only renewable carbon resource and should preferentially be used for the production of carbon containing products, when heat and electrical power can be efficiently be provided by other renewable energy sources. However, the annual upgrowth and sustainably available amount of biomass are limited and are not sufficient to cover the huge demand for all transportation fuels. Yet the much smaller carbon demand for organic chemistry and hydrocarbon aviation fuels could be supplied via biomass. Therefore, syngas is a valuable switch between energy and chemistry as well as between fossil feedstocks (coal and natural gas) and biomass.

In Table 10.4, a selection of pilot and demo projects is compiled, giving evidence to the range of technologies, conversion capacities, and state of development. Due to the importance of Scandinavians wood industries the main efforts can be observed in this region. Projects **A–C** are pilot plants in operation today, pilot plants **D** and **E** have been commissioned recently. Projects **F–H** are first of its kind demo plants planned or in design phase.

As an example, the bioliq® process, currently under development at Karlsruhe Institute of Technology (KIT) is described in more detail. It has been designed to overcome the problems in providing a large fuel production facility with sufficient amounts of residual biomass from agriculture and forestry like straw or wood residues. Therefore, the bioliq® concept firstly aims at pretreated biomass in a number of regionally distributed fast pyrolysis plants. An energy-dense intermediate, biosyncrude, is produced by mixing the pyrolysis char and liquid condensates, which is then collected at an industrial facility of reasonable, industrial size for further conversion to syngas and synfuels. Since syngas reactions principally demand for higher pressures from 30 to 80 bar (in case of those from Table 10.3), gasification in the bioliq® process is already conducted at the required pressure by entrained flow gasification. Liquids or slurries can more easily be fed to a gasifier and do not require an additional fluidization agent as in the case of feeding solid particles. As gasification agent, oxygen is used, which can be assisted by steam.

Table 10.4 Current synfuel pilot or demo projects

No.	Project	Feedstock	Pretreatment + gasification features	Synthesis, products
A	BioDME, S	Black liquor	Chemrec EF, 3 MW, 30 bar	DME, methanol
B	Güssing, A	Wood	Repotec dual fluid bed, 8 MW, atm.	CHP, SNG (1 MW) and FT laboratory plant
C	NSE Biofuels Varkaus, S	Forest biomass	CFB, 12 MW (5 MW for synfuel application)	Heat for a lime kiln, FT products
D	bioliq®/KIT	Lignocellulose	Fast pyrolysis 2 MW+ Lurgi EF, 5 MW, 80 bar	DME, gasoline
E	GoBiGas, S	Forest biomass	metso/repotec Dual bed, 20 MW	Biomethane
F	BioTfuel, F	Forest biomass	Torrefaction + Uhde Prenflow EF, 15 MW	FT products
G	Värmlandsmetanol, S	Forest biomass	Uhde-HTW gasifier, 111 MW	Methanol
H	ForestBtL, F	Wood, tall oil	Linde Carbo-V EF, 2 × 160 MW	FT products

On-site at the KIT, a 2–5 MW$_{th}$ pilot plant has been erected for process demonstration and as a platform for further research and development in regard to process improvement and optimization.

The separated process steps were commissioned in 2013; joint operation has been achieved in 2014 for the first time. Stage 1 of the pilot plant consists of a fast pyrolysis plant with a capacity of 500 kg biomass per hour (2 MW_{th}) in left building in Fig. 10.3, where biomass is contacted with hot sand at 500°C in a twin-screw reactor. While sand is recycled and reheated, the char formed after pyrolysis is separated from the hot pyrolysis vapors, which then are condensed to form liquids by two cooling stages. From wheat straw, two condensates are produced, which are converted to biosyncrude by adding the pyrolysis char in a colloidal mixer. Then, the biosyncrude is stored in the tank (middle of Fig. 10.3). For gasification, a 5 MW_{th} oxygen-blown high-pressure entrained flow gasifier has been put into operation. The slagging reactor is equipped with an internal cooling screen, particularly suited for the conversion of ash-rich feeds and fast start-up and shutdown procedures. The minerals sloughing down into the quench water reservoir are removed as solid slag. The tar-free, low-methane raw syngas passes through a water quench, which will later be extended to a water injection quench, by which the gas release temperature can be adjusted to the optimum gas cleaning temperature. Gasification pressures up to 80 bar can be applied, adjusted to the pressure required by the subsequent chemical synthesis processes. In case of the gasoline synthesis realized at the bioliq® pilot plant, 65 bar is required, which is applied on a separated syngas flow of 700 Nm^3/h (2 MW_{th}). A high-pressure, high-temperature process developed at the KIT for gas cleaning is used. Temperatures between 500°C and 800°C can be applied and will be adjusted to optimum operation temperatures, where energy savings can be expected compared to the conventional low temperature gas-cleaning processes. Fines are separated by a particle filter equipped with ceramic filter elements, sour gases (HCl, H_2S, COS) are retained by fixed-bed sorption. In a catalytic reactor, ammonia, HCN and organic trace compounds are decomposed. Another sorption bed is added as a safeguard. Prior to synthesis, carbon dioxide and water are separated from the purified gas. DME is produced in a single stage from syngas by simultaneous methanol formation, methanol dehydration, and a water gas shift reaction using a mixture of commercial catalysts for methanol production and dehydration. The formation of DME is thermodynamically favored at $CO:H_2$ ratios around 1:1, which are typically expected from biomass gasification. However, conversion is not complete and syngas recycling through a CO_2 separation unit can be done after the gasoline synthesis, which is subsequently following the DME reactor. The raw products of the fuel synthesis are separated from the nonconverted syngas and distilled to receive the main product, high octane gasoline, along with some light and little heavy hydrocarbon compounds.

Figure 10.3 Image of the bioliq® pilot plant at the Karlsruhe Institute of Technology (KIT). See also Color Insert.

10.5 Conclusions

A variety of pilot and demonstration projects are on the way to verify the technical and economic feasibility of biomass-based thermochemical syngas processes. Benefit may be taken from experiences of already established coal and gas conversion technologies. However, specific differences exist in biomass conversion, demanding for improved solutions. Issues are to be addressed in regard to the broad range of potential feedstocks differing in their composition, consistency, and availability. Many issues are related to the feed pretreatment, integration of gasifiers, gas cleaning, and conditioning according to the requirements of the chemical syntheses applied. It is nearly impossible to separate the fuels and chemicals business. Considering a BtL production complex, not only the most valuable mix of fuels and chemical has to be produced, but also coproduction of heat and power has to be included to achieve sufficient energetic efficiencies. Therefore, synthetic fuels production demands for cogeneration (polygeneration) of fuels, chemicals, heat, and power within an integrated biorefinery.

About the Authors

Nicolaus Dahmen studied chemistry at the University of Bochum, Germany, finishing his PhD in high-pressure thermodynamics in 1992. He started his professional work on the application of high pressure to chemical reactions and separation processes as a group leader and, since 2000, as the head of the High Pressure Process Technology division at the Research Centre Karlsruhe, which in 2010 merged into the Karlsruhe Institute of Technology (KIT) together with the University of Karlsruhe, Germany. In 2005, he became the project manager of the bioliq project, in which a large-scale pilot plant was installed at the KIT for synthetic fuels and chemical production from biomass. Shortly after, he also took over the Thermochemical Biomass Refining division at the Institute for Catalysis Research and Technology (IKFT), Germany, and after his habilitation on fundamentals for process developments with supercritical fluids, he became a lecturer on physical and technical chemistry at the University of Heidelberg in 2010. After commissioning the pilot plant in 2014, he now is responsible for the bioliq R&D program. At the end of 2014 he was announced as a professor at the KIT.

Jörg Sauer graduated in chemical engineering from the University of Erlangen-Nürnberg and received his PhD from the University of Karlsruhe in the group of Prof. Dr. Gerhard Emig. He started his industrial career at Degussa AG, Germany, and later at Evonik Industries AG, Germany, where he spent 18 years in different positions in research and development, production, process technology, and engineering and in different locations in Germany and the United States. His latest responsibility was as head of the department of chemical reaction technology at Evonik's Marl site in Germany. Since 2012 he is head of the Institute of Catalysis Research and Technology and speaker of the large-scale project Bioliq at Karlsruhe Institute of Technology.

Chapter 11

Energy from Biomass via Gasification in Güssing

Hermann Hofbauer

Vienna University of Technology, Institute of Chemical Engineering,
A-1060 Vienna, Getreidemarkt 9/166, Vienna, Austria
hermann.hofbauer@tuwien.ac.at

11.1   History and Background

Güssing, a small town with about 4000 inhabitants, is located about 200 km south of Vienna near the Hungarian border. For a long time until 1989 this border was called the "iron curtain," and therefore, no industry settled there, which in consequence led to a lack of jobs for the people living there. Many of them migrated to other regions forever or at least over weeks for working. The region was very poor until biomass as a source of energy was discovered.

About 25 years ago the mayor of Güssing and some other visionary people worked out an energy concept for the energy supply for the city of Güssing. The main aim of this energy concept was the substitution of fossil fuels by energy from renewable sources, mainly

Biomass Power for the World: Transformations to Effective Use
Edited by Wim van Swaaij, Sascha Kersten, and Wolfgang Palz
Copyright © 2015 Pan Stanford Publishing Pte. Ltd.
ISBN 978-981-4613-88-0 (Hardcover), 978-981-4669-24-5 (Paperback), 978-981-4613-89-7 (eBook)
www.panstanford.com

raw materials from the region as 40% of the region around Güssing is covered with wood. Therefore, sufficient raw material is available for the energy supply of the whole city (Fig. 11.1). This concept included mainly measures for energy saving as well as energy production by building new and innovative plants based on biomass.

Figure 11.1 Views of the city of Güssing. Left: Region around Güssing. Right: Castle of Güssing.

The first plant which was built was a biodiesel (rapeseed methyl ester [RME]) plant in the year 1990. This biodiesel plant produced RME until about 2005; afterward the production was stopped due to economic reasons. In 1998 a district heating system based on grate combustion using wood chips as fuel was commissioned which was the largest biomass based district heating system in Austria at that time. In the year 2001 a 2 MWel demonstration plant for combined heat and power (CHP) production based on a novel dual fluidized-bed (DFB) gasification system was realized in Güssing. Later another CHP plant was built on the basis of combustion of wood dust, which was available as residue from parquet factories settled in Güssing at that time. Today, these two CHP plants are able to produce the whole electricity which is consumed by the people living in Güssing, and together with the grate combustor, 95% of the city of Güssing and the industries around Güssing are supplied by district heat from biomass as well.

11.2 Short Description of the Güssing Plant

The innovative DFB-technology for CHP generation has been demonstrated in Güssing for the first time worldwide. Biomass is gasified in a DFB reactor. The producer gas is cooled, cleaned, and used in a gas engine. The most important data of the demonstration plant are summarized in Table 11.1. A detailed flow sheet of the whole plant is shown in Fig. 11.2.

Table 11.1   Characteristic data of the demonstration plant

Type of plant	Demonstration plant
Fuel power	8000 kW
Electrical output	2000 kW
Thermal output	4500 kW
Electrical efficiency	25.0 %
Thermal efficiency	56.3 %
Electrical/thermal output	0.44 kW/kW
Total efficiency	81.3 %

An overall view of the CHP plant in Güssing can be seen in Fig. 11.3. This photo was taken in 2003 after several months of full load operation of the plant. Wood is transported to the location of the plant as wood logs. These wood logs, which in terms of the quality cannot used for paper production, are chipped on-site (Fig. 11.4) and filled into the daily hopper, which is equipped with a walking floor at its bottom.

Biomass chips are transported from a daily hopper to a metering bin and fed into the fluidized-bed reactor via a rotary valve system and a screw feeder (Fig. 11.5, left side). The gasifier consists of two fluidized-bed reactors, a gasification reactor, and a combustion reactor. The gasification reactor is fluidized with steam which is generated by waste heat of the process in order to produce a nitrogen free producer gas. The combustion reactor is fluidized with air and delivers the heat for the gasification process via the circulating bed material.

The producer gas is cooled—mainly for steam production—and cleaned by a two-stage cleaning system (Fig. 11.5, right side). A water cooled heat exchanger reduces the temperature from 840°C–860°C to about 160°C–180°C. The first stage of the cleaning system is a

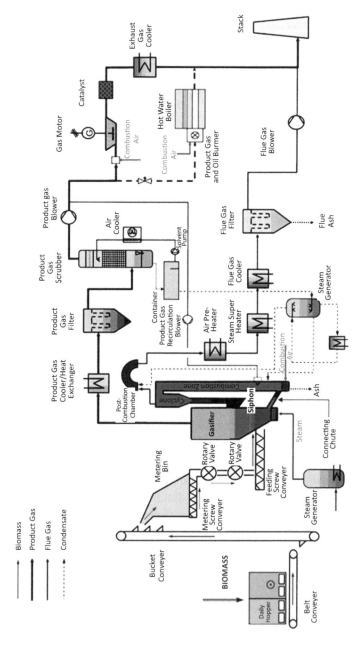

Figure 11.2 Flowsheet of the CHP plant in Güssing.

fabric filter to remove the particles and some of the tar from the producer gas. These particles are returned back into the combustion reactor of the gasifier. In the second stage the gas is liberated from tar by a RME scrubber.

Figure 11.3 Overall front view of the CHP plant in Güssing (castle of Güssing in the background). See also Color Insert.

Figure 11.4 On-site chipping of wood.

Figure 11.5 Views of the CHP plant in Güssing. Left: Side view. Right: Front view.

Spent scrubber liquid saturated with tar and condensate is vaporized and introduced into the combustion reactor of the gasifier. The scrubber is also used to reduce the temperature of the clean producer gas to about 40°C, which is necessary for the gas engine. The clean gas is finally fed into a gas engine to produce electricity and heat. If the gas engine is not in operation the whole amount of producer gas can be burned in boiler to produce heat only. The flue gas of the gas engine can be catalytically oxidized to reduce the CO emissions. The heat of the engine's flue gas is used to produce district heat.

The sensible heat of the flue gas from the combustion reactor is cooled in three stages. The heat is used for preheating the air, superheating the steam, and also delivering heat to the district heating system. A gas filter separates the particles before the flue gas is released via a stack to the environment.

11.3 Development of the Dual Fluid-Bed Gasification Process

11.3.1 Dual Fluid-Bed Gasification Reactor

The basic idea of the gasification process originates from Vienna University of Technology and was published first in 1982 for coal gasification [1]. The fundamental approach of the DFB (former fast internal circulating fluidized bed [FICFB]) gasification technology is to use a fluidized-bed system which physically separates the gasification reactions and the combustion reactions (Fig. 11.6) in order to obtain a largely nitrogen-free product gas and at the same time to avoid an air separation unit for oxygen production. In the gasification part of the system mainly devolatilization of the biomass takes place and only a small part of the charcoal is gasified.

Heat for the gasification reactions is supplied by the circulating bed material from the combustor. The fuel for the combustion is on the one hand remaining charcoal circulating with the bed material, and on the other hand additional fuel can be fed to the combustor in order to control the temperature in the gasification system. Both gas streams—producer gas and flue gas—are kept separated. With this novel system it is possible to produce a nitrogen free producer gas without the need of an expensive and energy-consuming air separation unit.

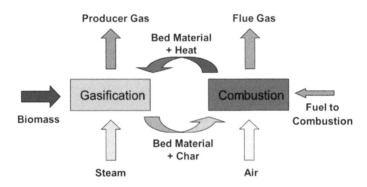

Figure 11.6 Basic principle of the DFB gasification process.

Starting from the original idea, several DFB designs were investigated in cold flow models to find the best solution. Laboratory-scale hot test rigs (10 kWth) and pilot plants (100 kWth) were further important tools to study the gasification process itself and to learn about the effect of important operation parameters. For scale-up purposes to the commercial size again cold flow models were applied to get a secure and optimal design for the large-scale fluidized bed.

Another important development tool is mathematical modeling and simulation. Mathematical modeling and simulation can help to understand the behavior of the gasifier itself and to optimize the performance of the overall process [2]. Several different modeling approaches were used during the development phase of the DFB process. Detailed modeling of the fluid dynamics and the chemical reactions were carried out for the gasification as well as for the combustion part of the DFB. Furthermore, CFD modeling of the gasifier delivered valuable insights of the fluid mechanic behavior [3].

From the fundamental work with cold flow models, mathematical modeling, and the operation of the pilot plants all the necessary information could be obtained to securely design an 8 MWth demonstration plant. Figure 11.7 summarizes all the development tools which were used in the case of scale-up of the DFB steam gasifier.

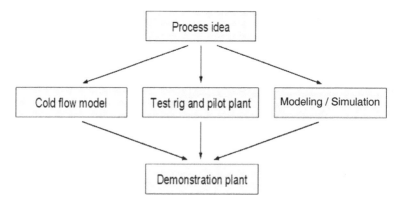

Figure 11.7   Tools for process development.

During the years since 1982 different reactor designs for the DFB gasifier were tested and are shown in Fig. 11.8. In 1993 a 10 kW test rig was built to study the fundamental behavior of an FICFB gasifier. The design at that time was a bubbling fluidized bed with a daft tube in its center. The basic behavior of such a system was intensively studied which was the base for further development.

In 1995 the first 100 kW pilot plant was built and here the combustor was located beside the gasification reactor and not in the center of the gasification reactor. This was due to scale-up reasons as in large-scale plants the combustor has to be reached without demounting too many parts, for example, from the gasification reactor. A rectangular cross section was used for this pilot plant. At that time the name FICFB remained although the design changed to an external circulating fluidized-bed system. Later the system was renamed to DFB gasifier.

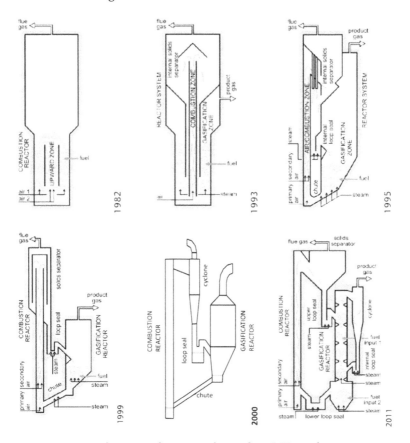

Figure 11.8 Development of a reactor design for a DFB gasifier.

After several years of investigations the next design of the pilot plant with the same capacity of 100 kW was built (1999). Now circular cross sections for the gasifier as well as combustor were used. Furthermore, another design for the particle separator at the exit of the combustor was tested. This type of particle separator was proposed to get a smooth separation without too much attrition for the particles.

For the demonstration plant in Güssing (year 2000) a configuration with circular cross sections but again a different solids particle separator was used. For large-scale plants a cyclone seemed to be a more reliable and well-known particle separator.

Since 2011, a new type of gasifier is under development. The basic idea is shown in Fig. 11.8. More details can be found elsewhere [4].

11.3.2 Gasification Process and Engineering of the Demonstration Plant in Güssing

Table 11.2 shows the main stages of the research and development process for the demonstration plant in Güssing. The project development started in 1995 with the installation of the first 100 kW pilot plant. The main aim of this pilot plant was to gasify different fuels (biomass, sewage sludge, coal, etc.) and to carry out comprehensive operation parameter studies. This pilot plant consisted not only of the gasifier but also of the same gas-cooling and gas-cleaning systems, which were later realized in a larger scale at the demonstration plant in Güssing.

Table 11.2 Development of the DFB gasification process

Stage	Institution	Installation/Tool	Period
1	TU Vienna	10 kW test rig	1993–1995
2	TU Vienna	100 kW pilot plant + gas cleaning	1995–2000
3	TU Vienna	Cold flow model for 8 MW plant	2000–2005
4	Competence Network Renet-Austria	8 MW demonstration plant, Güssing	2000–2007
5	BIOENERGY 2020+	Synthetic biofuel platform, Güssing	2008–2015

From the operation of the pilot plants a lot of design values for the demonstration plant could be achieved. A detailed description of the 100 kW pilot plant and the respective results can be found elsewhere [5, 6].

In 2000 a network of competence (Renet-Austria) was established to support the design, construction, commissioning, and demonstration phases of the CHP plant in Güssing. The members of Renet-Austria were the manufacturers of the plant (AE-Energietechnik, Jenbacher AG), the owner of the plant, and a research organization (TU Vienna). The work of the competence network was financially supported by the government and two federal states of Austria (Burgenland and Niederösterreich). Due to this support from Renet-Austria many questions during the design phase could be clarified by experiments at the 100 kW pilot plant or by mathematical modeling and simulation work [7].

The construction of the demonstration plant started in September 2000 and was finished again in September 2001. This very short time was possible, as no serious problems during this phase occurred. The official opening ceremony took place on September 20, 2001, together with officials from the Austrian government.

11.3.3 Commissioning and Demonstration Phases

The first test runs at the demonstration plant could be carried out in November 2001. First the control system of the plant has to be adjusted. This was necessary because no experience was available for such a type of gasification process. In the first months no constant and stable operation was possible. This unsteady operation led to deposits in the producer gas cooler due to bad gasification conditions. After this first period and with the experience collected during this time a good performance of the gasifier and gas-cleaning system could be obtained.

After more than 1500 hours of successful operation with the gasification and the gas-cleaning system the gas engine was coupled. Till the end of September 2002 more about 3000 hours of operation with the gasifier and gas cleaning and more than 900 hours of operation with the gas engine could be reached. Afterward, a two-year demonstration period was carried out to optimize the operation, to find the right process values for most of the operation parameters, especially to reduce the operation costs.

From 2005 ongoing a constant operation could be obtained (Fig. 11.9); however, there were some problems remaining which could not be solved due to financial reasons. The main two problems were:

- the lack of a biomass dryer as the mean water content of the biomass available is about 35%–40% over the year. This leads to bottlenecks in the combustion reactor and in the product gas scrubber;

• deposits in the air preheater in the flue gas line. This leads to shut down of the plant about every four to six weeks to clean the flue gas line.

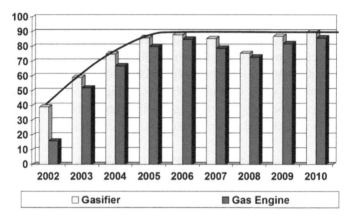

Figure 11.9 Availabilities of the gasifier and gas engine of the Güssing plant.

11.3.4 Operation Results of the DFB Plant in Güssing

At the end of the year 2014 the Güssing DFB gasification demonstration plant reached more than 80,000 full load hours of operation, and this number will be increased in the coming years. A lot of experience was obtained during these 13 years of continuous operation which was used for the design and operation of further industrial plants. Some main findings are discussed next.

Table 11.3 shows typical ranges for main components in the dry product gas of the DFB gasification plant in Güssing using wood chips as fuel. The highest concentration of all components is hydrogen with about 40%, followed by carbon monoxide and carbon dioxide, each about 20%–25%, and methane with about 10%. This composition can be obtained with gasification temperatures of about 830°C–850°C, a steam to carbon ratio of about 1.6 (kg steam + kg water in fuel/kg carbon in fuel), and olivine as bed material.

Table 11.3   Product gas composition (dry gas)

Component	Unit	Conventional	AER-Gas
Hydrogen	vol%	35–45	66–74
Carbon monoxide	vol%	22–25	5–8
Carbon dioxide	vol%	20–23	6–8
Methane	vol%	9–11	7–9
Ethene	vol%	2–3	1.1
H_2/CO	–	1.6–1.8	8–10

Table 11.3 contains also values for the adsorption-enhanced reforming (AER) process. This process can be realized with the same DFB concept using different temperatures and a different bed material. Instead of olivine calcite ($CaCO_3$) is used as bed material. Furthermore, the gasification reactor is operated at about 650°C (instead of 850°C) and the combustion reactor at 950°C. These temperatures can be obtained at low circulation rates of the bed material and—if necessary—by heat extraction from the gasifier.

Under these conditions calcination takes place in the combustor and CO_2 is released into the flue gas (Fig. 11.10). Calcium oxide CaO is then transported to the gasification reactor where carbonation is possible at 650°C and most of the CO_2 is removed from the producer gas. Due to the removal of CO_2 from the product gas additional H_2 is produced and CO is consumed by the water–gas shift reaction (Fig. 11.10). These effects lead to a composition with a high hydrogen content, as shown in Table 11.3. A lot of experimental runs in the AER mode were carried out in the 100 kW pilot plant and also one test in the Güssing demonstration plant. Calcite is an excellent catalyst for tar reforming, which leads to similar tar contents already at a gasification temperature of 650°C where in case of olivine 850°C are necessary. The main problem with calcite is the low resistance against attrition which may lead to a high consumption of bed material.

Cold gas efficiencies (chemical energy in product gas to chemical energy in solid fuel) ranges from 65% to 70% and carbon conversions (solid carbon to gaseous carbon) of the whole plant (gasification and combustion) is about 99.5% as the ash only contains a carbon content lower than 0.5%.

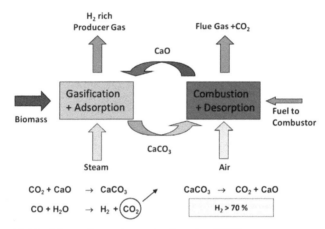

Figure 11.10 Adsorption-enhanced reforming (AER) process.

During conventional operation of the DFB gasification plant olivine has proven as suitable bed material. The attrition of olivine is low and it shows catalytic activity for tar reduction. It turned out that this catalytic activity is increased during operation. It was found that a calcium rich layer is built around each particle which is mainly responsible for this improvement of the catalytic activity. To check this hypothesis tests were carried out in the 100 kW pilot plant (Table 11.2) with fresh (unused) olivine and used olivine from the Güssing plant. The operational parameters at the pilot plant were chosen to be similar to those of the industrial-scale plant in Güssing (Table 11.4).

The gas composition of the two test series in the pilot plant and the industrial-scale plant is shown in Fig. 11.11. While the test run with the used olivine in the pilot plant shows similar results compared to the industrial-scale plant, the results for fresh bed material in the pilot plant show a significantly higher CO content and significantly lower CO_2 and H_2 contents. The values from the industrial-scale plant show a larger deviation compared to the pilot plant due to varying fuel quality (water content) and instability of inorganic flows (e.g., bed material consumption and ash circulation).

Comparison of the results of the test runs at the pilot plant with fresh (unused) and used bed materials reveals an enhancement of the water–gas shift reaction, with used bed material, which leads to higher H_2 and higher CO_2 contents, while the CO content decreases.

Table 11.4 Operational parameters at the industrial-scale plant and the pilot plant

	Unit	Pilot plant, fresh olivine from Güssing	Pilot plant, used olivine from Güssing	Industrial-scale plant in Güssing
Gasification temperature	°C	850	850	850
Combustion temperature	°C	897	872	~930
Fuel input	kW	97	97	8000
Steam to carbon ratio	–	1.8	1.8	1.2–1.6

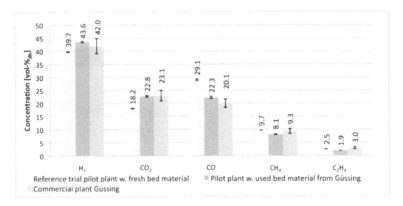

Figure 11.11 Product gas composition of unused and used bed material.

A comparison of gas compositions from the industrial plant in Güssing and the test run in the pilot plant with used bed material from Güssing presented in Fig. 11.11 shows very similar gas compositions of the pilot plant and the industrial-scale plant with the same bed material. Also, the results of tar analyses of the test run with the used olivine are in the range of other published values of the plant in Güssing. As shown in these results, the performance of the pilot plant can be easily transferred to the industrial-scale plant. A good scale-up of the 100 kW pilot plant to industrial-scale plants in terms of gas quality seems to be possible when the long-term interaction of ash and additives with the bed material is considered [8].

The tar content of the product gas in the pilot plant is shown in Fig. 11.12. It indicates that a substantially lower tar content was found with the used olivine from the plant in Güssing independently of the method of analysis. The gas chromatography–mass spectrometry (GCMS)-measured tars decreased by 82%, and the gravimetric tars decreased by 65%.

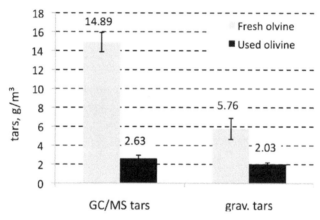

Figure 11.12 Tar contents with fresh, unused, and used olivine in the pilot plant.

11.4 Product Gas Applications

The favorable characteristics of the product gas—low nitrogen content, high hydrogen content, H_2:CO ratio of 1.6–1.8 (Table 11.3)—allow a lot of applications of the product gas. Besides CHP production in the demonstration plant research projects concerning the synthesis of synthetic natural gas (SNG), Fischer–Tropsch (FT) liquids, and mixed alcohols are ongoing since about 2005. Furthermore, hydrogen production from biomass via DFB gasification seems to be another attractive option. Figure 11.13 gives an overview about applications of the producer gas from a steam blown gasifier such as Güssing which have been investigated during the last years.

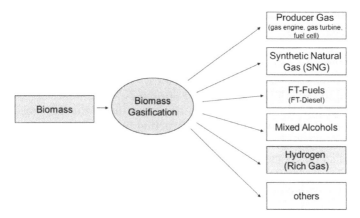

Figure 11.13 Overview about investigated product gas applications.

Figure 11.14 shows a view of several research facilities at the location of the 8 MW DFB gasifier in Güssing. These research facilities are operated from Renet-Austria and later from BIOENERGY 2020+ (see Table 11.2, stages 4 and 5). Generally, all products can be obtained from the synthesis gas as it is the case for syngas from coal or crude oil. All the necessary chemical pathways are well known since many decades. Therefore, in analogy to coal or oil chemistry one can say "green chemistry," as the original material is renewable (e.g., biomass).

Figure 11.14 Total view of the research facilities for product gas (syngas) applications.

11.4.1 Heat and Power Production

As already presented in Chapter 2 the product gas from the DFB gasifier in Güssing is used to produce heat and power. A gas engine with a nominal electrical output of 2 MW is applied for this purpose. Electrical efficiencies are dependent on the water content of the biomass. As no biomass dryer is installed at the Güssing plant the electrical efficiencies ranges from 21 (for water contents of about 40%) to 24% (for water contents around 20%).

The knowledge gained from the operation of the Güssing plant was used for designing and constructing further CHP plants. Table 11.5 (lines 1–4) gives an overview about these industrial installations so far. As can be seen, in addition to gas engines, also organic rankine cycles (ORCs) are applied to increase electrical efficiencies. A plant with a biomass dryer, a DFB gasifier, gas engines, and ORCs can be designed and operated with electrical efficiencies beyond 30%. Heat is normally delivered into a district heating grid, and the amount of heat depends so far (see Table 11.5) largely on the heat demand of connected consumers.

Table 11.5 Industrial installations with DFB gasifiers

Location	Usage/ Product	Fuel/Product MW, MW	Start up	Supplier	Status 2015
Güssing, AT	Gas engine	$8.0_{fuel}/2.0_{el}$	2002	AE&E, Repotec	Operational
Oberwart, AT	Gas engine/ ORC	$8.5_{fuel}/2.8_{el}$	2008	Ortner Anlagenbau	Operational
Villach, AT	Gas engine	$15_{fuel}/4.0_{el}$	2010	Ortner Anlagenbau	On hold
Senden/ Ulm DE	Gas engine/ ORC	$15_{fuel}/5.0_{el}$	2012	SWU/ Repotec	Operational
Göteborg, SE	BioSNG	$32_{fuel}/20_{BioSNG}$	2014	Metso Power/ Repotec	Operational

11.4.2 Bio-SNG Production

In this section a bio-SNG demonstration plant realized in Güssing, AT shall be presented shortly. This demonstration plant was built and operated as part of the EU project "Bio-SNG" in the years 2006 to 2009. A view of the demonstration plant can be seen from Fig. 11.14.

As it can be seen the demonstration plant is directly connected to the Güssing CHP plant. In Fig. 11.15 a simplified flow sheet of the bio-SNG demonstration plant is shown.

About 300 Nm^3/h of product gas is taken after the gas-cleaning section of the CHP plant. This gas stream is cleaned further (e.g., sulfur removal) and fed to the fluidized-bed methanation reactor to carry out the desired reaction.

$$CO + 3H_2 = CH_4 + H_2O$$

As side reaction also the water–gas shift reaction was observed.

$$CO + H_2O = CO_2 + H_2$$

The reactions are carried out in a bubbling fluidized at about 250°C–350°C and a pressure slightly above ambient pressure. A nickel catalyst is applied as bed material to facilitate the methanation reaction. Dry gas at the exit of the methanation plant contains about 40%–45% of methane and 45%–50% of CO_2 besides several minor impurities. Therefore, further upgrading of the gas is necessary. This upgrading consists of steps for a CO_2 removal, drying, and H_2 separation before the bio-SNG is fed into the fueling station for automotive utilization.

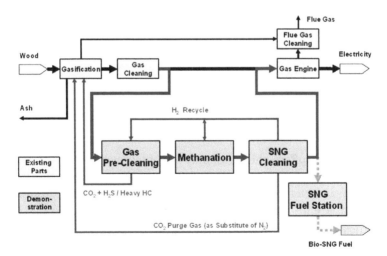

Figure 11.15 Flow diagram of the Bio-SNG demonstration plant at the Güssing CHP plant.

Table 11.6 shows some more details about the bio-SNG production out of syngas from the DFB biomass gasification and especially gas compositions along the gas path and some performance figures of the produced bio-SNG. The dry gas composition is presented at the entrance of the bio-SNG plant, after gas cleaning and the methanation reactor, and at the exit of the demonstration plant. Bio-SNG which is produced shows H gas quality and this quality would be sufficient for feeding it into the Austrian natural gas grid. As no natural gas grid is available in the region of Güssing, a bio-SNG fueling station was installed and the bio-SNG was tested successfully in natural gas cars.

Table 11.6 Gas transformation through a Bio-SNG demonstration plant

		Syngas after gasifier	After gas cleaning and methanation	After separation Bio-SNG
H_2	%	38–42	5–7	0.8
CO	%	23–27	0.1–0.4	0.1
CO_2	%	20–22	45–50	0.3
CH_4	%	9–11	40–45	94
$>C_2$	%	2–4	0.1–2.0	0.7
N_2	%	1	2–3	4
Sulphur	ppm	100	<0.1	n.d.
H_2O	%			<0.01
Rel density	–		*H*-gas quality $\eta > 60\%$	0.57
GCV	kWh/m^3			10.61
Wobbe index	kWh/m^3			14.00

Note: n.d.: not detectable; GCV: gross calorific value.

11.4.3 Fischer–Tropsch Synthesis

FT synthesis offers the possibility of obtaining clean fuels and chemical feedstock. The basic chemistry of the synthesis consists in the hydrogenation of CO molecules from syngas. Typical desired reactions are:

$$(2n+1)H_2 + nCO = C_nH_{2n+2} + nH_2O$$
$$2nH_2 + nCO = C_nH_{2n} + nH_2O$$

Several metals, especially cobalt and iron, have been shown to be active for FT reactions. A Co-based catalyst is the most preferred at the industrial scale due to its high activity, stability, and selectivity for the production of linear paraffins. The reaction is carried out typically at 230°C–300°C and pressures between 16 and 30 bars. In

Güssing a slurry reactor has been developed together with Vienna University of Technology and a commercial Co-based catalyst was used for most of the experiments carried out. Figure 11.16 shows the arrangement for FT-liquids production which is located in the Technikum at Güssing (Fig. 11.14). In the Technikum real syngas from biomass is available from the DFB gasification plant around the clock and all over the year. The capacity of the FT plant is about 5 Nm^3/h syngas.

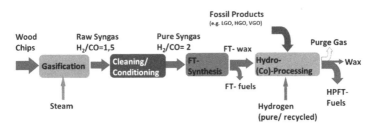

Figure 11.16 Flow diagram of the FT test rig at Technikum, Güssing.

Figure 11.17 shows product distributions obtained at different pressures between 16 to 24 bar. The diagram shows mass percent of the product dependent on the carbon number for direct run FT fuels. The main aim was to obtain a maximum amount of product between C_{10} and C_{20} which is the typical boiling range of diesel fuels.

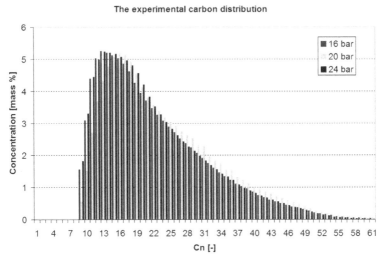

Figure 11.17 Product distribution of FT fuels at different operation pressures.

Produced FT waxes are further hydroprocessed to obtain additional so-called hydroprocessed Fischer–Tropsch (HPFT) fuels. Figure 11.18 shows the distribution of *n*-paraffins and *i*-paraffins for different carbon numbers for direct run FT fuels and HPFT fuels from hydrotreating of waxes. The cetane number of hydrotreated waxes is slightly lower but the cold behavior is excellent.

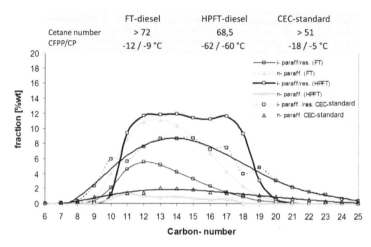

Figure 11.18 Comparison of produced FT fuels and HPFT fuels (hydroprocessed FT waxes).

In the years 2010 to 2012 an FT demonstration plant was installed and operated in the Technikum in Güssing for the production of 1 barrel/day of FT fuel. Figure 11.19 shows this plant consisting of a gas conditioning skid and an FT skid. The gas conditioning was designed by TU Vienna and built by Repotec. The FT multichannel technology originates from Velocys and Oxford Catalyst. The demonstration plant was operated by SGC Energia. After successful demonstration the Fischer–Tropsch demonstration plant was moved to another location for further improvement.

11.4.4 Mixed Alcohol Synthesis

Mixed alcohols from catalytic conversion of synthesis gas are valuable additives to gasoline to increase the octane number and reduce environmental pollution. Furthermore, a great benefit of mixed alcohol synthesis (MAS) is the high resistance of the catalysts against sulfur poisoning and the fact that the gas-cleaning facilities can be simpler as in other syntheses. Mixed alcohols can be also converted to high-quality fuels via dehydration and oligomerization.

Figure 11.19   One barrel per day Fischer–Tropsch demonstration plant.

Normally, the alkali-doped oxides (zinc and chromium oxides) and alkali-doped sulfides (molybdenum sulfides) are used as catalysts for mixed alcohols synthesis. Depending on process conditions and catalysts, the main primary products are generally methanol and water. Ethanol, propanol, butanol, etc., are gradually including methanol synthesis followed by its successive homologation.

The reaction mechanism of the mixed alcohols synthesis is following:

$$nCO + 2nH_2 = C_nH_{2n+1}OH + (n - 1)H_2O$$

Due to reaction stoichiometry the proposed CO/H_2 ratio is 2, but the optimal ratio is in practice closer to 1 because of water–gas shift reaction which occurs in parallel with the alcohol formation.

At the MAS pilot plant in Güssing a MoS_2 catalyst is used, as this type of catalyst is resistant against sulfur poisoning. So the gas treatment is much simpler which reduces the operation and investment costs. A simplified flow sheet of the research facility for mixed alcohol synthesis in Güssing can be seen in Fig. 11.20. The main parts of the mixed alcohol synthesis pilot plant are:

- a product gas blower to level out the pressure drop over the steam reformer

- a steam-reforming unit for converting the methane and aromatic components into hydrogen and carbon monoxide
- a glycol scrubber for gas drying
- a compression step for compressing the gas to a pressure between 90 and 300 bar
- a fixed bed reactor for the synthesis itself
- a condensation vessel for the separation of alcohols from the gas stream
- an expansion valve to reduce the pressure of the tail gas to ambient pressure

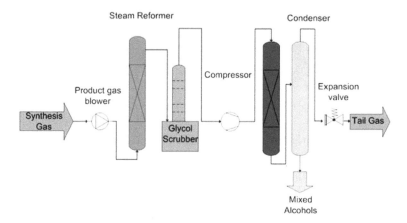

Figure 11.20   Flow sheet of the MAS pilot plant.

Since the first startup of the MAS plant during the summer of 2011, several experiments have been carried out. The parameters of temperature, pressure, space velocity, and gas composition were varied to investigate the impact on CO conversion, product distribution, and product yield. Also, side reactions were taken into consideration, especially reactions with sulfur components like mercaptanes. Typical pressures for the experiments are 110 bar and the temperature is set to about 320°C.

A further investigation of the experimental data showed that ethylene has a major influence on the product distribution and also on the distribution of the sulfur components.

In Fig. 11.21, a comparison between the yielded product with and without ethylene in the synthesis gas is obvious. It is clearly evident that ethylene in the inlet gas stream of the MAS reactor was responsible for the buildup of higher alcohols, predominantly propanol. The bars for MA30 show data from experiments with ethylene in the feed gas.

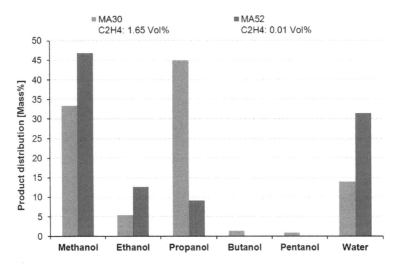

Figure 11.21 Comparison of product distribution with and without ethylene in the feed gas.

Ethylene reacts with H_2 and CO and predominantly forms propanol and some higher alcohols like butanol and pentanol. The bars for MA52 show data from a test run where ethylene was reduced to about 0.01 vol%$_{dry}$. Methanol was the major part and amounted to about 46 mass%. The amount of ethanol was about 12 mass% and the amount of propanol was about 9 mass%. The experiment also indicated that the composition of the alcohols was stable over a long time period. Further details can be found in literature [9].

11.4.5 Hydrogen Production

Hydrogen production from biomass is another application for the product gas from the DFB gasification process. Several theoretical studies were carried out for the production of hydrogen from biomass for industrial applications, for example, Ref. [10]. The process design was similar to that for the production of hydrogen from natural gas. Figure 11.22 shows a flow sheet for the process considered for industrial applications, for example, for refineries. Results show that an energetic efficiency from biomass to hydrogen of about 60% can be obtained.

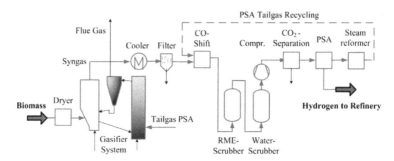

Figure 11.22 Process flow sheet for biohydrogen production for industrial applications.

Other activities on hydrogen production are based on a polygeneration strategy. This means that hydrogen besides electricity and heat is produced. A test rig was realized which takes a slip stream from an industrial dual fluid gasification industrial plant (Fig. 11.23).

From this slip stream hydrogen is enriched mainly by a membrane step (hydrogen enrichment) and cleaned by a pressure swing adsorption (carbon monoxide removal). The off-gas is led back to the gas engine line again. The aim is to produce pure hydrogen which can be fed into a polymer electrolyte membrane (PEM) fuel cell. Figure 11.24 shows a flow sheet of this hydrogen separation test rig. This test rig was mounted into a container and located at the industrial gasification plant in Oberwart (Fig. 11.23).

A lot of short- and long-term experiments (altogether 1500 hours) were carried out with this test rig. The hydrogen produced had a purity of 99.95% and could be fed into a PEM fuel cell without any problems. The hydrogen recovery was around 40%–60%. Figure 11.25 shows concentrations of hydrogen, carbon dioxide, and carbon monoxide after each main separation step included into this test rig. The gross electrical efficiency of the PEM fuel cell was beyond 60%. However, the hydrogen recovery can be increased and the power consumption decreased if a water–gas shift reaction is included into the process chain, perhaps also instead of the membrane step [11].

Preliminary tests show that the hydrogen recovery can be increased even to about 120% and the electrical power consumption reduced to half of that of the process chain shown in Fig. 11.24.

Figure 11.23 View into the container with the test rig for hydrogen separation and PEM fuel cell.

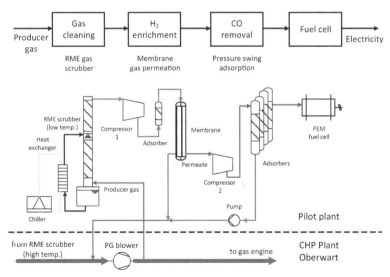

Figure 11.24 Process flow sheet of the test rig for hydrogen separation in the frame of a polygeneration strategy.

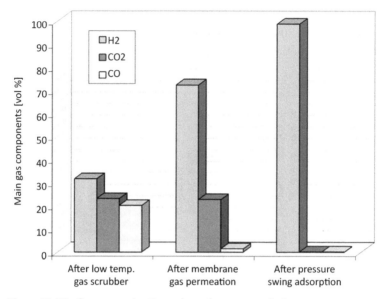

Figure 11.25 Gas concentrations along the process chain.

11.5 Visions for the DFB Biomass Gasification Process

In Fig. 11.26 a possible vision is shown for the DFB biomass gasification process as demonstrated in Güssing. This vision had been already in mind at the time of designing of the Güssing facility. Suitable feedstock is available in many regions in Austria as well as Europe. As suitable feedstock biomass and biogeneous residues from agriculture, industry, and municipalities can be considered. This very heterogeneous feedstock is gasified and a homogeneous product gas is obtained. For dry and/or lignocellulosic materials thermal gasification in a duel fluid gasifier is applied. For wet and easy biologically degradable material biogas plants can be combined with thermal gasification.

From these bio(product)gases several desirable final products can be obtained by suitable cleaning and upgrading and/or synthesis. The desired products are heat and electricity, transportation fuels, and in some cases also chemicals. The most common products are CHP production by gas engines (which is already realized in full scale), by gas turbines, and in the future also by efficient and environment-friendly fuel cells. For transportation fuels bio-SNG and FT fuels are under development which can be mixed with the existing fossil analogue up to 100% and therefore they are "drop in" fuels. The scale of plants available so far for these two products can be considered as demonstration plants. Further developments of products aim at methanol and mixed alcohols. which can be added to gasoline in some extent or used as a basis for chemicals.

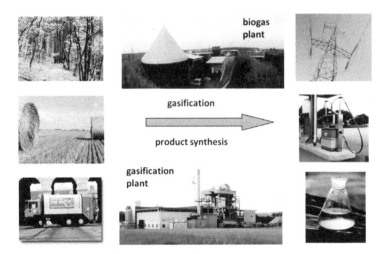

Figure 11.26 Vision for the DFB gasification process.

Finally, also biohydrogen can be produced very efficient from product gases out of biomass and biogeneous residues of the DFB gasification process. Hydrogen is considered as energy carrier for the future. Therefore, the DFB gasification process is suitable also for the energy supply of the future and based on long-term considerations.

References

1. Hofbauer, H. (1982). *Fundamental Investigations on an Internal Circulating Fluidized Bed with a Draft Tube*, PhD thesis, Vienna University of Technology.

2. Pröll, T., Rauch, R., Aichernig, C., Hofbauer, H. (2005). Fluidized bed steam gasification of solid biomass: analysis and optimization of plant operation using process simulation, *Proceedings of the 18th*

International Conference on Fluidized Bed Combustion, May 23–25, Toronto.

3. Zhang, K., Brandani, S. (2003). CFD modelling of a fast internally circulating fluidized-bed biomass gasifier, *Proceedings of the Sino-German Workshop on Energetic Utilization of Biomass*, October 7–9, Beijing, China.

4. Schmid, J.C., Pröll, T., Pfeifer, C., Hofbauer, H. (2011). Improvement of gas–solid interaction in dual circulating fluidized bed systems, *Proceedings of the 9th European Conference on Industrial Furnaces and Boilers (INFUB)*, Estoril, Portugal.

5. Fercher, E., Hofbauer, H., Fleck, T., Rauch, R., Veronik, G. (1998). Two years experience with the FICFB-gasification process, Proceedings of *the 10th European Biomass Conference*, Würzburg.

6. Hofbauer, H., Rauch, R. (2001). Stoichiometric water consumption of steam gasification by the FICFB gasification process, in *Proceedings of Progress in Thermochemical Biomass Conversion*, 199–208, Ed. Bridgewater, A.V., Blackwell Science, Oxford.

7. Kaiser, S., Weigl, K., Schuster, G., Tremmel, H., Friedl, A., Hofbauer, H. (2000). Simulation and optimization of a biomass gasification process, *1st World Conference and Exhibition on Biomass for Energy and Industry*, Seville.

8. Kirnbauer, F. (2013). *The Impact of Inorganic Matter on the Performance of Dual Fluidized Bed Biomass Steam Gasification Plants*, PhD thesis, Vienna University of Technology.

9. Weber, G., Rauch, R., Hofbauer, H. (2013). Production of mixed alcohols from biomass-derived synthesis gas using a sulfidized molybdenum catalyst, *Proceedings of ICPS13*, 137–145.

10. Müller, S., Kotik, J., Pröll, T., Rauch, R., Hofbauer, H. (2011). Hydrogen from biomass for industry: biomass gasification for integration in refineries, *Proceedings of ICPS11*, 225–231.

11. Fail, S., Diaz, N., Konlechner, D., Harasek, M., Rauch, R., Hackel, M., Bosch, K., Hofbauer, H. (2013). An experimental approach for the production of pure hydrogen based on wood gasification, *Proceedings of ICPS13*, 109–126.

About the Author

Hermann Hofbauer received a Dipl.-Ing. in mechanical engineering and a Dr. techn. in chemical engineering at the Vienna University of Technology, Austria. He was appointed as full professor for chemical engineering at the Vienna University of Technology in 1997. He is head of the Institute of Chemical Engineering and head of the research division for Chemical Process Engineering and Energy Technology. Since 2004 he is also dean for study affairs for chemical engineering at the Vienna University of Technology. Since 1990 he is lecturing on fuel technology, energy technology, fluidized-bed technology, basic engineering of combustion systems, thermal biomass utilization, and related topics at the Vienna University of Technology.

Prof. Hofbauer has more than 30 years of experience in thermal biomass utilization, and he has been coordinator of numerous European as well as national projects in this field. Especially in the field of biomass gasification he has developed a novel technology which is well known as the Güssing lighthouse project. He is key researcher in the competence center "BIOENERGY 2020+: Area Gasification" and head of the strategy board of this competence center. For more than 20 years he has been the Austrian representative in IEA-BIOENERGY, where he was active in several tasks (e.g., combustion, gasification) as well in the executive committee. He is author or coauthor of a large number of scientific publications, and he is frequently invited for keynote lectures in these fields. During the last few years he got several prestigious national as well as international awards, including the Linneborn Prize in the year 2012.

Chapter 12

Biomass Gasification Systems for Small-Scale Power Generation: Design Rules and Considerations for Systems Including the Down-Draft Gasifier

Frank G. van den Aarsen

Nesraad B.V., Meije 101, 2411 PM Bodegraven, the Netherlands

frank@nesraad.nl

12.1 Introduction

The availability of electrical power has become a prerequisite for the society at large, where power generation systems are basically fired with fossil fuels, both on the large as well as on the small scale. When in a given situation fossil fuels are only scarcely available, are no longer preferred for environmental reasons, or have become structurally expensive, biomass could become the obvious, sustainable fuel source. Biomass is thus used, for example, for cofiring operations in large-scale applications or as the principal fuel source in stand-

Biomass Power for the World: Transformations to Effective Use
Edited by Wim van Swaaij, Sascha Kersten, and Wolfgang Palz
Copyright © 2015 Pan Stanford Publishing Pte. Ltd.
ISBN 978-981-4613-88-0 (Hardcover), 978-981-4669-24-5 (Paperback), 978-981-4613-89-7 (eBook)
www.panstanford.com

alone operations; evidently the biomass source should in these cases be (or be made) readily and abundantly available.

The conventional and proven technology for generating power is through the steam cycle: combustion of the fuel (coal, oil, natural gas, biomass, or any other fuel) for generating high-pressure (HP) or medium-pressure (MP) steam, expanding the steam over a turbine, which drives an electrogenerator for power generation. Though proven and widely applied, the biomass-fired steam cycle generally requires a relatively large scale for satisfactory energy efficiency and financial feasibility. However, at favorable conditions and for relatively small-scale applications (<500 kWe) biomass gasification could be an interesting alternative by generating a combustible gas to replace the gasoline or diesel fuel in the internal combustion engine of generator sets. The gasification system can thus be considered the green fuel tank for the conventional genset. The boundary condition which has to be fulfilled here is that the gas be safely and reliably generated and the generator gas be of acceptable quality.

The use of generator gas is practiced since long; the gas has, for example, been widely applied since the middle of the 19th century in the steel production industry for the firing of furnaces. The idea of applying generator gas to fuel internal combustion engines came up briefly afterward but never got widely practiced until the outbreak of World War II in 1939. The shortage of liquid fuels which became apparent in those days fueled the interest in gas generators for automobile applications. The wood- or coal-fired generators at the time were generally of German or Swedish make and were built on traction vehicles to enable transport. Transport distances, and so operation times of the gasifier, were kept short to make the timely cleaning-up of the gasification system and the engine parts possible. The dirty and time-consuming operations with a relatively high maintenance cost were taken for granted as there was no alternative. After World War II the interest in small-scale gasifiers decreased rapidly but received revived interest after the first energy crisis in 1973 and the second energy crisis in 1978. In those days prices of liquid fuels rose sharply as a result of the policy of limited crude oil production by the oil-producing countries. Those high fuel prices were a drawback on the economic development of oil-importing countries, in particular developing countries, which boosted the demand for a cheap and reliable alternative for liquid fuels; here the gasification system was thought to be able to contribute to the answer. The desire to quickly produce a cheap gaseous fuel on the basis of locally abundantly available biomass (as generally was the case in developing countries) made many people revert to the World War II gasification technology. However, the necessity for improvement on the technical deficiencies of a "war technology" as well as the need for addressing the lack of awareness on how to properly operate and handle the gasification system were underestimated or insufficiently addressed. It is therefore understandable that the reintroduction of the gasification technology in the seventies was in many cases a source of miscalculation and financial loss.

Van Swaaij et al. [1] extensively reviewed the problems which might be encountered in operating gasification systems. Limiting ourselves here to what has been listed for the cocurrent down-draft gasifier, the identified problems range from poor gas quality to material problems and from hazardous operations to bunker flow problems. However, all problems have been related to poor design, wrong feedstock, improper material selection, or lack of trained operators. It emphasizes the importance of a well-thought-over design of the gasifier, as well as the necessity of creating awareness of how to go about with a gasification system, the latter not only in places where the technical culture is weak or nonexistent: the extensive training of operating personnel is a prerequisite for the safe and successful application of a gasification system anywhere.

To help prevent most of the frustrations of mishaps and malfunctioning of a gasification system in the future, we present in this chapter some of the most important topics in the design of a successful down-draft gasification system. When applying these rules it will contribute to the manufacture of a relatively cheap and well-functioning gasification system which is simple to operate. The design rules and considerations were drafted on the basis of a good understanding of what is happening inside the gasifier, after carefully analyzing the problems as experienced in biomass gasification and supported by practical experience gained during extensive testing.

The safe and reliable generation of a gas of good quality through biomass gasification in a properly designed down-draft gasifier was basically our starting point for the design of a robust gasification system back in the eighties. The choice for the down-draft gasifier as the central part of the gasification system was made because the cocurrent down-draft gasifier can be considered as the only small-scale gasifier in which biomass materials such as wood blocks, maize cobs, or coconut shells can be converted in one step into a good quality gas, virtually free of tar. Once the gas quality objective is achieved, the producer gas is only to be properly dedusted and cooled down prior to a trouble-free application in an internal combustion engine. On the basis of the design rules, as presented in the next sections, several biomass gasification systems have been manufactured for installation at different locations throughout the world, where they have been in island operations for many thousands of hours.

12.2 Gasification by Partial Combustion

The thermal conversion of biomass requires elevated temperatures and thus heat is required to arrive at the temperature necessary. This heat could be provided by external sources, but we consider the situation that the gasifier is in island operation, where the required heat is generated internally by the partial combustion of the biomass fuel. For accomplishing partial combustion, oxygen is required; the oxygen is supplied through injection of air into the gasifier; this air is often called primary air. After initial ignition of the biomass, the burning process is self-sustainable and the heat, generated by the partial combustion, fuels a number of physical and chemical processes inside the gasifier:

- Drying of the freshly fed biomass
- Pyrolysis of the biomass—thermal decomposition yielding noncondensable gases, water, tar, and char
- Partial combustion of the pyrolysis products
- Gasification of the char
- Thermal-catalytic steam cracking as well as gasification of higher gaseous molecules
- Homogeneous gas-phase reactions as well as the water shift reaction

Assuming that all tar components are fully converted into noncondensable gases and taking 10 wt.% moist cellulose as the feedstock, the resulting overall reaction for the complete gasification of biomass by partial oxidation could typically be represented as follows:

$$C_6H_{10}O_5 + H_2O + 1.5(O_2 + 3.76\,N_2) \rightarrow 2H_2 + 2H_2O + 2CO_2 + 3CO + CH_4 + 5.64N_2 \quad (12.1)$$

Through the water–gas shift reaction this gas composition will be somewhat altered depending on the equilibrium temperature.

From this simple representation of the gasification process and its related overall reaction, some conclusions can be drawn, relevant to the basis of design for the gasification reactor:

- Biomass and air have to be properly fed into the gasifier and the biomass is to be ignited to start the generation of heat, as required for lifting the temperature up to gasification conditions.
- The process proceeds at elevated temperatures, so heat-resistant materials have to be applied.
- The gasifier has to be thermally insulated to prevent significant heat losses and thus to prevent a larger share of biomass to be combusted.
- All tars evolving from pyrolysis have to be converted into noncondensable components prior to their leaving the reactor.

12.3 Mass and Energy Balances: Calculating the Proper Product Gas Composition and Quantity

Though an indication of the amount of gasification air required and of gas produced can simply be obtained from Eq. 12.1, accurate values can be obtained from a calculation model as initially introduced by Schläpfer and Tobler [2]. The basic equations for these calculations are:

- the elemental balances for
    o carbon,
    o oxygen,
    o hydrogen, and
    o nitrogen;

- the enthalpy balance;
- Dalton's law;
- the ideal gas law; and
- the homogeneous water–gas shift equilibrium: $K(T) = (CO_2)$ $(H_2)(CO)^{-1}(H_2O)^{-1}$.

The input data for this calculation model are:

- the absolute water–gas shift equilibrium temperature, T, which prevails close to the gas exit of the gasifier. The equilibrium constant can then be calculated from:
  $K(T) = -4.231 + 4.155 \times 10^4/(R \times T) - (2.155/R) \times \log T + (1.306 \times 10^{-2}) \times T/R - (2.64 \times 10^{-7}) \times T^2$,
  where R is the ideal gas constant, $8.134 \text{ J mol}^{-1} \text{ K}^{-1}$.
- related to the biomass fuel—the elemental composition, moisture content, ash content, and heating value. Some typical values for dry wood are C 48.6 wt.%, H 6 wt.%, and O 45.4 wt.% and lower heating value 17,824 kJ/kg dry wood. Fresh wood is typically half water half wood, but the moisture content may be altered through drying such that the preferred moisture content (roughly ranging from 10 to 25 wt.%) can be set.
- the gasification air composition, moisture content, and temperature at the gasifier air inlet.
- the estimated carbon loss (with the removal of ash) and the heat loss through the reactor wall, both typically taken at a few percent.
- the hydrocarbon content (predominantly methane) of the product gas; the methane value for a well-designed down-draft gasifier could be typically taken at 2 vol.%.
- the physical properties of the product gas components.

The gas composition as well as the amount of gas produced and the amount of gasification air required (per kilogram of biomass fed) result from solving the equations above. The gas and air quantities thus obtained are necessary for establishing the most critical dimensions of the down-draft gasifier.

A typical result of a calculation of such a material and energy balance, with 25 wt.% moisture in the fuel and an equilibrium temperature of 800°C, would be:

Gas composition (vol.%): H_2 15.1
                               CO 12.6
                               CO_2 14.0
                               CH_4 2.8
                               H_2O 16.5
                               N_2 39.0

Lower heating value gas (wet/dry, kJ/Nm^3)   4242/5080
Gas produced (Nm^3 gas/kg fuel)   2.13
Air requirement (Nm^3 air/kg fuel)   1.05

12.4   Down-Draft Gasifier Fuel Requirements

Though basically all biomass materials can be gasified, the suitability as a fuel for generating power through gasification in a down-draft gasifier is limited. The pre-requisite is that the biomass material be free flowing to allow proper bunker flow inside the gasifier. Next to this, the particle size/size distribution, moisture content, and ash content should be within limits to secure a successful application for power generation.

12.4.1   Particle Size

The maximum typical dimension of the biomass particle should be limited to approx. 10 cm. This will allow complete conversion at the prevailing conditions in the gasifier and within the time available. As such satisfactory pyrolysis of the larger biomass particles and gasification of the resulting char will result. On the other hand the biomass particles shouldn't be too small as this might cause plugging of the moving bed, thus inhibiting the gas to flow through. Therefore a minimum particle diameter of 1 cm is advisable, though a small percentage of fines (0.1–1 cm) is acceptable. In general the performance of the gasifier is best when the particle size is distributed between 1 and 10 cm.

12.4.2   Moisture Content

Fresh biomass exhibits a moisture content of 50–60 wt.% (on a wet basis); this means that more than half the mass of the fresh biomass is represented by water. If this fresh biomass would be fed into the gasifier, this water would absorb a considerable amount of sensible heat for evaporation and heating up inside the gasifier and as such adversely affect the product gas combustion value as more fuel has to be fully combusted to provide the required heat. Moreover, the maximum temperature inside the gasifier will be reduced, which affects tar and char conversion negatively. Predrying of the biomass to <25 wt.% moisture (w.b.) is therefore required.

On the other hand a little moisture in the biomass is favorable for satisfactory gasifier performance as water is used in the gasification and (steam) cracking reactions. We therefore advise a minimum water content of 10 wt.% (w.b.).

12.4.3 Ash Content

Biomass ashes, as it is meant here, include the minerals inside the biomass structure as well as the minerals and sand possibly attached to the biomass. Ashes may become weak and sticky at elevated temperatures around the oxidation zone in the gasifier and form clinkers, provided the ash is present in sufficient concentrations. Clinker formation has to be prevented as clinkers may plug the moving bed, the throat area, or the grid and as such hinder the proper progress of the gasification process. Our experience is that the formation of clinkers will be prevented if the fuel ash content is well below 5 wt.%, preferably <2 wt.%.

12.5 The Gasification Reactor

The primary objective of the gasifier is to fully convert the solid biomass into gaseous components, yielding a combustible gas. Because of the envisaged application of this gas in an internal combustion engine, not only should the gas be combustible, but the combustion value should be as high as possible and the gas should be virtually free of dust and tar. The true challenge in the design of a successful gasification system is to fulfil these requirements.

The gasifier, the vessel in which the actual conversion of biomass into a combustible gas is accomplished, forms the heart of each gasification system. Several reactor configurations exist for realizing the thermal conversion of biomass into a combustible gas. Packed-bed reactors, down-draft, up-draft, and cross-draft according to the direction of flow of gas and solids through the bed, bubbling or circulating fluidized-bed reactors, and entrained flow reactors are most commonly known.

The relative simplicity of the packed-bed reactors makes them most suitable for application on a small scale. Of these the down-draft reactor, with application of the throat concept, is the one reactor which can convert biomass in one step into a low-tar product gas [3].

12.5.1 Gas Tar Content

Tar in the producer gas may cause dramatic failures in the equipment downstream the gasifier due to tar condensation and subsequent deposition in tubes, control valves, engine inlet valves, etc. It therefore is of utmost importance to generate a combustible gas with as little tar as possible in it. The gas tar content should certainly be less than 50 mg/Nm^3 to guarantee lasting performance of the complete gasification system, including an engine-generator. This requirement can be fulfilled in the down-draft gasifier if properly designed.

12.5.2 Dimensioning the Down-Draft Gasifier

In Fig. 12.1 the down-draft gasifier is schematically represented. It is noted that biomass is fed from the top to feed the bunker section of the gasifier; air is being sucked into the gasifier to combust part of the combustible pyrolysis components, after which gas and solids flow concurrently down the gasifier.

Heat is continuously generated through partial combustion of the biomass close to the air inlet nozzles, locally resulting in high temperatures ranging from 1,100°C to 1,400°C [4]. Though most of the heat generated here is carried away with the combustion gases in the downward direction, some of the heat is transported in the upward direction, mainly through conductivity. Thus a temperature profile develops along the vertical axis of the gasifier: from room temperature to pyrolysis temperature (starting at approx. 250°C) in the top section, where also drying takes place, through the pyrolysis section directly above the combustion zone into the combustion zone.

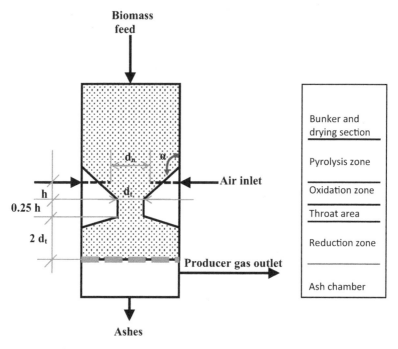

Figure 12.1 The dedicated down-draft gasifier.

The cylindrical space above the contracted area is serving as a bunker volume and drying space, dimensioned such that sufficient fuel can be kept, for example, for approximately one hour of operation at full load. The biomass moves slowly downward (bunkerflow) through the effect of the conversion of solid material into gas further down the gasifier. At the transfer from the cylindrical part into the conical part of the gasifier, a typical angle, α, of 140° is advised to enable smooth flow of the solid particles into the cone and subsequently into the throat. After the throat area, the diameter expands gradually from the throat diameter to the diameter of the cylindrical reduction zone.

On the way down, the biomass particles undergo a series of processes.

First through the conduction of heat from the lower hot oxidation zone the temperature slowly rises in the bunker section, causing the biomass to dry: free water evaporates from the biomass material. Then the biomass gradually descends to the conical part of the gasifier. In this inversed cone, the diameter of the gasifier becomes constricted from the cylindrical bunker diameter to a minimum diameter, which is called the throat. Prior to arriving in the oxidation zone, the biomass temperature has risen to a level where torrefaction and subsequently pyrolysis (the relatively slow thermal decomposition of the biomass into char, tar, and gaseous components) occur. The pyrolysis products are flowing further downward via the combustion zone into the throat area.

12.5.3   Air Inlet Nozzles

Though several other ways exist to introduce air into the gasifier, we choose air inlet nozzles, further named *tuyeres*, placed in a plane. This concept was earlier successfully applied for car and stationary engines in Sweden during World War II [5]. Here, the position of the tuyeres relative to the throat and the number of tuyeres are important design parameters to ensure a well-developed oxidation zone and related high temperatures extending in the throat at all (variable) loads.

The diameter of the tuyeres is calculated using the required air quantity at full load, which results from the mass and heat balances above, where a maximum air velocity of 25 m/s is allowed in the tuyeres. The number of tuyeres should be five for a lower-capacity gasifier (30–75 kWe) or seven for a larger (75–150 kWe) gasifier. These tuyeres should be evenly distributed in a nozzle plane, which is positioned such that the ratio of h/d_t is somewhere between 0.5 and 1.0, resulting in h to be typically 125 mm. The position of the nozzle outlet openings is to be such that the nozzles extend a little into the gasifier, where the ratio d_n/d_t is 2 ± 0.25.

Air is thus introduced into the combustion zone of the gasifier through the tuyeres. At the prevailing temperatures in and just below the oxidation zone (1,100°C–1,400°C) thermal cracking of the tarry components as well as char gasification commence. These temperatures require the throat to be shaped from a good-quality refractory material.

12.5.4   The Throat Concept

The throat concept in the design of the gasifier is of utmost importance to ensure that all pyrolysis products, especially tars, are forced through a hot zone in order to break down in smaller molecules which are noncondensable at room temperature. The diameter of the throat is determined by allowing a maximum superficial gas velocity (at standard conditions) of 2.5 m/s. The maximum gas quantity follows from the mass and heat balance calculations above. The height of the throat area is limited to a few centimeters, typically $0.25h$.

12.5.5   The Reduction Zone

After the throat area the diameter of the gasifier widens to the reduction zone diameter such that the gas velocity is well reduced (typically a factor 10) in order to allow sufficient residence time for the gas phase in the bed of glowing char. Char and gases flow concurrently from the throat area into the reduction zone, where the char gasification reactions continue. Next to the char reduction reactions, the remaining tar molecules will eventually be gasified and steam-cracked, catalyzed by the glowing char. To secure the maximum char and tar conversion in the reduction zone, it is of utmost importance that the temperature remains as high as possible for as long as possible. The gasifier should therefore be well insulated, especially around this area, to prevent the unnecessary loss of heat and to maintain the optimum reaction conditions. The shell of the gasifier is therefore to be lined with a well-insulating refractory material.

The height of the cylindrical part of the reduction zone can be taken at $2d_t$, thus allowing sufficient residence time for the char and gas to complete their reactions and yield a virtually tar-free product gas.

12.5.6   Gas and Ash Outlet

At the bottom of the gasifier the char bed is carried by a grid, where gas can flow through and the ashes and small char particles may fall through. We choose to have two grids on top of each other, of which the upper one could be rotated either manually or mechanically, to effect movement of the lower part of the char bed. As such ashes and some smaller char particles are stimulated to fall through the grid into the ash chamber, where these particles can accumulate. It is practical that ashes be removed from the ash chamber intermittently while the gasifier is in operation. This can be easily accomplished by a scraper at the bottom of the ash chamber, which is fixed on a grid-moving spindle. When rotating the grid, the scraper will push the ashes into the ash outlet, which is to be included in the bottom plate of the ash chamber. To prevent air to be sucked into the gasifier, ashes should then be removed via a two-valve sluice construction.

The producer gas leaves the gasifier at a temperature of 600°C–800°C through the gas outlet at the side of the gasifier. Ash particles which will be entrained by the gas have to be virtually fully separated from the gas downstream the gasifier. This will be dealt with in the next section.

The main design parameters for the down-draft gasifier are summarized in Table 12.1 below.

Table 12.1 The main design parameters for the down-draft gasifier

Angle α cylindrical bunker to inversed cone	typically 140°
Position nozzle plane above entrance throat	$0.5 < h/d_t < 1.0$, typically 125 mm
Maximum air velocity (at standard conditions) in tuyeres	25 m/s
Number of nozzles	5 for smaller capacity, 7 for larger capacity
Position of nozzle outlet openings relative to throat diameter (nozzles extending into the bed)	$d_n/d_t = 2 \pm 0.25$
Maximum superficial gas velocity (at standard conditions) through throat	2.5 m/s
Height of throat	typically $0.25h$
Height of reduction zone	$2d_t$

12.6 Dust Removal and Gas Cooling

As the producer gas is meant for application in internal combustion engines, any dust in the gas may cause severe wear and tear of moving parts or hamper smooth operation. Dust has therefore to be virtually completely removed from the gas prior to its downstream application.

The raw gas dust content can be as high as a few grams per Nm^3, which has to be brought down to <5, preferably <1 mg/Nm^3. We choose to accomplish this in a three-stage filtering section consisting of the following:

- A cyclone separator: Here mainly coarse dust, including possibly entrained sparks, is removed from the hot gases. The separation efficiency of the cyclone is determined by the gas inlet velocity under actual conditions. We based the design of the cyclone on a gas velocity of 25 m/s at full load of the gasifier. Certainly, with a turn-down ratio of approximately 1:4, the gas inlet gas velocity may get as low as 6 m/s, causing the separation efficiency to drop sharply. Though at lower gas quantities less dust will be entrained, it will mean that relatively more dust has to be filtered out in the next stages.
- An impingement separator: This separator is designed such that its optimum efficiency is obtained at the point where the gasifier will be operating at its minimum load.
- A cloth filter: Filter bags are made of Nomex quality, a material which is continuously applicable up to 260°C and which filters fine dust extremely well: the outlet dust content was generally measured at <1 mg/Nm^3.

As the maximum allowable gas temperature in the cloth filter is limited to 260°C, the raw producer gas has to cool down from its gasifier exit temperature to <260°C. The heat loss through the *noninsulated* carbon steel ducts and separators proved to be sufficient to arrive at this temperature under all conditions.

However, after the cloth filter, the now tar- and dust-free gas is still at a temperature well above its dew point (typically at 65°C–70°C). The gas combustion value can be considerably improved by removing as much water vapor as possible from this gas. To achieve this the gas has to be cooled down to a temperature well below its dew point. As such a considerable amount of water vapor can be condensed and be removed from the gas. To this end we included a simple air–gas tube heat exchanger, equipped with a condensate accumulator and a double-valve condensate removal tap. After the gas cooler and condensate removal the gas is at its optimum condition to be sucked in by the gas engine.

12.7 Starting and Stopping the Gasification System

To prevent tar formation at a cold start-up, the gasifier reduction zone and throat have to be filled with charcoal prior to the first start-up or after the gasifier has been fully emptied. After this has been secured, the biomass fuel can then be fed on top of that. Once the gasifier has been properly filled, the biomass is to be ignited for starting the gasification process. For this purpose an ignition pipe is foreseen, which extends through the wall of the gasifier into the oxidation zone. After starting the start-up fan a draught and slight vacuum are developing in the system. Then a burning paper or cloth is held at the opening of the ignition pipe, which will be sucked into the gasifier and will ignite the biomass and char. Once the ignition is visually confirmed, the ignition pipe should be closed and air will be sucked into the oxidation zone through the tuyeres only.

The gas is initially directed via the start-up fan and water lock (to prevent air to be sucked into the system after the fan stops) to the flare. A few minutes after ignition, the gas at the tip of the flare can be ignited. The flame should be colorless/light blue and can only be well distinguished in the dark. After 5 or 10 minutes the start-up fan can be stopped, the gas flow redirected to the engine, and the engine started.

Stopping the gasification is simply done by stopping the engine and redirecting the gas flow via the flare into the open while closing the air inlet at the gasifier. This will stop the gasification process and allow the excess gas to be evacuated via the flare pipe.

The full gasification system, as described above, is schematically represented in Fig. 12.2.

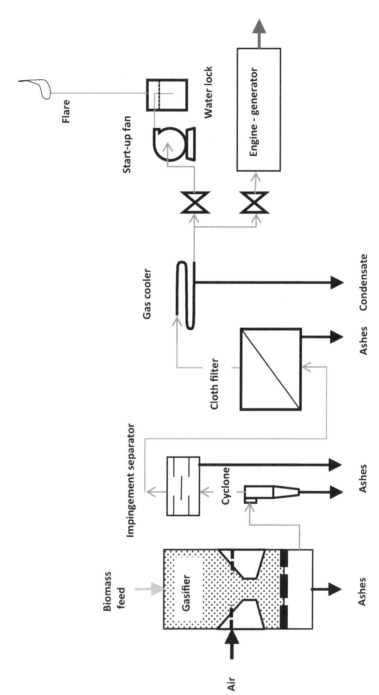

Figure 12.2 Schematic gasification system for power generation.

12.8 Some Safety Aspects

Producer gas contains carbon monoxide, hydrogen, and methane. These gases are combustible and toxic and may explode when in a proper mixture with air and at the availability of an ignition source. Utmost care and proper precautions should therefore be taken to prevent air to be sucked into the gasification system, especially at those spots at elevated temperatures, for example, where ashes or condensate has to be removed. Removal of ashes and condensate should only be done through a double-valve sluice to prevent any suction of air into the system.

During start-up the product gas should be vented in the open at a level such that the gas can easily mix up with the surrounding air. The gasification system as such should be placed in a well-ventilated area to prevent any accumulation of gas.

Feeding the biomass is to be done through a device which prevents too much air from being sucked into the gasifier while feeding the gasifier; too much air might disrupt the gasification process through excess oxidation, which should be prevented at all times.

A double-flap sluice on top of the gasifier will enable this, provided that both flaps cannot be in the open position at the same time.

12.9 The Biomass Gasification System as Installed at Giriulla Mills, Sri Lanka

One of our first gasification systems which were designed following the design rules and considerations above was installed at Giriulla Mills in Sri Lanka in November 1982. Coconut shells (not the husks!) were used as the feedstock to generate electricity and provide power to part of a coconut-desiccating mill. Since the beginning of January 1983, the unit has been operating 20 hours daily, 6 days a week, except on holidays. The unit has been in operation thousands of hours since.

The photographs in Figs. 12.3–12.5 (1983) show some pictures of the plant in operation.

Figure 12.3 Overview of the gasification plant as installed at Giriulla Mills, Sri Lanka, where the operator checks the gasifier air inlet hatch. To the right of the operator the water lock can be seen, to the left the steps to the fuel-feeding platform, and to the far left the engine-generator.

Figure 12.4 The operator on the feeding platform, feeding the gasifier with coconut shells via the twin-flap sluice.

Figure 12.5   Overview of the gasification plant with a clear view of the engine-generator to the left and the cloth filter to the right. At the far end the operator is seen to feed the gasifier.

References

1. van Swaaij, W.P.M., van den Aarsen, F.G., van den Bridgwater, A.V., Heesink, A.B.M. (April 1994). *A Review of Biomass Gasification*. A report to the European Community DG XII Joule Programme.

2. Schläpfer, P., Tobler, J. (1933). *Theoretische und Praktische Untersuchungen über den Betrieb von Motorfahrzeugen mit Holzgas*. Bern 1933, Bericht Nr. 3 (Separatabdruck aus "Der Motorlastwagen").

3. van den Aarsen, F.G. (1985). *Fluidised Bed Wood Gasifier: Performance and Modeling*, PhD thesis, University of Twente, Enschede.

4. Groeneveld, M.J. (1980). *The Co-Current Moving Bed Gasifier*, PhD thesis, University of Twente, Enschede.

5. Ingeniorsvetenskapsakademien (1950). *GENGAS*. Stockholm, Generalstabens Litografiska Anstalts Forlag. Original document translated by the Solar Energy Research Institute, Golden, Colorado (1972).

About the Author

Frank G. van den Aarsen (born 1950) studied mechanical engineering at the Technical University of Delft, the Netherlands. From 1973 to 1977 he was employed by Shell, where he worked on trickle-flow reactors and the development of oil processes at Shell Research Amsterdam. In 1977 he accepted an appointment at University of Twente, the Netherlands, to be sent out as senior lecturer (unit operations and petroleum technology) to Ahmadu Bello University, Zaria, Nigeria. In 1980 he became the project leader for the fluidized-bed biomass gasification program at University of Twente, and in 1985 he concluded his PhD thesis "Fluidised Bed Wood Gasifier: Performance and Modeling," where Prof. Van Swaaij was his promoter. In the same year Dr. van den Aarsen became the CEO of a midsize company (60 employees) in Almelo, the Netherlands, for the development, engineering, manufacturing, and marketing of industrial biomass combustion and gasification systems. In 1988 he established Nesraad BV, the Netherlands, for specialist advice as well as interim, process, and project management in the industrial sector. Since then he has been working with numerous institutions and (inter)national companies in the waste recycling, energy, and (petro)chemical business. Developing industrial projects and advising public and private parties throughout Western Europe, Dr. van den Aarsen is particularly involved in projects dealing with the thermochemical conversion of solid feedstock, amongst which are biomass and (the recycling of) residual materials.

Chapter 13

Torrefaction: Upgrading Biomass into High-Quality Solid Bioenergy Carriers

Robert W. Nachenius,[a] J. H. A. (Jaap) Kiel,[b] and Wolter Prins[a]

[a]*Department of Biosystems Engineering, Faculty of Bioscience Engineering, Ghent University, Coupure Links 653, 9000 Ghent, Belgium*
[b]*ECN Energy Research Centre, Westerduinweg 3, 1755 LE Petten, the Netherlands*
robert.nachenius@ugent.be

13.1 Introduction

Torrefaction is a thermochemical treatment applied to upgrade lignocellulosic biomass into a solid bioenergy carrier (torrefied biomass) with superior properties in view of logistics (handling, transport, and storage) and end use (combustion, gasification, and chemical processing). The word "torrefaction" is derived from the French verb *torréfier*, which means roasting (as in the roasting of coffee beans). As in most thermochemical treatments, torrefaction results in a combination of products, namely, solid torrefied biomass, condensable liquids, and permanent gases. While conventional slow pyrolysis operates at temperatures ranging from 450°C to 550°C, torrefaction employs milder conditions so that the mass yield

Biomass Power for the World: Transformations to Effective Use
Edited by Wim van Swaaij, Sascha Kersten, and Wolfgang Palz
Copyright © 2015 Pan Stanford Publishing Pte. Ltd.
ISBN 978-981-4613-88-0 (Hardcover), 978-981-4669-24-5 (Paperback), 978-981-4613-89-7 (eBook)
www.panstanford.com

and energy yield (in terms of the calorific value of the materials) of the solid, torrefied biomass are maximized. Torrefaction is typically carried out between 240°C and 320°C under nonoxidizing conditions. As the biomass is transformed to torrefied biomass it becomes brittle; moisture is evaporated; the biomass becomes more hydrophobic; and the calorific value increases (but the volumetric energy density decreases). Eventually, the grindability is largely improved and the material becomes more water resistant and less susceptible to biological degradation and self-heating. Torrefied biomass may be densified through pelleting or briquetting, thereby resulting in a high-quality solid bioenergy carrier with a high volumetric energy density. Torrefaction for upgrading biomass was first studied in the 1930s in France. In the 1980s, Pechiney operated an industrial torrefaction plant in Laval-de-Cere, France, for the production of torrefied wood as a reducing agent in metallurgical applications [1]. More recently, in the early twenty-first century, torrefaction gradually gained worldwide interest mainly as a technology to facilitate biomass cofiring in (pulverized) coal-fired utility boilers.

This chapter will focus on so-called dry torrefaction. It will not address wet torrefaction or hydrothermal carbonization: technologies under development where biomass (typically containing significant amounts of inherent moisture) is upgraded in water at elevated pressure.

13.2 Fundamentals of Torrefaction

13.2.1 Reactions and Heat and Mass Transfer

Heat transfer to the biomass lies at the heart of the torrefaction process as the temperature of the biomass must be increased sufficiently to allow the thermochemical degradation reactions to take place. If the biomass is not (or only partially) predried, moisture evaporation creates an additional heat demand. Since the heat requirement for drying is much larger than for torrefaction (already at low moisture levels), and the heat used in drying the biomass cannot be readily recovered, applying dry torrefaction for upgrading biomass feedstocks with high inherent moisture levels is typically less attractive unless low-value heat is available. The heat that can typically be recovered by combusting torrefaction gases and liquids (dependent on the severity of the torrefaction) may be roughly sufficient to dry biomass containing up to 40 wt.% moisture (dry basis). During torrefaction, as the temperature of the biomass is increased, the evaporation of physically bound water starts as the temperature approaches 100°C. At more elevated temperatures above about 160°C, the structural biopolymer constituents (cellulose, hemicellulose, and lignin) and the extractives within the biomass begin to degrade, forming gases and vapors.

Hemicellulose typically decomposes at temperatures above 220°C, whereas cellulose starts to decompose at a higher temperature, typically above 300°C. Lignin may already start decomposing at 160°C, but this takes place at a much slower rate than either hemicellulose or cellulose [2]. For energy applications, it is interesting to note than the higher heating value (HHV) of lignin is reported to range from 23.26 to 25.58 MJ/kg, while the HHVs of cellulose and hemicellulose are only approximately 18.6 MJ/kg [3]. The extractives that are present in biomass can consist of numerous nonstructural organic compounds which may be removed from the biomass by solvent extraction. The thermal behavior of the extractives may vary greatly and is difficult to predict. The degradation of the biomass constituents involves not only the removal of functional groups from the polymer chain but also depolymerization of the biopolymers and decomposition of the monomers. This is shown schematically in Fig. 13.1. The liquid reaction products may then be transferred to the bulk gas stream by evaporation or through aerosol formation. Finally, polymerization and recondensation/readsorption of evaporated degradation products may also occur.

The numerous types of reactions that can take place during torrefaction make it difficult to generalize about the overall reaction enthalpy for the torrefaction process and this is reflected in the variable results presented in the academic literature, ranging in the order of a few hundred kilojoules per kilogram (both endothermic and exothermic). Additionally, the net heat effects may include physical processes such as desorption or the evaporation of liquid torrefaction products, which would be sensitive to variations in the gas phase composition and pressure within the particle. Apparently, the observed overall reaction enthalpies are always found to be very small compared to the energy content of the biomass feedstock or the torrefied biomass product, as shown in Fig. 13.2.

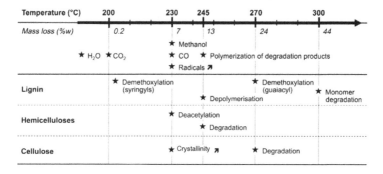

Figure 13.1 Overview of mass loss, volatile formation, and transformation of wood polymers as a function of torrefaction temperature. Stars indicate temperatures at which the processes have been clearly observed. Reprinted from Ref. [4], Copyright 2012, with permission from Elsevier.

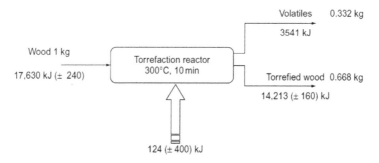

Figure 13.2 Typical mass and energy balance (including sensible heat) for wood torrefaction. Reprinted from Ref. [5], Copyright 2006, with permission from Elsevier.

In order to achieve the desired reaction temperatures heat must be transferred from an external source to the surface of the particle before heat is transferred further into the core of the biomass particle by conduction. Biot number analysis can indicate whether internal or external heat transfer is effectively limiting the heating of the biomass particle and this may simplify models for predicting transient heating if one mode of heat transfer is found to be limiting.

Furthermore, if it can be determined whether reaction kinetics or heat transfer limit the progress of the torrefaction within the particle, further simplification of the analysis is possible. Otherwise, both heat transfer (dependent on particle size) and reaction rates (dependent on the temperature within the particle) need to be taken into account simultaneously.

Mass transfer during torrefaction relates solely to the transport of product vapors and gases within the solid particle and eventually into the bulk, gas phase surrounding the biomass particles. The dominant mechanism of mass transport is convection, driven by the pressure built up inside the particle. Darcy's law, for the movement of fluids in porous media, is usually seen as being sufficient to describe convection if information regarding the pressure profile within the particle is available. Increasing the partial pressure of volatiles within the particle would also tend to suppress desorption of material into the vapor phase.

13.2.2 Reaction Models and Kinetics

Since torrefaction may be seen as a mild form of slow pyrolysis, many of the applied reaction models have been taken from slow pyrolysis literature. As indicated in Fig. 13.1, numerous individual reactions take place during torrefaction. As no exact model is possible, simplifications are necessary and this results in numerous, plausible approximations. Reactions may compete with one another, include intermediate products, or may take place independently. The initial reactant may be considered as one homogeneous material (raw biomass) or the model may be refined further to take into account the varying reactivities of the underlying biomass constituents (cellulose, hemicellulose, and lignin).

The thermal decomposition rates of biomass or the individual biomass constituents are frequently determined through thermogravimetric analysis (TGA), where the overall mass loss during the course of a predetermined temperature program indicates the extent to which these reactions take place. This approach has its limitations in that some overlapping reactions may not be easily distinguished. Furthermore, the presence of any heat transfer limitations during analysis may retard the observed mass loss thereby resulting in unreliable data. This necessitates the use of small biomass samples for TGAs. Care must also be taken to avoid systematic errors in temperature readings.

A thermogravimetric curve for the thermal decomposition of biomass constituents is given in Fig. 13.3. One can clearly see that the different individual biomass constituents start reacting at different temperatures and also produce different fractions of solid residues after the completion of the heating program. In addition, it is generally found that the behavior of the individual biomass constituents does not exactly add up to the behavior of the overall biomass; some interactions do occur.

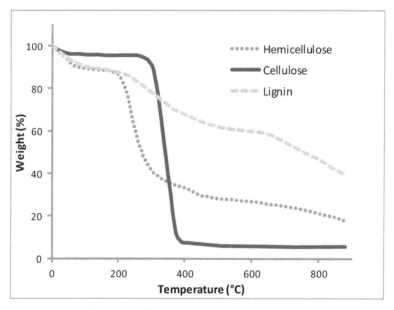

Figure 13.3 TGA curves for the thermal decomposition of cellulose, hemicellulose, and lignin for a linear heating rate of 10°C per minute [6].

Once a reaction model has been specified, then the thermogravimetric data can be analyzed to determine the rates of the reactions prescribed by the model. Typically, the temperature dependence of reaction rates is expressed through the Arrhenius equation and by evaluation of its parameters (pre-exponential factor, activation energy, and reaction order) for each reaction. There are "reservations regarding the application" of Arrhenius kinetics to solid state reactions, but given the robustness of this mathematical approach Arrhenius kinetics are widely used.

In some instances, Arrhenius plots can be used to determine the kinetic parameters. Alternatively, the kinetic parameters can be determined by numerical, model-fitting methods, whereby the square error between experimental and modeled results is minimized by manipulation of the parameter values. In addition to the model-based interpretation of the observed reaction rates, it is also possible to describe the process without explicitly defining an underlying model and by using Arrhenius reaction parameters which are not constant and which can vary based on some measurable property such as the extent of reaction or the reaction temperature. While such model-free methods are suitable for predicting the observed mass loss rates under various thermal conditions, by definition they do not offer the possibility to test the plausibility of any underlying reaction models against the observed results.

13.2.3   Properties of Torrefied Biomass

In literature, changes in product quality are generally reported as a function of reactor temperature and residence time, since these are parameters that may be controlled easily and indeed have a large impact on the torrefaction process. Increases in the torrefaction severities (due to increases in torrefaction temperatures and/ or residence times) consistently lead to a decrease in mass and energy yield of the torrefied biomass product. The effects of process parameters such as biomass particle size, operating pressure, and feedstock selection require more complex assessments.

The approach of comparing torrefaction behavior in different torrefaction processes based on torrefaction temperature and residence time can give useful initial indications. However, a detailed analysis of the torrefaction process needs to take into account the temperature–time history within the biomass particle (rather than just the peak temperature and the total residence time of the solid material within the reactor) since particle morphology, transient heat transfer, and temperature-dependent reaction rates all play a role in the overall conversion and product quality. To sum these various effects, attempts have been made to use the observed solid mass loss (to gases and vapors) during torrefaction as a characteristic parameter for predicting torrefied biomass quality with particular success for properties such as heating value, grindability, and proximate and elemental composition.

Typical properties of torrefied wood pellets and other solid fuels are compared in Table 13.1. It can be seen that, in many aspects, torrefied wood pellets have properties similar to (subbituminous) coal. In general, torrefaction facilitates the decoupling of biomass production and its end use in place, time, and scale. Torrefied biomass pellets or briquettes can be accommodated by existing solid infrastructure and end-use technology more easily than biomass. Moreover, these products may allow advanced trading schemes (comparable to coal) and have the potential, through standardization, to become commodity solid fuels.

In the next sections, individual properties will be discussed in more detail.

13.2.3.1 Elemental composition

Elemental analysis of biomass, before and after torrefaction, indicates that torrefaction preferentially removes oxygen- and hydrogen-rich compounds from the biomass. Some carbon also escapes with the produced volatiles (permanent gases and condensable vapors), but this amount is proportionately smaller than the original elemental composition of the biomass. This can be seen in Fig. 13.4. Such results are supported by Fig. 13.1, which indicates that the volatiles mainly consist of compounds such as carbon dioxide, carbon monoxide, water, methanol, and acetic acid.

The elemental composition of coal is frequently described by the elemental ratios of hydrogen and oxygen to carbon. This approach has also been adopted to describe the solid products of thermochemical biomass conversion. A plot which compares the relative O:C and H:C ratios (Van Krevelen diagram) is given in Fig. 13.5. Here it can be seen that torrefaction shifts the elemental composition in the direction of coal. Also, if one were to extrapolate the data obtained from torrefaction experiments, one would move toward an elemental composition typical of charcoal produced through slow pyrolysis.

Table 13.1 Typical properties of torrefied wood pellets and other solid fuels

	Wood chips	Wood pellets	Torrefied wood pellets	Charcoal	Coal
Moisture content (wt.% wet basis)	30–55	7–10	1–5	1–5	10–15
Lower heating value (MJ/kg)	7–12	15–17	18–24	30–32	23–28
Volatile matter (wt.% dry basis)	75–85	75–85	55–80	10–12	15–30
Fixed carbon (wt.% dry basis)	16–25	16–25	20–40	85–87	50–55
Bulk density (kg/L)	0.2–0.3	0.55–0.65	0.65–0.8	0.18–0.24	0.8–0.85
Volumetric energy density (GJ/m^3)	1.4–3.6	8–11	12–19	5.4–7.7	18–24
Hygroscopic properties	Hydrophilic	Hydrophilic	(Moderately) Hydrophobic	Hydrophobic	Hydrophobic
Biological degradation	Fast	Moderate	Slow	None	None
Milling requirements	Special	Special/Standard	Standard	Standard	Standard
Product consistency	Limited	High	High	High	High
Transportation costs	High	Medium	Low	Medium	Low

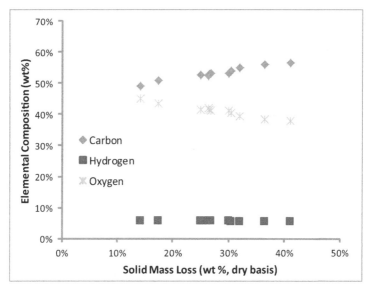

Figure 13.4 Elemental composition of torrefied pine versus solid mass loss due to torrefaction [7].

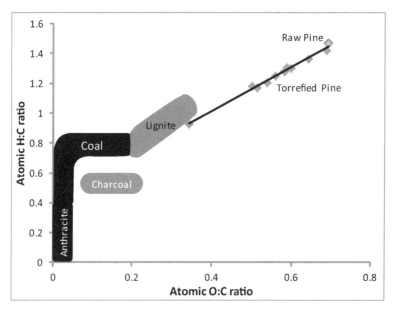

Figure 13.5 Van Krevelen diagram for torrefied pine produced in a screw conveyor reactor [7, 8].

13.2.3.2 Calorific value

The gases and vapors being released during torrefaction have a relatively low energy content. Typically, torrefaction results in the loss of 30% of the dry mass of the fuel (for woody biomass), while the solid product only has a 10% lower energy content (on a feed basis). Consequently, the specific energy content, or calorific value, of the solid product increases. This effect is stronger for increasing torrefaction severity. This is shown in Fig. 13.6. However, since the volume of the biomass particles stays largely the same, the apparent density and the volumetric energy density decrease. Torrefaction must be combined with a densification process (pelleting or briquetting) to achieve a high volumetric energy density necessary to provide a logistics benefit (reduced storage and transport costs on an energy equivalent basis).

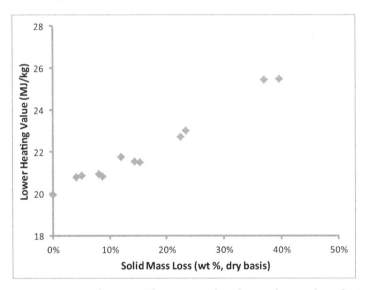

Figure 13.6 Lower heating value versus the observed mass loss during eucalyptus torrefaction for samples torrefied at various temperatures and residence times [9].

13.2.3.3 Grindability

Boilers fired with pulverized coal typically require the coal to be ground to a size of 50–100 µm to allow adequate pneumatic injection and sufficient burnout. Biomass is generally much more reactive which may allow for particle sizes of several hundreds of microns. However, grinding biomass to these small sizes is quite a challenge due to its fibrous structure and tenacious nature to which regular coal mills are not suited. It has been shown that through torrefaction, grindability can be greatly improved due to increased brittleness and a reduction of the cellulose fiber length. Practically, the grindability of densified, torrefied biomass becomes similar to that of coal. Densified, torrefied wood particles are also fairly spherical after grinding, which is in contrast to ground wood particles, which are generally needle shaped.

The classical Hargrove grindability index (HGI) method (applied for coal analyses) is inadequate for characterizing the grindabilities of biomass and torrefied biomass as it involves pregrinding to obtain a sample with a particle size in the range of 0.6–1.2 mm prior to the actual HGI test. This pregrinding energy, which is of great relevance for biomass and torrefied biomass, is not taken into consideration by the HGI method. To avoid this error, the grindability may instead be characterized by experimentally determining the energy requirement needed to achieve a certain particle size reduction, as initially developed by the Energy Research Centre of the Netherlands. Typical grinding results are given in Fig. 13.7 [10].

13.2.3.4 Hydrophobicity

The hydrophobicity of torrefied biomass is important for three reasons. Firstly, the drier the material remains, the higher its heating value will be on a mass basis, increasing its value as a fuel. Secondly, the likelihood of biological degradation is reduced when torrefied biomass remains dry, thereby increasing allowable storage times and reducing the risk of biochemically induced self-heating. Finally, hydrophobicity is essential for maintaining the durability of torrefied biomass pellets/briquettes in outdoor storage. Experiments involving the submersion of torrefied biomass in water indicated that torrefied biomass absorbs less moisture. This trend holds for increasing torrefaction severity but most of the increase in hydrophobicity can be achieved at relatively mild temperatures of approximately 225°C [11]. Analyses into the adsorption isotherms of biomass and torrefied biomass show similar improvements in hydrophobicity through torrefaction. Figure 13.8 indicates how the moisture uptake (defined as the mass of water absorbed per dry mass of torrefied biomass subjected to an immersion test) varies depending on the torrefaction severity. This improvement in hydrophobicity has been attributed to the removal of hydroxyl functional groups from the biomass (hemicellulose in particular), which reduces the number of potential sites at which water molecules can form hydrogen bonds.

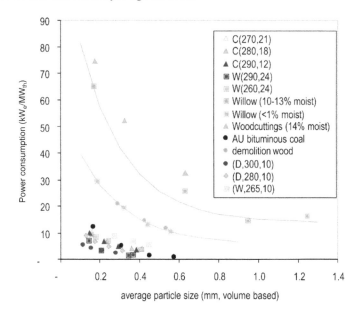

Figure 13.7 Grinding results for coal, biomass, and torrefied biomass, obtained by the ECN method. C = torrefied wood cuttings, W = torrefied willow, and D = torrefied demolition wood. The numbers within parentheses indicate torrefaction temperature and residence time, respectively [10].

Since allowing outdoor storage is an important potential feature of torrefied (and densified) biomass and hydrophobic behavior is an important precondition, the proper characterization of hydrophobicity and the design and execution of realistic outdoor storage tests have become important research topics. Major efforts are being made in a large European project focused on torrefaction (SECTOR) including the development and standardization of a small-scale hydrophobicity characterization test (based on climate chamber testing) and the execution of kilogram- and ton-scale outdoor storage testing [13].

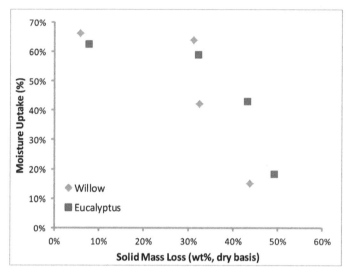

Figure 13.8 Reduction in moisture uptake (as measured by the water immersion experiments) with increasing torrefaction severity (increased solid mass loss) during the torrefaction of willow and eucalyptus [12].

13.2.3.5 Explosivity

Torrefied biomass is known to be more reactive than coal during combustion. Also, the increased brittleness that is achieved during torrefaction increases the possibility of forming fine particles which may form dust clouds during handling. The combination of increased reactivity and increased propensity to form dust clouds indicate that there is a definite risk of dust cloud explosion. While the information in academic literature is somewhat limited, initial results have shown that the reactivity and the flame speed of torrefied biomass dust explosions exceed those of coal [14] and any industrial application (including the production, handling, and utilization of torrefied biomass) needs to take this elevated risk into account to eliminate process and safety hazards.

13.2.4   Properties of Condensable Vapors

The composition of the volatiles produced during torrefaction is highly dependent on torrefaction severity. Furthermore, the efficiency of the separation step between the condensable vapors and the noncondensable gases will affect the composition of both these streams. Low-severity torrefaction will predominantly involve the removal of functional groups from the polymeric chains as indicated in Fig. 13.1. Increasing torrefaction severity results in increase in degradation products originating from the polymeric chains and their derivatives (cellulose degrades to levoglucosan derivatives, hemicellulose degrades to furfural derivatives, and lignin degrades to phenol derivatives). Typical yields for individual components within the condensable liquid stream are shown in Fig. 13.9. It can be seen, in the cases of willow and straw, that water, acetic acid, and formic acid are the most common compounds, while larch torrefaction shows comparatively greater formic acid production. The molecular structure of hemicellulose can vary between different plant species and this is the reason why certain biomass types favor either formic or acetic acid production [15]. The predominant use for the condensable liquids suggested in the literature is to combust this stream (in combination with the noncondensable gases) to partially meet the energy requirements of the torrefaction process and any biomass drying that is performed prior to torrefaction.

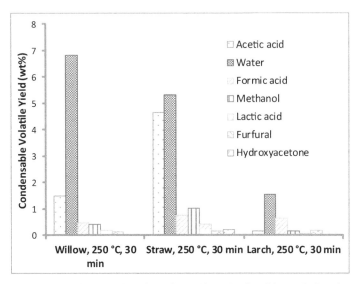

Figure 13.9 Product yields of condensable volatiles (dry, ash-free basis) formed during torrefaction at different conditions for (a) willow and (b) larch and straw [16].

13.2.5   Properties of Noncondensable Gases

Generally, the torrefaction gases are seen as a low-value stream. Most process designs suggest that these gases should be burnt on-site and that the heat produced should be used to reduce the energy requirements of the torrefaction and the predrying process. For mild torrefaction, carbon dioxide predominates and the calorific value of the stream is very low. Increasing the severity of the torrefaction favors the production of carbon monoxide and methane and small amounts of light hydrocarbons, resulting in an increase in the calorific value and size of this stream. If additional gas streams are introduced into the torrefaction reactor for purposes of direct heat transfer, then these will also be recovered with the torrefaction gases and will have a diluting effect on the calorific value. The composition of the gases produced during the torrefaction of pine is given in Fig. 13.10 for different torrefaction and slow-pyrolysis severities (indicated by solid mass loss). It can be seen that the gas product has a low calorific value (3–6 MJ/Nm^3), even before possible dilution effects are taken into account.

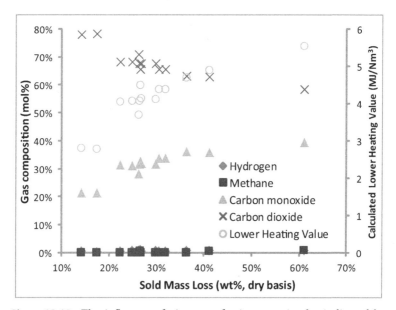

Figure 13.10   The influence of pine torrefaction severity (as indicated by the solid mass loss) on the composition and lower heating value of the torrefaction gases [7].

13.2.6   Feedstock Requirements

Lignocellulosic biomass, like woody biomass and straw, consists of varying proportions of the main constituents: cellulose, hemicellulose, lignin, and extractives. It has been shown previously in Fig. 13.1 and Fig. 13.3 that these constituents vary in terms of their thermal stability and that hemicellulose (and therefore also hemicellulose-rich biomass) is comparatively more susceptible to thermal degradation than cellulose and lignin. The products that are formed can also vary since hemicellulose is not a homopolymer such as cellulose, and its monomeric composition varies between different biomass sources. Generally speaking, hardwood hemicellulose consists largely of xylan (a pentose polysaccharide consisting of xylose), while softwood hemicellulose contains significantly more glucomannan (containing glucose and mannose) within the hemicellulose fraction [17].

While the moisture content of the biomass might not impact the torrefaction reactions themselves, it has a major impact on the overall energy balance. For a relatively wet biomass feedstock, the calorific energy available in the torrefaction vapors and gases may not be sufficient to cover the total heat demand of torrefaction and drying, while for relatively dry biomass there may be an energy surplus. The moisture content, for which the calorific energy in the torrefaction vapors and gases is just sufficient to cover the internal heat demand (autothermal operation), depends on the torrefaction severity. Figure 13.11 shows this relation, where the torrefaction severity is expressed in terms of energy yield. If predrying is not separated from the torrefaction step, the evaporated water will combine with the torrefaction vapors and gases and will reduce the calorific value of this stream although there will also be a commensurate increase in the size of this stream.

The inorganic matter that is present in biomass (reported as ash when determined through proximate analysis) consists of alkali metals, heavy metals, chlorine, and silicon-based compounds. Various metals may have a catalytic effect on secondary reactions involving degradation products derived from cellulose, hemicellulose, and lignin monomers, but since the mild temperatures involved in torrefaction should not generate these reactants in high concentrations, such catalytic activity will not be significant (in comparison to pyrolysis). The inorganic matter present in the biomass will largely remain in the solid torrefied biomass, although a small fraction may be released with the vapors and gases (especially organically bound inorganics). Inorganic compounds that remain in the torrefied biomass may have a negative impact during the eventual conversion of the torrefied biomass in the form of slagging, fouling, agglomeration, or corrosion of boilers and gasifiers. In this respect, alkali metals and chlorine are particularly problematic [19]. Leaching the minerals from the biomass through water-washing may be an effective way of removing these inorganics, but this will increase the overall costs of biomass upgrading.

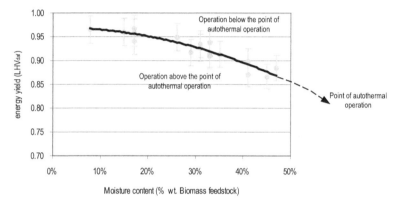

Figure 13.11 Relation between energy yield of torrefaction and moisture content of the biomass feedstock with respect to autothermal operation of the torrefaction process [18].

The particle size distribution of the biomass should be suited to the selected reactor type. Incorrect particle size selection can lead to undesirable flow patterns within the torrefaction reactor or excessive gas-phase pressure drop over the reactor. Large particles will require relatively longer residence times to allow the core of the particle to reach the desired reaction temperatures, and variations in surface and core temperatures that arise from using large biomass particles can result in nonuniform conversion within the same particle. Significant particle size reduction upstream of the torrefaction reactor should be avoided, however, as otherwise the benefit of the improved grindability of torrefied biomass would be effectively negated by increased upstream costs.

13.3   Torrefaction Technology

Many research groups and companies are involved in the research and development, and market implementation of torrefaction technologies. Rather than developing entirely new reactor designs, most developers rely on the re-engineering of proven technologies developed for applications, such as drying, pyrolysis, combustion, and ore roasting.

Torrefaction was primarily conceived as a technology for application in coffee processing. The philosophy of designing biomass torrefaction reactors, however, is conceptually dissimilar, since unlike coffee, biomass fuel is not a high-value product and the quantities to be processed are several orders of magnitude larger. However, just as in coffee bean roasting, good process control in terms of temperature, residence time, and mixing is essential for achieving consistency in product quality. Variations in temperature generally have a larger impact on product quality than variations in residence time. Energetic efficiency is the key to economic viability and to the overall sustainability of torrefied biomass-based value chains. This necessitates a proper integration between the several heat sources and sinks in the process. Although the torrefaction process itself is not very endo- or exothermic, predrying does require substantial energy inputs. Hence, proper design of the predrying process is also important to achieve good overall torrefaction process efficiencies (especially for a biomass feedstock with a high moisture content). Figure 13.12 gives an example of heat integration, where predrying of the biomass and torrefaction are carried out by utilizing three heat sources, namely, the torrefaction gases, supplementary fuel combustion (either fossil or biomass), and heat recovered from cooling the solid product.

When designing the heat integration strategy, a major design choice concerns the means of heating biomass and cooling the torrefied product. For instance, it is possible to use the heat contained in the torrefaction gas or flue gas by direct contact with the biomass or indirect heat transfer may be preferred: the choice will be largely based on the specification of the torrefaction reactor and its geometry. Convective, conductive, and radiative (infrared, microwave) heat transfer have all been applied for torrefaction process design.

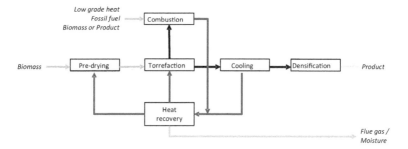

Figure 13.12 Example of heat integration in a torrefaction process.

In addition to high efficiency and low cost, a proper design should also aim at avoiding operational and safety problems resulting from possible blockages (due to solid agglomeration) and uncontrolled temperature excursions (due to excessive exothermicity and ineffective heat removal). Using an inert convective medium instead of flue gas might be safer but would incur the cost of reduced energy efficiency. The recirculation of torrefaction gases for effective heat integration might lead to condensation or polymerization of organic constituents, thereby fouling or blocking pipes.

The main reactor concepts, currently considered by the various torrefaction developers, are discussed below. They are all continuous reactors as these allow high throughputs and more efficient heat and process integration. For each reactor type, some current, relevant industrial initiatives are mentioned despite the fact that the torrefaction scene is still rather turbulent. Industrial initiatives come and go, and most of them are still at the pilot or semi-industrial scale. Different developers may also have changed their preferred reactor concept along the development path. Therefore, the overview of industrial initiatives bears the risk of already being outdated at the time of print.

13.3.1 Moving-Bed Reactor

A moving-bed torrefaction reactor consists of a vertical column filled with biomass particles, where biomass is continuously fed to the top and torrefied biomass extracted at the bottom. Heat transfer may occur across the wall of the reactor, but at larger reactor diameters, directly contacting the biomass with hot gas (possibly produced by the combustion of torrefaction gases and vapors) is more efficient and allows better temperature control. However, the temperature distribution can still be uneven due to a maldistribution of the gas flow due to internal variations in bed porosity, and the pressure drop over the bed may be considerable when the biomass contains a significant proportion of millimeter-size particles. The slowly moving bed of biomass results in relatively long reactor residence times but variations in particle residence time may arise from nonuniform particle velocity profiles (similar to what is seen in the flow of bulk materials in hoppers). Therefore, much care should be given to the solids extraction device at the bottom of the reactor. The major benefit of the moving bed reactor is its simplicity and the relatively high biomass hold-up (ratio of contained mass to reactor volume), which make it cheaper than other alternatives. AREVA (formerly Thermya, France) is developing a torrefaction process based on this reactor type under the TORSPYD trademark [20–22].

13.3.2 Screw Reactor

To decrease the risk of nonuniform biomass flow, mechanical force may be applied to the biomass by a rotating, helical screw so that the biomass particles are pushed through the torrefaction reactor. This is the principle behind screw (auger) reactors. These modified screw conveyors can handle a wide range of feed materials, provided that the biomass particles sizes are significantly smaller than the pitch (the axial distance between the blades) of the screw. Indirect heating may be applied via the outer reactor wall or via a hollow screw. Since this surface-to-volume ratio decreases for larger reactors, the potential for scaling up such reactors is limited. The rotational speed of the screw provides an effective means of controlling the particles' residence time within the reactor due to the (near) plug-flow behavior. An undesirable effect of the rotating screw is that particles may be crushed and this attrition is exacerbated during torrefaction as the biomass becomes more brittle. Any entrainment of fine particles with the torrefaction vapor stream requires more effective downstream separation steps. While screw reactors can accommodate feed materials with a range of morphological properties, those materials which are likely to cause fouling of the heat transfer surfaces are to be avoided [23]. The first torrefaction demonstration plant of Pechiney in France was based on a twin-screw reactor with a production capacity of 12,000 tons of torrefied biomass per year. Despite its technical suitability, the process was not economically successful and the operations were ended in the 1990s [1]. Currently, screw reactor technology is applied by Biolake in the Netherlands, CENER in Spain (reactor developed by LIST of Switzerland) and BioEndev in Sweden [20, 21, 24].

13.3.3 Rotary Drum Reactor

Rotary drum reactors are adapted from the rotary kilns and dryers that are extensively used in the mineral processing and solid handling industries. Biomass is fed to a long drum which is slightly inclined. As the drum rotates, the material is lifted until it tumbles downward and forward within the drum. By increasing the rotational speed, the residence time can be reduced (for a given reactor length) but since the tumbling action cannot be perfectly controlled, ideal plug-flow is not achieved. Heat transfer to the biomass can be either direct or indirect. Particle attrition is increased due to the tumbling motion and these fine particles need to be dealt with in the separation processes downstream of the reactor. Rotary drum reactors have a limited scalability, necessitating a modular approach for higher capacities. Companies applying this type of reactor include TSI (USA), Torr-Coal (the Netherlands), and Andritz (Austria), although the development of the so-called ACB process of the latter has apparently been discontinued.

13.3.4 Multiple-Hearth Furnace Reactor

Multiple-hearth furnaces are well known for ore refining and smelting. They consist of multiple trays (hearths) stacked vertically. Biomass is fed to the top tray and driven by rotating arms (or rotating trays with stationary arms) to a hole in the tray, where it drops to the tray below. Direct heat transfer is accomplished by hot gas which is either fed at the bottom of the reactor and flowing through the (porous) trays or fed per tray through the side wall. The top trays may be used for (final) drying and the bottom trays for torrefaction. If the hot gas is supplied to each tray individually, the temperature at each layer of the reactor can be set independently to control the torrefaction process. This reactor allows flexibility of operation due to accurate control concerning the residence time. Also, the relatively low agitation ensures limited production of dust. Higher costs due to the complexity of these reactors and higher maintenance requirements may be a drawback [25]. Amongst others, CMI-NESA in Belgium and Wyssmont in the United States have been involved in applying this reactor technology [24]. Recently, Andritz and the ECN have successfully demonstrated a tray-type reactor design in Denmark. The design involves multiple stationary trays with rotating scrapers and separate gas loops for final drying and torrefaction. The reactor is operated at elevated pressure to improve gas–solid heat transfer and allow better scale-up.

13.3.5 Torbed Reactor

The Torbed reactor is a proven reactor concept for several applications, including drying and combustion. It involves entraining the solid particles in a high-velocity swirling flow of gas. The "tor" in the name of the reactor refers not to torrefaction but to the resulting toroidal motion of the fluidized solid particles within the bed. In this reactor, the heat transfer to the solid particles is efficient and residence times are very short (several minutes). The biomass input must consist of adequately small particles which can be fluidized in the bed. The reactor typically has a low biomass hold-up. The reactor technology has been developed by Torftech in the United Kingdom and it has been applied for torrefaction by (the now-defunct) Topell Energy (the Netherlands).

13.3.6 Microwave Reactor

In this relatively new concept, electromagnetic radiation in the microwave range (300 MHz–300 GHz) is used to heat up the biomass. This allows heat to be transferred directly to the core of the particles (principally the water molecules), thereby avoiding internal heat transfer limitations and allowing particles with differing sizes to react equally [26]. The torrefaction temperature is controlled by adjusting the power of the applied microwave radiation. Microwave torrefaction relies on electrical power, which is more expensive than the heat which could be recovered from the process or from the combustion of torrefaction vapors and gases. Microwave torrefaction technology is being developed by Rotawave of the United Kingdom [21].

13.4 Torrefaction Development Status and Outlook

Initially, the main driver for torrefaction development has been the increasing demand for solid biofuels in Western Europe as a result of incentives promoting biomass cofiring in large power utilities. Given the limited availability of woody biomass in this region, other regions in the world (rich in woody biomass) have been considered as biomass sources. In this respect, the focus has been on North America (USA and Canada) in particular, where the logistic benefits of torrefied (and densified) biomass can have a significant positive impact on the economics of such long-distance supply chains. Consequently, torrefaction technology has found several developers in Europe and North America. Currently, many technologies (>10) are in the demonstration phase, while some have entered early industrial application. The latter includes the Torbed technology of Topell, with a 60,000 ton/a plant in Duiven, the Netherlands, and the rotating drum technology of Torr-Coal, applied in a 32,000 ton/a plant in Dilsen-Stokkem, Belgium. Many of these plants are able to produce high-quality torrefied biomass pellets, which are able to meet the requirements of white wood pellets, for instance, in terms of durability. The biomass feedstocks used in the processes mentioned above mainly consist of forestry residues (mixtures of soft- and hardwood).

The larger-scale torrefaction plants now also allow for industrial-scale application trials to confirm the technical feasibility of using of torrefied biomass pellets as a feedstock for coal cofiring. Recent cofiring trials include the following:

- A trial in the NOUN-/Vattenfall-owned integrated gasification combined cycle plant in Buggenum (the Netherlands) in 2012, where 1200 tons of torrefied pellets were comilled with coal and cofired in proportions of up to 70% (on an energy basis).

- A trial at the Amer 9 power plant of RWE/Essent in Geertruidenberg (the Netherlands) at the end of 2013, where 2300 tons of Topell torrefied biomass pellets were comilled with coal (5%–25% comilling) and then fed to the burners, leading to 1%–4% cofiring (on an energy basis).
- A trial in the Studstrup 3 power plant of DONG Energy (Denmark) in March 2014, where 200 tons of Andritz pellets were fed to a dedicated mill, leading to up to 33% cofiring (on an energy basis).
- A trial at Helsingin Energia Hanasaaren (Finland) in March 2014), in which 140 tons Torr-Coal torrefied biomass pellets were tested at cofiring percentages of up to 14% (on an energy basis).

In general, these trials showed no big technical challenges during handling, (co-)milling and feeding. Moreover, during the combustion of the pulverized-fuel, stable, attached flames and low levels of carbon in the ash (indicating proper burnout) could be maintained. From the Buggenum 70% cogasification trial, it was estimated to be possible to achieve 90% of the nominal plant capacity without major modifications in the fuel-feeding system. Dust formation during the handling and transport of torrefied biomass pellets and a high explosivity of this dust may be an issue, but dust formation could be effectively suppressed, and explosivity reduced by fine-tuning the production process. These findings help to create end-user confidence in the torrefaction technology and resulting products, which is essential for a broader market introduction of the technology. Moreover, the industrial-scale experience in handling, storage, milling, feeding, and conversion (combustion and gasification) of torrefied biomass pellets provides valuable inputs for further product quality optimization and the developments of product standards.

Despite the significant technical progress, commercial market introduction of torrefaction technology appears slower and more difficult than initially anticipated. An important reason is that many (European) utilities are no longer very eager to invest in torrefaction plants or engage in long-term off-take agreements for torrefied material. This is mainly attributable to the current lack of incentives for cofiring in many European countries. Alternative thermal conversion processes, such as the industrial production of heat and steam and the production of (entrained-flow) gasification-based biofuels and biochemicals, may appear to be more attractive initial routes for launching torrefaction. In addition, interest in torrefied biomass as a renewable fuel is growing rapidly in other parts of the world such as Brazil, South Africa, and in several Asian countries like Malaysia, Indonesia, Japan, and South Korea. Therefore, there are still promising prospects for torrefaction to become a mature, commercially proven technology in the coming years.

References

1. Bergman, P.C.A., Kiel, J.H.A. (2005). Torrefaction for biomass upgrading, *Proceedings of the 14th European Biomass Conference Exhibition*.

2. Yang, H., Yan, R., Chen, H., Lee, D.H., Zheng, C. (2007). Characteristics of hemicellulose, cellulose and lignin pyrolysis, *Fuel*, **86**, 1781–1788.

3. Demirbaş, A. (2001). Relationships between lignin contents and heating values of biomass, *Energy Conversion Management*, **42**, 183–188.

4. Melkior, T., Jacob, S., Gerbaud, G., et al. (2012). NMR analysis of the transformation of wood constituents by torrefaction, *Fuel*, **92**, 271–280.

5. Prins, M.J., Ptasinski, K.J., Janssen, F.J.J.G. (2006). More efficient biomass gasification via torrefaction, *Energy*, **31**, 3458–3470.

6. Yang, H., Yan, R., Chen, H., Zheng, C., Lee, D.H., Liang, D.T. (2006). In-depth investigation of biomass pyrolysis based on three major components: hemicellulose, cellulose and lignin, *Energy and Fuels*, **20**, 388–393.

7. Nachenius, R., van de Wardt, T., Ronsse, F., Prins, W. (2014). Torrefaction of biomass in a continuous rotating screw reactor, in *Proceedings of the EU BC&E 2014 22nd European Biomass Conference Exhibition*, 1018–1024, Eds. Hoffmann, C., Baxter, D., Maniatis, K., Grassi, A., Helm, P., ETA-Florence Renewable Energies.

8. Van der Stelt, M.J.C., Gerhauser, H., Kiel, J.H.A., Ptasinski, K.J. (2011). Biomass upgrading by torrefaction for the production of biofuels: a review, *Biomass and Bioenergy*, **35**, 1–15.

9. Almeida, G., Brito, J.O., Perré, P. (2010). Alterations in energy properties of eucalyptus wood and bark subjected to torrefaction: the potential of

mass loss as a synthetic indicator, *Bioresource Technology*, **101**, 9778–9784.

10. Bergman, P.C.A., Boersma, A.R., Zwart, R.W.R., Kiel, J.H.A. (2005). *Torrefaction for Biomass Co-Firing in Existing Coal-Fired Power Stations "BIOCOAL"*. ECN report ECN-C-05-013.

11. Tapasvi, D., Khalil, R., Skreiberg, Ø., Tran, K.-Q., Grønli, M.G. (2012). Torrefaction of Norwegian birch and spruce: an experimental study using macro-TGA, *Energy and Fuels*, **26**, 5232–5240.

12. Ibrahim, R.H.H., Darvell, L.I., Jones, J.M., Williams, A. (2013). Physicochemical characterisation of torrefied biomass, *Journal of Analytical and Applied Pyrolysis*, **103**, 21–30.

13. EU-FP7 project, *The Production of Solid Sustainable Energy Carriers by Means of Torrefaction (SECTOR)*, https://sector-Project.eu/.

14. Huéscar, M.C., Phylaktou, H.N., Andrews, G.E., Gibbs, B.M. (2014). Comparison of explosion characteristics of torrefied and raw biomass, in *Proceedings of the EU BC&E 2014 22nd European Biomass Conference Exhibition*, 1025–1033, Hoffmann, C., Baxter, D., Maniatis, K., Grassi, A., Helm, P., Hamburg, ETA-Florence Renewable Energies.

15. Wang, X., Kersten, S.R.a., Prins, W., van Swaaij, W.P.M. (2005). Biomass pyrolysis in a fluidized bed reactor. Part 2: experimental validation of model results, *Industrial and Engineering Chemistry Research*, **44**, 8786–8795.

16. Prins, M.J., Ptasinski, K.J., Janssen, F.J.J.G. (2006). Torrefaction of wood: part 2. Analysis of products, *Journal of Analytical and Applied Pyrolysis*, **77**, 35–40.

17. Ciolkosz, D., Wallace, R. (2011). A review of torrefaction for bioenergy feedstock production, *Biofuels, Bioproducts, and Biorefining*, **5**, 317–329.

18. Bergman, P.C.A. (2005). *Combined Torrefaction and Pelletisation: The TOP Process*. ECN report ECN-C-05-073.

19. Dayton, D.C., Jenkins, B.M., Turn, S.Q., et al. (1999). Release of inorganic constituents from leached biomass during thermal conversion, *Energy and Fuels*, **13**, 860–870.

20. Chew, J.J., Doshi, V. (2011). Recent advances in biomass pretreatment: torrefaction fundamentals and technology, *Renewable and Sustainable Energy Reviews*, **15**, 4212–4222.

21. Kleinschmidt, C.P. (2011). Overview of international developments in torrefaction, *2011 Century European Biomass Conference Torrefaction Workshop*, Graz.

22. Ratte, J., Fardet, E., Mateos, D., Héry, J.-S. (2011). Mathematical modelling of a continuous biomass torrefaction reactor: TORSPYD column, *Biomass and Bioenergy*, **35**, 3481–3495.

23. Waje, S.S., Thorat, B.N., Mujumdar, A.S. (2006). An experimental study of the thermal performance of a screw conveyor dryer, *Dry Technology*, **24**, 293–301.

24. Kleinschmidt, C.P. (2010). Statusoverzicht en impactanalyse van torrefactie in Nederland.

25. Acharya, B., Sule, I., Dutta, A. (2012). A review on advances of torrefaction technologies for biomass processing, *Biomass Conversion and Biorefinery*, **2**, 349–369.

26. Ren, S., Lei, H., Wang, L., Bu, Q., Chen, S., Wu, J. (2013). Thermal behaviour and kinetic study for woody biomass torrefaction and torrefied biomass pyrolysis by TGA, *Biosystems Engineering*, **116**, 420–426.

About the Authors

Robert Nachenius completed his undergraduate studies in chemical

engineering at Stellenbosch University (South Africa) and subsequently obtained his masters in chemical engineering and a professional doctorate of engineering (process technology) at the University of Twente, the Netherlands, in 2006. After completing his studies, Robert worked in the industry as a process engineer for Sasol Synfuels and Sasol Technology at their Secunda site (South Africa), which is focussed on the production of fuels and chemicals from coal via the Fischer–Tropsch process. Since 2011, Robert has returned to academic research and has taken up the position of doctoral researcher at the Department of Biosystems Engineering at Ghent University, Belgium, where his field of interest incorporates torrefaction and slow pyrolysis.

Jaap Kiel holds an MSc in mechanical engineering from the

University of Twente and has completed a PhD at the same university on simultaneous SOx/NOx removal from flue gases of coal-fired boilers.

In 1989, Jaap joined the ECN (Energy research Centre of the Netherlands) and became involved in the execution and later organization and management of a broad range of R&D projects in the field of thermal conversion of solid fuels. These include biomass pretreatment and upgrading, combustion and gasification of biomass and coal, and biomass cofiring. As such, he has played a pioneering role in the development of torrefaction technology, and he has coordinated the torrefaction activities at the ECN from early on.

Jaap currently is the development manager of the biomass program at the ECN, covering R&D on various thermochemical biomass conversion technologies, including dry and wet torrefaction, combustion and gasification, and biorefinery concepts for lignocellulosic biomass and seaweed. He also is a part-time professor of thermochemical conversion of biomass at Delft University of Technology, the Netherlands, and the coordinator of the subprogram on thermochemical processing within EERA Bioenergy, an alliance of European energy R&D centers.

Wolter Prins (1951) got his master's degree in chemical engineering from the University of Groningen, the Netherlands, and his PhD from the University of Twente. In 1984 he was appointed as an assistant professor in the Department of Chemical Technology at the University of Twente. Since 1992 he combined his work at the University of Twente with a position as head of R&D in BTG Biomass Technology Group B.V., Enschede, the Netherlands. While working for the University of Twente and BTG, Wolter has participated in many European FP6 and FP7 projects. In 2008 he was appointed as a professor for bioresources processing in the bioscience engineering faculty of the University of Ghent, Belgium.

Wolter has published around 200 papers in the area of novel gas–solid reactors, heat and mass transfer in fluidized beds, and thermochemical conversion of biomass. He has participated in the European Network for Pyrolyis (Pyne) for many years and was invited by NEDO and the Chinese Academy of Sciences to present his work in Japan and China, respectively. Since September 2010 he is an editor of the Elsevier journal *Biomass and Bioenergy*.

Wolter is a member of the Scientific Committees of the European Biomass Conference, the International Conference on Renewable Resources and Biorefineries, and of the International Conference on Chemical Reactors. For the European Biomass Conference he is responsible for the annual publication of a special issue in *Biomass and Bioenergy*.

Chapter 14

Current Issues in Charcoal Production and Use

Hubert E. Stassen

H.E.M. Stassen Holding B.V., Langenkampweg 107, 7522 LM Enschede, the Netherlands
huubstassen@netscape.net

The chapter strives to present an overview of important issues which currently affect the charcoal business. To illustrate matters, two fundamentally different markets and production environments were chosen for analysis, that is, the charcoal markets and industries of sub-Saharan Africa and Europe.

In Africa, for large parts of the population, charcoal is the most important cooking fuel. Charcoal production and consumption are huge and likely to increase in the decennia to come. This has raised doubts about the sustainability of charcoal production on the continent. The chapter goes into the biomass resources which are used for charcoal production. Alternative resources are identified and evaluated. Production technologies are described and technical, economic, and ecologic issues of the different methods are considered. Attention is paid to the prospects of charcoal briquette production in sub-Saharan Africa, and a potentially beneficial alternative

Biomass Power for the World: Transformations to Effective Use
Edited by Wim van Swaaij, Sascha Kersten, and Wolfgang Palz
Copyright © 2015 Pan Stanford Publishing Pte. Ltd.
ISBN 978-981-4613-88-0 (Hardcover), 978-981-4669-24-5 (Paperback), 978-981-4613-89-7 (eBook)
www.panstanford.com

application of charcoal, that is, replacement for petroleum fuels in small-scale power production at isolated sites, is briefly discussed. In the section which looks into the future of the charcoal industry in Africa an alternative to the prevailing fragmented, inefficient, and unsustainable charcoal value chain is proposed.

The section on charcoal production and use in Europe discusses the most important applications of charcoal on this continent. The major part of the charcoal which is sold is used for leisure cooking and grilling. This market is briefly discussed. Aspects of the charcoal markets for important industrial applications (in metal ore reduction and for activated carbon) are considered as well. Different proven production technologies for industrial lump charcoal and charcoal briquette production are described and their merits and limitations discussed. The section on the future of charcoal production and use in Europe discusses a number of market developments which ultimately may seriously alter the structure of the charcoal business on the continent.

14.1 Introduction

Charcoal is usually produced by a process known as slow pyrolysis—the heating of wood or other biomass in the absence of oxygen. Mankind has made and used charcoal from wood since prehistoric times. Stone Age people, in their cave dwellings, roasted meat on charcoal fires. Bronze and Iron Age artisans and craftsmen produced tools and weapons from metals extracted from ores through reduction by charcoal. In 1991 the mummy of an Ice Age man who lived around 3500 BC was discovered in the Tyrolean Alps. Among other things the man was carrying a birch-bark box filled with charred wood pieces wrapped in maple leaves. As the man did not carry a flint stone for starting a fire it appears that he may have carried smoldering charcoal instead [1]. From the Middle Ages in Europe charcoal was used as a reducing agent in iron production. Large-scale application resulted in massive deforestation in parts of England and Central Europe. Due to scarcity and high costs, in blast furnaces and forgeries it was gradually replaced by coke. Charcoal, however, is superior to coke because it contains no sulfur. Therefore, today in Brazil, charcoal from (eucalyptus) plantations is still used in the production of pig iron and quality steel.

According to the Food and Agriculture Organization (FAO) of the United Nations about 51.3 million tons of wood charcoal were produced worldwide in 2012. In that year Africa, Latin America– (including the Caribbean) and Asia-dominated production with 60%, 21% and 17% of total world output, respectively (Table 14.1).

Table 14.1   Wood charcoal production by region, 2004–2012

Region	Production in 2012 (10^6 tons)	Share 2004 (%)	Share 2012 (%)	% Change 2004–2012
Africa	30.6	52	60	28.6
Asia	8.6	17	17	1.8
Europe	0.5	1	1	5.4
Latin America and the Caribbean	10.7	27	21	−2.9
Northern America	0.9	2	2	0.0
Oceania	0.0	0	0	4.5
World	51.3	99*	101*	13.7

Source: FAOSTAT-ForesSTAT.
*due to rounding

Demand for charcoal continues to increase in Africa, where the charcoal sector has been growing by 3% to 4% per year in the period of 2004 to 2012 (exceeding the annual population growth rate of 2.6%). Nigeria and Ethiopia jockey for leadership in Africa, with a production of about 4.0 million tons each in 2012. Production seems to decrease somewhat in Latin America. Brazil is the world's largest charcoal producer, with a production of 7.6 million tons in 2012. Statistics on charcoal in Brazil only report in the context of the production of pig iron; therefore annual data on charcoal production fluctuate considerably depending on actual production levels in the national iron and steel industry [2]. Longer-term trends seem to indicate that the charcoal consumption of the steel industry in Brazil is slowly decreasing. However, this trend may be reversed following Brazilian law changes in 2010 to reduce carbon emissions as part of the president's commitment to make "green" steel. Production in Europe, Asia, and North America remains more or less stable. In the industrialized countries recreational use is concentrated over the summer months. Europe accounts for only 1% of the world production. The largest European producer is Ukraine (100,000 tons in 2012). Within the E27 Poland and France are the largest producers, with productions in 2012 of, respectively, 88,700 and 50,000 tons. The United States remains a major manufacturer of charcoal, producing in 2012 just less than 1 million tons. This results in a ranking amongst the 10 major charcoal-manufacturing countries in the world. In 2012 wood charcoal exports totaled at 1.9 million tons, or only 3.7% of the global production. Five countries, that is, Somalia, Indonesia, Myanmar, Paraguay, and Argentina, in that year accounted for half the global wood charcoal exports. The top five wood charcoal–importing countries in 2012 were Germany, followed by, respectively, the People's Republic of China, Japan, Thailand, and the Republic of Korea. Together those countries accounted for 37% of the global import market.

Official European Union (EU) quality standards for charcoal and charcoal briquettes for cooking, grilling, and barbecuing are filed in EN-1860-2. The directive for barbecue charcoal specifies parameters like fixed-carbon content (minimum 75% by weight), ash content (maximum 8% by weight), total moisture (maximum 8% by weight), size (maximum 10% by weight may exceed 80 mm in size, at least 80% by weight shall be greater than 20 mm, 0–10 mm shall not exceed 7% by weight), and bulk density (at least 130 kg/m^3). It is notable that no maximum or minimum limits are set for volatiles. The directive also describes standardized methods and protocols for determination and analysis of those parameters. Tests can only be performed in qualified laboratories. Charcoal manufacturers and commercial and individual charcoal buyers in and outside of the EU seldom employ industrial laboratories to check on the above specifications. Buyers of charcoal for cooking and grilling as a rule employ just visual and sensual means for judging the quality of the charcoal for sale. They look for a hard charcoal which cannot easily be broken. The lumps or sticks should be of a shiny black color at the outside, and there should be an absence of (yellowish) discoloring at the inside, indicating a (too) high level of volatiles. The best grilling charcoal produces a crisp metallic sound when tapped.

Industrial charcoal users like the steel industry, the ferrosilicon smelting industry, and the activated carbon industry stipulate charcoal specifications and quality tests which relate to the characteristics of the equipment which is used in their production processes. Quite often those companies have their own quality control laboratories, which perform company-specific unpublished tests. To establish the suitability of a charcoal from a new supplier, before entering into an agreement large industrial users will always run pilot tests in their industrial-scale production equipment, even when all laboratory tests before have shown satisfactory results.

14.2 Issues of Charcoal Production and Use in Africa

14.2.1 Urban Cooking Fuel

Especially for households dwelling in the growing (peri)-urban parts of sub-Saharan Africa, charcoal is the most important cooking fuel. Although data should be considered with care due to difficulties with reliability, it is plausible that as a result of urbanization and economic development, the last three decades have seen a shift from the use of wood to charcoal [3]. According to the World Bank [4] in 2007 charcoal was in sub-Saharan Africa a US$ 8 billion industry, employing more than 7 million people in the subregion. Some development economists argue that 70% of charcoal is not accounted for in national statistics. The bank expects that despite rising incomes, charcoal consumption in the region will remain at high levels or even increase in absolute terms over the next decades, due to the factors mentioned above, as well as a result of growing population and in some cases because of rising prices of alternatives like liquefied petroleum gas (LPG) and kerosene. The International Energy Agency (IEA) forecasts that by 2030 charcoal will become a US$ 12 billion industry employing 12 million people [5].

The FAO estimates that 2.4 billion people still rely on wood and charcoal as fuel for daily cooking. In the (peri)-urban centers of sub-Saharan Africa charcoal is the dominant cooking fuel of the poor and middle-income segment of the population. Urban women from different African regions, when interviewed in household energy surveys, indicated that they found wood fuels difficult to control, heavy to transport, smoky, messy, and dangerous for children [6]. Charcoal is preferred (Fig. 14.1) because it is lightweight, has double the energy content of wood fuel, is easy to store without the risks of rot and mold, and burns relatively clean and even. Charcoal is cheap and reliable and available through a large network of wholesalers and retailers contrary to more "modern" alternatives like kerosene, electricity, and LPG, which are sometimes in short supply and often more expensive (especially when the costs of appliances and stoves are taken into account).

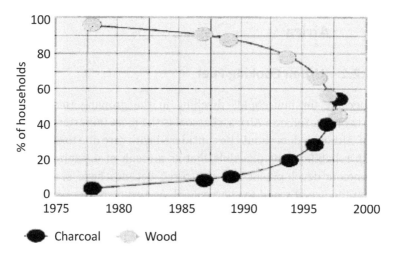

Figure 14.1 Domestic fuel use in Bamako, Mali: 1975–1998 (% of households using wood or charcoal) [3].

14.2.2 Environmental Issue: Deforestation

The use of charcoal as a cooking fuel in developing countries, particularly sub-Saharan Africa, has been cause of much controversy. Charcoal production has been linked to forest degradation and deforestation, which is the reason that in most countries the charcoal sector is viewed negatively. This attitude has caused governments to focus on restricting and regulating the sector as much as possible, trying to move away to other energy sources. Substitution programs and discouragement of charcoal production, however, can have a negative impact on the employment opportunities in rural areas, thus increasing migration to towns and cities and augmenting the demand for charcoal in urban centers. The FAO reported already in 1993 that banning charcoal production and trade drove the sector underground and resulted in cartel-like operations and corruption [7]. A recent study by the Kenya Forest Service revealed that bribes can amount to 10%–13% of the final charcoal price, resulting in low returns for producers and high prices for consumers [8].

Already in 1985 Foley found that clearance for agriculture is the major cause of deforestation in most regions of sub-Saharan Africa. The charcoal which is produced as a by-product from wood from clear-felling is, however, often blamed for this [9]. Early predictions that supply sources for firewood and charcoal would be unable to meet the growing demand, with steeply rising prices for urban consumers, have been proven wrong. There is significant regional variation in demand and supply of charcoal in sub-Saharan Africa. Wood-harvesting techniques vary from selective cutting and coppicing to clear-cutting or selective species felling, with different consequences for forest cover and biodiversity. However, charcoal production for urban markets in rural areas with weak supply sources (particularly dry land forests) is still a major cause of forest degradation and therefore is of great concern. The ongoing shift from firewood to charcoal could have major adverse environmental consequences if not properly managed or mitigated, especially in countries with weak institutions and a lack of financing and skilled personnel. Characteristics of the national charcoal sector and charcoal value chain were studied in depth in many African countries with support from international and national organizations and agencies. Insights and results gained from those efforts relating to raw materials for charcoal production, charcoal production technologies, including emissions, and charcoal use are discussed below.

14.2.3 Charcoal Sources

Charcoal in sub-Saharan Africa is mainly produced from wood harvested from natural forests or bush lands. Preferably hardwood species from dry savannah forests are used, which yield a dense, slow-burning charcoal. In many places those slow-growing species are endangered by overexploitation. Three types of wood sources can be distinguished:

- Wood residues resulting from another forestry-related (thinning, cutting, sawmilling) activity
- Wood from forest clearing for agricultural purposes (including slash and burn agriculture or removal of invasive species)
- Forest and bush wood specifically harvested for charcoal production

A particular (niche) source of sustainable charcoal production exists in Southern African savannah regions which are affected by bush encroachment. This phenomenon has decreased the grazing capacity in large areas, thus rendering many livestock properties no longer economically viable. Removal of some or most of the woody plants will result in increased grass production and grazing capacity. In Namibia 10–12 million hectares of farmland and rangeland suffer from bush encroachment [10], which is responsible for the reduction of the total number of livestock in Namibia from 2.5 million in 1958 to 800,000 in 2001. Other studies [11] show that about 10 tons per hectare of excess woody biomass is available, representing a total of about 100 million tons of raw material. Assuming a bush reproduction cycle of 10 years, the total biomass potential is 10 million tons per year (moisture content 20 wt.%). At an assumed (low) conversion efficiency of 10%, Namibia only could sustainably produce about 1 million tons of high-quality charcoal, or about 2% of the current world or 3.3% of the current total African charcoal production. The Namibian charcoal industry has grown from exporting 50,000 tons in 2004 to 85,000 tons in 2012. Exports are mainly (70%) destined for European countries, while the remainder is sold in the South African market.

Dedicated wood plantations for charcoal production are uncommon in sub-Saharan Africa, although many different species of relatively fast-growing and coppicing (eucalyptus) trees (*E. saligna, E. torreliana, E. tereticornis*, etc.) are available. The wood is suitable for charcoal production and the available range of tree species allows fitting to local conditions. Mean annual increments of 15–20 m^3 wood per hectare over 15–20 years rotations are quoted in the literature [12]. Eucalyptus charcoal is less dense than charcoal from savannah hardwood [13], but in mass terms the energy content is the same. In local sub-Saharan African markets charcoal is sold by volume and not by weight, which explains the unpopularity of lightweight charcoal. A major shift from hardwood charcoal to plantation eucalyptus wood-based charcoal can therefore only occur when the sale-by-volume retail system is replaced by a sale-by-weight system. About 1.2 million hectares of commercial wood plantations are found in the Republic of South Africa, mainly in Kwazulu-Natal and Mpumalanga. Wood is used for pulp and timber but plantation waste and sawmill residues are carbonized on a (semi-)industrial scale and for the larger part exported by sea containers to Europe and Asia.

Many agricultural residues have been considered as substitute raw materials for wood charcoal. A preferred material for making superior-quality charcoal for industrial and recreational use is coconut shells. In the literature proposals can be found for charcoal making based on agricultural residues like cotton stalks, soy bean stems, rice husks, coconut husks, sugarcane bagasse, and processing residues from coffee, tea, and oil press cake. Those materials produce a powdery high-ash-content charcoal, which must be briquetted before it can be used in traditional domestic stoves. The heating value per unit weight is lower than from wood charcoal, which may lead to acceptance problems in the domestic market. For the production to be profitable there must be a market with sufficient buying power for powdery charcoal or charcoal briquettes. Charcoal and charcoal briquettes from residues can be costly. Residues are sometimes dispersed in the field, which can result in (too) high collection and transport costs. Residue supply is sometimes seasonal, which, unless the residues can be easily stored, calls for a very-low-cost-processing technology which can operate commercially only part of the year. Also the carbonization costs may be expensive in terms of both equipment and manpower. Unless factors like this are carefully taken into account projects are bound to give out on economic terms. Fast-growing bamboo has been proposed [14] as a renewable alternative for domestic fuel wood and charcoal production in sub-Saharan Africa. In an European Commission (EC)-funded project, activities are implemented at pilot sites in Ethiopia and Ghana. Different bamboo species were planted for testing. Bamboo charcoal technology was acquired in China [15] and fitted to local conditions and charcoal makers were trained. To date (2014) more than 1000 tons of bamboo charcoal have been produced.

14.2.4 Production Techniques

In sub-Saharan Africa by far the largest part of charcoal is produced by (sometimes part-time) small- and medium-scale independent charcoal burners. The producers as a rule use traditional methods which are characterized by low investments and limited efficiencies. Large scale (semi-)industrial type operations are uncommon outside the Republic of South Africa and Namibia. More advanced technologies (see the next chapter) call for a better infrastructure in finance, technical support, human resources, and logistics than is usually available in the charcoal sector in sub-Saharan Africa [4]. A traditional charcoal production technique (earth pit kiln), which has been practiced for centuries (Fig. 14.2), is to start a wood fire in a dug-out hole in the ground, to fill-up this pit with wood until the fire burns well, and subsequently to cover the pit with leaves and loose earth. Combustion of part of the wood produces enough heat to carbonize the remainder. After one or two days the pit can be opened and the charcoal recovered. Small earth pit kilns may show very low efficiencies (below 8%–10% air-dried weight basis), especially as they are often used by less experienced part-time charcoal makers. However, also large and deep pits in which big pieces of wood can be carbonized are used by professional charcoal producers and can be encountered in sub-Saharan Africa. Due to a lack of process controls the technique produces batches of variable-quality charcoal characterized by wide differences in fixed-carbon content and volatile content.

Figure 14.2 A traditional small earth pit kiln with the wood charge already lit but before the pit is covered by leaves and loose earth (http//www.greenyourhead.com/biochar/).

Alternatively a wood pile (traditional mound kiln) is covered with grass or sod and loose earth. A number of air inlets (up to 10) are made at the bottom of the mound, and a vent is provided in the top to allow exit of smoke during burning. The mound is lit through the opening in the top. By judiciously opening and closing the air inlet holes in the outer cover, the air supply to the kiln can be checked, thus allowing more control over combustion and carbonization of the load than with the earth pit method. When the smoke from the top opening becomes clear, the charcoal is ready and all air inlets and the vent are closed and the mound left to cool. Depending on the size of the mound this may take 3–12 days. Experienced charcoal makers can manage mass yields of about 20%–25% (200–250 kg of charcoal from 1 ton of air-dried wood) with well-constructed and well-tended traditional mounds. In practice poor craftsmanship (like the use of green wood and insufficient supervision) often results in much lower yields of 8%–12% [16]. Both techniques persist to this day in many African countries because they require little or no investment. Since the 1970s and 1980s efforts are made to introduce improved traditional charcoaling techniques. One technique developed in Senegal (Casamance kiln) (Fig. 14.3) equips a traditional mound kiln with one or more simple steel chimneys made from oil drums, which allow a better control of air flow and hot flue gas circulation in the mound.

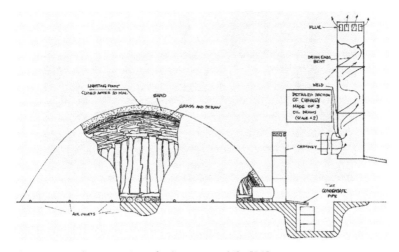

Figure 14.3   Cross section of a Casamance kiln [12].

As a result carbonization is faster and more uniform than in a traditional mound. Casamance kilns, when properly tended, can achieve air-dried wood efficiencies up to 30% [17]. Care should be taken when comparing yield data from different references. Although the concept of mass yield or efficiency is simple and unequivocal (ratio of mass charcoal produced and wood used) the outcome of trials depends heavily on the characteristics of the wood (moisture content, ash content, sand content) and the charcoal (fixed-carbon content, volatile matter content, ash content). Other critical factors which influence the production efficiency are craftsmanship in construction and operation of the kiln and climate and weather conditions. All these factors have led to some controversy in the literature about the relation between efficiency and production technology. Some authors conclude from the very scattered data that the traditional and improved kilns do achieve more or less the same efficiency [18]. However, all other factors being equal, the Casamance kiln (Fig. 14.4) is found to produce about 1.4 times as much charcoal from the same amount of wood as the traditional earth mound [19, 20].

Figure 14.4 Casamance kiln in operation in Sambande Forest, Senegal [19].

Other (semi-industrial) technologies which were introduced and tested for charcoal production in sub-Saharan Africa comprise brick and steel kilns (Fig. 14.5). Brick kilns, unlike traditional mounds or

Casamance kilns, are stationary once built. Therefore they are only practical and economic at sites were an easy, long-term wood supply is guaranteed. A typical brick kiln like the Argentine "half orange" with a 6 m diameter can produce up to 7 tons of charcoal in a 14-day carbonization cycle with an air-dried wood efficiency of up to 30% [21]. Different types and sizes of steel kilns are the workhorses of the charcoal industry all over the world. They are designed for dismantling and transport and show reasonable to good efficiencies of 25% to 30%, when operated properly. Because of high investment (typical US$ 1000–1500 for a steel kiln manufactured in low-labor-cost countries and producing 100–150 tons per year), they are out of reach of small-scale traditional producers but may be an option for larger production enterprises.

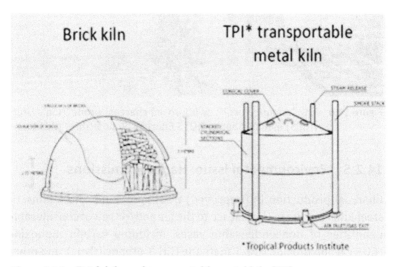

Figure 14.5 Brick kiln and transportable metal kiln [21].

An interesting development is the improved charcoal production system (ICPS), or "adam-retort"® [22] (Fig. 14.6). This stationary device burns inferior waste wood and residues in an external fire box to generate heat for drying charcoal wood inside the retort. Steam from the drying wood is emitted to the air through a chimney. Once steam emission stops and the wood starts carbonizing (as indicated by the color of the smoke from the chimney), the combustible gases which evolve are redirected into the fire box, where they burn and the heat generated is recycled. An air-dried wood efficiency of 30%–

40% is claimed. Depending on the local labor and material costs, a unit producing 200–350 kg of charcoal per day can be built for €300–€1200.

Figure 14.6 The "adam-retort,"® or improved charcoal production system (ICPS) in Mali. *Source*: NOTS Energies Renouvables (2011).

14.2.5 Environmental Issue: Harmful Emissions

Charcoal production in (improved) traditional kilns, brick kilns, or steel kilns results in the release to the air and/or soil of considerable quantities of noncondensable gases, including carbon monoxide (CO), carbon dioxide (CO_2), methane (CH_4), ethane (C_2H_6), and other volatile organic compounds (VOCs); condensable organic compounds like acetic acid, pyroacids, heavy oils, and tars; and particulate matter. Indeed, areas and sites where traditional charcoal making is in progress can often be identified from a distance by the dark clouds of smoke in the air. The emissions from charcoal making are a definite health hazard and a source of different types of respiratory problems and diseases to charcoal workers and villagers living near the charcoaling sites. Retort kilns like the "adam-retort,"® which burn organic compounds in the smoke, reduce the emissions (except CO_2) by 90%–99% [23]. Emissions from traditional technologies expressed in grams per kilogram of charcoal are in the order of 450 to 550 g for CO_2, 450 to 650 g for CO, 700 g for CH_4, and 10 to 700 g for nonmethane hydrocarbons (NMHCs) [24]. CO_2, CH_4, and some NMHCs are greenhouse gases which contribute to global warming.

14.2.6   Charcoal Briquettes

During handling and transport, about 10% to 20% by weight of wood charcoal is reduced to powder ("fines"). Charcoal fines have a much higher ash content (10%–30%) than lump charcoal (5%) because they contain sand ("tramp ash") which is inevitably present on the surface and in the bark of the wood. Powdered fine charcoal produced from agricultural and processing residues has also a (much) higher ash content than lump charcoal. To be used as household fuel in the usual African charcoal cooking stove this charcoal powder needs to be briquetted. Briquetting requires addition of (hot) water and a binder to the charcoal fines, a machine or device which turns the mixture into lumps, balls, cakes, or other type of briquettes and a (solar) drying device or (wood) heated drying furnace to cure and strengthen the product. Usually starch is used as a binder; also molasses from sugar production and tar and pitch recovered from charcoal production are used. Some types of plastic clays are suitable and cheap, although they are incombustible and lower the heating value of the briquette. There is briquetting equipment available for village- and small-scale industrial application, in the form of (hand) molds, simple extruders, agglomerators [25], and small roller presses producing many different forms of cakes, balls, and briquettes. In the United States and Asia briquettes are universally used for home barbecue and grilling and cooking in restaurants. In sub-Saharan Africa the uptake of charcoal briquettes has been much more limited in scale. Only a few African entrepreneurs (among others Chardust Ltd., Nairobi, Kenya) have been successful in the charcoal briquetting business. Perceptions of inferior quality compared to regular lump charcoal (difficult to light, higher fuel consumption, lower combustion temperature) can be due to briquette quality or may be caused by insufficient adjustment to briquette fuel of the domestic stove or oven. Difficulties in ensuring a reliable, cheap supply of fines in urban sites far away from the rural charcoal production areas can make it hard to generate sufficient sales volume and revenue. Scarcity of drying space in urban areas and drying problems in the rainy season may lead to low productivity or even temporary stoppage of the factory. The upfront costs to start and operate a briquette factory seem often prohibitive to potential entrepreneurs. Most producers in East Africa obtained start-up financing partially or fully through donor funding and grants [26]. This suggests that successful industrial briquette operations depend on a confluence of factors which are usually absent in sub-Saharan Africa, that is:

- adequate capital for good equipment and facilities and a sufficient turnover to cover the amortization;
- competent commercial and technical manpower;
- a product of adequate quality at an acceptable and competitive price to the domestic user; and
- ability to produce/acquire charcoal dust/fines at very low cost, close to the major market, and in steady quantities throughout the year.

14.2.7 Charcoal as a Fuel for Small-Scale Power Generation

Charcoal (and other carbonaceous solid fuels) have a potential to replace petroleum fuels in internal combustion engines by means of an old technology called (small-scale) gasification. So-called producer gas is made by passing air through a bed of hot glowing charcoal. The air burns the charcoal to produce carbon dioxide. This substance is largely reduced to carbon monoxide on the hot charcoal next to the burning zone. Any water present in the fuel or the air is reduced by the hot charcoal to hydrogen and carbon monoxide. The result is a combustible gas consisting of carbon monoxide, carbon dioxide, hydrogen, and nitrogen (from the air), which after cooling and filtering is suitable for use in petrol, diesel, and gas engines. Producer gas can be made from many different fuels like mineral coal, brown coal, peat, biomass, etc. There are many types of gasifiers which are adapted to specific fuels [27]. Charcoal is a very good fuel because it produces a tar-free gas and has a relatively low ash content. The gas from a typical "cross draft" small-scale charcoal gasifier has a lower heating value (LHV) of 4.0–5.2 MJ/Nm^3 dry gas, which is about 25% of the LHV of Dutch natural gas. Nevertheless it can be readily used in internal combustion engines, although a considerable derating of the maximum engine power output (in the order of 50%) has to be taken into account.

Small-scale gasifiers for stationary and mobile power applications have a long history. They were used commercially in the 19th century and have seen extensive use in the 20th century, mainly as a means to compensate for war-time shortages of petroleum fuels. The energy crises of the 1970s and 1980s have rekindled interest in small-scale biomass gasification. At the end of the century this has specifically led to a number of initiatives to demonstrate the potential benefits of the introduction of biomass (wood or charcoal) gasifiers in developing countries. It was seen that dissemination of the gasifiers in developing countries could reduce fuel costs for small-scale power generation in remote areas and that it could improve the reliability of fuel supply by making isolated rural industries or communities more self-reliant. By the early 1980s, more than 15 manufacturers (mainly in Europe and North America) were offering wood and charcoal power gasifiers in capacities up to about 250 kWe. In addition, agencies such as Directoraat Generaal voor Internationale Samenwerking (DGIS) of the Netherlands, Gesellschaft fur Technische Zusammenarbeit (GTZ) of Germany, Agence Française pour la Maitrise de l'Energie (AFME), the Swedish International Development Authority (SIDA), and the DGI of the European Community were financing the installation of test and demonstration biomass power systems in many developing countries. In addition, at least six developing countries (Brazil, China, India, Indonesia, Philippines, and Thailand) had started power gasifier implementation programs of their own, based on locally developed technologies. In Brazil, India, Paraguay, and Thailand, the technology was promoted largely by local entrepreneurs and manufacturers [27].

With this revival of small-scale gasification the World Bank and the UNDP decided to assess the technology thoroughly before endorsing further dissemination in developing countries. An intensive worldwide monitoring program of existing operating systems resulted in the following conclusions:

- The economics of charcoal gasifiers are depend highly on three factors, that is, the on-site cost of diesel fuel, the on-site cost of charcoal, and amortization costs of the equipment.

Small-scale (10 kW) charcoal gasifiers require on-site diesel fuel costs of US$ 500 per ton or more and (good-quality, low-volatile) charcoal costs below US$ 50–75 per ton on-site when the equipment is in constant use. If the engine is used only sporadically or for a short time (as is often the case in isolated locations in developing countries) charcoal gasification is not economic.

- There are some disadvantages in the use of charcoal as compared to the convenience of liquid fuel. The advantage of replacing imported expensive liquid fuels by local charcoal is obvious. However, operation of a gasification system requires training and experience. Operator motivation and discipline are necessary. There are some success stories of replacement of liquid fuel by producer gas. However, the operational history of the gasifiers which were monitored shows that not all operators could master the required competences. Gasifier operations appeared to require a well-structured operator training program and operator backup for a prolonged period.

14.3 Issues of Charcoal Production and Use in Industrial Countries

14.3.1 Applications

Even in industrial countries the major part of charcoal is used as a fuel for cooking meat either in leisure activities like barbeque grilling or in grill restaurants. In sharp contrast to the United States, most European users prefer lump charcoal over charcoal briquettes. Price is the overruling factor in the choice for a specific supplier or brand, although especially in the restaurant sector there is some awareness about differences in heating value, burning behavior of softwood and hardwood charcoal, smoke and smell production, and ease of ignition. Relatively high prices are paid for two specific types of grill charcoal. "White" or Binchotan charcoal is a very-high-carbon-content (95% C) dense charcoal made from oak wood. It is a prized product used in high-class Japanese grill (yakitori) restaurants and increasingly also popular in specific restaurants in Europe. It is produced by slow carbonization at low temperature (350°C–400°C), a short recarbonization at high temperature (1000°C), and fast cooling under a sand layer. It is completely odorless and burns very slowly. Traditionally the best Binchotan comes from the Kishu region in Japan and can be recognized by the metallic sound charcoal pieces make when tapped together. Nowadays an almost comparable quality can be achieved by expert carbonization of high-quality binderless extruded sawdust briquettes [28]. Also charcoal from coconut shells usually fetches a higher price in the restaurant grill charcoal market than ordinary lump charcoal. Quality coconut shell charcoal is preferred because it burns slow, flameless, and homogenous, is odorless, and produces minimal ash.

Another major application of charcoal is as a reductant in various metallurgical industries like iron production and copper refining, production of ferrosilicon and ferromanganese, or silica smelting. Silicon in different grades of purity is used in the electronics industry, in which there is a fast growing demand for semiconductor and "solar grade" silicon. The reduction of 2.5 tons of good-quality quartzite ore containing about 1 ton of silica (SiO_2) to silicon (Si) requires about 1 ton of charcoal. The most important property of charcoal in metallurgical applications is the fixed-carbon content, since it is the fixed carbon which reduces the oxide to pure metal. Metallurgical-grade charcoal has a high fixed-carbon content (75%–80% or higher), low ash content (3%–4%), good crushing strength, uniform size (20–50 mm), low-volatile matter content (below 20%), and a high ignition temperature. A typical silicon smelter may consume 30,000 tons of charcoal per year. Large industrial users know from research and experience which exact qualities they seek in the charcoal they purchase and have methods to check whether it conforms to specifications. Usually they produce at least part of their charcoal consumption themselves. Recently (2013) a South African silicon-smelting company [29] announced the establishment of about 10,000 ha of dedicated wood plantations in order to cover its charcoal demand in a sustainable manner.

A sector which has fast grown in importance and volume over the last 30 years is activated carbon production. Activated carbon is a solid, porous, carbonaceous material prepared by carbonizing and activating organic substances. It is marketed in granular and in powder form. Activated carbon can be made from raw materials such as wood or agricultural residues but also from peat, lignite, petroleum coke, and coal. Exceptionally good activated carbon [30] is made from coconut shells mainly due to high strength and a fine macropore structure which renders it more effective, especially in gas or vapor adsorption applications. Activated carbon is used for many purposes like refining and bleaching of vegetable oils and beverages, water purification, air purification (toxic gases and mercury), pharmaceuticals, semiconductor and electronics manufacture, recovery of solvents and vapors, filters, and gas masks, as well as mining and gold recovery. The activation process is carried out in two stages. Firstly the feedstock is carbonized in some sort of charcoal kiln or retort to result in a charcoal of at least the following specifications: fixed carbon minimum 82%, ash maximum 4%, and volatiles around 10% [31]. Secondly the charcoal is activated with steam at a temperature of 750°C–1100°C under controlled atmosphere in a rotary kiln. Sometimes the two stages are combined in one piece of equipment, usually a rotating drum or kiln. The reaction between steam and charcoal takes place at the internal surface, creating active sites for adsorption. The activation temperature is important. Below 750°C the reaction becomes very slow. Above 1100°C the reaction becomes diffusion controlled and therefore takes place on the outer surface of the charcoal, resulting in charcoal losses. The global market for activated carbon was valued at about US$ 2 billion in 2012 and is expected to reach US$ 4 billion by 2019 [32]. In terms of volume, activated carbon demand was about 1.2 million tons in 2012 and this amount is expected to grow by 10.2% annually from 2013 to 2019. The market is dominated by large industry participants (among others Calgon Carbon Corporation, Carbon Resources LLC, Haycarb PLC, Cabot Norit Activated Carbon, and Siemens Water Technologies Corporation).

14.3.2 Industrial Technologies

Industrial charcoal kilns can be roughly classified according to the way in which the wood is heated. The oldest method is to burn part of the wood inside the kiln in order to generate heat for carbonization of the rest. Advantage of the system is that heat is generated on the spot where it is needed, thus limiting problems in heat transfer. However, adequate control of air admittance to the kiln is essential to achieve efficiency. Specifically in traditional equipment this is difficult because it requires experience, craftsmanship, and dedication. The air pollution associated with large-scale charcoal production in internal combustion units can be alleviated by burning the (condensable) gases and smoke with additional fossil fuel in afterburners. Acceptance of this expensive technology requires strict emission legislation and control. Still internal combustion is the technique by which to this day, most of the world's charcoal is produced in mound, brick, or steel kilns. The last half of the 19th and the first half of the 20th century saw a breakthrough in wood carbonization equipment. This development is closely linked to the emergence of a chemical industry which was interested in the process of "wood distillation." Different batch and continuous retort systems for wood carbonization were developed which were mainly focused on the bulk chemicals (like acetic acid, acetone, and crude methanol) and some special chemicals (like butyric acid) which can be reclaimed from wood vapors. Charcoal was produced as a by-product. In a retort system heat is generated outside the kiln through combustion of secondary wood and/or wood gas. Heat is transferred to the wood in the retort vessel either by radiation/conduction through its walls or by circulating hot inert flue gas inside. This causes the wood to carbonize and as a result (combustible) gases/vapors and charcoal are produced. The former can be (partly) condensed and/or burnt after leaving the retort vessel. The remaining charcoal in the vessel is recovered after cooling. The wood distillation industry was the precursor of the petrochemical industry. Before the mineral oil era started in the 1920s, all important organic chemicals were produced from wood. By the middle of the 20th century the wood distillation industry had completely vanished due to the competition of the petrochemical industry. In the industrial countries, however, the demand for quality charcoal to use as a reductant in the metallurgical industry or to make activated charcoal remained. This resulted in the development of charcoaling technologies which were more efficient and less polluting and labor intensive than the kilns based on internal combustion. Some systems with a proven commercial track record are commented upon hereunder. It is important to understand that each system arose out of a specific need or situation and therefore may not automatically be successfully employed in different circumstances.

14.3.2.1 The Missouri kiln

The Missouri kiln is a stationary industrial unit for batch-wise carbonization of sawmill slabs and edgings or low-quality round wood. The kiln was developed in Ozark County area in Missouri, USA, over some period of time and is based on the principle of internal combustion. Around 1960 a more or less standard design emerged (Fig. 14.7). This unit is constructed in brick or poured concrete which must be able to withstand the stress of high heat. The internal volume is 150–200 m^3. The ground plan is rectangular in shape, the walls are 2.5 m high and 250 mm thick, and the vaulted roof has a maximum height of about 4 m above floor level. Steel doors allow entrance to a tractor, small lorry, or front-end loader for mechanically loading the wood and unloading the charcoal. The kiln is equipped with air inlets (in boiler pipe steel quality) in the bottom, front, and side walls and has ventilation ports in the roof and the upper walls. A number of stoneware pipe chimneys (usually four on each side) allow smoke to escape to the atmosphere.

Figure 14.7 A battery of Missouri kilns in Ozark County, Missouri, USA. The kilns are connected to an afterburner (center picture) for pollution abatement. *Source*: Missouri Department of Natural Resources, Jefferson City (MO), USA.

The kiln is lighted through the air inlets or special lighting ports by means of an oil burner with the ventilation ports in the open position. The wood inside is allowed to burn for some time (about 0.5–1 hour). Large amounts of steam are released from the ventilation ports. Once the kiln has reached the required temperature, the ventilation ports are closed. The bottom air inlets to the center of the kiln may be partially closed, as judged necessary by the operator, and hot gases move through the center to the top of the kiln and down on to the outer edge, where they are released through the smoke chimneys. Missouri kilns are usually equipped with a number of thermocouples which indicate any cold or hot spots inside the kiln which require action. Nevertheless charcoal making in a Missouri kiln is still more an art than a science and requires extensive experience and craftsmanship from the operator. Kiln damage due to overheating is quite common with inexperienced operators [33]. Carbonization is complete when the smoke from the chimneys becomes transparent as hot air. At this point the air inlets in the bottom of the kiln are closed and the smoke chimneys are plugged, thus starting the cooling phase. The thermocouples are useful for checking the progress of the cooling. When the kiln is sufficiently cool the metal doors are opened and the kiln content discharged. A crew of two workers with front-end loader and lorry is employed for loading and unloading, one operator per shift is sufficient to control the carbonization phase and one operator can supervise a number of kilns in the cooling phase. Loading takes about 2 days, carbonization about 6 days, cooling 20 days or more, and, depending on weather conditions, unloading about 2 days. The total cycle time is in the order of 1 month or slightly more. To optimize labor costs it is advantageous to operate Missouri kilns in groups of three or more. A well-operated average Missouri kiln has a dry weight efficiency of about 33% [18]. A production of 500–600 tons of lump charcoal per average kiln and per year from 1500–1800 tons of air-dried hardwood (like white oak, red oak, or hickory) is feasible in a temperate climate. The big advantage of the Missouri kiln is the possibility of mechanical loading and unloading and the associated labor cost savings. Disadvantages are the high investment costs requiring a long amortization period and the huge amount of wood which must be sustainably available for a period of at least 10 years within economic hauling distance of the charcoaling site. In the 1960s about 50,000 tons per year of lump charcoal mainly for metallurgical purposes were produced with Missouri kilns in the Ozark Mountains. For decades before and thereafter the area had to live with the columns of black acrid smoke which rose from the charcoaling sites, leaving behind a residue of gritty soot. In 1997, after five years of discussions with the Missouri Department of Natural Resources and a toughened stand by the US Environmental Protection Agency, charcoal makers agreed to close down at least 64 kilns which were too old to modify and equip the remaining 310 and any new kilns with afterburners (Fig. 14.7) before 2005 [34].

14.3.2.2 Carbo twin-retort carbonizer

The Carbo twin-retort carbonizer [35, 36] is an example of a successful stationary batch (or semicontinuous) medium-size industrial system. It is a proven technology for carbonizing air-dried (10–30 cm) waste lump-wood pieces and extruded briquettes from sawdust or agricultural residues. A twin-retort system consists of a steel sheet and frame module, which inside is heavily lined with refractory bricks (Fig. 14.8). Inside the module are two stainless-steel retorts. Each retort can hold a 4.5 m^3 steel carbonization vessel containing the wood charge. The outside retort wall is heated by hot flue gas from the wood gas afterburner. Heat is transferred from the hot retort wall to the vessel wall by radiation. The wood pieces inside the vessel are heated by the convection caused by the hot vessel wall. A cold or hot vessel can be moved in and out of the hot retort by means of an overhead crane. The two vessels are effectively operated counterwise. When the wood inside a vessel is heated, as described above, steam and gas are consecutively released through holes in the bottom of the vessel, and these flow under pressure to the afterburner furnace, where the combustible components are completely burned with excess air. The resulting hot flue gas is directed by means of a damper system to a chimney via the outside of the other retort. The radiation from this retort causes a fresh cold wood-filled vessel to heat up and eventually reach carbonization temperature. Once a vessel stops releasing gas, to be judged from the flame pattern in the afterburner, the charcoal is ready and the hot vessel is removed from the retort and replaced with a fresh cold one. The other vessel should now have reached carbonization temperature and by means of a damper system the flow of hot wood gas from this vessel to the afterburner furnace and of hot flue gas from the furnace via the outside of a retort to a second chimney is reversed. Initial start-up of the kiln is achieved by means of an oil or gas burner. Once the process temperature is reached the cycle can be continued interminably by burning wood gas without need for another fuel. Because of changes in heating value of the wood gas over the time of the carbonization cycle sufficient refractory material inside the kiln is necessary to act as a heat store and dampen out temperature differences caused by variations in heat production over time.

Figure 14.8 Carbo twin-retort kiln with a carbonization vessel in the foreground to the right. *Source*: Ekoblok B.V. See also Color Insert.

The carbonization time of the wood is influenced in part by the temperature of the flue gas from the furnace. The latter can be adjusted by control of the excess combustion air flow to the furnace. It takes about 8–12 hours for the wood inside a vessel to carbonize sufficiently, depending on wood properties (moisture and morphology), wood species, and charcoal requirements (fixed-carbon content). Therefore every 4–6 hours a (hot) vessel must be removed and a (cold) vessel uploaded. Hot vessels are left to cool for about 24 hours, sealed by sand locks to prevent air entering through the gas release holes and set fire to the charcoal inside. Hot (350°C–450°C) flue gas from the chimneys can be used to predry the wood. In view of loading, predrying, carbonization, and cooling requirements each twin-retort system should be equipped with at least 10 vessels. The Carbo twin-retort has an efficiency of 30% or more. The capacity depends on the carbonization time. A twin-retort can produce about 900–1000 tons charcoal per year from about 3000 tons per year of air-dried hardwood. One operator per shift can operate (load and discharge) and supervise a battery of ten twin-retorts, provided wood-charged vessels are available. One worker equipped with a front-end loader and conveyor belt needs about 8 hours to load 16 vessels. Advantages of the twin-retort technology in comparison to most other charcoaling equipment of comparable size and capacity are:

- The system is equipped with an internal afterburner furnace operated with an excess of air which completely burns all organic components (like CH_4, NMHCs, and CO) in the wood gas vapors, converting them to carbon dioxide and water and resulting in a very low level of environmentally harmful emissions to the air.
- The low gas flow rates in the furnace and chimneys result in low particulate emissions which comply with strict Dutch emission directives.
- The labor requirements compare favorably with systems of comparable capacity.
- The possibility to remove the hot vessel from the module separates the charcoal cooling from the rest of the carbonization process, thus allowing maximum flexibility in process procedures and control.
- The modularity of units allows a relatively easy expansion of production capacity.
- The system is able to produce a uniform product in accordance with the fixed-carbon specifications of the client by adjustment of the carbonization time and control of the carbonization temperature.

Under circumstances a weakness of the systems is the need for a relatively big uniform-sized lump wood or briquetted feed in order to enable sufficient convective gas flow for good heat transfer inside the vessel. A sufficiently low feed moisture content of the feedstock (below 25%-w) is necessary in order to allow effective balancing of the drying/heating time and the carbonization/vaporization time of the two alternating vessels. A disadvantage is the comparatively high system investment cost, as besides the twin-retort proper also the building and auxiliary equipment (vessels, hoist and rail, front-end loader and rotator) must be taken into account. Carbo twin-retorts are in use, among others, in China (silicon smelter charcoal), Estonia, France, the Netherlands, Senegal (husk briquettes charcoal), Singapore, and South Africa. This suggests that the system is competitive and applicable in situations of, among others, strict emission directives and emission control, high labor costs, specific charcoal quality requirements, and high charcoal prices.

14.3.2.3 Herreshoff, or multiple-hearth, furnace

The Herreshoff, or multiple-hearth, furnace is a proven large-scale technology for continuous carbonization of small-particle-size material like sawdust, bark, straw, and husk. It consists of a series (4–10) of circular hearths or plates located one above the other inside a refractory lined steel shell (Fig. 14.9). The raw material enters continuously at the top of the carbonizer and is spread onto the surface of the first hearth. A rotary shaft in the center of the furnace slowly (1–2 rpm) moves air-cooled radial agitators ("rabble arms"), which circulate the feedstock in spiral lines across the hearth until it drops through exit holes to the hearth directly below. The material circulates in this way over each of the hearths, descending progressively through the entire furnace until the product is discharged through one of more ports in the bottom hearth. As the material progresses through the furnace, it is gradually dried, heated, and carbonized. The unit is started by heating up each hearth to a temperature of 500°C–600°C using external gas or oil burners. This causes the feedstock to ignite and the combustible volatiles which are released are burnt with controlled amounts of combustion air. This provides the heat for the process and so enables for the burners to be switched off. The temperature is controlled by the combustion air admission rate. The heat transfer is achieved through direct contact of the hot gases with the thin bed of material, which is constantly agitated. The hot gases circulate upward in the furnace in countercurrent to the feedstock. Upon discharge, these gases, which are still rich in combustible components, are burned in a boiler to produce process steam or power. The powder charcoal from the bottom hearth is cooled before storage and briquetting. Herreshoff furnaces are large industrial-size factories (Fig. 14.10) with feedstock capacities in the order of 4–10 tons of sawdust/bark per hour. The charcoal yield is about 25% by weight, resulting in a production capacity of about 1.0–2.5 tons of charcoal powder per hour per furnace or a production of about 8,000–20,000 tons of powder charcoal per year per Herreshoff plant.

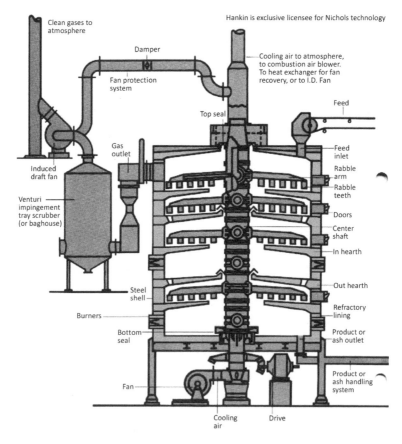

Figure 14.9 Schematic cross section of a Herreshoff, or multiple-hearth, furnace [37].

2000-1130

Figure 14.10   Hankin–Nichols multiple-hearth furnace [37].

The major advantage of a Herreshoff furnace is its ability to efficient and flexible use fine-grained feedstock material of little commercial value. A disadvantage is the need for briquetting of the charcoal powder before it can be used in barbecuing and the associated added cost of this process to the final product. Under circumstances another disadvantage is that charcoal production costs are markedly influenced by unit size [37]. Units at the top of the capacity range can produce charcoal at almost half the cost of the smallest equipment. Because of high investment cost and a long amortization time a Herreshoff unit needs a guaranteed supply at an economic transport distance of very large quantities of suitable feedstock for at least a 10-year period. Herreshoff kilns are successful in the southern United States, where the availability of large amounts of sawdust and bark from big plywood mills and sawmills and the proximity of a sophisticated urban barbecue market for charcoal briquettes provide a profitable combination [38].

14.3.2.4 The Lambiotte retort

The SIFIC/CISR Lambiotte retort[1] is an example of a successful stationary continuous large-scale industrial system. It is a proven technology for the production of conventional all-purpose lump charcoal from logs and pieces of lump wood of relatively uniform size (length about 25–35 cm and diameter about 8–15 cm) and a maximum moisture content of 25 wt.% (preferred moisture content for SIFIC plants is about 10–15 wt.%).[2] Although there are some differences in design between the SIFIC unit (equipped with a system for recovery of by-products from the pyroligneous vapors) and the CISR unit (no by-product recovery), the working principle of both retorts is the same (Fig. 14.11). Precut and (when necessary) predried wood is entered by means of a conveyor belt and skip over the top of the vertical shaft furnace through a lock hopper to prevent gases escaping. The level of feedstock in the retort is monitored by means of a gamma-ray detector and the loading controlled to keep the level constant. Inside the furnace the charge moves slowly down under gravity. Typical residence time of the charge inside the retort is around 11 hours. However, the operator can control the residence time by adjusting the intervals between the removals of charcoal at the base of the retort. A Lambiotte retorts employs two closed gas loops, one for drying and carbonization of the wood charge and the other for cooling of the charcoal. On its way down the wood charge is dried and raised to carbonization temperature by hot flue gas flowing in countercurrent to the charge, injected more or less at the middle of the furnace and drawn by a fan. Addition of controlled amounts of combustion air for burning some of the vapors inside the retort compensate for deviations in feedstock moisture and provide the possibility of temperature control. The flue gas and remaining carbonization vapors which leave over the top of the furnace may be (partly) condensed and the remainder is burnt. The hot flue gas from this burner is partly re-injected for heating the charge and partly cooled. The cooled inert gas is injected in the bottom of the furnace and flows upward thus cooling the charge which at this spot is converted to charcoal. The (now hot) gas is removed from the furnace just below the carbonization zone. The

---

[1]www.lambiotte.com.
[2]Personal information J.G. Mares, Société Usines Lambiotte, Prémery, France (2001).

cold charcoal is removed from the furnace in intermittent intervals. The correct circulation of the two gas streams is controlled by a pressure sensor and damper valve. The system is started by means of oil and gas burners which also can be used to heat inert gas in case of heat deficiency due to moist wood. Lambiotte retorts are large industrial equipment (typical height of retort 18 m, diameter 3 m) with feedstock capacities in the order of 2000 to 6000 tons per year.

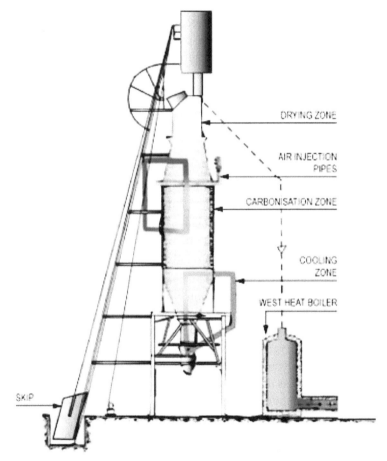

Figure 14.11   Diagram of a Lambiotte retort (info@livanucarbon.lv).

The dry wood efficiency can be very high and reaches 35% or even more. Balt Carbon Ltd.[3] (supplier of Lambiotte retorts for

---

[3]www.baltcarbon.lv (accessed 28/03/2014).

Russia and a number of East European and Central Asian countries) claims that for the same capital investment continuous Lambiotte systems give an increase in output compared to industrial batch systems. The surplus combustible vapors constitute a considerable energy resource which can be utilized for production of process steam or generation of electricity. Balt Carbon Ltd. is developing two Lambiotte charcoal production projects which combine charcoal production and power generation. The company reports that the 2000 tons charcoal per year project in Kaplava (Eastern Latvia) and the 8000 tons charcoal per year project in Ugale (Western Latvia) will be equipped with power-generating equipment of, respectively, 250 kW and 1 MW installed capacity. The Lambiotte system has a number of advantages over batch systems and traditional kilns:

- A continuous and fully automated process resulting in a high labor efficiency
- A very high charcoal yield
- A homogeneous and controlled quality product (fixed-carbon content control by adjustment of the intervals of char removal and the feedstock residence time)
- At remote sites (especially in sub-Saharan Africa) the possibility of charcoal making with cogeneration of electricity probably having significant potential [39]

A disadvantage of the system is the marked influence of the moisture content on the operation characteristics and the installed capacity. Increased moisture content reduces the capacity and in extreme cases can result in a need for burning auxiliary (oil) fuel to keep the process going. The vertical movement of the load causes increased fines formation in comparison to batch systems. Vertical retorts are especially prone to corrosion by acetic acid, because there is always some area in the retort where the temperature is right for acid corrosion. The problem can be solved at a cost by using stainless steel in those areas. Lambiotte retorts of the SIFIC type (with chemicals recovery) were operating in a number of European countries, among others, France and Austria. The last plant at Prémery, Bourgogne in France closed down in 2002. At present (2014) two CISR retorts build by Balt Com Ltd. are operated by Livanu Karbons Ltd. in Livani, Latvia. The annual production capacity is about 2500 tons of charcoal per retort (Fig. 14. 12).

Figure 14.12 Lambiotte retort, Livanu Karbons Ltd., Livani, Latvia. *Source*: Livanu Karbons Ltd., Latvia.

14.3.3 Industrial Charcoal Briquettes

In the United States the transition of charcoal from largely a heating and industrial fuel to a recreational cooking material took place around 1920 when Henry Ford invented the charcoal briquette. His original idea was to make profitable use of sawdust and waste wood from his car factory, but soon it became clear that the briquette business encouraged recreational use of cars for picnicking. Bags of Ford Charcoal (now Kingsford Charcoal) and barbecue grills were promoted and sold at Ford car dealerships and became an integral part of the stock [40]. Industrial briquette production methods have changed little in the past several decades. The most significant innovation in recent years has been the development of "instant light" briquettes. In a briquette factory charcoal fines and lumps are fed to a hammer mill by means of a screw feeder or conveyor belt and ground to a powder with a particle size below 3–4 mm.

The ground charcoal goes to a storage bin from where a measured flow of material goes to a paddle mixer which is usually situated directly below the bin. Inside the mixer a measured amount of binder and water is added. Although it is expensive starch is still the most preferred binder. Addition of 5–7 kg of starch and 30–35 kg of (warm) water to 100 kg charcoal powder results in a satisfactory briquette quality. The use of other binders (like molasses, china clay, and Arabic gum) is reported in the literature [41], but they are less common. Sometimes materials like wax or sodium nitrate are added to improve the ignition and/or burning quality of the briquette. By means of a feed screw a controlled flow of charcoal/binder mixture is fed to the briquette press. Roll-type briquette presses are most common. Those machines apply pressure to the feedstock particles by squeezing them between two rolls rotating in opposite directions. Cavities or indentions cut into the surface of the rolls form the briquettes (Fig. 14.13). Briquettes can be oval shaped, square shaped, or pillow shaped. Because charcoal is an abrasive, high-quality material, and tight tolerances are necessary to ensure a consistent end product.

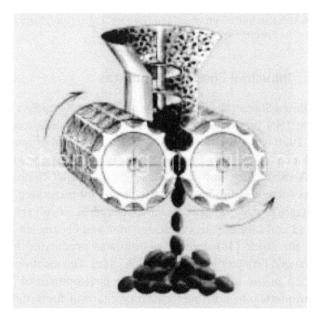

Figure 14.13 Principle charcoal roller briquette press. *Source*: Changzhou Shengong Granulation Equipment Co., Ltd.

The wet or "green" briquettes are treated in a continuous belt dryer at a temperature of about 80°C for about three to four hours. This causes the starch to set and produces a briquette which can be used like lump charcoal in barbecue and grilling appliances. For reasons of economics, industrial briquette plants should operate at three shifts per day. The relative unpopularity of charcoal briquetting and charcoal briquettes in Western Europe may be partly caused by the relatively small production volumes of many local charcoal producers. Taking into account investment and operation costs, a commercial industrial briquette plant in Western Europe (the Netherlands) must have a minimum production capacity of about 15 tons of briquettes per day.[4] This would require at least 375 tons of charcoal fines/dust monthly. In Europe (unlike in the United States) such an amount of raw material is not likely available at one charcoaling site. Therefore a briquette factory must draw on feedstock from geographically dispersed production sites, which adds unfavorably to raw material costs because of hauling and therefore in many cases results in insufficient profitability.

14.4 The Future of Charcoal Production and Use: Africa

One undeniable trend in the 21st century is the rapid urbanization which takes place worldwide. It is expected that the proportion of the world's urban population will increase from about 47% in 2000 to about 57% by 2050. In Africa the share of the urban population in 2010 was about 36% of the total, and this will increase to about 50% and 60% by 2030 and 2050 [42]. This rapid expansion is going to change the continent's demographic landscape and will have consequences for the household energy sector in sub-Saharan Africa. There are no indications that urbanization is going to bring about economic growth and relief of urban poverty. Therefore only the better-off urban households (about 20% of the total) will have the option to switch from traditional to more expensive alternative fuels like gas, kerosene, or electricity. And even those groups, due to sociocultural factors will continue to consume charcoal for particular dishes or at special occasions. Because of income constraints the

---

[4]Author's valuation.

poor and middle-income households (about 80% of the total) will remain dependent on charcoal as their most important fuel for cooking. The World Bank estimates that in sub-Saharan Africa one% increase in urbanization results in 14% increase in charcoal production [4]. This means that in many sub-Saharan African countries the demand for domestic charcoal will remain quite high and most likely will even increase in the future and that charcoal will continue to play an important role in the energy mix of many African countries. By the end of the 20th century international and national donor organizations as well as African national governments slowly became aware of those facts. Many activities and studies were undertaken to essentially answer the question, What measures are needed and practical to produce in a sustainable manner sufficient amounts of (wood) charcoal at an affordable cost to the fast growing sub-Saharan African poor and middle-income segments of the urban population? Studies have come up with proposals and policy recommendations for interventions. National governments are urged to stimulate efficiency and regulate and control the players [43, 44] in the successive stadia of the charcoal value chain (Fig. 14.14). However, the structure and the basic characteristics of the traditional chain remain unchallenged.

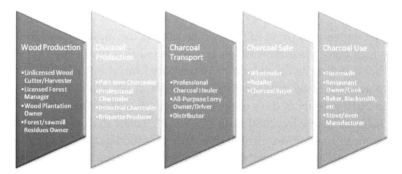

Figure 14.14 Successive stadia and major players in the traditional charcoal value chain.

The future will teach whether and in which circumstances and/or regions the implementation of the proposed actions and recommended activities is going to be practical and successful. Where this is the case, increased affordable supply and sustainability in the short run are possible. It seems very unlikely, however, that the existing natural resources, even with the implementation of (improved) forest management techniques, the large-scale introduction of more efficient improved traditional charcoaling techniques and more efficient charcoal stoves, the establishment of an improved regulatory framework and measures to banish corruption, etc., will everywhere be able to meet in a sustainable manner the massive growth of urban charcoal demand expected in the coming two decennia.

To take on this challenge and avoid the environmental and socioeconomic problems which follow a rapid depletion of natural forest resources, a fresh look at the situation and an innovative approach are needed. It is unlikely that hardwood charcoal from natural forests, especially but not exclusively from semiarid and wet savannah lands, will be able to meet the future demand. The only practical and economic way to sufficiently increase the wood resource in a sustainable manner is the establishment of dedicated wood plantations. Fast-growing (coppicing) tree species can be adapted to different soils and rainfall patterns and have been successfully planted, among others, in the Republic of South Africa. The problem is that for economic reasons those plantations produce (relatively) fast-growing softwoods which (on a weight basis) are more expensive to carbonize in (semi-)industrial equipment then hardwoods. Also, depending on distances, infrastructure, and logistics the cost of transport is an important, and often the most important single, factor in the cost of charcoal to the end user. Transport costs of charcoal are calculated per unit volume and not per unit weight, and for this reason per unit energy the transport costs for softwood charcoal are (much) higher than for hardwood charcoal. Every charcoal producer and trader knows that cost-wise plantation softwood charcoal cannot compete with natural forest hardwood charcoal if there is a need for transport over some distance. The only practical way to increase the density of softwood charcoal is by briquetting. A properly equipped briquette factory should be of certain minimum capacity which partly depends on local factors. Probably 3–5 tons briquettes per day are an absolute minimum for economic operation. The capacity of the briquette plant determines the size of the charcoal kiln(s) (minimum about 1000 tons per year) and the size of the wood plantation. The plantation/

charcoal/briquetting factory should be situated as close as possible and practical to the major urban market(s). Adequate infrastructure to connect the production site(s) to the market(s) has a high priority.

The present charcoal value chain is in all stadia very fragmented, inefficient, and unmanageable and unable to produce and distribute in a sustainable and cost-effective manner the charcoal quantities which soon will be needed. Therefore in (parts of) Africa the future of the charcoaling industry could be one of industrial production by competing relatively large scale vertically integrated private silvicultural/industrial charcoal companies with adequate control over their resources, production processes, products, and markets. Governments should aim at fostering conditions which would favor the establishment of such companies. The structure of such a charcoal sector is amenable to government guidelines and controls in the areas of ecology, environment (including emissions), labor, safety, health, etc. Companies of this type and size will be able to strike an economic balance between efficiency, investments, and costs and will structure their processes accordingly. Economic and technical efficient co-generating of electric power with charcoal production for off-grid applications in rural areas becomes a realistic possibility. All this will create quality sustainable rural jobs in forestry and (agro) forestry related products, agriculture (cassava or corn starch binder production), industry, and transport. Companies will not depend on scattered donor-financed consultancy efforts but have the size and financial means to improve through tests and research on the quality and the characteristics of their product (like "instant light" briquettes) and to actively develop, promote, and market a better and adapted briquette stove.

14.5 The Future of Charcoal Production and Use: Europe

The three traditional markets for charcoal in industrial countries are:

- the barbecue market;
- the metallurgical market; and
- the active coal market.

Part-time and leisure consumption account for more than 95% by volume of the barbecue charcoal sold in Western Europe. The remainder is high-quality grill charcoal used by professionals mainly in restaurants in Northwestern Europe. Consumption in the E27 is more or less stable (Fig. 14.5), but the Russian market, although still relatively small, grows at a phenomenal rate.[5] The gradual opening for Western trade of Eastern and Southeastern Europe culminating in the fall of the Iron Curtain in 1990 brought profound changes to the West European charcoal Industry. Traditionally the West European consumer barbecue charcoal market was supplied with lump hardwood charcoal by large and medium-size producers from, among others, France, Germany, Austria, Spain and Italy. They used raw materials like low-quality oak and beech wood or residues from sawmills and furniture factories. After 1995, because of relentless price competition from lump hardwood charcoal originating from Central and Southeastern Europe (Poland, Ukraine, Croatia, Serbia, Romania, and Southern Russia) and from African countries (Nigeria, Namibia, and the Republic of South Africa), gradually many Western European producers were forced to close down or went into bankruptcy.[6] Charcoal production in Western Europe became largely uneconomic partly because of the need for high cost investment in new equipment due to the gradual enforcement of strict emission norms. The high cost of labor which cannot be sufficiently offset by efficiency measures is another factor. This situation is likely to continue as long as the uptake and upholding of environmental, labor, safety, and health legislation, etc. in the exporting countries is not comparable to the practice in Western Europe. Therefore the prospects for barbeque lump charcoal production from hardwood in Western Europe are not sunny in the near future. Today most lump charcoal used for barbeque and grilling in the EU27 is imported from Eastern Europe or Africa (Fig. 14.15). Many former charcoal producers became importers, charcoal packers, and distributors to supermarkets etc., some diversified into other fuels and related products, a few specialized in high-quality charcoal for demanding users and special (niche) markets and applications like for *hookah* or *shisha* smoking.

---

[5]Barbeque Industry Association Grillverband (BIAG) e.V. (2013) [45].
[6]The last big Lambiotte plant in W-Europe (Prémery, France) ceased operation in 2002.

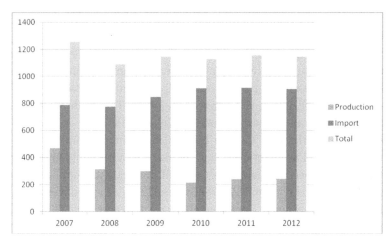

Figure 14.15 Charcoal production and import in the EU27: 2007–2012 (tons × 1000). *Source*: FAOSTAT-ForesSTAT.

A new development which could ultimately make a profound change to the European barbeque charcoal scene is the establishment of very big state-of-the-art sawmills and plywood mills in (Northwest) Russia. Those factories are strategically located in the proximity of the softwood (pine, larch, etc.) resource and with good road, rail, and waterway connections to the major markets in inside Russia and abroad. The mills produce large amounts of residues in the form of sawdust, bark, and off-cuts. At present those residues are often used to produce wood pellets which are mainly exported to the EU.

The EU imports large amounts of wood pellets which are partly used for substitution of coal in base load power plants. Other applications are domestic heating and combined heat and power (CHP) generation in plants. The rational is greenhouse gas (CO_2) emission mitigation by substituting fossil carbon by short-cycle carbon. Major importers of wood pellets in the EU are Belgium, Denmark, Italy, the United Kingdom, and the Netherlands. The latter country imports mainly pellets for cofiring. Wood pellets are traded in a global market. In addition to the 10 million tons which were produced internally, the EU imported in 2010 about 2.8 million tons of wood pellets from Canada and the United States. Co-combustion of pellets in coal-fired power plants is heavily subsidized. Between 2003 and 2006 the Dutch government granted long-term subsidies for cofiring of biomass of up to €0.06 and €0.07/kWh(e) [46]. This is a subsidy equal to about €120–€135 per ton of pellets, and it has given an enormous boost to pellet cofiring. A comparison of the value of the subsidy with the price of pellets at the time (between €115 per ton and €140 per ton in the years 2007 to 2010) shows the reason why. The Dutch subsidy scheme ends in 2015 and nobody knows what will happen thereafter. Most experts predict continued growth of the pellet market but how this is going to come about without the co-combustion subsidies is not clear at all. Plywood/sawmills in Russia produce sufficient amounts of sawdust and bark for economic exploitation of multiple hearth carbonization technology. In case of a collapse of the co-combustion wood pellet market those mills could choose to opt for the production of charcoal briquettes for the European barbeque market instead. Price is the most important factor in this market and in present market circumstances it is likely that briquettes from West Russia can out-compete the hardwood lump charcoal exported from Africa to Europe.

Expensive metallurgical charcoal (€420–€460 per ton) is used in high-value applications as reductant in silicon smelting or as recarburizer in steelmaking. Silicon smelting requires relatively large amounts of charcoal (30,000 tons per year for an average smelter) and therefore the smelting companies often own short-rotation wood plantations to use the wood as an input for dedicated metallurgical charcoal production. Little information is available about the volume of charcoal consumption by the European metallurgical industry. In Norway charcoal from South America and Asia is used to a small extent in the production of silicon (Si) and ferrosilicon (FeSi) [47]. There is mention[7] in the market of several European charcoal companies having been approached over the last 10 years by European metallurgical companies for test delivery of charcoal batches of defined specifications. There are no indications that this has resulted into regular supply contracts. It seems that in current market circumstances West European producers are unable to match price conditions of metallurgical charcoal from South America and elsewhere.

The activated carbon market is dominated by large industrial companies which produce (active) charcoal from different raw

[7]Author's information.

materials through proprietary processes and technologies. Some active carbon companies have production facilities in Western Europe, among others, in the Netherlands, and the United Kingdom. Charcoal buying from third parties is basically limited to coconut shell charcoal. A profitable market niche in the activated coal business may exist for specialized European companies producing high-quality heavy charcoal from high-density extruded sawdust briquettes.

References

1. Castor-Perry, S. (2009). "The discovery of Oetzi the Iceman," *The Naked Scientists (University of Cambridge)*, 21 Sept., 2009, 4.

2. Steierer, F. (2011). *Highlights on Wood Charcoal: 2004–2009*, Rome, Italy, FAO Forestry Department.

3. World Bank. *ESMAP Household Energy Strategy Leaflet*, 2000.

4. Sander, K., et al. (2011). *Wood-Based Biomass Energy Development for Sub-Saharan Africa*, World Bank Group, Washington DC.

5. IEA (2010). *World Energy Outlook 2010*, OECD/IEA, Paris, France.

6. Madon, G. (2000). An assessment of tropical dry-land forest management in Africa, what are its lessons? *World Bank Seminar Village Power 2000, Empowering People and Transforming Markets*, World Bank Group, Washington DC.

7. FAO (1993). *A Decade of Wood Energy Activities within the Nairobi Programme of Action*, FAO forestry paper no. 108, Rome, Italy.

8. Kenya Forestry Service (2013). Charcoal value chain analysis; analytical study on corruption, *Sustainable Developmentin Kenya Draft SESA Road Map*, Nairobi, Kenya.

9. Foley, G. (1985). *Exploring the Impact of Conventional Fuel Substitution on Woodfuel Demand*, Earthscan Working Paper, International Institute for Environment and Development, London, U.K.

10. De Klerk, J.N. (2004). *Bush Encroachment in Namibia*. Windhoek, Ministry of Environment and Tourism, Government of Namibia, Namibia, 254.

11. Leinonen, A. (2007). *Wood Chip Production Technology and Costs for Fuel in Namibia*, VTT Technical Research Centre of Finland, Jyväskylä, Finland.

12. FAO (1987). *Simple Technologies for Charcoal Making*, FAO forestry paper no. 41, FAO Forestry Division, Rome, Italy.

13. Girard, P. (2000). Charcoal production and use in Africa: what future?, *Unasylva*, **2002**, 30–34.

14. INBAR (2010). *Bamboo as Sustainable Biomass Energy: A Suitable Alternative for Firewood and Charcoal Production in Africa*, Inbar, Beijing, China.

15. Zhang, Q., et al. (2013). *Bamboo Charcoal Unit*. Nanjing Forestry University, Nanjing, China.

16. Stassen, H.E. (2002). Developments in charcoal production technology, *Unasylva 211*, **53**, 34–35.

17. Nturanabo, F., et al. (2011). Performance appraisal of the Casamance kiln as a replacement to the traditional charcoal kilns in Uganda, *Second International Conference on Advances in Engineering and Technology (ICAET)*, Makerere University, Kampala, Uganda, 530–536.

18. Kammen, D., Lew, D. (2005). *Review of Technologies for the Production and Use of Charcoal*, Renewable and Appropriate Energy Laboratory Report, University of California, Berkeley, USA.

19. Mundhek, Ph., et al. (2010). *Comparison des Rendements de Production de Charbon de Bois Entre la Meule Traditionelle et la Meule Casamance dans la Foret Communautaire de Sambande*, Peracod, Dakar, Senegal.

20. Nahayo, A., et al. (2013). Comparative study on charcoal yield produced by traditional and improved kilns: a case study of Nyaruguru and Nyamagabe districts in southern province of Rwanda, *Energy and Environment Research* **3**(1), 40–48.

21. Brown, R.C. (2009). *Biochar for Environmental Management: Science and Technology*, Earthscan, London, U.K.

22. Adam, J.C. (2009). Improved and more environmentally friendly charcoal production system using a low-cost retort-kiln (eco-charcoal), *Renewable Energy*, **34**(8), 1923–1925.

23. Halouani, K., et al. (2003). Depollution of atmospheric emissions of wood pyrolysis furnaces, *Renewable Energy*, **28**, 129–138.

24. Domac, J., Trossero, M. (2008). *Environmental Aspects of Charcoal Production in Croatia*, North-West Croatia Regional Energy Agency, Zagreb, Croatia.

25. Hood, A. (1998). *Briquettes de Charbon au Mali: Valorisation des Dechets*, SED-SEED-CIRAD-AFRITECH-BTG.

26. Mwampamba, T., et al. (2013). Opportunities, challenges and way forward for the charcoal briquette industryin Sub-Saharan Africa, *Energy for Sustainable Development*, **17**(2), 158–170.

27. Stassen, H.E. (1995). *Small-Scale Gasifiers for Heat and Power: A Global Review*, World Bank technical paper no. 296, World Bank, Washington DC.

28. Grover, P.D., Mishra, S.K. (1996). *Biomass Briquetting: Technology and Practices*, Field document no. 46, Food and Agriculture Organization of the United Nations, Bangkok.

29. O'Hanlon, J. (2013). FerroAtlantica: the metals of energy, *BE Mining*, 1 May 2013.

30. Woodroof, J. (1970). *Coconuts: Production, Processing, Products*, AVI, Westport, CT, USA.

31. Emrich, W. (1981). *The Charcoal Markets in Industrialized Countries and the Impacts of Charcoal Exports in Developing Countries*, FAO, Rome, Italy.

32. Transparency Market Research (2013). *Global Activated Carbon Market-Industry Analysis, Size, Share, Growth, Trends and Forecast 2013-2019*, Transparency Market Research, Albany, New York, USA.

33. Emrich, W. (1985). *Handbook of Charcoal Making*, D. Reidel, Dordrecht/Boston/Lancaster.

34. Braun, S. (1997). Missouri charcoal makers agree to clean their kilns, *Los Angeles Times*, 14 Aug. 1997.

35. Reumerman, P.J., et al. (2002). Charcoal production with reduced emissions, *Proceedings of the 12th European Conference on Biomass for Energy, Industry and Climate Protection*, European Commission, Brussels, Belgium, 4.

36. Siemons, R., et al. (2008). *Industrial Charcoal Production*, FAO TCP 3101, FAO, Rome, Italy.

37. Hankin Environmental Systems, The production of charcoal with the Hankin/Nichols carbonizer, *Hankin Highlights*.

38. Baker, A.J. (1985). Charcoal production in the U.S.A., *Symposium on Forest Product Research International*, 5, South African Council for Scientific and Industrial Research, National Timber Research Institute, Pretoria RSA, 15p.

39. Carneiro de Miranda, R., et al. (2013). Co-generating electricity from charcoaling: a promising new advanced technology, in *Energy Sustainable Development*, Elsevier, 6.

40. Maykuth, A. (2014). Charcoal: the old fashioned way Henry Ford and E.G. Kingsford invented briquettes to rid Ford's factory of wood waste, *The Philadelphia Inquirer*, 1 April 2014.

41. FAO (1985). *Industrial Charcoal Making*, FAO Forestry Department, Rome, Italy.

42. Ncube, M. (2012). Urbanization in Africa, *AFDB: Championing Inclusive Growth across Africa*, 12 Dec. 2012.

43. Vos, J., et al. (2010). *Making Charcoal Production in Sub Sahara Africa Sustainable*, Netherlands Programmes Sustainable Biomass, NL Agency, Utrecht.

44. Peter, C., et al. (2009). *Environmental Crisis or Sustainable Development Opportunity? Transforming the Charcoal Sector in Tanzania*, Policy note, World Bank, Washington DC.

45. Barbecue Expo 2013 in Moscow, www.biag.org.

46. Sikkema, R. (2010). The European wood pellet markets: current status and prospects for 2020, *Biofuels, Bioproducts and Biorefining*, **2010**, 250–278.

47. Monsen, B., et al. (2004). Use of charcoal in silicomanganese production, *Proceedings of the 10th International Ferroalloys Congress*, Document Transformation Technologies, Kaapstad, South Africa, 392–404.

About the Author

Hubert E. Stassen (1942) studied chemical engineering at the Technical University of Eindhoven, the Netherlands. He worked as a process researcher and developer with National Defence Laboratory in Rijswijk and Philips, Eindhoven, the Netherlands. From 1975 till 1979 he acted as head of the department of the (then) recently founded Department of Chemical Engineering at Ahmadu Bello University in Zaria, Nigeria, where he became interested in sustainable energy, more specifically energy from biomass. Upon his return to the Netherlands he worked with University of Twente, the Netherlands, but was soon seconded by the Netherlands Foreign Office to the New York Secretariat of the United Nations Conference on New and Renewable Sources of Energy which was held in Nairobi, Kenya, in August 1981. During this conference he acted as the key European Union speaker in the conference's technical panel. He returned to the University of Twente in 1982 and founded a small research and consultancy group in the area of biomass gasification within the Department of Chemical Engineering. In 1985 this Biomass Technology Group (BTG) B.V. was privatized, and he became the first director of the commercial company. Spin-offs from the research and development activities at the BTG led to participations in international charcoal and briquette production companies. Major production facilities were located in the Netherlands, South Africa, and Namibia.

Stassen has expertise in a wide range of energy-related areas, in particular in the field of renewable energy technologies and environmental aspects of energy generation, and specific expertise in design and implementation of small-/medium-scale biomass (co)gasification projects for decentralized power production and cofiring projects in medium-/large-scale coal-fired power stations. He has a strong background in small-, medium-, and large-scale thermochemical biomass/coal/gas conversion processes (cofiring, gasification, combustion, carbonization, pyrolysis). He has worked in many countries as an energy consultant for organizations like the World Bank, the United Nations Development Programme (UNDP), the Food and Agriculture Organization (FAO), the United Nations Industrial Development Organization (UNIDO), the Commission of the European Communities, the European Bank for Reconstruction and Development (EBRD), the Ministry of Development Co-operation (the Netherlands), the Ministry of Economic Affairs (the Netherlands), the Netherlands Organization for Energy and Environment (NOVEM), and the Ford Foundation, as well as for a large number of multinational and national private corporations.

In 2001 Stassen was awarded the Netherlands National Energy Award ("DOW Energieprijs"), and in 2002 HM Queen Beatrix of the Netherlands appointed him "Knight in the Order of the Dutch Lion." Also in 2002 he retired from the position of director of BTG. Since then he has been studying, writing, traveling, and working as a private consultant for multinational and national organizations and companies.

Chapter 15

Pyrolysis of Biomass

Anthony Victor Bridgwater

Bioenergy Research Group, European Bioenergy Research Institute,
Aston University, Birmingham B4 7ET, UK
a.v.bridgwater@aston.ac.uk

15.1   Introduction

15.1.1   Background

Pyrolysis has become of major interest due to the flexibility in operation, versatility of the technology, and adaptability to a wide variety of feedstocks and products. Pyrolysis operates in anaerobic conditions where heat is usually provided externally and the constituents of biomass are thermally cracked to gases and vapors, which usually undergo secondary reactions, thereby giving a broad spectrum of products. There are a number of conditions and circumstances which have a major impact on the products and the process performance. These include feedstock, technology, reaction temperature, additives, catalysts, hot-vapor residence time, solid residence time, and pressure.

Biomass Power for the World: Transformations to Effective Use
Edited by Wim van Swaaij, Sascha Kersten, and Wolfgang Palz
Copyright © 2015 Pan Stanford Publishing Pte. Ltd.
ISBN 978-981-4613-88-0 (Hardcover), 978-981-4669-24-5 (Paperback), 978-981-4613-89-7 (eBook)
www.panstanford.com

Pyrolysis has been applied for thousands of years for charcoal production but it is only in the last 35 years that fast pyrolysis for liquids has been developed. This operates at moderate temperatures of around 500°C and very short hot-vapor residence times of less than 2 seconds. Fast pyrolysis is of considerable interest because this directly gives high yields of liquids of up to 75 wt.% which can be used directly in a variety of applications [1] or used as an efficient energy carrier. Intermediate pyrolysis and slow pyrolysis focus on production of solid char as the main product, with the liquids and gases usually as by-products, although increasing attention is being paid to maximizing the value of the products. Pyrolysis has also been used for over 50 years to reduce quantities of waste which require disposal as well as making the residues less harmful to the environment. These processes often employ slow pyrolysis as the core technology.

15.1.2 Science

Pyrolysis is thermal decomposition occurring in the absence of oxygen. Lower process temperatures and longer hot-vapor residence times favor the production of charcoal. Higher temperatures and longer residence times increase biomass conversion to gas, and moderate temperatures and short vapor residence time are optimum for producing liquids. Three products are always produced, but the proportions can be varied over a wide range by adjustment of the process parameters. Table 15.1 and Fig. 15.1 indicate the product distribution obtained from different modes of pyrolysis, showing the considerable flexibility achievable by changing process conditions. Fast pyrolysis for liquid production is currently of particular interest commercially as the liquid can be stored and transported, and used for energy, transport fuels, chemicals or as an energy carrier.

Table 15.1 Typical product weight yields (dry wood basis) obtained by different modes of pyrolysis of wood

Mode	Conditions	Liquid	Solid	Gas
Fast	~ 500°C, short hot-vapor residence time < 2 s	75 wt.% (bio-oil)	12 wt.% char	13 wt.%

Mode	Conditions	Liquid	Solid	Gas
Intermediate	~ 500°C, moderate hot-vapor residence time 5–30 s	50 wt.% in 2 phases	25 wt.% char	25 wt.%
Carbonization (slow)	~ 400°C, long hot-vapor residence time hours → days	30 wt.%	35 wt.% char	35 wt.%
Gasification (allothermal)	~ 750–900°C, moderate hot-vapor time 5 s	3 wt.%	1 wt.% char	96 wt.%
Torrefaction (slow)	~ 280°C, solid residence time ~10–60 min	0 wt.% unless vapors are condensed, then up to 15%	80 wt.% solid	20 wt.%

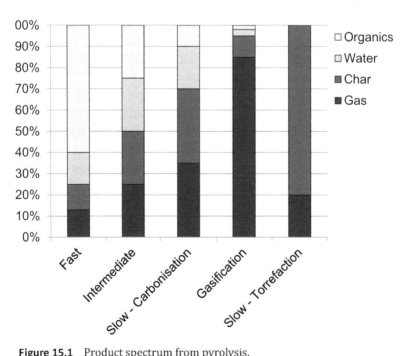

Figure 15.1 Product spectrum from pyrolysis.

15.2　Fast Pyrolysis

In fast pyrolysis, biomass decomposes very quickly to generate mostly vapors and aerosols and some charcoal and gas. After cooling and condensation, a dark brown homogenous mobile liquid is formed which has a heating value about half that of conventional fuel oil. This is referred to as bio-oil and is the basis of the recent ASTM standard [2]. A high yield of liquid is obtained with most low ash biomass. Yields of up to 75 wt.% on dry biomass feed can be achieved. The essential features of a fast-pyrolysis process for producing liquids are:

- Small particle sizes of typically less than 5 mm for high heating rates and rapid devolatilization.
- Feed moisture content of less than 10 wt.%, since all the feed water reports to the liquid phase along with water from the pyrolysis reactions.
- Very high heating rates and very high heat transfer rates at the biomass–particle reaction interface, usually requiring a finely ground biomass feed of typically less than 3 mm, as biomass generally has low thermal conductivity. The rate of particle heating is usually the rate-limiting step.
- Carefully controlled pyrolysis reaction temperature of around 500°C to maximize the liquid yield for most biomass.
- Short hot-vapor residence times of typically less than 2 seconds to minimize secondary reactions.
- Rapid removal of product char to minimize cracking of vapors.
- Rapid cooling of the pyrolysis vapors to give the bio-oil product.

As fast pyrolysis for liquids occurs in a few seconds or less, heat and mass transfer processes and phase transition phenomena, as well as chemical reaction kinetics, play important roles. The critical issue is to bring the reacting biomass particles to the optimum process temperature and minimize both their exposure to lower temperatures which favor formation of charcoal and their exposure to higher temperatures which accelerate thermal cracking. One way this objective can be achieved is by using small particles, for example in the fluidized-bed processes which are described later. Another possibility is to transfer heat very fast only to the particle surface which contacts the heat source which is used in ablative pyrolysis.

Several comprehensive reviews of fast pyrolysis for liquid production are available such as [3–4].

15.2.1   Feedstocks

Before considering catalytic upgrading of bio-oil, it is important to appreciate firstly that biomass contains very active catalysts within its structure. These are the alkali metals which form ash and which are essential for nutrient transfer and growth of the biomass. The most active is potassium followed by sodium. These act by causing secondary cracking of vapors and reducing liquid yield and liquid quality, and depending on the concentration, the effect can be more severe than char cracking.

Ash can be managed to some extent by selection of crops and harvesting time but it cannot be eliminated from growing biomass. Ash can be reduced by washing in water or dilute acid, and the more extreme the conditions in temperature or concentration, respectively, the more complete the ash removal. However, as washing conditions become more extreme firstly hemicellulose and then cellulose is lost through hydrolysis. This reduces liquid yield and quality. In addition, washed biomass needs to have any acid removed as completely as possible and recovered or disposed of and the wet biomass has to be dried.

So washing is not often considered a viable possibility, unless there are some unusual circumstances such as removal of contaminants. Another consequence of high ash removal is the increased production of levoglucosan which can reach levels in bio-oil where recovery becomes an interesting proposition, although commercially markets need to be identified and/or developed.

15.2.2   Technology

A conceptual fast-pyrolysis process is depicted in Fig. 15.2 from biomass feed to collection of a liquid product. Each process step has several or even many alternatives such as the reactor and liquid collection but the underlying principles are similar.

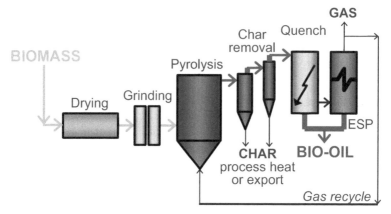

Figure 15.2   Conceptual fast-pyrolysis process.

At the heart of a fast-pyrolysis process is the reactor. Although it probably represents only about 10–15% of the total capital cost of an integrated system, most research and development has focused on developing and testing different reactor configurations on a variety of feedstocks, although increasing attention is now being paid to improvement of liquid collection systems and improvement of liquid quality. The rest of the fast-pyrolysis process consists of biomass reception, storage and handling, biomass drying and grinding, product collection, storage, and, when relevant, upgrading.

15.2.2.1   Bubbling fluid-bed reactors

Bubbling fluid beds have the advantages of a well-understood technology which is simple in construction and operation, good temperature control, and very efficient heat transfer to biomass particles arising from the high solid density. These are incompletely depolymerized lignin fragments which seem to exist as a liquid with a substantial molecular weight. Evidence of their liquid basis is found in the accumulation of liquid in the ESP which runs down the plates to accumulate in the bio-oil product. Demisters for agglomeration or coalescence of the aerosols have been used but published experience suggests that this is less effective.

Heating can be achieved in a variety of ways and scaling is well understood. However, heat transfer to bed at large scales of operation has to be considered carefully because of the scale-up limitations of different methods of heat transfer. Fluid-bed pyrolyzers give good and consistent performance with high liquid yields of typically 70–75 wt.% from wood on a dry-feed basis. Small biomass particle sizes of less than 2–3 mm are needed to achieve high biomass heating rates, and the rate of particle heating is usually the rate-limiting step.

Vapor and solid residence time is controlled by the fluidizing gas flow rate and is higher for char than for vapors. As char acts as an effective vapor-cracking catalyst at fast-pyrolysis reaction temperatures, rapid and effective char separation is important. This is usually achieved by ejection and entrainment followed by separation in one or more cyclones so careful design of sand and biomass/char hydrodynamics is important. The high level of inert gases arising from the high permanent gas flows required for fluidization result in very low partial pressures for the condensable vapors and thus care is needed to design and operate efficient heat exchange and liquid collection systems. In addition the large inert gas flowrates result in relatively large equipment thus increasing cost.

The by-product char is typically about 15 wt.% of the products but about 25% of the energy of the biomass feed. It can be used within the process to provide the process heat requirements by combustion or it can be separated and exported, in which case an alternative fuel is required. Depending on the reactor configuration and gas velocities, a large part of the char will be of a comparable size and shape as the biomass fed. The fresh char is pyrophoric, that is, it spontaneously combusts when exposed to air so careful handling and storage is required. This property deteriorates with time due to oxidation of active sites on the char surface.

15.2.2.2 Circulating fluid-bed and transported-bed reactors

Circulating fluid-bed (CFB) and transported-bed reactor systems have many of the features of bubbling beds described above, except that the residence time of the char is almost the same as for vapors and gas, and the char is more attritted due to the higher gas velocities. This can lead to higher char contents in the collected bio-oil unless more extensive char removal is included. An added advantage is that CFBs are potentially suitable for larger throughputs even though the hydrodynamics are more complex as this technology is widely used at very high throughputs in the petroleum and petrochemical industry.

Heat supply is usually from recirculation of heated sand from a secondary char combustor, which can be either a bubbling bed or a CFB. In this respect the process is similar to a twin fluid-bed gasifier except that the reactor (pyrolyzer) temperature is much lower and the closely integrated char combustion in a second reactor requires careful control to ensure that the temperature, heat flux, and solid flow match the process and feed requirements. Heat transfer is a mixture of conduction and convection in the riser. One of the unproven areas is scale-up and heat transfer at high throughputs.

All the char is burned in the secondary reactor to reheat the circulating sand, so there is no char available for export unless an alternative heating source is used. If separated the char would be a fine powder.

15.2.2.3 Ablative pyrolysis

Ablative pyrolysis is substantially different in concept compared with other methods of fast pyrolysis. In all the other methods, the rate of reaction is limited by the rate of heat transfer through the biomass particles, which is why small particles are required. The mode of reaction in ablative pyrolysis is like melting butter in a frying pan—the rate of melting can be significantly enhanced by pressing the butter down and moving it over the heated pan surface. In ablative pyrolysis, heat is transferred from the hot reactor wall to "melt" wood which is in contact with it under pressure. As the wood is moved away, the molten layer then vaporizes to a product very similar to that derived from fluid-bed systems.

The pyrolysis front thus moves unidirectionally through the biomass particle. As the wood is mechanically moved away, the residual oil film both provides lubrication for successive biomass particles and also rapidly evaporates to give pyrolysis vapors for collection in the same way as other processes. There is an element of cracking on the hot surface from the char which is also deposited. The rate of reaction is strongly influenced by pressure of the wood onto the heated surface, the relative velocity of the wood and the heat exchange surface, and the reactor surface temperature. The key features of ablative pyrolysis are therefore as follows:

- High pressure of particle on hot reactor wall, achieved by centrifugal force or mechanically

- High relative motion between particle and reactor wall
- Reactor wall temperature less than 600°C

As reaction rates are not limited by heat transfer through the biomass particles, larger particles can be used and in principle there is no upper limit to the size which can be processed. The process, in fact, is limited by the rate of heat supply to the reactor rather than the rate of heat absorption by the pyrolyzing biomass, as in other reactors. There is no requirement for inert gas, so the processing equipment is smaller and the reaction system is thus more intensive. In addition the absence of fluidizing gas substantially increases the partial pressure of the condensable vapors leading to more efficient collection and smaller equipment. However, the process is surface-area-controlled so scaling is less effective and the reactor is mechanically driven, and is thus more complex. The char is a fine powder which can be separated by cyclones and hot-vapor filters as for fluid-bed reaction systems.

15.2.2.4 Screw and augur kiln reactors

There have been a number of developments which mechanically move biomass through a hot reactor rather than using fluids, including screw and augur reactors. Heating can be with recycled hot sand as at the Bioliq plant at KIT (FZK until 2009) [5, 6], with heat carriers such as steel or ceramic balls such as Haloclean also at KIT (e.g., Ref. [7]), or external heating. The nature of mechanically driven reactors is that very short residence times comparable to fluid beds and CFBs are difficult to achieve, and hot-vapor residence times can range from 5 to 30 seconds, depending on the design and size of reactor. Examples include screw reactors and more recently the Lurgi LR reactor at Karlsruhe Institute of Technology (KIT) [10, 11] and the Bio-oil International reactors which have been studied at Mississippi State University [8]. Screw and augur reactors have also been developed as intermediate pyrolysis systems.

Screw reactors are particularly suitable for feed materials which are difficult to handle or feed, or are heterogeneous. The liquid product yield is lower than fluid beds and is usually phase-separated due to the longer residence times and contact with by-product char. Also the char yields are higher. KIT has promoted and tested the concept of producing a slurry of the char with the liquid to maximize liquid yield in terms of energy efficiency [13], but this would requires an alternative energy source to provide heat for the process.

15.2.2.5   Heat transfer in fast pyrolysis

There are a number of technical challenges facing the development of fast pyrolysis, of which the most significant is heat transfer to the reactor. Pyrolysis is an endothermic process, requiring a substantial heat input to raise the biomass to reaction temperature, although the heat of reaction is insignificant. Heat transfer in commercial reactors is a significant design feature, and the energy in the by-product charcoal would typically be used in a commercial process by combustion of the char in air. The char typically contains about 25% of the energy of the feedstock, and about 75% of this energy is typically required to drive the process. The by-product gas only contains about 5% of the energy in the feed, and this is not sufficient for pyrolysis. The main methods of providing the necessary heat are listed below:

- Through heat transfer surfaces located in suitable positions in the reactor.
- By heating the fluidization gas in the case of a fluid-bed or CFB reactor, although excessive gas temperatures may be needed to input the necessary heat resulting in local overheating and reduced liquid yield, or alternatively very high gas flows are needed resulting in unstable hydrodynamics. Partial heating is usually satisfactory and desirable to optimize energy efficiency.
- By removing and reheating the bed material in a separate reactor as used in most CFB and transported-bed reactors.
- By the addition of some air, although this can create hot spots and increase cracking of the liquids to tars.

There are a variety of ways of providing the process heat from by-product char or gas; or from fresh biomass. This facet of pyrolysis reactor design and optimization is most important for commercial units and will attract increasing attention as plants become bigger. Examples of options include:

- combustion of by-product char, all or part;

- combustion of by-product gas which normally requires supplementation for example with natural gas;
- combustion of fresh biomass instead of char, particularly where there is a lucrative market for the char;
- gasification of the by-product char and combustion of the resultant producer gas to provide greater temperature control and avoid alkali metal problems such as slagging in the char combustor;
- use of by-product gas with similar advantages as above, although there is unlikely to be sufficient energy available in this gas without some supplementation;
- use of bio-oil product; and
- use of fossil fuels where these are available at low cost, do not affect any interventions allowable on the process or product, and the by-products have a sufficiently high value.

15.2.3 Products

The liquid is formed by rapidly quenching and thus 'freezing' the intermediate products of flash degradation of hemicellulose, cellulose, and lignin. The liquid thus contains many reactive species, which contribute to its unusual attributes. Bio-oil can be considered a microemulsion in which the continuous phase is an aqueous solution of holocellulose decomposition products, which stabilizes the discontinuous phase of pyrolytic lignin macromolecules through mechanisms such as hydrogen bonding. Aging or instability is believed to result from a breakdown in this emulsion.

Pyrolysis oil typically is a dark brown, free-flowing liquid and approximates to biomass in elemental composition. Depending on the initial feedstock and the mode of fast pyrolysis, the color can be almost black through dark red-brown to dark green, being influenced by the presence of microcarbon in the liquid and chemical composition. Hot-vapor filtration gives a more translucent red-brown appearance owing to the absence of char. High nitrogen content can impart a dark green tinge to the liquid.

It is composed of a very complex mixture of oxygenated hydrocarbons with an appreciable proportion of water from both the original moisture and the reaction product. Solid char may also be present. Typical organics yields from different feedstocks and their variation with temperature is shown in Fig. 15.3, and Fig. 15.4 shows the temperature dependence of the four main products from a variety of feedstocks [9]. Similar results are obtained for most biomass feedstocks, although the maximum yield can occur between 480°C and 525°C, depending on feedstock. Grasses, for example, tend to give maximum liquid yields of around 55–60 wt.% on a dry feed basis at the lower end of this temperature range, depending on the ash content of the grass. Liquid yield depends on biomass type, temperature, hot-vapor residence time, char separation, and biomass ash content, the last two having a catalytic effect on vapor cracking. It is important to note that maximum yield is not the same as maximum quality, and quality needs careful definition if it is to be optimized. Bio-oil quality and quality management and improvement have also been reviewed [10].

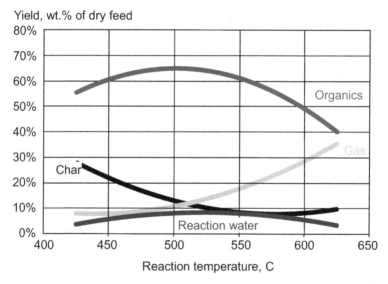

Figure 15.3 Variation of products from Aspen Poplar with temperature [3].

The liquid contains varying quantities of water, which forms a stable single-phase mixture, ranging from about 15 wt.% to an upper limit of about 30–50 wt.% water, depending on the feed material, how it was produced and subsequently collected. A typical feed material specification is a maximum of 10% moisture in the dried feed material, as both this feed moisture and the water of reaction from pyrolysis, typically about 12% based on dry feed, report to the liquid product. Pyrolysis liquids can tolerate the addition of some water, but there is a limit to the amount of water which can be added to the liquid before phase separation occurs, in other words the liquid cannot be dissolved in water. Water addition reduces viscosity, which is useful; reduces heating value which means that more liquid is required to meet a given duty; and can improve stability. The effect of water is therefore complex and important. It is miscible with polar solvents such as methanol, acetone, etc., but totally immiscible with petroleum-derived fuels. This is due to the high oxygen content of around 35–40 wt.%, which is similar to that of biomass, and provides the chemical explanation of many of the characteristics reported. Removal of this oxygen by upgrading requires complex catalytic processes which are described below.

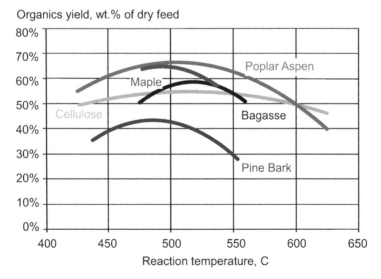

Figure 15.4 Organics yield from different feedstocks [3].

The density of the liquid is very high at around 1200 kg t^{-1}, compared with light fuel oil at around 0.85 kg/litre. This means that the liquid has about 42% of the energy content of fuel oil on a weight basis, but 61% on a volumetric basis. This has implications for the design and specification of equipment such as pumps and atomizers in boilers and engines.

Viscosity is important in many fuel applications [11]. The viscosity of the bio-oil as produced can vary from as low as 25 m^2 s^{-1} to as high as 1000 m^2 s^{-1} (measured at 40°C) or more, depending on the feedstock, the water content of the bio-oil, the amount of light ends collected and the extent to which the oil has aged.

Pyrolysis liquids cannot be completely vaporized once they have been recovered from the vapor phase. If the liquid is heated to 100°C or more to try to remove water or distil off lighter fractions, it rapidly reacts and eventually produces a solid residue of around 50 wt.% of the original liquid and some distillate containing volatile organic compounds and water. While bio-oil has been successfully stored for several years in normal storage conditions in steel and plastic drums without any deterioration which would prevent its use in any of the applications tested to date, it does change slowly with time, most noticeably there is a gradual increase in viscosity. More recent samples which have been distributed for testing have shown substantial improvements in consistency and stability, demonstrating the improvement in process design and control as the technology develops.

Aging is a well known phenomenon caused by continued slow secondary reactions in the liquid which manifests as an increase in viscosity with time. It can be reduced or controlled by the addition of alcohols such as ethanol or methanol. In extreme cases phase separation can occur. It is exacerbated or accelerated by the presence of fine char. This has been reviewed by Diebold [12].

A fast-pyrolysis liquid has a higher heating value of about 17 MJ/ kg as produced with about 25 wt.% water which cannot readily be separated. While the liquid is widely referred to as bio-oil, it will not mix with any hydrocarbon liquids. It is composed of a complex mixture of oxygenated compounds which provide both the potential and challenge for utilization. There are some important properties of this liquid which are summarized in Table 15.2 and Table 15.3. There are many particular characteristics of bio-oil which require consideration for any application [6]. Oasmaa and Peacocke have reviewed physical property characterization and methods [13, 14].

Table 15.2   Typical properties of wood-derived crude bio-oil

Physical property		Typical value
Moisture content		25%
pH		2.5
Specific gravity		1.20
Elemental analysis	C	56%
	H	6%
	O	38%
	N	0%–0.1%
HHV as produced		17 MJ/kg
Viscosity (40°C and 25% water)		40–100 MPa s
Solids (char)		0.1%
Vacuum distillation residue		up to 50%

Table 15.3   Characteristics of bio-oil

Characteristic	Cause	Effects
Acidity or low pH	Organic acids from biopolymer degradation	Corrosion of vessels and pipework
Aging	Continuation of secondary reactions, including polymerization	Slow increase in viscosity from secondary reactions such as condensation Potential phase separation
Alkali metals	Nearly all alkali metals reporting to char so not a big problem High ash feed Incomplete solid separation	Catalyst poisoning Deposition of solids in combustion Erosion and corrosion Slag formation Damage to turbines
Char	Incomplete char separation in process	Aging of oil Sedimentation Filter blockage Catalyst blockage Engine injector blockage Alkali metal poisoning

(Continued)

Table 15.3 *(Continued)*

Characteristic	Cause	Effects
Chlorine	Contaminants in biomass feed	Catalyst poisoning in upgrading
Color	Cracking of biopolymers and char	Discoloration of some products such as resins
Contamination of feed	Poor harvesting practice	Contaminants, notably soil, acting as catalysts and can increase particulate carry-over
Distillability is poor	Reactive mixture of degradation products	Bio-oil not be distilled— maximum 50% typically Liquid beginning to react below 100°C and substantially decomposing above 100°C
High viscosity		High pressure drop, increasing equipment cost High pumping cost Poor atomization
Low H:C ratio	Biomass low H:C ratio	Upgrading to hydrocarbons more difficult
Materials incompatibility	Phenolics and aromatics	Destruction of seals and gaskets
Miscibility with hydrocarbons is very low	Highly oxygenated nature of bio-oil	Will not mix with any hydrocarbons so integration into a refinery more difficult
Nitrogen	Contaminants in biomass feed High nitrogen feed such as proteins in wastes	Unpleasant smell Catalyst poisoning in upgrading NO_x in combustion
Oxygen content is very high	Biomass composition	Poor stability Nonmiscibility with hydrocarbons
Phase separation or Inhomogeneity	High feed water High ash in feed Poor char separation	Phase separation Partial phase separation Layering Poor mixing Inconsistency in handling, storage, and processing

Characteristic	Cause	Effects
Smell or odour	Aldehydes and other volatile organics, many from hemicellulose	While not toxic, the smell often objectionable
Solids	See also char Particulates from reactors such as sand Particulates from feed contamination	Sedimentation Erosion and corrosion Blockage
Structure	Unique structure caused by rapid depolymerization and rapid quenching of vapors and aerosols	Susceptibility to aging such as viscosity increase and phase separation
Sulphur	Contaminants in biomass feed	Catalyst poisoning in upgrading
Temperature sensitivity	Incomplete reactions	Irreversible decomposition of liquid into two phases above 100°C Irreversible viscosity increase above 60°C Potential phase separation above 60°C
Toxicity	Biopolymer degradation products	Human toxicity positive but small Ecotoxicity negligible
Viscosity	Chemical composition of bio-oil	Fairly high and variable with time Greater temperature influence than hydrocarbons
Water content	Pyrolysis reactions Feed water	Complex effect on viscosity and stability: Increased water lowering heating value, density, and stability and increasing pH Effect on catalysts

15.2.4 Liquid Collection

The gaseous products from fast pyrolysis consist of aerosols, true vapors, and noncondensable gases. These require rapid cooling to minimize secondary reactions and condense the true vapors, while the aerosols require additional coalescence or agglomeration. Simple indirect heat exchange can cause preferential deposition of lignin-derived components leading to liquid fractionation and eventually blockage in pipelines and heat exchanges. Quenching in product bio-oil or in an immiscible hydrocarbon solvent is widely practiced.

Orthodox aerosol capture devices such as demisters and other commonly used impingement devices are not reported to be as effective as electrostatic precipitation, which is currently the preferred method at both laboratory- and commercial-scale units. The vapor product from fluid-bed and transported-bed reactors has a low partial pressure of condensable products due to the large volumes of fluidizing gas, and this is an important design consideration in liquid collection. This disadvantage is reduced in the rotating cone and ablative reaction systems, both of which exclude inert gas, which leads to more compact equipment and lower costs [15].

15.2.5 By-Products

Char and gas are by-products, typically containing about 25% and 5% of the energy in the feed material, respectively. The pyrolysis process itself requires about 15% of the energy in the feed, and of the by-products, only the char has sufficient energy to provide this heat. The heat can be derived by burning char in orthodox reaction system design, which makes the process energy self-sufficient. More advanced configurations could gasify the char to an low heating value (LHV) gas and then burn the resultant gas more effectively to provide process heat with the advantage that the alkali metals in the char can be much better controlled.

The waste heat from char combustion and any heat from surplus gas or by-product gas can be used for feed drying and in large installations could be used for export or power generation.

An important principle of fast pyrolysis is that a well-designed and well-run process should not produce any emissions other than clean flue gas, that is, CO_2 and water, although they will have to meet local emissions standards and requirements.

15.2.5.1 Char

Char acts as a vapor-cracking catalyst, so rapid and effective separation from the pyrolysis product vapors is essential. Cyclones are the usual method of char removal; however, some fines always pass through the cyclones and collect in the liquid product where they accelerate aging and exacerbate the instability problem which is described below. Some success has been achieved with hot-vapor filtration which is analogous to hot-gas cleaning in gasification systems (e.g., [16–17]). Problems arise with the sticky nature of fine char and disengagement of the filter cake from the filter.

Pressure filtration of the liquid for substantial removal of particulates (down to <5 μm) can be difficult because of the complex interaction of the char and pyrolytic lignin, which appears to form a gel-like phase which rapidly blocks the filter. Modification of the liquid microstructure by addition of solvents such as methanol or ethanol which solubilize the less soluble constituents can improve this problem and contribute to improvements in liquid stability.

15.2.5.2 Gas

The gas contains a small proportion of the initial energy of the biomass feed and is insufficient to provide all the necessary process heat. The heating value depends on the process technology and the extent to which the off-gas is diluted by inert and/or recycle gas.

15.2.6 Environment, Health, and Safety

As bio-oil becomes more widely available, attention will be increasingly placed on environment, health, and safety aspects. A study was completed in 2005 to assess the ecotoxicity and toxicity of 21 bio-oils from most commercial producers of bio-oil around the world in a screening study, with a complete assessment of a representative bio-oil [18]. The study includes a comprehensive evaluation of transportation requirements as an update of an earlier study [19] and an assessment of the biodegradability [20]. The results are complex and require more comprehensive analysis but the overall conclusion is that bio-oil offers no significant health, environment or safety risks.

15.3 Bio-Oil Upgrading

Bio-oil can be upgraded in a number of ways—physically, chemically, and catalytically. This has been extensively reviewed (e.g., Refs. [1, 21–22]) and only the more significant features and recent developments are reported here. A summary of the main methods for upgrading fast-pyrolysis products and the products is shown in Fig. 15.5.

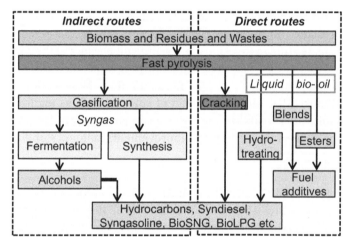

Figure 15.5 Overview of fast-pyrolysis upgrading methods.

15.3.1 Physical Upgrading of Bio-Oil

The most important properties which may adversely affect bio-oil fuel quality are incompatibility with conventional fuels from the high oxygen content of the bio-oil, high solid content, high viscosity, and chemical instability.

15.3.1.1 Filtration

Hot-vapor filtration can reduce the ash content of the oil to less than 0.01% and the alkali content to less than 10 ppm, much lower than reported for biomass oils produced in systems using only cyclones.

This gives a higher quality product with lower char [23]; however, char is catalytically active and potentially cracks the vapors, reduces yield by up to 20%, reduces viscosity, and lowers the average molecular weight of the liquid product. There is limited information available on the performance or operation of hot-vapor filters, but they can be specified and perform similar to hot-gas filters in gasification processes.

Diesel engine tests performed on crude and on hot-filtered oil showed a substantial increase in burning rate and a lower ignition delay for the latter, due to the lower average molecular weight for the filtered oil [24]. Hot-gas filtration has not yet been demonstrated over a long-term process operation. A little work has been done in this area by NREL [32], VTT, and Aston University [25], and very little has been published.

Liquid filtration to very low particle sizes of below around 5μm is very difficult due to the physicochemical nature of the liquid and usually requires very high pressure drops and self-cleaning filters.

15.3.1.2 Solvent addition

Polar solvents have been used for many years to homogenize and reduce the viscosity of biomass oils. The addition of solvents, especially methanol, showed a significant effect on the oil stability. Diebold and Czernik [26] found that the rate of viscosity increase ("aging") for the oil with 10 wt.% of methanol was almost twenty times less than for the oil without additives.

15.3.1.3 Emulsions

Pyrolysis oils are not miscible with hydrocarbon fuels but they can be emulsified with diesel oil with the aid of surfactants. A process for producing stable microemulsions with 5%–30% of bio-oil in diesel has been developed at CANMET [27]. The University of Florence, Italy, has been working on emulsions of 5%–95% bio-oil in diesel [28, 29, 39] to make either a transport fuel or a fuel for power generation in engines which does not require engine modification to dual fuel operation. There is limited experience of using such fuels in engines or burners, but significantly higher levels of corrosion/erosion were observed in engine applications compared to bio-oil or diesel alone. A further drawback of this approach is the cost of surfactants and the high energy required for emulsification.

15.3.2 Catalytic Upgrading of Bio-Oil

15.3.2.1 Upgrading to biofuels

Upgrading bio-oil to a conventional transport fuel such as diesel, gasoline, kerosene, methane, and LPG requires full deoxygenation and conventional refining, which can be accomplished either by integrated catalytic pyrolysis as discussed above or by decoupled operation as summarized above in Fig. 15.7. There is also interest in partial upgrading to a product which is compatible with refinery streams in order to take advantage of the economies of scale and experience in a conventional refinery. This area has been well reviewed (e.g., Ref. [15]). Integration into refineries by upgrading through cracking and/or hydrotreating has been reviewed by Huber and Corma [40]. The main methods are:

- hydrotreating;
- catalytic vapor cracking;
- gasification to syngas followed by synthesis to hydrocarbons or alcohols.

15.3.2.2 Hydrotreating

Hydroprocessing rejects oxygen as water by catalytic reaction with hydrogen. This is usually considered as a separate and distinct process to fast pyrolysis, which can therefore be carried out remotely. The process is typically high pressure (up to 20 MPa) and moderate temperature (up to 400°C) and requires a hydrogen supply or source [41]. Most attention is now focussed on multiple step processes with increasingly severe conditions. Full hydrotreating gives a naphtha-like product which requires orthodox refining to derive conventional transport fuels. This would be expected to take place in a conventional refinery to take advantage of know-how and existing processes. A projected typical yield of naphtha equivalent from biomass is about 25% by weight or 55% in energy terms excluding provision of hydrogen [30]. Inclusion of hydrogen production by gasification of biomass reduces the yields to around 15 wt.% or 33% in energy terms. The process can be depicted by the following conceptual reaction:

$$C_1H_{1.33}O_{0.43} + 0.77\ H_2 \rightarrow CH_2 + 0.43\ H_2O$$

A key aspect is production of hydrogen. Since the hydrogen requirement is significant, it should be renewable and sustainable. Few refineries have a hydrogen surplus, so this has to be provided. There are many ways of providing hydrogen such as gasification of biomass, shifting CO to H_2 followed by scrubbing CO_2, or bio-oil or the aqueous phase from a phase separated product can be steam reformed to hydrogen; or hydrogen can be generated locally by electrolysis of water. Supply of hydrogen from external sources is unlikely to be feasible due to very high cost of storage and transport. The necessary purity of hydrogen is unknown, but some CO shifting may take place in the hydroprocessing reactor removing the need for dedicated shift reactors.

15.3.2.3   Zeolite cracking

Zeolite cracking rejects oxygen as CO_2, as summarized in the conceptual overall reaction below:

$$C_1H_{1.33}O_{0.43} + 0.26\ O_2 \rightarrow 0.65\ CH_{1.2} + 0.34\ CO_2 + 0.27\ H_2O$$

Cracking takes place at atmospheric pressure in a close coupled process. There is no requirement for hydrogen or pressure. The projected yield is around 18 wt.% aromatics and the process is now being commercially developed by Kior. ZSM-5 has attracted most attention due to its shape selectivity to aromatics, with promoters such as Ga or Ni. A key disadvantage is that the catalyst rapidly cokes which requires frequent regeneration as in an FCC unit. Oxygen is thus removed as CO and CO_2 compared to H_2O in hydroprocessing. Production of aromatics is also likely to be of significant interest to the chemicals sector.

15.3.2.4   Gasification for synfuels

A recent concept which has attracted much interest is the decentralized production of bio-oil or bio-oil–char slurries for transportation to a central process plant for gasification and synthesis of hydrocarbon transport fuels, by, for example, Fischer–Tropsch synthesis, or alcohols. Although there is a small energy penalty from the lower pyrolysis energy efficiency, transportation energy and additional bio-oil gasification stage, this is more than compensated by the economies of scale achievable on a commercial-sized gasification and transport fuel synthesis plant [42].

Although the concept of very large gasification plants of 5 GW or more has been promoted [43] on the basis of importation of biomass on a massive scale to an integrated plant, based for example at Rotterdam, there are significant obstacles to be overcome. Decentralized fast-pyrolysis plants of up to 100,000 t/y or 12 t/h are currently feasible and close to being commercially realized. Bio-oil gasification in an entrained-flow oxygen-blown pressurized gasifier is also feasible, such as a Texaco or Shell system, with the added advantage that feeding a liquid at pressure is easier than solid biomass, and offers lower costs, and the gas quality under such conditions is likely to be higher than from solid biomass. Finally transport fuel synthesis at 50,000 to 200,000 t/d is also commercially realisable from the extensive gas-to-liquid plants currently operational around the world. Future Energy (now Siemens) has successfully conducted pressurized oxygen blown gasification tests on both bio-oil and bio-oil/char slurries [30, 45]

15.3.2.5 Other methods and routes

A variety of methods and catalysts have been investigated in recent years as listed below. Some are combined or integrated processes and some employ more traditional chemistry in new applications. It is important to emphasize the importance of maximizing yield and minimizing unwanted reactions especially minimizing residues since these will have to be disposed of at a potential cost as well as lower efficiency.

- Acid cracking in supercritical ethanol
- Aqueous-phase reforming + dehydration + hydrogenation
- Blending
- Dicationic ionic liquid $C_6(mim)_2$–HSO_4
- Esterification of pyrolysis vapors
- Esterification of liquid bio-oil
- Hydrogenation–esterification over bifunctional Pt catalysts
- Reactive distillation
- Solid-acid catalysts $40SiO_2/TiO_2$–SO_4^{2-}
- Solid-base catalysts $30K_2CO_3/Al_2O_3$–NaOH
- Steam reforming
- ZnO, MgO and Zn-Al, and Mg-Al mixed oxides

15.4 Applications of Bio-Oil

Bio-oil can substitute for fuel oil or diesel in many applications, including boilers, furnaces, engines, and turbines for electricity generation, which was been thoroughly reviewed in 2004 [1]. Although many aspects have not changed very much, the most significant changes since then include:

- an appreciation of the potential for fast pyrolysis to be a pretreatment method, that is, for bio-oil to be an effective energy carrier;
- greater interest in bio-oil as a precursor for second-generation biofuels for transport;
- greater awareness of the potential for fast pyrolysis and bio-oil to offer more versatile processes routes to a wider range of products and contribute to biorefinery concept development; and
- considerably greater interest in upgrading bio-oil sufficiently for it to be used for heat, power, and other applications with greater confidence by users.

Figure 15.6 summarizes the possibilities for applications for bio-oil and the main developments are expanded below.

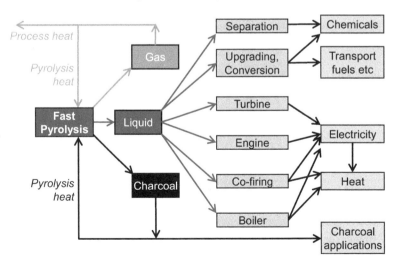

Figure 15.6 Applications for fast-pyrolysis products. See also Color Insert.

15.4.1 Pretreatment Method for Energy Carriers

Biomass is a widely dispersed resource which has to be harvested, collected, and transported to the conversion facility. The low bulk density of biomass, which can be as low as 50 kg/m^3, means that transport costs are high and the number of vehicle movements for transportation to a large-scale processing facility is also very high, with consequent substantial environmental impacts. Conversion of biomass to a liquid by fast pyrolysis at or near the biomass source will reduce transport costs and reduce environmental concerns as the liquid has a density of 1.2 kg/m^3—nearly 10 times higher than low-density crops and residues. This not only reduces the number of vehicle movements and costs by up to 87%, it also reduces costs of handling and transportation by virtue of it being a liquid which can be pumped. This leads to the concept of small decentralized fast-pyrolysis plants of 50,000 to 250,000 t/y for production of liquids to be transported to a central processing plant. It is also possible to consider mixing the by-product char with the bio-oil to make a slurry to improve the energy content of the product, but the pyrolysis process will then require that its process energy needs are met from another source.

Adoption of decentralized fast pyrolysis with transportation of the resultant liquid to a central gasification and fuel synthesis plant has both technical and economic advantages and disadvantages, as summarized in Table 15.4. The impact of inclusion of fast pyrolysis as a pretreatment step on biomass-to-liquid (BtL) cost and performance has been analyzed [42].

Table 15.4 Comparison of solid biomass gasification vs. bio-oil gasification

Impact from using liquid bio-oil	Capital cost	Performance	Product cost
Transport costs	Lower	Higher	Lower
Very low-alkali metals	Lower	Higher	Lower
Handling and transporting costs	Lower	None	Lower
Liquid feeding to a gasifier, particularly pressurized	Lower	Higher	Lower
Lower gas-cleaning requirements	Lower	Higher	Lower
Higher costs for fast pyrolysis	Higher	Lower	Higher
Lower efficiency from additional processing step	Higher	Lower	Higher

Source: Ref. [42].

15.4.2   Cofiring

Coprocessing of biomass with conventional fuels is potentially a very attractive option which enables full economies of scale to be realized as well as reducing the problems of product quality and clean-up. Most current cofiring applications are those where the biomass fuels are added to the coal feed and this is widely practiced at up to 5% on the energy demand of the power station. A few applications involve conversion to a fuel gas via gasification followed by close coupled firing to the power station boiler. There are also some successful examples of cofiring fast-pyrolysis liquids in coal-fired and natural gas–fired power stations.

15.4.3   Fast Pyrolysis–Based Biorefinery

The large majority of chemicals are manufactured from petroleum feedstocks. Only a small proportion of the total oil production, around 5%, is used in chemical manufacture but the value of these chemicals is high and contributes comparable revenue to fuel and energy products. There is a clear economic advantage in building a similar flexibility into the biofuels market by devoting part of the biomass production to the manufacture of chemicals. In fact, this concept makes even more sense in the context of biomass because it is chemically more heterogeneous than crude oil and conversion to fuels, particularly hydrocarbons, is not so cost effective. Figure 15.7 shows fast pyrolysis at the heart of a biorefinery, while Fig. 15.8 shows some potential roles of fast pyrolysis within a biorefinery

A key feature of the biorefinery concept is the coproduction of fuels, chemicals, and energy. As explained earlier, there is also the possibility of gasifying biomass to make syngas, a mixture of hydrogen and carbon monoxide for subsequent synthesis of hydrocarbons, alcohols, and other chemicals. However, this route is energy intensive and much of the energy content of the biomass is lost in processing; therefore, electricity generation may be the most efficient use of biomass [46].

Since the empirical chemical composition of biomass, approximately $(CH_2O)_n$, is quite different from that of oil $(CH_2)_n$, the range of primary chemicals which can be easily derived from biomass and oil are quite different. Hence, any biomass chemical industry will have to be based on a different selection of simple 'platform' chemicals than those currently used in the petrochemical industry. Since the available biomass will inevitably show major regional differences, it is quite possible that the choice of platform chemicals derived from biomass will show much more geographical variation than in petrochemical production.

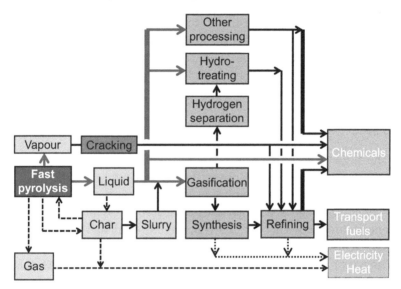

Figure 15.7 Fast pyrolysis–based biorefinery.

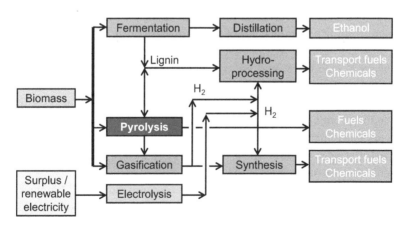

Figure 15.8 Possible roles of fast pyrolysis in a biorefinery.

While biorefineries are not new, the recognition of their strategic and economic potential is recent. A biorefinery can be defined as the optimized performance of the use of biomass for materials, chemicals, fuels, and energy applications, where performance relates to costs, economics, markets, yield, environment, impact, carbon balance, and social aspects. In other words, there needs to be optimized use of resources, maximized profitability, maximized benefits, and minimized wastes.

15.5 Intermediate Pyrolysis

Intermediate pyrolysis has more characteristics of slow pyrolysis than fast pyrolysis and the descriptor was devised to provide the connotation of a modern and advanced technology.

15.5.1 Technology

Processes include rotary kilns, screw reactors, and auger reactors. As processes are scaled up, heat transfer becomes an increasing challenge and over the years many devices have been developed such as circulating steel or ceramic balls. The other challenge is provision of effective and long-life seals around the moving parts which are at high temperature. Early systems based on rotary kilns claimed that a third of the product gas and vapor was required to be burned to provide the process heat. Overall this is comparable to fast pyrolysis. The attraction of the technology is its ability to handle difficult to handle feed materials and these are usually mechanically processed through the reactor.

15.5.2 Products

As shown in Table 15.1 at the beginning of this chapter, liquid yields are lower than fast pyrolysis and are usually in two phases, although a third phase has sometimes been mentioned. The organic phase is more condensed and has a higher energy value than fast-pyrolysis liquid, reaching up to 35 MJ/kg. It has been claimed that this will successfully run a compression ignition engine. The aqueous phase has very high organics and a very high chemical oxygen demand (COD) and biological oxygen demand (BOD) and would require utilization and/or disposal.

The most significant product is charcoal which is produced in yields comparable to slow pyrolysis and typically accounts for around 50% of the energy in the products. The charcoal may tend to have a reduced particle size, depending on the feedstock, the reactor design, and solid handling.

Finally the gas may have a significant heating value if inert gas use is minimized for purging etc. It can be successfully used in engine both compression ignition and spark ignition, where it can be cofed with the organic liquid. The attraction of decoupling liquid and gas production from char is that most of the ash in the feed is retained in the char and thus provides an effective fractionation and isolation of the inorganics.

15.6 Slow Pyrolysis

Slow-pyrolysis characteristics are summarized and compared in Table 15.1 at the beginning of this chapter. Slow pyrolysis has been practiced for production of charcoal for thousands of years and is still used throughout the world. The rapid growth of interest in biochar has created a new level of interest as slow pyrolysis is usually the preferred technology for production of biochar for agricultural applications.

15.6.1 Technology

Technologies can be broadly classified as batch, which are still used throughout the world to provide charcoal for cooking and for export, and high-capacity continuous reactors (such as Lambiotte and Lurgi), which are still used for charcoal production for iron and steel in Brazil, for silicon in Australia, and for specialist charcoal applications in Europe. Feed size and shape are important, especially in moving-bed reactors. Heating can be direct through air addition as in Lambiotte kilns or indirect as in rotary kilns, often using by-product gases and vapors to provide the fuel.

15.6.2 Products

Charcoal is mostly in lump form with smaller particles and dust, which depend on the technology, feedstock, and extent of handling. Recent attention has focussed on use of char for carbon sequestration and soil conditioning—biochar. Char recycles potassium in biomass, provides a microbial base for soil, and improves soil texture. Gases, vapors, and liquids are seldom collected or processed. Exceptions include Usine Lambiotte (now shut down) and proFagus (Chemviron, Degussa) in Germany (still operating). Table 15.5 summarizes performance and financial data from Usine Lambiotte operation in France around 2002. Of significance is the very small production of speciality chemicals at 0.056 wt.% of the feed which provided over 35% of the plant revenue. Chemical sales provided 62% of the total income compared to 31.5% from charcoal. This is a good illustration of how integrated processing in a biorefinery context can maximize benefits. It is also important to appreciate that what is not recovered as a valuable product or by-product has to be disposed of, often at a cost, and often by incineration in the case of charcoal production.

Table 15.5 Financial data from Usine Lambiotte charcoal operation around 2002

	t/year	€/t$^{\#}$	k€/y	%
Biomass input	100,000			
Charcoal	25,000	100*	2500	31.5
Total pyroligneous liquid	40,000			
Water	30,000			
Organics	10,000			
Acids and alcohols	3830	452	1732	21.8
Oils	310	1258	390	4.9
Fine chemicals	56	49,732	2785	35.2
Fuel	5804	90	522	6.6
Total organics	10,000	543	5429	
Total income			7929	100.0

#Average
*Assumed

15.7 Torrefaction

Torrefaction is a recent development and is effectively very low-temperature slow pyrolysis at around 250°C–320°C. This results in complete drying and loss of a fraction of the hemicellulose and also some cellulose, depending on the final temperature reached by the biomass. This results in a hydrophobic, slightly-higher-density, slightly-higher-heating, and much more brittle material. This is well suited to the grinding requirements of a pulverized coal-fired power station or an entrained flow gasifier for synfuel production. Torrefaction enhances the properties of the biomass by:

- removing water;
- reducing hemicellulose;
- improving heating value by up to 15%, depending on the temperature;
- improving storability by inhibiting biological degradation; and
- improving the friability of the product for grinding for cofiring and entrained flow gasification.

The process requires an external heat source which provides around 15% of the energy in the biomass, although the exact figure depends on the technology and the temperature control. Torrefaction is costly in both energy and financial terms and a key question is whether the improvement in product value justifies these costs. The production cost has to be compared to alternative methods of size reduction for large-scale cofiring and gasification in entrained flow reactors.

15.7.1 Technology

The technology is similar to slow pyrolysis. Most reactors are based on continuous processing in moving beds, both horizontal and vertical, and on mechanically driven systems such as auger or screw or rotary kilns. A key requirement is temperature control since excessive local temperatures in the biomass will cause excessive pyrolysis leading to excessive loss of weight and energy, while temperatures which are too low will detract from the claimed advantages listed above. Since most types of biomass are very heterogeneous, temperature control of the heating medium is important to avoid excessive localized temperatures, and thus small temperature gradients are needed between the heat source and the biomass. This results in large reactors with low throughputs. There is interest in pelletization of the torrefied material for improved consistency.

15.7.2 Products

The products are the torrefied biomass and an off-gas containing water vapor and decomposition products from hemicellulose and cellulose. In addition there may be solid particles of biomass and solid product entrained with the off-gas which will require filtration for emissions control. Vapors can either be burned to provide some process heat for torrefaction, to effect waste disposal of the volatiles, or may be collected to yield potentially valuable chemicals arising from the low-temperature pyrolysis or thermal degradation of the hemicellulose.

15.8 Conclusions

15.8.1 Fast Pyrolysis

- The liquid bio-oil produced by fast pyrolysis has the considerable advantage of being storable and transportable, as well as having the potential to supply a number of valuable chemicals. Fast pyrolysis provides a liquid as an energy carrier. The liquid bio-oil is alkali metal free, which has advantages in downstream utilization and/or processing
- Fast pyrolysis has some basic and essential requirements if good yields of good quality bio-oil are to be derived. These are mostly well understood.
- Decentralized pyrolysis plants offer system improvements with environmental and logistical advantages but at a small efficiency and cost penalty. In centralized systems, there is a small cost penalty for using fast pyrolysis for pretreatment.
- Bio-oil from fast pyrolysis can be used for fuel, chemicals or biofuels.

- Fast-pyrolysis technology needs to be improved to reduce costs and increase liquid yield and improve liquid quality.
- Fast-pyrolysis liquid upgrading needs to be further developed and demonstrated. There has been a rapid growth in research into improving bio-oil properties, particularly for dedicated applications and for biofuel production. Much research is still at a fundamental scale, even to the use of model compounds and mixtures of model compounds which purport to represent whole bio-oil. It is doubtful if a limited component mixture can adequately represent the complexity of bio-oil. Some of the most interesting and potentially valuable research is on more complex and more sophisticated catalytic systems, and these will require larger-scale development to prove feasibility and viability.

15.8.2   Intermediate Pyrolysis

The organic part of the liquid from intermediate pyrolysis can be used as fuel, but there is so far insufficient data for other applications. Intermediate pyrolysis is a relatively young technology, and more research and development is needed to evolve an optimal system.

15.8.3   Slow Pyrolysis

The aqueous faction of liquids from slow pyrolysis offers substantial potential for valuable chemicals recovery, but considerable market knowledge is required to exploit the opportunities. Biochar is of great interest but the economics are questionable.

15.8.4   Torrefaction

Torrefaction to pretreat biomass is currently very fashionable and attracting considerable attention. The technology is still being developed with no commercially proven technologies available and uncertain process performance and economics.

15.8.5   Biorefineries

Biorefineries offer considerable scope for optimization of fast-pyrolysis-based processes and products, and these will require development of component processes in order to optimize an integrated system. They will necessarily include provision of heat and power for at least energy self-sufficiency.

There is an exciting future for both fast-pyrolysis and bio-oil upgrading as long as these are focussed on delivering useful and valuable products.

15.8.6 Challenges and Generic Issues

The key challenges to be addressed include:

- feed selection and modification;
- reactor development;
- heat transfer;
- improved modeling;
- char separation;
- bio-oil quality;
- norms and standards; and
- integration with upgrading.

The key generic issues to be addressed include:

- scale-up;
- cost reduction;
- more operational plants;
- development of opportunities;
- health and safety; and
- lack of familiarity with process and products.

References

1. Czernik, S., Bridgwater, A.V. (2004). Overview of application of biomass fast pyrolysis oil, *Energy and Fuels*, **18**, 590–598.

2. ASTM standard D7544-12 for fast pyrolysis liquid (bio-oil).

3. Bridgwater, A.V. (2012). Review of fast pyrolysis of biomass and product upgrading, *Biomass and Bioenergy*, **38**, 68–94.

4. Mohan, D., Pittman, C.U., Steele, P. (2006). Pyrolysis of wood/biomass for bio-oil: a critical review, *Energy and Fuels*, **20**, 848–889.

5. Kersten, S.R.A., Wang, X., Prins, W., van Swaaij, W.P.M. (2005). Biomass pyrolysis in a fluidized bed reactor. Part 1: literature review and model

simulations, *Industrial and Engineering Chemistry Research*, **44**, 8773–8785.

6. Bridgwater, A.V. (2003). Renewable fuels and chemicals by thermal processing of biomass, *Chemical Engineering Journal*, **91**, 87–102.

7. Bridgwater, A.V., Czernik, S., Piskorz, J. (2002). The status of biomass fast pyrolysis, in *Fast Pyrolysis of Biomass: A Handbook*, Vol. 2, 1–22, Ed. Bridgwater, A.V., CPL Press, UK.

8. Bridgwater, A.V., (2009). Fast pyrolysis of biomass, in *Thermal Biomass Conversion*, Eds. Bridgwater, A.V., Hofbauer H, van Loo, S., CPL Press, UK.

9. Venderbosch, R.H., Prins, W. (2010). Fast pyrolysis technology development, *Biofuels, Bioproducts and Biorefining*, **4**(2), 178–208. DOI:10.1002/bbb.205

10. Raffelt, K., Henrich, E., Steinhardt, J., Dinjus, E. (2006). Preparation and characterisation of biomass slurries: a new feed for entrained flow gasification, in *Science in Thermal and Chemical Biomass Conversion*, Eds. Bridgwater, A.V., Boocock, D.G.B., CPL Press, UK.

11. Henrich, E., Dahmen, N., Dinjus, E. (2009). Cost estimate for biosynfuel production via biosyncrude gasification, *Biofuels, Bioproducts and Biorefining*, **3**, 28–41.

12. Hornung, A., Apfelbacher, A., Richter, F., Seifert, H. (2007). Thermo-chemical conversion of energy crops: haloclean; intermediate pyrolysis, *6th International Congress on Valorisation and Recycling of Industrial Waste (VARIREI, 2007)*, L'Aquila, Italy, 27–28 June 2007.

13. Ingram, L., Mohan, D., Steele, P., Strobel, D., Mitchell, B., Mohammad, J., et al. (2008). Pyrolysis of wood and bark in an auger reactor: physical properties and chemical analysis of the produced bio-oils, *Energy and Fuels*, **22**(1), 614–625.

14. Toft, A.J. (1998). PhD thesis, Aston University.

15. Bridgwater, A.V. (2011). Upgrading biomass fast pyrolysis liquids, in *Thermochemical Processing of Biomass: Conversion into Fuels, Chemicals and Power*, Ed. Brown, R.C., Wiley Series in Renewable Resources, Wiley-Blackwell. ISBN: 978-0-470-72111-7

16. Diebold, J.P., Milne, T.A., Czernik, S., Oasmaa, A., Bridgwater, A.V., Cuevas, A., et al. (1997). Proposed specifications for various grades of pyrolysis oils, in *Developments in Thermochemical Biomass Conversion*, Eds. Bridgwater, A.V., Boocock, D.G.G, Springer Netherlands, 433–447.

17. Diebold, J.P. (2002). A review of the chemical and physical mechanisms of the storage stability of fast pyrolysis bio-oils, in *Fast Pyrolysis of*

Biomass: A Handbook, Vol. 2, 243–292, Ed. Bridgwater, A.V., CPL Press, UK.

18. Oasmaa, A., Peacocke, G.V.C. (2001). *A Guide to Physical Property Characterisation of Biomass Derived Fast Pyrolysis Liquids*, VTT Technical Research Centre of Finland, Espoo Finland, VTT publication, 450.

19. Oasmaa, A., Peacocke, G.V.C. (2010). *Properties and Fuel Use of Biomass Derived Fast Pyrolysis Liquids*, VTT Technical Research Centre of Finland, Espoo Finland, VTT publication, 731.

20. Peacocke, G.V.C., Bridgwater, A.V. (2004). Techno-economic assessment of power production from the Wellman Process Engineering and BTG fast pyrolysis processes, in *Science in Thermal and Chemical Conversion*, 1785–1802, Eds. Bridgwater, A.V., Boocock, D.G.B., CPL Press, UK.

21. Diebold, J.P., Scahill, J.W., Czernik, S., Philips, S.D., Feik, C.J. (1996). Progress in the production of hot-gas filtered biocrude oil at NREL, in *Bio-Oil Production and Utilisation*, Eds. Bridgwater, A.V., Hogan, E.N., CPL Press, UK, 66–81.

22. Hoekstra, E., Hogendoorn, K.J.A., Xiaoquan, W., Westerhof, R.J.M., Kersten, S.R.A., van Swaaij, W.P.M., et al. (2009). Fast pyrolysis of biomass in a fluidized bed reactor: in situ filtering of the vapors, *Industrial and Engineering Chemistry Research*, **48**(10), 4744–4756.

23. Vivarelli, S., Tondi, G. (2004). Pyrolysis oil: an innovative liquid biofuel for heating the COMBIO project, *International Workshop Bioenergy for a Sustainable Development*, Casino Viña del Mar, Chile, 8–9 November 2004.

24. Sitzmann, J., Bridgwater, A.V. (2007). Upgrading fast pyrolysis oils by hot vapour filtration, *15th European Energy from Biomass Conference*, Berlin, 7–11 May 2007.

25. http://ec.europa.eu/energy/renewables/bioenergy/doc/pyrolysis/biotox_publishable_report.pdf, accessed 16 February 2011.

26. Peacocke, G.V.C. (2002). Transport handling and storage of fast pyrolysis liquids, in *Fast Pyrolysis of Biomass: A Handbook*, Vol. 2, 293–338, CPL Press, UK.

27. Blin. J., Volle, G., Girard, P., Bridgwater, A.V., Meier, D., (2007). Biodegradability of biomass pyrolysis oils: comparison to conventional petroleum fuels and alternatives fuels in current use, *Fuel*, **86**, 2679–2686.

28. Maggi, R., Elliott, D. (1997). Upgrading overview, in *Developments in Thermochemical Biomass Conversion*, 575–588, Eds. Bridgwater, A.V., Boocock, D.G.G., Blackie, A.C.

29. Bridgwater, A.V. (1966). Production of high-grade fuels and chemicals from catalytic pyrolysis of biomass, *Catalysis Today*, **29**, 285–295.

30. Bridgwater, A.V. (1994). Catalysis in thermal biomass conversion, *Applied Catalysis A*, **116**, 5–47.

31. Zhang, Q., Chang, J., Wang, T., Xu, Y. (2007). Review of biomass pyrolysis oil properties and upgrading research, *Energy Conversion and Management*, **48**, 87–92.

32. Diebold, J.P., Czernik, S., Scahill, J.W., Philips, S.D., Feik, C.J. (1994). Hot-gas filtration to remove char from pyrolysis vapours produced in the vortex reactor at NREL, in *Biomass Pyrolysis Oil Properties and Combustion Meeting*, 90–108, Ed. Milne, T.A., National Renewable Energy Laboratory, Boulder, CO.

33. Shihadeh, A.L. (1998). *Rural Electrification from Local Resources: Biomass Pyrolysis Oil Combustion in a Direct Injection Diesel Engine*, PhD thesis, Massachusetts Institute of Technology.

34. Sitzmann, J., Bridgwater, A.V. (2007). Upgrading fast pyrolysis oils by hot vapour filtration, *15th European Energy from Biomass Conference*, Berlin, 7–11 May 2007.

35. Diebold, J.P., Czernik, S. (1997). Additives to lower and stabilize the viscosity of pyrolysis oils during storage, *Energy and Fuels*, **11**, 1081–1091.

36. Ikura, M., Slamak, M., Sawatzky, H. (1998). Pyrolysis liquid-in-diesel oil microemulsions, US patent 5820640.

37. Baglioni, P., Chiaramonti, D., Bonini, M., Soldaini, I., Tondi, G. (2001). BCO/diesel oil emulsification: main achievements of the emulsification process and preliminary results of tests on diesel engine, in *Progress in Thermochemical Biomass Conversion*, Ed. Bridgwater, A.V. Blackwell Sciences, UK, 1525–1539.

38. Baglioni, P., Chiaramonti, D., Gartner, K., Grimm, H.P., Soldaini, I., Tondi, G., et al. (2003). Development of bio crude oil/diesel oil emulsions and use in diesel engines: part 1; emulsion production, *Biomass and Bioenergy*, **25**, 85–99.

39. Baglioni, P., Chiaramonti, D., Gartner, K., Grimm, H.P., Soldaini, I., Tondi, G., et al. (2003). Development of bio crude oil/diesel oil emulsions and use in diesel engines: part 2; tests in diesel engines, *Biomass and Bioenergy*, **25**, 101–111.

40. Huber, G.W., Corma, A. (2007). Synergies between bio- and oil refineries for the production of fuels from biomass, *Angewandte Chemie, International Edition*, **46**(38), 7184–7201.

41. Elliott, D.C., Baker, E. (1983). In *Energy from Biomass and Wastes X*, Ed. Klass, D., IGT 765–782.

42. Bridgwater, A.V. (2009). *Technical and Economic Assessment of Thermal Processes for Biofuels*, Report to NNFCC, June 2009, COPE, UK, available at www.copeltd.co.uk.

43. Calis, H.P.A., Haan, J.P., Boerrigter, H., van der Drift, A., Peppink, G., van den Broek, R., Faaij, A.P.C., Venderbosch, R.H. (2002). Preliminary techno-economic analysis of large-scale synthesis gas manufacturing from imported biomass, in *Proceedings of Pyrolysis and Gasification of Biomass and Waste*, Ed. Bridgwater, A.V., CPL Scientific Press, Strasbourg.

44. Morehead, H. (2008). *Siemens Gasification and IGCC Update, Gasification Technologies*, Washington, DC, 7 October 2008.

45. Volkmann, D. (2004). *Update on Technology and Projects, Gasification Technologies Conference*, Washington, DC, October 2004.

46. Kamm, B., Gruber, P.R., Kamm, M. (2006). *Biorefineries: Industrial Processes & Products*, Vols. 1–2, 45, Wiley-VCH, Weinheim, Germany.

About the Author

Tony Bridgwater is a professor of chemical engineering at Aston University, Birmingham, UK. He has worked at Aston University for most of his professional career and is currently director of the European Bioenergy Research Institute. He has a world-wide research portfolio focussing on fast pyrolysis as a key technology in thermal biomass conversion for power, heat, biofuels, and biorefineries. He is a fellow of the Institution of Chemical Engineers and a fellow of the Institute of Energy.

Of particular note was Bridgwater's technical leadership of United Kingdom's SUPERGEN Bioenergy Consortium—the United Kingdom's virtual center of excellence in biomass and bioenergy research with 14 academic partners and 10 companies, which ran for 8½ years until the end of 2011. In addition he has led and coordinated nine major European Commission (EC) research and development projects in bioenergy and has an active current involvement in six further research and development projects. He has attracted funding from national research funding councils in Canada, Holland, Norway, and the United States. He has been responsible for raising over £28 million during his research career. He formed and led the IEA Bioenergy Pyrolysis Task—Pyne— from 1994 to 2008 with parallel European networks on pyrolysis, gasification, and combustion, which included the EC-sponsored ThermoNet and ThermalNet networks.

Bridgwater was awarded the following:

- The European Johannes Linneborn Prize in 2007 for "Outstanding Contributions to Bioenergy"
- The North American Don Klass Award in September 2009 for "Excellence in Thermochemical Biomass Conversion"
- The Aston University Chancellor's medal in July 2010 for "Outstanding Service to the University"

- Green Leader in the West Midlands in 2011 in recognition of individuals who are having a positive environmental impact in helping the West Midlands achieve its vision for a low carbon economy.
- The UK Environmental Capital Peterborough Clean Energy Award in 2012 to recognize outstanding contributions to the development, implementation, and promotion of clean-energy technologies.

Chapter 16

Coprocessing of (Upgraded) Pyrolysis Liquids in Conventional Oil Refineries

R. H. Venderbosch[a] and H. J. Heeres[b]
[a]Biomass Technology Group (BTG) B.V., Enschede, the Netherlands
[b]University of Groningen, Groningen, the Netherlands
venderbosch@btgworld.com

Fast pyrolysis is an efficient technology to convert biomass to a high yield of "pyrolysis liquids" and as such is a significant concentration step to reduce storage space and transport costs. In addition, it allows feeding of biomass derived liquids in existing refineries, with or without suitable pretreatment. The focus of this chapter is to provide an overview of the state of the art regarding the (co-)feeding of (upgraded) pyrolysis liquids in existing refineries.

16.1 The Gap between Agriculture and Petrochemistry

Environmental concerns and possible future shortages have boosted research on alternatives for fossil-derived products and biomass is

Biomass Power for the World: Transformations to Effective Use
Edited by Wim van Swaaij, Sascha Kersten, and Wolfgang Palz
Copyright © 2015 Pan Stanford Publishing Pte. Ltd.
ISBN 978-981-4613-88-0 (Hardcover), 978-981-4669-24-5 (Paperback), 978-981-4613-89-7 (eBook)
www.panstanford.com

considered an attractive option. Compared to fossil fuels, however, biomass is much more reactive, mainly due to its high content of a wide range of oxygen containing biopolymers (mainly cellulose, hemicellulose, and lignin) Besides, it is much more difficult to handle due to its huge variability in composition, depending among others on the biomass source (e.g., wood, grass, and straw). Conversion of biomass presents challenges for implementation and radical changes are required to improve the state of the art and particularly (i) to increase the overall carbon conversion efficiency, (ii) to broaden the biomass feedstock basis, and (iii) to improve the economic, environmental, and social impacts relative to fossil fuels and currently available biofuels.

Fast pyrolysis, as elucidated elsewhere in this book, produces a pyrolysis liquid (PL) with a higher energy density than the original biomass, which allows for simpler and cheaper transport of biomass resources. With a high energy yield (70+% for just PLs from woody biomass but nearly 90% if production of electricity and heat are also taken into account), conservative estimates for the GHG emission savings of PLs for heat and electricity are high [1].

Future perspectives for PLs are to transport them over long distances (just like crude oil) to refining units. However, and similar to biomass, PLs have a low heating value per kg (14–17 MJ/kg), are not miscible with mineral oils, are polar and acidic, and have a high oxygen content. Furthermore, the liquids have a high coking tendency due to possible polymerization of the organics. Possibilities to convert such liquid biomasses into new generation biofuels for the existing transport sector include various routes, schematically outlined in Fig. 16.1 below. A straightforward option is fast pyrolysis of biomass to PLs, followed by (co-)hydrotreating of the oxygenated components and possibly (co-)processing with fluid catalytic cracking (FCC) catalysts. The corefining (FCC or hydrotreating) can occur by mixing (emulsification) the oil with crude oil derivatives. Alternatives include the use of an intermediate fraction of the PL obtained by a fractionation step (e.g., separation of the lignitic fraction from the carbohydrate fraction) and separate upgrading (FCC or hydrotreating) of one or more of these fractions. A specific option includes the aqueous-phase processing (APP) of liquids (or fractions thereof) for the generation of hydrogen for hydro treatment.

Another approach indicated in Fig. 16.1 is reduce the oxygen content upfront, for example, by catalytic pyrolysis of biomass to yield liquids containing less oxygen (catalytic pyrolysis liquids [CPLs]). It is suggested that the oils from catalytic pyrolysis have higher heating values and low acid and water content and even may yield so-called drop-in biofuels miscible with crude oil derivatives. Likely though, also these liquids require further upgrading.

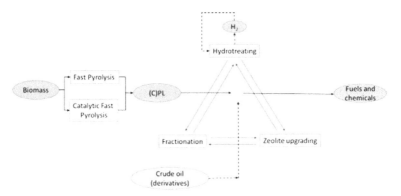

Figure 16.1   Routes from biomass to fuels and chemicals through (catalytic) pyrolysis.

16.1.1   This Chapter

This chapter provides an overview of the possibilities to use the product liquids from (catalytic) fast pyrolysis, directly or after a suitable upgrading processing step, in existing refineries. The main emphasis will be on a two-step approach where a PL obtained by noncatalytic fast pyrolysis technology is (co-)fed in an FCC unit, as such or after prior treatment (e.g., by catalytic hydrotreatment). In addition, product compositions and properties of the various products will be discussed and compared.

16.2   Integration of Pyrolysis Liquids in a Refinery

An option for introducing biomass-derived liquids into the fuels marketplace is to use PLs or upgraded PLs as a (co-)feed in a petroleum refinery as a replacement of fossil fuels. This would, in principle, facilitate the introduction of renewable carbon into the existing fuels infrastructure. Moreover, it is economically advantageous for the biofuels industry to use the multitrillion dollar refining and distribution infrastructure as it is already in place. From a refiner's perspective, however, important properties include the boiling-range distribution obtained from the main crude oil fractionator and the hydrocarbon types and heteroatom (sulfur, nitrogen, oxygen) contents of each of the resulting primary distillation and process-derived intermediate fractions. The boiling-range distribution from the main fractionator impacts all of the major downstream unit operations, which are in turn designed to optimize the refinery product slate to produce the most profit per barrel of feedstock. Before detailing possibilities to introduce biomass-derived liquids in refineries, it is important to understand that there are five major types of hydrocarbons in petroleum feedstocks, viz., paraffins, iso-paraffins, aromatics, naphtenens, and olefins (PIANO). The main objectives of refineries are (i) to steer the product slate to fractions with same carbon number (the number of carbons atoms in a particular molecule) by distillation, (ii) to decrease density by increasing hydrogen-to-carbon ratio, and (iii) to process and control the various hydrocarbon isomers within a fraction. For these specific purposes, crude oil in a refinery is first fractionated (distilled) in atmospheric and vacuum columns, where after a relatively large fraction (referred to as gas oil and residue) are further processed, usually to reduce the molecular weight by cracking and to increase the hydrogen-to-carbon ratio by coke extraction. The key conversion technology for the latter is FCC (Figs. 16.2 and 16.3). Worldwide significant changes exist in the demand for fuel types and crude oil available, resulting in different refinery complexities. The installed capacities (in bbl/y) of atmospheric distillation, vacuum distillation, cracking, and hydrotreating roughly correspond to the different demands in the US and Europe. Indicatively, in 2011 in the US approx. 50% of the crude oil processed in the atmospheric distillation unit was further processed in an FCC unit and 33% in hydrotreaters, while these values were around 40% and 20% respectively in Europe [3].

Detailed overviews of refinery units are presented elsewhere; a quick review (with perspectives for the introduction of oxygenated PLs) was presented recently [4].

The ability to profitably process complex hydrocarbon mixtures suggests that refineries are well suited to handle PLs. However, this is definitely not the case, as the oxygen content in petroleum usually is, in contrast to typical PLs, far below 1 wt.% (and treated as a contaminant). Feeding of PLs in such complexes introduce significant uncertainties, and related to this, quite some potential problems and pitfalls:

- The high oxygen content is not accommodated for by the refineries, usually dealing with oxygen contents in the crude oil far below <1 wt.%.

Fluid Catalytic Cracking

Over more than 60 years Fluid Catalytic Cracking (FCC) is effectively used to upgrade heavy gas oils (HGO) and residue fuels to mainly gasoline and diesel blend components. The history of process design is well-related to catalyst development, with the earlier catalysts being activated natural clays with low catalytic activity and poor stability being used in packed bed. With the advent of (much) more active and stable synthetic silica/alumina catalysts in the 1940's, and in the 1960's the introduction of zeolite Y, the process shifted to the use of circulating fluid bed reactors. Cracking is nowadays achieved by bringing the

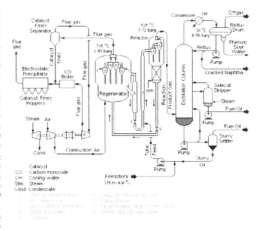

liquid in contact with powdered catalysts at temperatures of around 600 - 700°C. The reaction is usually carried out in an up-flowing vertical reactor ('the riser reactor'), where oil vaporizes and cracks to lighter products in a few seconds. Licensors for the FCC technology include ABB Lummus, Exxon Mobil, Kellog Brown & Root, KBR, Shaw Stone & Webster – Axens, Shell and UOP. A typical example of an FCC process is presented below.

A preheated petroleum feedstock (about 315 to 430 °C) consisting of long-chain hydrocarbon molecules is combined with recycle slurry oil from the bottom of the distillation column and injected into the catalyst riser (7). Here it is vaporized and cracked into smaller molecules by contact and mixing with the hot catalyst from the regenerator. The hydrocarbon vapors

Typical FCC operating conditions

wt.%	FCC
Feed	
Temperature	
Reactor (°C)	510 - 500
Regenerator (°C)	670 - 730
Reactor pressure (barg)	1.4 – 2.1
Reaction time (s)	2
Catalyst:Oil ratio (wt./wt.)	5 – 8

"fluidize" the powdered catalyst and the mixture of hydrocarbon vapors and catalyst flows upward to enter the reactor at a typical temperature of 535 °C. The cracking reactions take place in the catalyst riser within seconds, where after the cracked product vapors are separated from now 'spent' catalyst by means of cyclones (5). The spent catalyst flows downward through a steam stripping section to remove any hydrocarbon vapors before the catalyst returns to the catalyst regenerator. In the cracking reactions solid carbonaceous material is produced ('coke') that deposits on the catalyst and very quickly reduces the catalyst reactivity. This coke is removed by burning, hereby regenerating the catalyst. The regenerator typically operates at a temperature above 700 °C and a pressure slightly above atmospheric (up to 2 barg).

FCC units are often 'heat balanced' and the amount of coke produced is an essential design parameters of the unit. The flow of regenerated catalyst to the feedstock injection point below the catalyst riser is regulated by a slide valve, and a catalyst flow in an FCC unit processing 75,000 barrels per day (12,000 m³/d) is typically in the range of 60,000 to 120,000 ton/day. General FCC operating conditions are listed above.

Figure 16.2 Fluid catalytic cracking (FCC). (http://en.wikipedia.org/wiki/Fluid_catalytic_cracking).

- Oxygenated compounds show boiling points typically much higher than the hydrocarbons with the same carbon number.
- Densities of the various oxygenated compounds are much more diverse than for hydrocarbons.
- The number of possible isomers of oxygenated components is far higher than for hydrocarbons.
- Due to its polar nature, PLs do not mix with the apolar crude oil (derivatives).
- The high water content of PLs is a disadvantage; water is considered a contaminant in conventional refining processes.
- The acidity of PLs is (much) higher than of crude oil.
- The presence of the various reactive functionalities results in thermal polymerization reactions, resulting in a high coking tendency.
- PLs contain large content of different metals (merely alkali and alkaline earth metals) relative to crude oil (heavy metals).

Theoretically possibilities for the introduction of biomass-derived liquids by the fast pyrolysis platform into petroleum refineries is schematically presented in Fig. 16.4. Introduction of pure PLs in a refinery is questionable, as (too) large amounts of coke/char can be expected. The latter points at (co-)FCC options for slightly upgraded PLs. The FCC unit is merely the heart of the refinery complex (see text box), and is available at very large scales in the order of 200 t petroleum derivative/h. The application of such a cofeeding (co-FCC) concept can in principal be rapidly integrated into an existing refinery, the more so as FCC is a rather flexible process unit operation, be readily tuned to accommodate different feedstocks (by operating conditions and catalyst). On the other hand, it must be noted that the FCC unit is optimized for the production of gasoline, which is a potential disadvantage as the demand for diesel is growing.

A major drive for coprocessing is to leverage the capital costs by using the existing refineries' infrastructure and logistics rather than construct an independent route from biomass to fuels. Pioneering work started with the introduction of pure PLs in FCC units, and this will be the topic of the next paragraph. Thereafter, co-FCC of mildly hydrotreated PLs will be discussed.

Products from FCC

A typical product slate obtained from vacuum gas oil (VGO) are usually defined as: fuel gas (ethane and lighter hydrocarbons), olefinic and saturated C_3/C_4 liquefied petroleum gas (LPG), gasoline, light cycle oil (LCO, comparable with diesel), fractionator bottoms (slurry oil+ bottoms), and coke (combusted in regenerator). Preferred products are gasoline, LCO and LPG. One of the strengths of FCC is its versatility in producing these individual components by adjusting the operating parameters. Three modes of FCC operation are consequently a) the Gasoline mode, b) the Distillate mode and c) The light olefin mode.[3] In the latter different catalysts and operating conditions are required. More or less representative yields for the various regimes though are listed in Table below.

Most units have been designed for gasoline production, wherein the reaction severity is high enough to convert a substantial portion of the feed, but no so high as to destroy the gasoline produced. Usually no recycle or unconverted components is needed. In a high severity system, light olefins and a higher octane gasoline is produced, sometimes referred to as a liquefied petroleum gas mode or even as petrochemical FCC because of the increased amount of light material. Finally production of distillates can be optimized in a distillate mode (or maximum LCO) if the reaction severity is limited.

Table 4. Typical product slate from an FCC (Meyers, 1997).

wt.% Feed	wt.% Other references	Middle distillate mode Full range	Undercut	Gasoline mode	Light olefin mode
H_2S		0.7	0.7	1.0	1.0
Dry gas	fuel gas;	2.6	2.6	3.2	4.7
$C_3+ C_4$	Light olefins	16.7	16.7	26.1	36.6
C_3 olefins				8.2	
C_4 olefins				13.1	
C_5^+ (-225°C)	gasoline; naphtha	43.4	33.3	60.0	55.2
C_5 olefins					
LCO (220-344°C)	Diesel; light cycle oil	37.5	47.6	13.9	10.1
Bottoms (> 344°C)	(clarified slurry oil, CSO)	7.6	7.6	9.2	7.0
Coke	char	4.9	4.9	5.0	6.4

FCC Catalysts: State-of-the-Art FCC catalysts are fine powders (particle diameter from 50 to 80 µm) constituting essentially a silica-alumina matrix and a very active material (zeolite). The catalyst contains (Brönsted and Lewis) acid sites able to crack heavy hydrocarbons to gasoline and others lighter products without excessive coke production. In the 1980's zeolite ZSM-5 was introduced, and in the early 1990s new alumina technologies allowed processing of heavier crude sources with higher tolerance to Ni and V. FCC catalyst usually contain three components: zeolites, active matrix, and a binder. The overall activity of the catalyst increases and the coke yield decreases as the zeolite content is increased. The effect is evident with easier-to-crack paraffin feeds, as aromatic feeds are less susceptible to higher zeolite activity. The large pore structure of active matrix portion of catalyst provides for easy access of large, heavy oil molecules, providing also their effective conversion and improving the stripping of hydro carbonaceous product from the solid structure. Feed composition and contaminants such as sulfur, nitrogen, and trace of metals (i.e., Ni and V) affect the yield and quality of the products, the catalyst consumption of the unit but also the heat balances (by the formation of coke). By default, more sulfur in the feed leads to (cracked) products containing sulfur, with a distribution of sulfur over the products varying widely (and not controlled by operating parameters nor catalyst design). Nitrogen compounds cause reversible catalytic deactivation and low conversion levels, the latter requiring higher reactor temperatures. Ni and V cause the most serious problems, with Ni increasing the coke and gas yield and V causing irreversible catalytic deactivation.

Figure 16.3   Products from FCC [2].

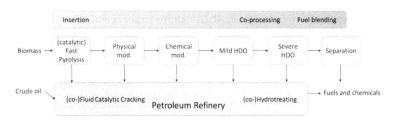

Figure 16.4   Pathways for the introduction of biomass in petroleum refineries. Adapted from Ref. [5].

16.2.1 Fluid Catalytic Cracking of Pure Pyrolysis Liquids

Conversion of pure PLs over conventional FCC catalysts dates back to the nineties [6–8]. The idea generally referred to is that similar to methanol over ZSM-5 catalysts, oxygen can be rejected from the oil as CO_2. Indeed, when pure PLs are led over such catalysts (amongst others HZSM-5, H-Y, and other silica-alumina-based materials, used in batch-wise-operated packed-bed systems) at temperatures ranging from near 300°C to over 400°C hydrocarbons are produced. The product slate includes char (material deposited on the wall before the catalytic reactor), coke (deposited on the catalyst), gas, tar (the "acetone soluble" fraction of the solids remaining in the reactor), residue (the "nonvolatile" liquid product fraction), water and a preferred organic distillate fraction (ODF). However, in all cases hydrocarbon yields (from PLs to ODF) were well below 30 wt.%, with maximum values for HZSM-5 (near 28 wt.%) and the lowest (<15 wt.%) for silica-alumina, respectively. These values correspond to maximum carbon yield to liquids far below 40%, the remaining carbon lost in char, coke, and gas. The product liquids still contain oxygen, and additional treatments of the liquids seemed required to produce transport fuels. In such batch-wise-operated system the catalysts are deactivated rapidly, similar to conventional FCC catalysts.

Due to the large loss of carbon as coke and char, blending of PLs with conventional crude-derived liquids (usually vacuum gas oil [VGO]) before introduction in FCC is a logical step (see Fig. 16.5). With the interest of oil companies for this concept, standardized techniques for the simulation of co-FCC of PLs were introduced in the form of small FCC test units (referred to as ACE or MAT). In these units, representing the actual behavior of commercially operated FCC units quite well, various parameters as cat-to-oil ratio and temperature, conversion, and product composition are systematically investigated. At the same time, terminology applied in the petrochemical world was introduced, with reference to conventional product slates. First results for such blending tests in ACE units were reported by UOP [9]. Table 16.1 shows typical results for pure VGO as a feed and a comparion with results obtained for blends of PLs and VGO (20:80). These findings also imply high carbon losses in the form of coke. On an incremental basis, the "pure" PL formed has more than 80% coke + water + CO_2, with roughly 40% of the carbon ending up in the coke. The results are consistent with those for FCC of pure PLs (*vide supra*). Significant amounts of carbon are also transferred to gas (fuel gas and liquefied petroleum gas [LPG]) but suprisingly also a larger fraction than expected to gasoline (here referred to as naphta), and less to diesel (or light cycle oil [LCO]) and bottoms (here clarified slurry oil [CSO]). Indicatively, replacement of 20% of the crude oil by PLs effectively reduces the total amount of carbon fed to the FCC unit with approx. 13% (due to the oxygen in the PL), but the apparent yield of naphtha is reduced only less than 5%. The PL appears to increase the crackability of the VGO and shifts the VGO product portfolio towards light ends. For LCO and CSO, a 20% substitution by PLs is reflected by an almost 40% reduction in carbon transformed into LCO and more than a 75% reduction for CSO, which can be an economically attractive outcome. On the other hand 25% more LPG is formed. All this indicates that synergetic effects between the crude oil derivative and PLs take place, and VGO seems to act as a hydrogen donor to the PL. Overall, the cofeeding of PLs to FCC units is not beneficial, with only an estimated 10% of the carbon from the liquids ending up in useable products (LPG and liquids).

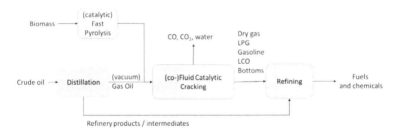

Figure 16.5 PL corefining in an FCC unit. Adapted from Refs. [10, 11].

Much of the recent advances to obtain a better understanding of the coprocessing of untreated PLs in oil refineries have been the result of the research carried out under the European 6th Framework program BIOCOUP [12]. Data on the use of pure PLs or co-FCC PLs with VGO are not published, but it is mentioned that despite a lower oxygen content, PLs upgraded without hydrogen (decarboxylated oils [DCOs]) or without catalyst or hydrogen (high-pressure thermal treatment oils [HPTTs]) could not be successfully coprocessed. An important criteria to allow successful coprocessing of such oils is to ensure a low-coking tendency (measured as Micro Carbon Residue Testing or MCRT), high H/C ratios, and a lower average molecular weight [13].

Table 16.1   Yields from co-FCC of VGO and PL

wt.%	VGO	20 wt.% PL and VGO
Ethylene	2.0	3.3
Propane	1.2	2.1
Propylene	5.9	5.9
Butane	11.1	13.5
Gasoline	42.7	40.6
LCO	14.8	9.1
CSO	18.5	4.8
Coke	3.9	7.1
Water and CO_2	0.0	13.4

Source: Ref. [9].

Results for the cofeeding of untreated PLs were published by Grace [14], who coprocessed a blend of 3 wt.% pure PLs and VGO in a DCR system (see Table 16.2). Small amounts of PLs already affect the yields significantly, again causing higher yields of coke. In these experiments, however, more bottoms were formed, at the expense of gasoline and LCO yields. Less PG is formed, and yields of useable products (LPG and liquids) are around 40%.

Table 16.2   (Interpolated) yields at a constant conversion of 82.5%, 520°C, and C/O = 10.6

wt.%	VGO	3 wt.% PL and VGO
Dry gas		
H_2	0.05	0.05
C_1 and C_2	3.1	3.2
LPG ($C_3 + C_4$)		
C_3	8.6	8.2
C_4	15.1	14.4
Gasoline (−225°C)	49.0	47.9
Diesel (220°C–344°C)	13.4	13.3
Bottoms (residues; >344°C)	4.1	4.2
Coke	6.40	6.65
Water		1.27
$CO + CO_2$	0	0.64

The complete set of data is presented in Fig. 16.6. It is shown that already at small 3 wt.% substitution ratios, the PL causes a significant yield shift in product slate. In addition, a synergetic effect of the PL on the VGO is suggested. Obviously the oxygen bound to the PL is converted to water, CO and CO_2, while yields of coke and bottoms increase at the expense of the preferred gasoline and LCO. The overall conclusion from these experiments likely precludes the processing of untreated PLs in FCC units.

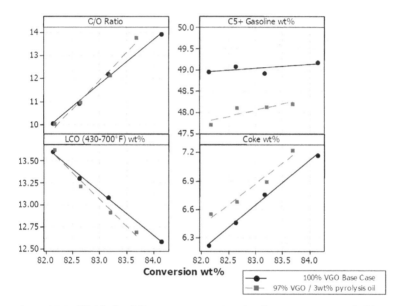

Figure 16.6   Yields (wt.%) versus conversion for the coprocesisng of a PL in an FCC unit. (C/O: cat-to-oil ratio; LCO: light cycle oil, or diesel fraction).

Data for pure PLs can be derived assuming that the PLs do not show any synergetic effects on the VGO. By doing so, yields for gasoil, LCO, and bottoms which can be traced back to the PL only are substantially lower than for VGO but surprisingly higher compared to earlier reported data from UOP. Yields for the higher molecular weight fraction (diesel and LCO) are rather high, and a significant decrease in the LPG yield is observed. Oxygen is traced back in water (3/4) and in (CO_x) (1/4). In these experiments, synergetic effects between the PL and the VGO must be accounted for to explain the reduced yield in LPG due to the PL. However, as with the experimentally observed synergetic effect of PLs on diesel yield in UOP's experiments, explanations are missing and fundamental understanding is absent.

It is generally accepted that small-scale ACE reactors provide data which substantially deviate from continuous operated systems, amongst others due to the complexity of feeding (very) small amounts of feedstock. Specifically for VGO these problems are solved, but now reappear, while introducing even smaller fractions of PLs (not miscible with VGO). To overcome this artefact, Petrobras deployed a near-commercial FCC vacuum gasoil unit ("Six" located in São Mateus Sul in Brazil) to coprocess pure PLs [10, 11]. The untreated PLs were converted into biofuels via catalytic cracking, and, the cracker naphtha was successfully hydrotreated in a subsequent step. The test unit (Fig. 16.7) consists of a vertical riser reactor (50 mm ID and 18 m riser height), a fluid-bed regenerator reactor, a stripper, and a lift line, with a total catalyst inventory of 450 kg. VGO and pure PLs were injected at two different heights into the riser reactor, at a combined feed rate of 150 kg/h at conditions close to that used in commercial FCC units.

* Catalytic Cracking Demonstration Unit
* Location: São Mateus do Sul (PR)
* Feed Rate: 200 kg/h

* Purchase of 3 tons of bio-oil from BTG/Netherlands
* Refining/Cracking: 80% gasoil (fossil) plus 20% bio-oil

Figure 16.7 Catalytic cracking unit used for cofeeding experiments of VGO with PLs (80:20 and 90:10) by Petrobras [10, 11].

An overview of the results is presented in Table 16.3; detailed results for higher temperature, different catalyst circulation rates, and cracking severities are presented elsewhere [11].

Table 16.3 Co-FCC data for pyrolysis liquids by Petrobras at 540°C

wt.%

Feed	Gasoil	90% Gasoil and 10% PL	80% Gasoil and 20% PL
Fuel gas	3.9	2.8	2.5
LPG (C_3 + C_4)	15.2	12.9	9.9
Naphta (C_5: –225°C)	40.4	40.7	37.7
LCO (220°C–344°C)	18.1	17.4	16.5
Decanted oil (+344°C)	14.8	14.0	13.7
Coke	7.4	7.5	8.5
CO	0.1	1.9	3.1
CO_2	0.1	0.5	0.8
Water	0	2.3	7.3

By separating the contributions of PLs and VGO to the product slate (ignoring any synergetic effects) an estimation of the carbon yields for the PL only can be made. Though the liquid yield from the VGO-PL mixture is not dramatically low, a significant decrease is noticed in the fuel/LPG yield (as shown in the Grace runs as well), and again, a dramatic decrease in the (preferred) gasoline fraction. Also at these milder conditions—relative to the Grace experiments— the oxygen in the oil (in water and bound to carbon) is primarily released as water, but at a slightly lower ratio of 2:1 for the carbon oxides. The gasoline yield obtained in coprocessing is approximately the same when 10% PL is added to the fossil feedstock, but decreases when 20% PL is introduced. The coke yield obtained at 10% substitution level is rather similar to the pure VGO test, but increases with the 20% substitution level. Nevertheless, this increase is not as high as found in the data of Grace and UOP, both at the bench scale with the same or even lower PL addition levels.

The experimental studies reported by UOP, Grace, and those in BIOCOUP reported technical problems while cofeeding pure PLs, specifically due to nozzle plugging. but this could be corrected by modifications to the liquid injection system. In addition, being an alient component in petroleum processing, difficulties were encountered in addressing the correct water content and yield in these experiment (especially as a large dilution ratio requires high accuracy). These problems were indeed solved in the Petrobras's experiments.

It can be concluded that there are substantial differences in the product slates obtained at lab-scale ACE or MAT (UOP), pilot plant DCR (Grace), and semicommercial scale (Petrobras). Partially these may be related to differences in the manner the PL is introduced in the reactor system, the complications in dispersion of the liquid at a smaller scale (e.g., in MAT and ACE), but also in artefacts as cat-to-oil ratios, type of VGO applied (e.g., a high coke-forming tendency for the fossil feed used Petrobras's experiments versus a low coke forming tendency for UOP's feed) etc.

Only limited research has been carried out on the use of catalytically produced PLs, to be further cofed in FCC units [15]. In these tests, the most recalcitrant component from the liquids are reportedly destroyed in the initial catalytic pyrolysis step (significantly lowering the overall carbon yield as well), but rendering the liquid to resemble more closely the VGO. Recent data reported the use of liquids from a catalytic pyrolysis process in a further coprocessing in an FCC unit, and results compared with those obtained from hydrotreated PLs, as further detailed below [16].

16.2.2 Cofluid Catalytic Cracking of Hydrotreated Pyrolysis Liquids

On the basis of the experimental findings described in the previous paragraph for untreated PLs, an integrated strategy is the coprocessing of treated PLs in an FCC unit. Prior hydrotreatment offers the best possibilities, and the first successful results date back to the end of the nineties [17]. This strategy entails the use of a refinery's FCC unit to introduce (partially) hydrotreated PLs (see Fig. 16.8). Various treatment of liquids were tested in a European project BIOCOUP: liquids catalytically upgraded without hydrogen (DCOs) or upgraded without catalyst or hydrogen (HPTTs) showed more coke and char than pure PLs. Interestingly, substantial deoxygenation to low oxygen levels in the hydrotreated product (requiring ~800 NL hydrogen/kg pyrolysis oil) was not needed, as liquids with a relatively high remaining oxygen content in between ~10 and 30 wt.% (and associated low hydrogen use required for preparation, order of magnitude ~100–300 NL/kg) could be coprocessed successfully together with VGO.

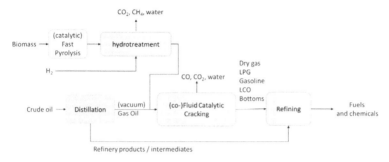

Figure 16.8 Hydrotreated PL corefining in an FCC unit.

Important objectives for the hydrotreatment step in this integrated hydroprocessing—FCC approach (HT/FCC) is (i) to reduce the overall hydrogen consumption, (ii) to maintain a high overall carbon yield, and (iii) to meet specific product properties targets for further FCC processing (such as low water content, low acid number, volatility, miscibility with a-polar component, low charring tendency, and high yields of useable products). The first results following this strategy were published as an outcome of BIOCOUP [18]. PLs were first treated over Ru/C in a batch-wise-operated autoclave for four hours at pressures up to 290 bar. The oil's characteristics are indicated in Table 16.4 below.

Table 16.4 Oil properties of PLs treated over a 5 wt.% Ru/C catalyst at 290 bar and for four hours in a batch-wise-operated autoclave (on as-received basis)

HDO temperature (°C)	230	260	300	330	340
Carbon	54.4	60.3	64.0	71.9	71.8
Hydrogen	9.3	9.5	9.8	10.3	9.8
Oxygen	36.4	30.3	26.2	17.8	18.4
Water	15.9	10.1	5.7	3.2	2.1
MCRT	11.7	9.1	4.7	1.8	2.2

Source: Ref. [18].

The FCC product slates derived from these hydrotreated oils (HDOs) in co-FCC using long residue in a standardized MAT-5000 fluid bed reactor are listed in Table 16.5.

Table 16.5 Product yield in the MAT unit (20 wt.% HDO with long residue at 60% conversion); CO_x (<0.5 wt.%) included in the bottoms yield

Yield	VGO	230°C	260°C	300°C	330°C	340°C
C/O ratio	3.1	4.3	3.4	3.4	3.7	3.8
Dry gas	1.5	2.3	2.1	1.8	1.9	2.0
LPG	8.5	10.1	9.4	9.6	8.9	9.2
Liquids						
Gasoline	44.0	40.2	41.7	43.4	43.5	43.0
LCO	25.2	21.3	22.2	22.5	23.8	24.0
Bottoms + CO_x	14.8	10.8	11.0	11.6	12.2	12.0
Coke	5.9	7.2	6.6	5.2	5.5	5.8
Water	–	7.9	6.7	5.7	3.9	3.9

The liquid products are defined as gasoline (C_5^+: –221°C), light cycle oil (LCO) (221°C–370°C), and bottoms as a mix of heavy cycle oil (370°C–425°C) and slurry oil (>425°C). Further conditions: $T_{reaction}$ = 520°C (long residue), $T_{regeneration}$ = 650°C, $T_{oil\ injection}$ = 40°C, stripping time = 11 minutes, regeneration time = 40 minutes, catalyst amount = 10 g, feed intake = 3.33–1.25 g, cat/oil ratio range = 3–8.

The respective yields from the PLs can also be calculated by subtracting the yields attributed to the long residue fraction only. Reliable mass balances are quite difficult to obtain though (due to the small amounts used in MAT units), and the amount of water and CO_x produced are difficult to measure.

As was expected, the temperature applied during the hydrotreatment of PLs appears to have a major effect on the product slate from the co-FCC with the long residue. Whereas relative high and constant gasoline yields are obtained already at low HDO temperatures, the overall liquid, or better the contribution of LCO and bottoms, is limited due to high LPG and coke yields. More specifically, the bottom yield which can be determined from the contribution of the PL is negative if the HDO temperatures are low. This indicates that there is an overall negative impact on the bottom yield, probably in favor of the coke yield. It suggests that synergetic effects between the larger components in the hydrotreated PL cause the bottoms produced from only the long residues to form coke as well. The lower liquid yields are clearly associated with higher coke and LPG yields.

The oils hydrotreated upfront at the higher temperatures, where more dehydration and cracking reaction take place, are clearly better suited for such co-FCC processing. Typically overall liquid yields are obtained close to those obtained for the standard FCC feed, but interestingly less bottom product are produced, and more gasoline and LCO. In practically all cases, all oxygen in the feed is exiting the FCC reactor as water (and very low amounts of CO and CO_2, the latter representing less than 5 wt.% of oxygen).

A systematic analysis of the cat-to-oil ratio and type of cracking catalysts on the co-FF of a hydrotreated PL (330°C over Ru/C) indicates rather limited differences in product slates between VGO and mixtures of VGO and this particular hydrotreated PL (80:20) [19]. The structure of the zeolites, in terms of pore size and Si/Al ratio, affects the conversion level and product distribution, and it is believed that a large part of the (possibly oligomeric) oxygenated compounds present in the hydrotreated PLs do not enter the zeolite pores, but are partially converted on the silica-alumina matrix. The latter explains the higher coke yields, and indicates a much stronger absorption of the more polar oxygenates in the hydrotreated PL on the active sites of the catalyst than the less polar hydrocarbons. In addition, the cracking of these oxygenated molecules on the outer part of the zeolite framework can consume part of the hydrogen from the paraffins in VGO, and thus the H_2 formation is decreased upon addition of the treated PL. This latter observation, however, conflicts with increased hydrogen yields, as found by others.

Nevertheless, all these data definitely show good potential for cofeeding of such hydrotreated PLs in an FCC unit. A mild hydrotreatment (with roughly 20^+ wt.% residual oxygen in the products) appears already suitable for co-FCC purposes. Results from other studies on this co-FCC feed approach are rather identical, despite large variations in PL characteristics, treatment procedures, the FCC process conditions and alike: co-FCC is technically possible, and no unexpected deviation in the product spectrum derived from standard VGO is observed. There are no substantial difference between the products derived from pure VGO and its cofeeding mode, and small deviations can be attributed to the significant lower amounts of carbon fed as the upgraded liquid contains water and oxygen. Such "upgraded" PLs are thus suitable for corefining and successful corefining of upgraded oils has been demonstrated at the lab scale.

Recent data report that coprocessing in the FCC of these hydrotreated liquids is rather similar to the coprocessing of liquids from catalytic pyrolysis, however, with similar yields but significant difference in quality, with more aromatics and residual oxygenates in the products derived from the catalytically produced PLs. In the latter case a (rather expensive) hydrogenation step can be avoided, while overall yields are suggested to be higher than for the HDO route [16].

16.3 Product Composition of the Co-FCC Products from (Upgraded) PLs

16.3.1 Co-FCC of Pure PL

Samples obtained from the various liquids in the Petrobras's FCC experiments with untreated PLs were analyzed to calculate the PLs contribution to the yields profile by measuring the renewable carbon content by ^{14}C analysis. The liquid products contain 3% to 5% of renewable carbon when 20% of a PL is added to the feed [11]; for 10% substitution the renewable carbon in the total liquid product is around 2%. These values are close to the theoretical predictions based on the carbon input of VGO and the PLs. The water contents of both, the liquid effluents produced from PLs and those from pure VGO, are around 400 ppm, with no detectable difference between the two products. Likewise, elemental analysis did not indicate oxygen (obtained by difference) in the liquid effluents, however, a relatively high level of phenolic compounds could be measured in the liquid product (ranging from 12,000 up to 16,000 ppm). It was expected that this high phenolic content can be reduced rather straightforward in a naphtha-hydrogenating step (a common process step carried out under moderate pressure using a sulfided cobalt-molybdenum catalyst to reduce the sulfur level down to 50 ppm) but complete conversion could not be achieved yet, rendering relatively high levels of phenolic in the naphtha (order of several hundred ppm). The effect of such compounds on naphtha stability requires specific attention in the future. Though some differences were noted (e.g., some loss on octane numbers), no problems were detected for sulfur removal, and the products still comply with local gasoline specification.

16.3.2   Co-FCC of Hydrotreated Products

First results on the product characteristics of cofeeding mildly hydrotreated PLs in FCC units are presented within BIOCOUP. Cracking of conventional hydrocarbons over FCC catalysts leads to smaller hydrocarbons and coke as a result of cracking, hydrogen production, and hydrogen consumption and to larger molecules by C–C bond formation (e.g., by Diels–Alder reactions). Cracking of oxygenates in the mildly hydrotreated PLs over an acid FCC catalyst yields, in addition to hydrocarbons and coke, also smaller oxygenated molecules such as CO_2, H_2O, and CO, while also dehydration occurs. Decomposition of oxygenated hydrocarbons starts by breaking of the C–C and C–O bonds by β-scission. Over an acid catalyst, C–C and C–O bond cleavages are competing. Decarboxylation and decarbonylation reactions proceed more rapidly than the C–C bond cleavage of saturated hydrocarbons, yielding CO and CO_2 as the main oxygenated compounds and a mixture of hydrocarbons. Large(r) amounts of coke are formed, attributed to the polar nature of the oxygenated compounds, which results in a stronger adsorption on the acid sites where the coke precursors are formed. Besides the production of some oxygenated compounds, coprocessing also induces changes in the hydrocarbon product distribution as well:

- In the dry gas less hydrogen and more methane and ethylene are observed.
- In the LPG a slight tendency is seen in lower olefinicity for both the C_3 and C_4 components, as propane and butanes are increasing, while propylene and butenes decrease.
- In both, the gasoline as well as in the LCO, more aromatics are found.

Co-FCC of hydrotreated products obtained at higher hydrotreatment severity (indicated by a lower residual oxygen content) leads to a gradual shift of the distillate product slate towards the lighter fractions (naphtha and light ends). For hydrotreated products from low severity processing, the residual oxygen (here 8 wt.%) after co-FCC is concentrated in the lighter distillate fractions, which are seemingly easier to remove than the oxygenates in the heavy fractions [4].

Interestingly, coprocessing of hydrotreated PLs results in a higher amounts of aromatic compounds in the gasoline, HCO, LCO, and bottom ranges. This is likely due to the fact that the mixed feed already contains more aromatics than the pure feed, and such aromatics are more refractory to cracking than a paraffinic feedstock. More detailed compositions of the gasoline at an isoconversion of 85 wt.% are summarized in Fig. 16.9 for VGO as well as for the VGO/ HDO mixture. Coprocessing favors branched paraffin formation at the expense of linear paraffins compared to VGO cracking, which is also indicated in Fig. 16.10. Short-alkyl-chain (C_1–C_3) benzene derivatives are more typical for the VGO/HDO feed than for the pure VGO cracking.

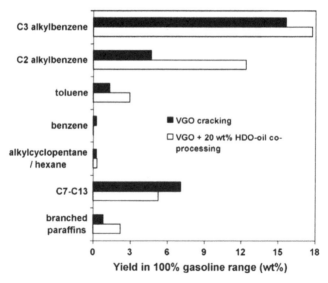

Figure 16.9 Detailed gasoline range composition by compounds at a 85% conversion level. Reprinted from Ref. [19], Copyright 2010, with permission from Elsevier.

Nevertheless, even at a conversion level of over 85%, some oxygenated compounds remain, possible due to the presence of the stable alkylphenols in the HDO cofeed. Detailed analyses of the products after co-FCC by GC×GC analysis confirmed the presence of such alkylphenols (Fig. 16.9) among the other coprocessing products.

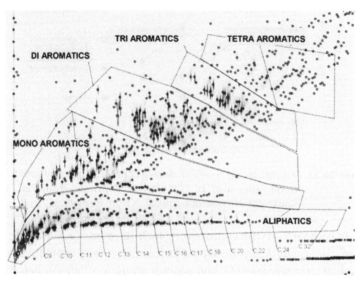

Figure 16.10   Contour plot of volatile part of the liquid phase products from VGO/HDO cracking. Reprinted from Ref. [19], Copyright 2010, with permission from Elsevier.

The presence of phenol and alkylphenols with side chains up to C_5 was observed. The decomposition of alkylphenols as a function of the conversion is shown in Fig. 16.11. Clear reactivity differences were observed. Phenol and alkylphenols with up to C_2 side chains were identified among the coprocessing products, whereas the conversion of C_3-, C_4-, and C_5-side-chain phenolics is complete at 71% conversion.

The concentration of C_2 and C_1 alkylphenols decreases much slower, that is, at 86% conversion still ~30% of the initial C_1 alkylphenols are present. Phenol concentrations show an optimum: an increase at low conversion compared to its initial concentration and only at a higher conversion a reduction is observed. At 500°C dealkylation and dehydroxylation are considered to be the most important reactions. Dealkylation leads to C_1 and C_2 alkylphenols and phenol, and dehydroxylation to benzene and alkylbenzenes.

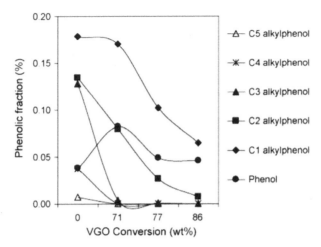

Figure 16.11 Alkylphenol conversion during coprocessing at different cat-to-oil ratios over an FCC catalyst. Reprinted from Ref. [19], Copyright 2010, with permission from Elsevier.

In comparison with coprocessing hydrotreated PLs, the products from the catalytically derived PLs contain more aromatics and residual oxygeanates [16], and a larger amount of (alkyl)phenols is present in the gasoline fraction.

16.4 Conclusions

An attractive option for the introduction of biomass into the fuels marketplace is to use PLs as a (co-)feed in an existing petroleum refinery as a (partial) replacement of fossil sources. The proof of principle is shown on the lab scale for both PLs and treated PLs, but also on the semidemonstration scale by introducing pure PLs in a larger FCC unit. In the latter case pure PLs lead to significantly lower overall carbon yields compared to the use of hydrotreated liquids.

The concept of corefining would, in principle, facilitate the introduction of renewable carbon into the transportation fuel infrastructure within a short time frame. In addition it could also be economically advantageous for the biofuels industry as the capital costs of thermochemical processes could be reduced by leveraging existing process units available in petroleum refineries. The challenges of processing such bio-based-derived feeds in an archetype oil refinery are significant. From a refiner's perspective important properties include the boiling-range distribution obtained from the main crude oil fractionator and the hydrocarbon types and heteroatom (sulfur, nitrogen, oxygen) contents of each of the resulting primary distillation and process-derived intermediate fractions.

The main unknown is also to identify the best refinery insertion points for bioliquids and this can be directly translated to what extent the bioliquid must be upgraded (deoxygenated) prior to insertion and to what extent should the refinery be adapted to accept less upgraded, oxygen-containing intermediates. The high oxygen content of biofeeds will translate to risks of corrosion and (more) extensive coking of catalyst surfaces as well as downstream contamination risks and requirements for venting of oxygenated gases (CO, CO_2, and H_2O). To mitigate these challenges, the blending rate of biofeeds in petroleum feeds can be limited, and the insertion point is preferably located at the end of the refinery processing.

References

1. Fan, J., et al. (2011). Life cycle assessment of electricity generation using fast pyrolysis bio-oil, *Renewable Energy*, **36**(2), 632–641.

2. Meyers, R.A. (1997). *Handbook of Petroleum Refining Processes*, McGraw-Hill Education, ISBN: 9780071391092.

3. http://www.gasandoil.com/image_store/Global%20Refining%20 Capacity%20table1_0.PNG/view, accessed on Nov. 17, 2014.

4. Talmadge, M.S., et al. (2014). A perspective on oxygenated species in the refinery integration of pyrolysis oil, *Green Chemistry*, **16**(2), 407.

5. Zacher, A.H., et al. (2014). A review and perspective of recent bio-oil hydrotreating research, *Green Chemistry*, **16**(2), 491.

6. Adjaye, J.D., Bakhshi, N.N. (1995a). Catalytic conversion of a biomass-derived oil to fuel and chemicals I: model compound studies and reaction pathways, *Biomass and Bioenergy*, **8**(3), 131–149.

7. Adjaye, J.D., Bakhshi, N.N. (1995b). Production of hydrocarbons by catalytic upgrading of a fast pyrolysis bio-oil. Part I: conversion over various catalysts, *Fuel Processing Technology*, **45**(3), 161–183.

8. Adjaye, J.D., Bakhshi, N.N. (1995c). Production of hydrocarbons by catalytic upgrading of a fast pyrolysis bio-oil. Part II: comparative catalyst performance and reaction pathways, *Fuel Processing Technology*, **45**(3), 185–202.

9. Marinangeli, R., Marker, T., Petri, J., Kalnes, T., McCall, M., Mackowiac, D., Jerosky, B., Reagan, B., Nemeth, L., Krawczyk, M., Czernik, S., Elliott, D., Shonnard, D. (2006). *Opportunities for Biorenewables in Oil Refineries*, Report No. DE-FG36-05GO15085, UOP.

10. De Rezende Pinho, A., Almeida, M.B.B., Mendes, F.L. (2014). Production of lignocellulosic gasoline using fast pyrolysis of biomass and a conventional refining scheme, *Pure and Applied Chemistry*, **86**(5), 859–865.

11. De Rezende Pinho, A., Almeida, M.B.B., Mendes, F.L., Ximenes, V.L., Casavechia, L.C. (2014). Co-processing raw bio-oil and gasoil in an FCC unit, *Fuel Processing*, http://dx.doi.org/10.1016/j.fuproc.2014.11.008.

12. http://www.biocoup.com/, accessed on Nov. 17, 2014.

13. Karatzos, S., McMillan, J.D., Saddler, J.N. (2014). The *Potential and Challenges of Drop-in Biofuels*, IEA bioenergy task 39, http://task39.org/2014/01/the-potential-and-challenges-of-drop-in-fuels-members-only/.

14. Habib, E.T., Bryden, K., Weatherbee, G. (2013). http://www1.eere.energy.gov/bioenergy/pdfs/biomass13_habib_2-d.pdf, accessed on Nov. 17, 2014.

15. Agblevor, F. A., Mante, O., McClung, R., Oyama, S.T. (2012). Co-processing of standard gas oil and biocrude oil to hydrocarbon fuels, *Biomass and Bioenergy*, **45**, 130–137.

16. Thegarid, N., et al. (2014). Second-generation biofuels by co-processing catalytic pyrolysis oil in FCC units, *Applied Catalysis B: Environmental*, **145**, 161–166.

17. Samolada, M.C., Baldauf, W., Vasalos, I.A. (1998). Production of bio-gasoline by upgrading biomass flash pyrolysis liquids via hydrogen processing and catalytic cracking, *Fuel*, **14**, 1667–1675.

18. De Miguel Mercader, F., et al. (2011). Hydrodeoxygenation of pyrolysis oil fractions: process understanding and quality assessment through co-processing in refinery units, *Energy and Environmental Science*, **4**(3), 985.

19. Fogassy, G., et al. (2010). Biomass derived feedstock co-processing with vacuum gas oil for second-generation fuel production in FCC units, *Applied Catalysis B: Environmental*, **96**(3–4), 476–485.

About the Authors

R. H. (Robbie) Venderbosch (1966) carried out his PhD research  at the University of Twente, the Netherlands, on the role of clusters in gas–solid reactors. He then carried out a postdoc at the same university on the fast pyrolysis of biomass, after which he joined the Biomass Technology Group (BTG) B.V., the Netherlands, in 1998. Here he is working on a range of topics related to fast pyrolysis, hydrogenation reactions, sugar-based chemistry, supercritical gasification in water, production of chemicals from biomass (derivatives), and alike. His research interests concern bio-based applications, with emphasis on heterogeneous catalysis in relation to chemical reaction engineering and specific focus on all kinds of possibilities to bridge the gap between biomass-related activities and petrochemical engineering. Dr. Venderbosch is a coauthor of around 30 papers in international peer-reviewed journals and around. 10 book chapters. He has approximately 10 patents in the field of biomass (applications) to his credit.

For more information, please visit www.btgworld.com.

H. J. (Erik) Heeres (1963) carried out his PhD research at  the University of Groningen, the Netherlands, on the development of novel homogeneous lanthanide catalysts for the conversion of unsaturated hydrocarbons and graduated in 1990. Afterward, he performed a postdoc at the University of Oxford, UK, in the group of Prof. J. M. Brown on asymmetric catalysis. From 1991 to 1999, he was employed at Shell Research B.V. (Amsterdam and Pernis, the Netherlands) and worked on a range of applied catalysis topics. He joined the chemical engineering department of the University of Groningen in 1999 as an assistant professor. In 2003 he was appointed professor in green chemical reaction engineering, with interest in acid- and metal-based catalytic biomass conversions, with emphasis on biofuels (pyrolysis oil upgrading), platforms chemicals), and performance materials from biomass. Heeres is the (co-)author of 145 papers in international peer-reviewed journals and 11 patents in the field of (applied) catalysis and chemical reaction engineering. He is an editorial board member of the journal *Sustainable Chemical Processes*.

For more information, please visit http://www.rug.nl/staff/ h.j.heeres/research.

Chapter 17

Options for Adaptation of the Biogas Sector to Serve Evolving Electricity Grids

David Baxter

Institute for Energy and Transport, Joint Research Centre (JRC),
European Commission, Westerduinweg 3, 1755 LE Petten, the Netherlands
david.baxter@ec.europa.eu

This chapter addresses the challenges facing the evolving biogas sector and integration into electricity grids fed with increasing amounts of energy generated by intermittent wind and solar sources. Biogas is a readily storable gas which can be used to generate electricity, when needed, and therefore to make up for periods of time when the wind is not blowing or the sun is not shining. Various options for biogas storage are considered, both locally at the site of biogas plants and in centralized locations where bigger and more efficient gas engines can be used for electricity generation. Upgrading to biomethane allows gas storage in the natural gas grid and later use for conversion to electricity. Additional biomethane can also be produced by a range of power-to-gas concepts, which also allow storage. Finally, while biogas plants are normally operated

Biomass Power for the World: Transformations to Effective Use
Edited by Wim van Swaaij, Sascha Kersten, and Wolfgang Palz
Copyright © 2015 Pan Stanford Publishing Pte. Ltd.
ISBN 978-981-4613-88-0 (Hardcover), 978-981-4669-24-5 (Paperback), 978-981-4613-89-7 (eBook)
www.panstanford.com

under constant conditions, recent research has shown the anaerobic digestion process may not be adversely affected by substantial changes in operating conditions over a period of a few hours or days, and therefore biogas production could in principle be regulated and used on demand when the electricity grid is needed from the power.

17.1 Introduction

It has been known for centuries that a flammable gas, nowadays referred to as biogas, is formed naturally under anaerobic conditions from the decomposition of organic matter. The first biogas plant built to capture biogas was in Mumbai, India, around 1859. The technology was transferred to the United Kingdom toward the end of the 19th century and the gas used to light street lamps. Subsequently, in the 20th century the technology was developed firstly for sewage treatment at a time when urban drainage systems were being built. By the time of the Second World War biogas was being used as a vehicle fuel produced at sewage treatment plants. Extension of biogas plant technology to other feedstocks started in the 1930s, again in India, as a way of providing rural areas with fuel for domestic use. China followed a similar route to build millions of small-scale biogas plants based on septic tanks by the 1980s. Since that time the biogas sector has expanded in many ways in terms of technologies employed, feedstocks used, size of installations, and utilization of the biogas. And of course all kinds of legislation have been introduced to ensure minimum health and safety standards and quality.

Nowadays, biogas plants are built to address a range of environmental, economic, and social needs. From the environmental side, treatment of wastes and residues is necessary to avoid health problems, contamination of the ground and of ground water, infestation by insects and rodents, and air pollution. Environmentally acceptable treatment of waste also avoids unsightly waste dumps and associated odors. In recent times, biogas has also come to be seen as a source of renewable energy which can contribute to reducing fossil-derived carbon emissions. The economic side of the biogas sector has always been an economic challenge and remains the case to the present time. In many cases some kind of government support, or support from local taxes, is needed for biogas plants to be economically viable. All the time, industry and researchers are working to improve biogas plant performance in order to improve economic performance.

From early in the 21st century renewable energy from biogas has become recognized as a significant contributor to reducing fossil emissions through use as bioheat, bioelectricity, and biofuel. For this reason in many developed countries biogas plants are set up to be as energy efficient as possible and export energy to heat, electricity, and gas distribution grids. A range of feed-in tariffs, quotas, and green certificates have been put in place to meet targets and compensate operators for relatively high production costs compared to equivalent fossil energy generators. Over time, substantial renewable energy markets have emerged. In most developed countries biogas plants are therefore operating in markets where there is competition with other energy providers, both renewable and nonrenewable. This introduces both constraints and opportunities and possibly additional responsibilities.

This also brings us to the next challenge in the history of the biogas sector so far—full integration in complex energy supply infrastructures. Whereas biogas is widely used as fuel for electricity generation, and the electricity is fed into power grids as base load, the rapidly increasing contribution of intermittent wind and solar electricity to power grids has led to increasing demand for a flexible power supply which is controllable so that grids can both meet electricity demand and remain stable. In principle, the biogas sector has the ability to export energy in a flexible manner, because gas can be stored and used when needed, but how this can be done in a cost-effective and efficient manner still has to be established. At the time of making this publication, various options were being considered for managing biogas plants in integrated energy systems. This high-tech demand for biogas plant operation of course does not mean that small-scale biogas plants in developing countries will change in any significant way. The scope of biogas production will simply become ever more diverse.

17.2 Drivers for Biogas Plants

From its origin in the treatment of sewage the anaerobic digestion (AD) process for biogas production has been adopted and adapted for use in technologies applied in many sectors. While in the earlier years, waste was the primary resource treated by AD, dedicated energy crops have been used increasingly in the 21st century. In terms of waste treatment, application of AD has been promoted by policy makers who have recognized that AD technologies address a number of challenges simultaneously. Application of the technology in an appropriate manner will reduce greenhouse gas (GHG) emissions, for example, from manure storage. Biogas is a suitable renewable energy carrier for heat, electricity, and vehicle fuel production, and the technology can be used to avoid ground, water, and air pollution. In addition, the process can be used to upgrade waste into products; for example, it can be used to make high-grade organic fertilizer in which vital nutrients are preserved and recycled back to the soil. With respect to soil nutrients, AD is often applied for agricultural residue treatment to negate the impacts of nutrient overload in areas of intense agricultural activity, for example, concentrated rearing of animals for meat or dairy production where not only the agricultural land itself is affected but also freshwater and coastal waters can be badly affected by growth of algae.

In Europe, a raft of legislation has been introduced over the years to tackle the adverse environmental impacts of waste. The most notable European legislation involves the control of nitrates in agriculture [1] and landfilling of biodegradable waste [2].

The nitrates directive obliges EU countries to implement nitrate action programs aimed at taking measures to protect waters against pollution from agricultural sources, including both nitrogen and phosphorus. This means farmers have to adopt practices which minimize risks of pollution of water courses and local authorities have to inspect and monitor action plans and report to the EU. The directive also demands strengthened enforcement for better farm management, including provisions relating to times of the year, weather, and soil conditions when fertilizers may be applied and minimum storage capacity for manures. Manure storage and use as a fertilizer is a main driver for adoption of AD on farms. Uptake of

AD by farmers has been made easier by the availability of subsidies in many countries for renewable energy which can be exported to national energy grids.

In Europe and elsewhere, a raft of legislation has led to energy crops being used for renewable energy, both in the form of biofuels and in the form of bioenergy. In the EU, biofuels were first promoted through implementation of the biofuels directive [3], while in 2005 in the United States a statutory standard [4] set the requirement for 2.78% of the gasoline sold or dispensed in 2006 to be renewable fuel. This was followed by the renewable fuel standard [5], which set ambitious targets for biofuels to be reached by 2022 in order to reduce environmental impacts and to reduce reliance on imported fossil fuels for transport. The 2003 EU directive set a goal of reaching a 5.75% share of renewable energy in the transport sector by 2010. The target was superseded by the more general renewables directive in 2009 [6], which set the target at 10% renewables (rather than only biofuels) in the transport sector by 2020. Earlier EU legislation in 2001 set targets for the generation of electricity from renewable sources [7]. The latter led to strong incentives for renewable electricity generation from energy crops in some countries, particularly Germany.

Since the middle of the first decade of the 21st century questions have been raised concerning the ethics of using energy crops for energy. What came to be known as the "food versus fuel" debate led to a discussion of using agricultural land for vehicle fuel production. In Europe, the expansion of land used for energy crops was due in part to the availability of land set aside because of overproduction of food in preceding years and a payment subsidy for that land for keeping it out of food production. Consideration of the wider environmental impacts of bioenergy and biofuels led to the imposition of strict rules on environmental sustainability in the 2009 EU renewables directive. Subsequently, environmental sustainability rules have been adopted in many countries to ensure minimum standards for GHG savings, land use, and the impacts of land use change. Environmental impacts seem set to provide the main drivers for AD and biogas plants in the future. Moreover, environmental impacts are likely to emerge in the future as the way by which the performance of biogas plants, both individually and collectively, will be judged.

17.3 Biodegradable Feedstocks for Biogas Production

Feedstocks suitable for biogas production come from a wide range of biomass and typically have high contents of sugar, starch, proteins, and fats. Starch is relatively easy and quick to break down biologically, whereas cellulose and hemicellulose break down more slowly under AD conditions. Lignin is generally considered to be resistant to biological breakdown by AD. As a consequence, in many cases it is necessary to provide some kind of pretreatment in order for the key components to be available for methane formation. Methane yield varies widely between biodegradable feedstocks suitable for AD as shown in Table 17.1.

Table 17.1 Methane yields of some common feedstocks used in biogas plants

Feedstock	Methane yield (m^3/ ton of volatile solid)
Cow manure	200
Pig manure	300
Chicken manure	300
Separated food waste	500–600
Fish oil	800
Sewage sludge	400
Maize (whole crop)	205–450*
Maize crop residue	<450
Sorghum	295–372*
Wheat (grain)	384–426*
Barley (grain)	353–658*
Straw	242–324*
Potatoes	276–400*
Sugar beet	236–381*
Grass	298–467*
Clover	345–350*
Ryegrass	390–410*

*Braun, 2007 [8].

Feedstock pretreatment is carried out using a range of mechanical, thermal, chemical, and biological methods, which may be combined [9]. No single pretreatment technology is suitable for all feedstocks and for all AD systems and so some knowledge is needed before selection is made. For example, feedstocks with very high solid matter content are better suited to mechanical milling or extrusion, whereas chemical treatments are appropriate for feedstocks with high lignin contents. A number of pretreatments have been developed for lignocellulosic materials, like straw for the production of bioethanol (so-called lignocellulosic ethanol) since the late 20th century [10], and these can also be applied to feedstocks for biogas. Thermal treatment usually involves heating under pressure in order to disrupt plant matter. This can sometimes be done with the addition of chemicals, although this adds cost and the chemicals must be taken into account in the subsequent AD process. Biological technologies involve a range microbial, fungal, or enzyme treatments. Enzyme pretreatments have been developed intensively for lignocellulosic biofuel production and the knowledge is available for application in the biogas sector. Some of the advantages and disadvantages of pretreatment processes are summarized in Table 17.2.

The big challenge for the biogas sector is selecting a pretreatment technology which achieves a level of improved biogas yield or kinetics at acceptable capital and operating costs and without a significant impact on the overall process energy balance. Research and development on pretreatment technologies has revealed that fibrous feedstocks can be effectively transformed into readily digestible materials like animal slurry and maize and used to produce biogas in the most commonly available continuously stirred tank reactor (CSTR) design of biogas plants. However, there is evidence that a different biogas reactor design would be more suited to lignocellulosic feedstocks in the future. For feedstocks with very high solid content, dry digester designs could be used, whereas for wet lignocellulosic feedstocks a kind of leach bed or percolator could be used, combined with a high-rate reactor like an up-flow anaerobic sludge blanket (UASB) [9]. It is possible that if different phases of the AD process are separated, pretreatment could be integrated into one of the process steps to achieve greater and/or faster methane production. The idea of separating the AD process into separate steps or phases is not at all new. Over the period of rapid expansion of the biogas sector since the 1980s single-step AD became commonplace, but this might change in the future, particularly because separation of the individual steps also presents the possibility to manage biogas production in a flexible manner in order to meet market demand for exported energy. Feedstock pretreatment is playing an increasingly important role in biogas plants in the search of better performance and this trend will almost certainly play a substantial role in the future.

Table 17.2 Advantages and disadvantages of pretreatment processes. Adapted by Montgomery and Bochmann (2014) [9] from Taherzadeh and Karimi (2008) [11]

Process	Advantages	Disadvantages
Milling	-Increases surface area -Makes feedstock easier to handle -Usually improves fluidity in digester	-Increases energy demand -Has high maintenance costs/is sensitive to foreign objects (e.g., stones)
Hot water TDH (from the German, thermo-druck-hydrolyze)	Increases enzyme accessibility	-Has high heat demand -Is only effective up to a certain temperature
Alkali	Breaks down lignin	-Leads to high alkali concentration in digester -Has high cost of chemical
Microbial	Consumes less energy	-Is a slow process -Involves no lignin breakdown
Enzymatic	Consumes less energy	-Requires continuous addition -Involves high cost of enzyme
Steam explosion	Breaks down lignin and makes hemicellulose soluble	-Has high heat and electricity demand -Is only effective up to a certain temperature

Process	Advantages	Disadvantages
Extrusion	Increases surface area	-Has increased energy demand -Has high maintenance costs/is sensitive to foreign objects (e.g., stones)
Acid	Makes hemicellulose soluble	-Involves high cost of acid -Can lead to corrosion problems -Involves formation of inhibitors, particularly with heat

17.4 Biogas Plant Technologies and Their Application

The simplest technology for recovery of biogas from the process of AD involves simply placing an airtight cover over the digesting biomass and collecting the gas as it is released. This method is still used in some situations and in principle it is the concept adopted to recover residual methane from digestate storage tanks or lagoons. In effect, landfills are little more than airtight containers (in the ideal case), albeit on a very large scale, for the collection of methane in the landfill gas, while at the same time avoiding contamination of the ground and ground water. Many small-scale and very simple biogas plants have been, and are still being, used to generate methane for use for domestic purposes in developing countries.

Regardless of the size of the biogas plant and its primary purpose, it is important that investment and operating costs be kept as low as possible, and in the case of commercial enterprises it is important that a profit can be achieved, even if this needs the help of schemes for environmental protection or promoting of alternative, renewable forms of energy. As a consequence, substantial effort has been invested to optimize the performance of AD technologies applied in many situations to produce specific products and to address a spectrum of market conditions and environmental challenges (Table

17.3). Given that AD is a biological process which occurs naturally under the right conditions, the aim of biogas plant designers and operators is to understand how best to optimize conditions so that the quantity and quality of saleable products is maximized. This of course means maximizing output without causing undue strain on the AD biological environment, which could result in complete system failure. A totally failed AD process can lead to months of lost production and substantial costs.

Table 17.3 Summary of situations where AD and biogas plant technologies are used for different reasons

Type	Where used	Main products	Drivers
Household	Developing countries	Biogas for cooking	Rural development
Landfill	Worldwide	-Landfill gas for electricity and heat -Biomethane	Renewable energy, GHG emissions
Sewage treatment	Worldwide	-Biogas for electricity and heat -Biomethane	Sludge volume reduction and renewable energy
Municipal waste treatment	Worldwide	-Biogas for electricity and heat -Low-grade compost	Waste treatment and GHG emissions
Source-separated waste treatment	Worldwide	-Biogas for electricity and heat -Biomethane -Biofertilizer	Waste treatment, renewable energy, and nutrient recycling
Animal manure and slurries	Worldwide	-Biogas for electricity and heat -Biofertilizer	Water pollution, GHG emissions, and nutrient recycling
Crop residues	Worldwide	-Biogas for electricity and heat -Biomethane -Biofertilizer	GHG emissions and farm enterprise
Energy crops	Developed countries	-Biogas for electricity and heat -Biomethane	Renewable energy and biofuels

Type	Where used	Main products	Drivers
Industry	Worldwide	-Biogas for electricity and heat -Biomethane -Biofertilizer -Feedstock for further processing	Waste treatment, environment protection, and business opportunities
Biorefinery	Worldwide	-All energy products -Biofertilizer and chemicals	Additional value-added products and waste treatment

The main criteria influencing biogas plant design are the composition and characteristics of the feedstock and the quantity of feedstock to be treated. Feedstock characteristics such as dry matter content, methane potential, rate of methane production, and the presence of contaminants like foreign objects (stones, soil, plastic, etc.) and chemical and biological properties (e.g., concentrations of nitrogen, sulfur, and pathogens) are key parameters. Taking into account these key parameters as well as the amount of feedstock and how much feedstock will be codigested, the basics of plant design can be established [12]. Technologies developed over the years fall into three main types—continuous wet, continuous dry, and batch (Fig. 17.1)—and operate within one of two temperature ranges, 30°C to 40°C, for mesophilic digestion and above 55°C for thermophilic digestion. The fundamental science and engineering of the AD process for biogas production have been described by many authors (see, for example, Ref. [13]).

The vast majority of biogas plants are of the CSTR type, which are designed to run under constant operating conditions (see the example in Fig. 17.1d). The key word here is "stability," with the aim to maintain constant temperature and chemistry (mainly pH) by careful loading of new feedstock which avoids disturbances to the AD process. Traditional wisdom says that maintaining stability under optimum conditions should achieve maximum biogas production over very long periods of time, typically years. Substantial effort has been directed toward AD process control. Process monitoring is vital in any well-controlled process and extensive work has been carried out in recent years to better understand various aspects of the AD process and to develop and apply methods for measuring a range of process parameters mainly in biogas plants of the CSTR type [14].

(a)   (b)

(c)   (d)

Figure 17.1   Examples of biogas plant technologies. (a) Simple small-farm biogas plant with covered manure modules; (b) series of batch digesters designed to give relative even biogas output; (c) very small-scale farm digester which in this example in Brazil is linked by a biogas pipeline to a centralized CHP unit; and (d) commonly used continuous biogas plant on a farm in Europe. See also Color Insert.

Among the various process control parameters for CSTR biogas plants, there are two different groups. The first group includes methods to provide early indications of process imbalance which should allow the plant operator to take appropriate and timely action. In the second group are methods which can often help to better identify the cause of the imbalance. Through application of a raft of monitoring techniques it has been possible to establish general guidelines for stability limits for different process parameters. It has nevertheless been found that the guidelines must be adapted for use on individual biogas plants. It is also necessary to adapt the process-monitoring strategy to each biogas plant and its particular feedstock. It is clear from the biogas sector that process-monitoring technology needs to be developed further. One of the challenges for the future will be the application of process monitoring to biogas plants which work in a flexible mode to meet fluctuating demands from electricity grids. Under such flexible operating conditions, the need for process monitoring will likely increase considerably.

Batch processes are used predominantly for low-moisture-content feedstocks like municipal waste. When individual batches of typically five or six vessels (see the example in Fig. 17.1b) are run in parallel, but with different start and end times, biogas output can be maintained at a relatively constant level. The inertia of such systems is substantial and as such they are only suitable for constant energy output, unless gas storage facilities are available or biogas is upgraded to biomethane, which can be readily stored locally or injected into a natural gas grid. Flexible operation of the AD process is limited with such systems.

Landfills were the preferred method for disposal of waste in many countries and remained so in some, until the late 20th century when greater emphasis was placed on environmental impacts from the waste management sector. The EU Landfill Directive (1999) set the course for a wholesale shift from landfill to reuse, recycling, and energy recovery from waste. The main thrust behind the EU landfill legislation is to ensure that methane emissions to the environment are minimized. According to the Intergovernmental Panel on Climate Change (IPCC) [15], methane has a global warming potential 34 times greater than that of carbon dioxide. Therefore, biodegradable waste, and that includes most of the waste with a methane production potential under anaerobic conditions, is preferably treated to recover methane in a separate facility, or it is composted aerobically so that methane formation is avoided. Either way, the chance of methane loss to the environment is reduced to a level well below that in a traditional landfill.

Most existing landfills in developed countries and increasing numbers in developing countries are now equipped to collect the methane produced. The landfill gas, containing typically 45% to 55% methane, is widely used as a fuel to generate electricity for export to the grid. In some cases, landfill gas is cleaned and upgraded to give biomethane which is of the same quality as natural gas in the natural gas grid. In some cases, landfill gas is simply flared to ensure that the emission is carbon dioxide and not methane. One key problem with landfills is that they are not sufficiently leak tight and some methane inevitably escapes. The amount of methane which can be recovered from a typical landfill compared to the methane potential is typically in the region of 60%, with the result that even with gas recover for energy, losses are still substantial. As a consequence, it seems that in the future landfilling of biodegradable waste is coming to an end.

17.5 Biogas Utilization

After simple cleaning in particular to remove sulfur (H_2S), water, and solid particles biogas can be used as a fuel in internal combustion engines to generate electricity. This form of biogas utilization is adopted for the majority of commercial farm-scale and municipal of industrial biogas plants. In most cases a combined heat and power (CHP) unit comprising a gas engine with spark ignition is used from which the heat is recovered. Gas cleaning is necessary to avoid corrosion of engine components and consequent costly maintenance. Gas engines have become standard equipment since the 1980s, and while conversion efficiency is related to engine capacity, it has been gradually increased by improvements in engine technology [16]. Conversion efficiency to electricity is generally in the range from 30% to more than 40%, depending on capacity and quality of the biogas, with higher methane content enabling higher conversion efficiencies to be achieved. While gas engines continue to be improved, new technologies have been tried and tested to achieve higher conversion efficiencies to electricity. Table 17.4 summarizes data for the commonly used gas engines and alternatives. There are of course a number of factors which will influence selection of electricity-generating technology, not least the capacity. For example, gas turbines can achieve conversion efficiency to electricity (power efficiency) up to 40%, and around 60% in a combined cycle configuration with a steam turbine, but their size is normally too large for most individual biogas plants. For gas turbines to be viable, gas would need to be collected from a number of biogas plants connected by a pipeline. Gas turbines can reach maximum power output in short times, making then perfectly suited for supplying peak load to the electricity grid. Microturbines have been adapted to work on small-scale biogas production but they have lower power efficiency. The main challenge for the future is to fully utilize residual heat from the engines or turbines used to generate electricity. However, if nearby heat customers are not available, the residual heat will usually be wasted. Due, in part, to this challenge, upgrading of biogas to natural-gas-quality biomethane is an obvious option.

Table 17.4 Summary of biogas power generation technologies

Parameter	Gas engine	Gas turbine	Microturbine	Fuel cell
Power efficiency	30%–42%	25%–40%	25%–30%	40%–45%
Overall efficiency with heat utilization (HHV)	70%–80%	70%–75%	55%–65%	75%–80%
Typical capacity (kWe)	110–3000	500–250,000	25–500	300–1500
Power to heat ratio control	Not possible	Very good	Very good	Very good
Start-up time	10 sec	10 min–1h	60 sec	3 h–2 d
Capital cost (€/kWel))	400–1100	900–1500	600–1200	3000–4000

Source: Adapted from Ref. [16].

Biogas upgrading to biomethane has been practiced for many decades but has seen major growth only since the beginning of the current century. A number of upgrading technologies (pressure swing adsorption, physical or chemical absorption using water or a range of inorganic or organic solvents and membranes) are in commercial operation and continue to be refined [17]. The main driver for expanded biomethane production began with the desire to reduce vehicle emissions from the use of diesel in urban areas. Sweden started this trend and saw a rapid rise in the use of methane, and biomethane in particular, as illustrated in Fig. 17.2. In Sweden, biomethane is supplied as a transport fuel close to the place of production, which is commonly at municipal wastewater or solid waste treatment sites close to cities. As a consequence, the pressurized gas is commonly supplied in tanks (>200 bar pressure). To achieve higher storage density, biomethane can be liquefied and used for heavy duty engines in road transport (e.g., Ref. [18]) as well as trains and ships. Growth in the production, distribution, and use of liquefied natural gas (LNG) provides an impetus for utilization of both fossil methane and biomethane liquid in the heavy-duty engine sector.

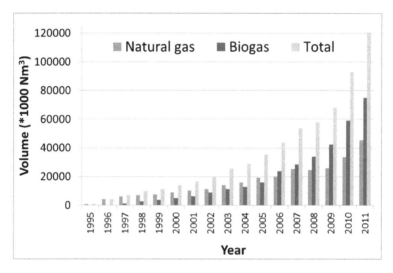

Figure 17.2 Growth in the use of biomethane as a vehicle fuel in Sweden [19].

The availability of a natural gas grid permits easy transport of biomethane as long as it meets quality requirements of gas in the grid. While international standards have been slow to be developed for injection of biomethane into natural gas grids, grid injection was made easier after introduction of the renewable energy directive in Europe in 2009 [6]. With increasing supply of intermittent power from solar and wind renewables to the electricity grid there is a growing need to use storable energy or energy carriers for balancing electricity supply and demand. The fact that biomethane is a stable and readily storable gas means demand for its supply will increase as a green alternative to fossil fuels. The storage potential of methane is clearly illustrated in Fig. 17.3.

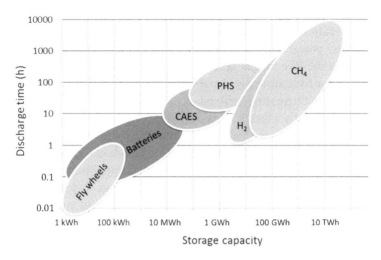

Figure 17.3 Comparison of various energy storage systems with respect to discharge time and storage capacity. Modified by Persson et al. [25] from original Ref. [20]. (CAES, compressed air energy storage; PHS, pumped hydrostorage; H_2, hydrogen; CH_4, methane).

17.6 Developments in the 21st Century

There are many thousands of biogas plants around the world, providing significant amounts of renewable energy from wastes, residues, and energy crops. Biogas is used to generate electricity for supply to national grids, heat to local and district heating systems, and fuel for transport after upgrading. What form of energy is exported from a biogas plant is determined by local circumstances and local markets. The majority of biogas plants use a CHP unit to generate electricity, which is exported and contributes to base load in the electricity grid. However, in very many cases the residual heat from the CHP unit is not fully utilized and this represents around 50% of the original energy content of the biogas. Hence, one key challenge is to try and find ways to utilize the residual heat which is otherwise wasted.

One other key challenge for the biogas sector is the rapidly changing way in which electricity grids are fed from generators. Decades ago a small number of fossil and nuclear electricity generators dominated supply to the grids and load was managed in a relatively straightforward, although somewhat wasteful manner. That situation is changing as intermittent suppliers from the wind and solar power sectors provide increasing shares of electricity to power grids. The biogas sector has been identified as a potential contributor to ensuring grid stability by providing balancing power according to fluctuating demand.

Until recently, electricity supply to grids mostly provided base load. However, with increasing contributions of intermittent renewable electricity to the grids, from wind and solar photovoltaics, there are greater demands for technologies capable of providing flexible supply to grids and for energy storage capabilities. In principle, biogas can be used in a flexible manner to support grid balancing. In addition to having many possible uses, biogas can be readily stored as a gas until needed. The main question is whether flexible energy delivery can be effectively achieved within the evolving smart grid energy supply infrastructure. If it can, extra demands will likely be placed on the way biogas plants are operated and possibly on the anaerobic process itself. While most efforts over the last two or three decades have concentrated on optimizing stable biogas production, mainly because of the tight financial situations in which most biogas plants operate, some attention has turned to making biogas output more flexible in a fully controlled manner. The main question is to what extent it will eventually be possible to manage the anaerobic process to provide biogas output according to demand. Alternatively, it might be that biogas storage could facilitate flexible energy export to the electricity grid. In practice it is likely to be a combination of the two approaches.

17.6.1 Biogas Storage

Biogas can be stored, provided that storage capacity is available either locally or via pipelines connecting several small biogas plants. In the case of temporary biogas storage, CHP units can be used to serve demand for periods of insufficient intermittent renewable electricity generation or for peak load demand. However, there will likely be additional investment cost involved because the CHP capacity would no longer be selected to operate on the amount of steady gas production from the biogas plant but on larger volumes of gas. Hochloff and Braun (2014) [21] examined the situation for a biogas plant operating in the German market with an annual average electricity output of 600 kW. They showed that increasing the electricity-generating capacity to 1 MW would lead to the best cost–benefit ratio when taking into account additional costs on the one side and additional market revenues on the other. Costs of additional on-site biogas storage have been considered by Hahn et al. (2014a) [22] to be in the range of 25 to 67 €/m^3 (estimated costs in 2014), depending on the storage design. The main design considered is a double-membrane system in which an inner biogas container, usually spherical and made from flexible material, is surrounded by an air envelope to maintain a constant biogas pressure. Variable-speed electric fans are used to maintain the air pressure in the outer envelope. Biogas losses during storage have been considered and may amount to typically 1% per day [23], although technology improvements should be able to reduce this amount. In addition to the environmental impact of methane losses during storage, potential risks associated with storing large volumes of explosive gas have to be taken into account.

In a review of possible approaches in Germany and taking into account national laws, Hahn et al. (2014b) [24] concluded that on-farm storage of biogas could provide balancing power to the electricity grid using gas stored for up to approximately six days for an average biogas plant with 500 kW generating capacity. This time period would clearly be able serve daily and weekend fluctuations of electricity demand but not long periods of lack of wind, which can persist for weeks in some regions. For this reason, other or additional approaches are needed.

17.6.2   Storage Using the Gas Grid

Given that the natural gas grid itself has very large storage capacity (Fig. 17.3), production of biomethane from biogas immediately after production would appear to be an attractive option. Biomethane production from biogas is a mature technology, and injection into the natural gas grid is carried out in a number of countries. In principle, biomethane can be taken from the grid at any time after injection, and in displacing natural gas it contributes to GHG savings. It should be a matter simply of certification and accounting for biomethane use in order that such an approach to storage for later conversion to electricity is auditable. An added benefit of upgrading biogas to biomethane is that residual heat from electricity generation with a CHP unit is not wasted.

17.6.3 Power to Gas

Power-to-gas concepts are being demonstrated to make use of excess grid electricity which is used for hydrogen production by electrolysis and subsequent reaction with CO_2 from biogas upgrading to produce additional methane. Persson et al. (2014) [25] have reviewed a range of options for using H_2 from the electrolysis of water for the production of methane. These processes include:

- Direct injection of H_2 into the liquid phase of the biogas digester and in situ conversion of CO_2 to additional CH_4, while ensuring that the extra H_2 addition is carefully managed so that the AD process is not adversely affected.
- Injection of H_2 into a separate vessel with hydrogenotrophic methanogens to combine with CO_2 to produce CH_4. An optimized process has been shown to achieve methane-rich gas with gas grid injection specification by this method [26].
- Catalytic methanation to promote an exothermic reaction of H_2 with either CO_2 or CO to produce CH_4. The heat which is extracted to maintain a high rate of conversion is reused in the biogas process. This process was scaled up in 2013 to megawatt capacity for demonstration. Conversion of CO means the process is also suitable for conversion of syngas from thermal gasification of biomass.

In the best case, power-to-gas technology could replace or be an alternative to traditional biogas-upgrading technologies and in principle could eventually save biomethane production costs. It is clear that power-to-gas technologies have the potential to increase methane production from a given amount of feedstock used in biogas plants. This is a positive result on its own. Future developments and cost will determine to what extent power-to-gas technologies will be adopted in this way. If used as part of a strategy to balance electricity grids some form of gas storage will be needed, like the natural gas grid.

17.6.4   Flexible Biogas Production

The standard approach to continuously operating biogas plants is to maintain relatively constant conditions with respect to temperature, acidity (pH), rate of feedstock supply, and removal of digested material in order to ensure high biogas yield and minimal methane potential in the digested material. The microbiological community is complex and far from fully understood, and hence there is generally a conservative approach to management of the AD process. Nevertheless, various configurations of AD processes to achieve variable biogas production are receiving increasing attention in the research area. The challenge is to produce biogas on demand with minimal inertia and without having a significant short- or long-term negative impact on the performance of the AD process.

The main parameters which can be varied with the commonly used CSTR concept are frequency and rate of feedstock supply to the process, pretreatment of the feedstock, and use of alternative feedstocks which degrade in a predictably rapid manner. Many aspects of feedstock selection, a better understanding of AD microbiological communities, management of individual steps of the AD process, and process monitoring and control are being investigated for their potential to allow variable and controlled rates of biogas production. Research in the laboratory has shown that flexible feeding of a CSTR with selected combinations of feedstocks with different biological degradation rates can result in biogas production volume varying by a factor of 6 over a period of one day and that the AD process performance is not adversely affected over long-term operation [27]. The response of a continuously running AD process to the addition of a feedstock depends on the feedstock composition, in particular the accessible amount of easily degradable material, which can be influenced by pretreatment. Mauky et al. (2014) report using a mixture of maize silage and cow slurry fed into a primary digester to produce one stream of biogas and sugar beet silage fed to a secondary digester along with digestate from the primary digester (Fig. 17.4). Careful management of up to six individual injections of feedstocks to the primary and secondary digesters each day enabled peak biogas production to be achieved at times of maximum demand for power.

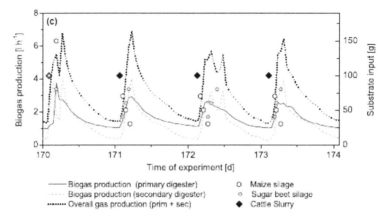

Figure 17.4   Variable biogas production from a two-step anaerobic digestion process with feeding of maize silage and cow slurry and sugar beet silage into primary and secondary steps, respectively. Results obtained from four days in a long-term laboratory experiment carried out with a CSTR under mesophilic conditions (38°C). Reprinted from Ref. [27], Copyright 2014, with permission from Elsevier.

Flexible biogas production would, in principle, require additional time for daily process management, while saving on the need for biogas storage capacity. In practice, a combination of biogas storage and flexible AD operation should enable the maximum flexibility for supply of power to the electricity grid to meet fluctuating demand. This concept would need to use the highest possible power conversion efficiency, particularly if effective use of residual heat from the CHP unit cannot be ensured.

17.6.5   Biogas Plant Networks and Virtual Power Plants

If numerous biogas plants are linked by biogas pipelines to centralized gas utilization economies of scale and higher conversion efficiencies can be achieved with CHP or with biogas upgrading. This concept of sharing to reduced costs has already been introduced in numerous locations, for example, in Brazil, where 33 farms have been linked to a central CHP unit which can additionally make better use of waste heat than many small distributed CHP engines [28]. Another example of a biogas pipeline is in the Netherlands, where desulfurized and dried biogas is transported from one farm to a local community about 5 km away for use in a CHP unit [29]. The CHP unit provides electricity to the grid and injects waste heat into a district heat grid, thereby making maximum use of the available energy in the biogas. In the latter case, the farm maintains a small CHP unit for its own power and heat needs. This project helps the farmer to achieve higher income with the sale of both electricity and heat and to contribute to reducing use of fossil fuel for energy. Local gas grids have been introduced in places like Sweden, where the natural gas grid has limited coverage [18]. In the Swedish case the local grid is used to distribute biomethane to vehicle-filling stations.

An ambitious concept has been developed in Bavaria to pool the resources of approximately 2300 biogas plants in order to create one virtual power plant which can be managed in a central, better-coordinated manner than if plants are operated independently [30]. For the concept to have a significant impact, biogas storage capacity has to be increased. In the Bavaria example, flexible operation would provide 450 MW additional power, which is approximately 15% of the fluctuating demand.

17.7 The Future

There will likely be four main types of biogas plants in the future, one operated to provide power on demand to the electricity grid, one optimized for biogas upgrading to biomethane and injection into the natural gas grid, one operated independent of the national grid and providing energy for local use, and finally the humble small-scale biogas plants of the sort deployed in developing countries. A choice will need to be made between electricity generation from biogas and injection into the electricity grid and biogas upgrading to biomethane and injection into the natural gas grid for subsequent use on demand for generation of electricity. There is already competing demand for biomethane from the transport sector, which is slowly evolving to a low-carbon future. There are a number of possible scenarios of how the relative contributions of biofuels, including biomethane, electricity, and eventually hydrogen, will develop after the fairly well-planned period to 2020 is passed. There is growing potential to utilize CO_2 emitted from biomethane production in order to increase biomethane production capacity from excess renewable electricity generated from the intermittent wind and solar sources.

Many countries consider biogas and biomethane attractive clean fuels for heat, electricity, and biofuel use and have planned expansion of the biogas sector to address demand. The versatility of the AD process means that is already established in the waste treatment, agriculture, wastewater treatment, and a range of process industries. The big challenge is to make biogas production economically viable in these sectors. With the exception of some waste treatment processes and in a number of process industries, some form of financial support is needed to ensure economic viability of biogas plants. As a consequence, substantial effort is being made to maximize biogas output from individual feedstocks or feedstocks codigested in terms of both methane potential and reaction kinetics. Selling electricity to the power grid in a flexible manner to meet peak demand presents an opportunity to plant operators to increase income, while at the same time providing vital grid-balancing capability.

Acknowledgments

The author is grateful for support of the European Commission through the JRC work program and to members of IEA Bioenergy Task 37 for their support in the collection and assessment of information on emerging technologies in the biogas sector.

References

1. EU Nitrates Directive (1991). Council Directive 91/676/EEC of 12 December 1991 concerning the protection of waters against pollution caused by nitrates from agricultural sources.

2. EU Landfill Directive (1999). Council Directive 99/31/EC of 26 April 1999 on the landfill of waste.

3. EU Biofuels Directive (2003). Council Directive 2003/30/EC of 8 May 2003 on the promotion of the use of biofuels or other renewable fuels for transport.

4. US EPA (2005). Statutory Default Standard, 2005. In December 2005, the US Environmental Protection Agency (EPA) set a statutory default

standard which required that 2.78% of the gasoline sold or dispensed in calendar year 2006 be renewable fuel.

5. US EPA (2007). Renewable Fuel Standard, 2007. Regulations for a Renewable Fuel Standard (RFS) Program for 2007 and beyond.

6. EU Renewables Directive (2009). Council Directive 2009/28/EC of 23 April 2009 on the promotion of the use of energy from renewable sources.

7. EU Renewable Electricity Directive (2001). Directive of the European Parliament and of the Council of 27 September 2001 on the promotion of electricity from renewable energy sources in the internal electricity market.

8. Braun, R. (2007). Anaerobic digestion: a multi-faceted process for energy, environmental management and rural development, in *Improvement of Crop Plants for Industrial End Users*, Ed. Renalli, P., Springer, ISBN 978-1-4020-5486-0.

9. Montgomery, L.F.R., Bochmann, G. (2014). Pretreatment of feedstock for enhanced biogas production, in *IEA Bioenergy Task 37 Technical Brochure, February 2014*, ISBN 978-1-910154-04-5.

10. Sun, Y., Cheng, J. (2002). Hydrolysis of lignocellulosic materials for ethanol production: a review, *Bioresource Technology*, **83**, 1–11.

11. Taherzadeh, M.J., Karimi, K. (2008). Pretreatment of lignocellulosic wastes to improve ethanol and biogas production, *International Journal of Molecular Sciences*, **9**, 1621–1651.

12. Bachmann, N. (2013). Design and engineering of biogas plants, in *The Biogas Handbook, Science, Production and Applications*, Ed. Wellinger et al., Woodhead, 191–211, ISBN 978-0-85709-498-8.

13. Murphy, J., Thamsiriroj, T. (2013). Fundamental science and engineering of the anaerobic digestion process for biogas production, in *The Biogas Handbook, Science, Production and Applications*, Ed. Wellinger et al., Woodhead, 104–130, ISBN 978-0-85709-498-8.

14. Drosg, B. (2013). Process monitoring in biogas plants, in *IEA Bioenergy Task 37 Technical Brochure, December 2013*, ISBN 978-1-910154-02-1.

15. Intergovernmental Panel on Climate Change (2013). *Intergovernmental Panel on Climate Change (IPCC), Fifth Assessment Report, 2013*.

16. Kaparaju, P., Rintala, J. (2013). Generation of heat and power from biogas for stationary applications: boilers, gas engines and turbines, combined heat and power (CHP) plants and fuel cells, in *The Biogas Handbook, Science, Production and Applications*, Ed. Wellinger et al., Woodhead, 191–211, ISBN 978-0-85709-498-8.

17. Beil, M., Beyrich, W. (2013). Biogas upgrading to biomethane, in *The Biogas Handbook, Science, Production and Applications*, Ed. Wellinger et al., Woodhead, 191–211, ISBN 978-0-85709-498-8.

18. Persson, Svensson (2014). Non-grid biomethane transportation in Sweden and the development of the liquefied biogas market, in *IEA Bioenergy Task 37 Case Study, September 2014*. http://www.iea-biogas. net/case-studies.html.

19. Persson, T. (2012). *Task 37 Sweden Country Report 2012*. http://www. iea-biogas.net/country-reports.html.

20. Specht, M., Zuberbuhler, U., Baumgart, F., Feigl, B., Frick, V., Sturmer, B., Sterner, M., Waldstein, G. (2011). *Storing Renewable Energy in the Natural Gas Grid: Methane via Power-to-Gas (P2G): A Renewable Fuel for Mobility, October 2011*. http://www.etogas.com/nc/en/information/ publications/.

21. Hochloff, P., Braun, M. (2014). Optimising biogas plants with excess power unit and storage capacity in electricity and control reserve markets, *Biomass and Bioenergy*, **65**, 125–135.

22. Hahn, H., Ganagin, W., Hartmann, K., Wachendorf, M. (2014). Cost analysis for a demand orinted biogas supply for flexible power generation, *Bioresource Technology*, **170**, 211–220.

23. Liebetrau, J., Clemens, J., Cuhls, C., Hafermann, C., Friehe, J., Weiland, P., Daniel-Gromke, J. (2010). Methane emissions from biogas-producing facilities within the agricultural sector, *Engineering in Life Sciences*, **10**, 595–599.

24. Hahn, H., Krautkremer, B., Hartmenn, K., Wachendorf, M. (2014b). Review of concepts for a demand-driven biogas supply for flexible power generation, *Renewable and Sustainable Energy Reviews*, **29**, 383–393.

25. Persson, T., Murphy, J., Jannasch, A-K., Ahern, E., Liebetrau, J., Toyama, J. (2014). A perspective on the potential role of biogas in smart energy grids, *IEA Bioenergy Task 37*, ISBN 978-1-910154-12-0 (http://www. iea-biogas.net/technical-brochures.html).

26. Luo, G., Angelidaki, I. (2012). Integrated biogas upgrading and hydrogen utilization in an anaerobic reactor containing enriched hydrogenotrophic methanogenic culture, *Biotechnology and Bioengineering*, **109**, 2729–2736.

27. Mauky, E., Jacobi, H.F., Liebetrau, J., Nelles, M. (2014). Flexible biogas production for demand-driven energy supply: feeding strategies and types of substrates, *Bioresource Technology*, **178**, 262–269.

28. Bley, C., Amon, D. (2013). Bioenergy in family farming: a new sustainable perspective for the rural sector in Brazil, in *IEA Bioenergy Task 37 Case Study, September 2013*. http://www.iea-biogas.net/case-studies.html.

29. Task 37 Success Story, 2011. Biogas pipeline for local heat and power production in a residential area; Zeewolde, NL, in *IEA Bioenergy Task 37 Case Study, September 2014*. http://www.iea-biogas.net/case-studies.html.

30. Schmidt, M., Schäfer, R., Ortinger, W. (2013). Bayernplan: an innovative business concept for biogas, in *21st European Biomass Conference and Exhibition, June 2013*, ISBN 978-88-89407-53-0.

About the Author

David Baxter is associated with the European Commission, Joint Research Centre, Institute for Energy and Transport, Petten, the Netherlands. He obtained a PhD in the topic of high-temperature-material integrity in power generation processes in 1981. Afterward, he spent five years at the Argonne National Laboratory in the United States on laboratory research on coal gasification and the integrity of heat exchange systems for energy recovery, looking at innovative alloying combinations for steel and coating systems as a way of increasing corrosion resistance in a cost-effective manner.

Dr. Baxter worked in the United Kingdom for the next five years in the manufacturing industry as a technology support and production manager at International Nickel Company (INCO). The industrial role involved working on the development of new and the improvement of existing manufacturing processes, including composition and microstructure control during metal melting, forging processes for gas turbine alloys, powder metallurgy applied to both mechanical alloying and powder atomization, environmental management in a factory, and metal recycling.

Dr. Baxter joined his current employer, the European Commission, in 1991 and has been concerned with laboratory and desktop research in the area of waste and biomass. The laboratory research involves bioenergy and biofuel conversion processes, while desktop research focusses mainly on environmental sustainability assessments. Since 2000, most of the research in the fields of biomass, biofuels, and bioenergy has been directed toward providing scientific support to European Union policy makers.

Dr. Baxter is leader of IEA Bioenergy Task 37 (Energy from Biogas); chairman of the Scientific Programme Committee of the European Biomass Conference; member of SET-Plan bioenergy industrial initiative (EIBI) management team; member of the Institute for Materials, Minerals and Mining; and a chartered engineer (C.Eng.).

Chapter 18

The Ethanol Program in Brazil

José Roberto Moreira

Institute of Energy and Environment, University of Sao Paulo, Brazil

rmoreira69@hotmail.com

The ethanol program in Brazil started in 1975 and is evolving continuously. Initially triggered by the high price of oil and the difficulty to conciliate it with the limited amount of hard currency, it became a significant source of liquid fuel for transportation, an important driver for the country leadership of the world sugar market, an stimulus for large amount of renewable electricity generation and provide technology development for the industrial and transport sectors. The chapter tries to describe the evolution of all these sugarcane-based actions, as well as to speculate about the future, trying to infer potential results which may be achieved, The forecast is strongly tied to the 38 years of history, where sequences of huge success and serious barriers were observed, but whose average consequences was the capacity of supplying almost half of the liquid fuels consumed by 38 million cars, generation of 6% of the more than 500 TWh/yr of electricity consumed in the country. In parallel with such technical outcomes, the program has contributed significantly for GHG emission reduction, as demonstrated by being classified

Biomass Power for the World: Transformations to Effective Use

Edited by Wim van Swaaij, Sascha Kersten, and Wolfgang Palz

Copyright © 2015 Pan Stanford Publishing Pte. Ltd.

ISBN 978-981-4613-88-0 (Hardcover), 978-981-4669-24-5 (Paperback), 978-981-4613-89-7 (eBook)

www.panstanford.com

as advanced ethanol by the US EPA. On the basis of such long-term success, in the large amount of renewable energy produced, and in the continuous capacity of expansion due to technology advances and availability of soil, water, and manpower, the text speculates about the ethanol program to become more economically attractive than offshore oil expansion over deep water, another much bigger liquid and gas fuel program under execution in Brazil.

18.1   Introduction

Fuel ethanol replacing gasoline has been used continuously since 1975. In the period 1975–1980 it was used only blended in gasoline. Starting in 1980 and motivated by the fast increase in ethanol availability due to the success of the PROALCOOL program driven by private entrepreneurs but with very attractive financing mechanisms from the federal government, neat-ethanol cars started to be commercialized. These cars are powered by hydrous ethanol (93% ethanol and 7% water) as opposed to the ones which use gasohol, which is gasoline blended with neat ethanol (99.7% ethanol). Since 2003, with the manufacturing of flex-fuel cars, which are able to use gasohol (blend of gasoline with 25% neat ethanol, which is the only gasoline available in service stations), neat hydrous ethanol, or any blend of gasohol and hydrous ethanol, the market share of such cars has increase to 61% of all cars in service by 2013[1] (see Fig. 18.1). Thus, potential demand for ethanol has exceeded supply, and is only limited by economic reason. Drivers, when attending any service station can choose between gasohol and hydrous ethanol on the basis of the lower-cost option.

Ethanol production grew very quickly in the first 10 years of the program (1976–1986) (shown in Fig. 18.2), followed by a decline in the period 1999–2003 and retake growth from 2003 up to 2009. After 2009 until 2013 a decline in the production of ethanol and even sugarcane was observed, and even after a recent retake (2012–2013) on these activities the sugarcane production has not yet exceeded the maximum value achieved in 2010 (see Fig. 18.2). This peculiar behavior can be explained by the following major drivers [2]:

---

[1]Flex-fuel motorcycle production started in 2009, and they are becoming common, and they already represented 12.7% of such fleet in 2012 [1].

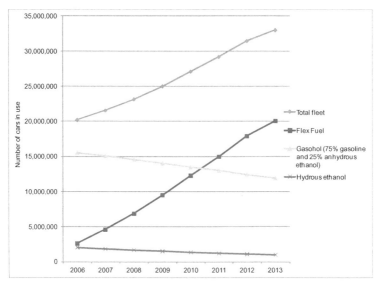

Figure 18.1 Evolution of the car fleet according to fuel use in the period 2006–2013. Note that 2013 data is only up to the month of September. Flex-fuel cars use hydrous ethanol, gasohol, or any blend of both. No neat gasoline is sold for cars. Prepared by the author on the basis of Ref. [1].

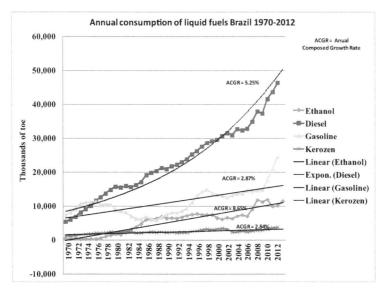

Figure 18.2 Annual consumption of liquid fuels in Brazil between the years 1970 and 2012 [3].

- 1976–1986: For the first four years, ethanol was only used as a blend for gasoline. In 1980 with the introduction of neat-ethanol cars in the market, national energy legislation required that all services stations have, at least, one dedicated ethanol pump. With the low price of ethanol, the supply availability, compulsory consumption through the blend in gasoline, and voluntary consumption by neat-ethanol cars the demand rose very quickly. At the end of the period 6,900 ktoe was produced, which was 27 times bigger than the 1976 value (see Fig. 18.2). It is worthwhile to note that neat-ethanol cars represented 95% of all new cars manufactured in 1985. Strong institutional infrastructure was created, price mechanisms designed to keep ethanol price below the one from gasoline, and taxes differentiation between both fuels drive up ethanol production. With favorable price, demand also increased due to the production and sales of neat-ethanol cars.

- 1986–1994: This was the phase of stagnation (see Fig. 18.2). Since 1985 oil prices declined significantly to the level of US$ 12–20/bbl. In Brazil, this effect was noticed only by 1988 when shortage of public money to subsidize programs of alternative energies occurred, driving down the volume of investments in further ethanol supply expansion. The low price of the fossil fuel also reduced private interest on new investment in the ethanol sector, limiting production expansion since 1986. On the demand side sales of neat-ethanol cars continued. Nevertheless, the strong demand was not fulfilled by the installed capacity and a significant shortage of ethanol for the voluntary neat-ethanol consumers showed up. The problem was partially mitigated through importation of ethanol and methanol, and the replacement of the compulsory 20% of ethanol blended in gasoline. Due to this fact ethanol producers decided to dedicate a significant share of sugarcane to sugar production (see Fig. 18.3), while the ethanol fuel market was supplemented by imported ethanol.

During the two phases described it was necessary to subsidize ethanol since it had its price controlled by the government and set as a fraction of the gasoline price. Total accumulated amount of subsidy was high, reaching more than US$ 30 billion by 1998 [5]. Fortunately, this accumulated amount started to decline in the last 10 years, when the government price control was removed (see next paragraph).

- 1995–2000: This was the phase of redefinition. The government removed price control on fuels letting the ethanol to rely in market competition to survive. During this time neat-ethanol annual car sales dropped to 1% of the total annual cars production. It is interesting to note that in this period sugarcane production continue to grow, essentially due to the expansion of sugar for exportation (see Fig. 18.3).

- 2001–2009: This is a phase where significant expansion of sugarcane production occurred and fuel ethanol production almost achieve the same level of gasoline consumption in 2009, when accounted in terms of its energy content (see Fig. 18.2). New planting areas occurred in regions of the state of São Paulo and in other neighboring states like Minas Gerais, Mato Grosso do Sul, and Goias, which were modest sugarcane producers. Most of the motivation to increase ethanol production was the introduction of the flex-fuel cars [6]. Simultaneously, the government understood the significant electricity potential associated with the use of an ethanol by-product—the bagasse—and set policies in favor of its promotion. The promotion is being carried out not only through the PROINFA,[2] but also through public bidding specially designed for biomass-based electricity plants using the installed electric grid.

- 2010–Today (end of 2013): This is the recent phase, where ethanol and even sugarcane production declined for two years and only started to retake growth by 2012 and 2013, but not enough to exceed the year 2010 production. National gasoline prices were frozen for most of the period, impacting in the financial return of the ethanol business. In parallel to that, significant amount of debt accumulated by the sugar/ethanol sector, during the preceding expansion phase, had a negative impact on ethanol producers due to an increase in the

---

[2]PROINFA, an incentive program for alternative energy sources of electricity in Brazil, is a federal government program, created in 2002, to foster and/or promote the use of biomass, wind, and small hydropower as sources of electricity.

international interest rate charged to developing countries, triggered by the 2008 world economic crisis. International sugar prices were kept at a high value up to 2012, providing, from one side, some revenue to the sugarcane sector, but, on the other side, reducing ethanol production since more than the usual share of the sugarcane was directed to be used as sugar feedstock. Furthermore, climatic conditions were quite poor for sugarcane crops. In the last two years, with better climatic conditions and with the international sugar price decline, it became obvious that the ethanol market deserved better care and the sugarcane producers, and the cane mills owners recognize the necessity of investments to renew the plantation and expand again the ethanol market. Expectation for the ending of gasoline price frozen policy has helped the retake. Unfortunately, due to past debit and the still unrealistic national gasoline price, new investments in further expansion are being modest. New green cane projects are very few and new bioelectricity producing plants have not been built in this period.

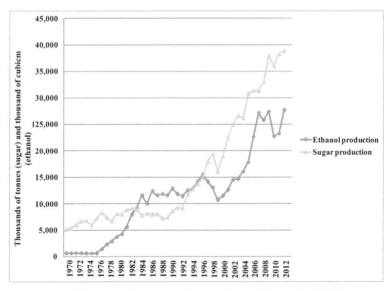

Figure 18.3   Ethanol and sugar production from 1970 to 2013 [3]. Forecast for 2013 from Ref. [4].

Considering all these five phases, Fig. 18.2 shows the annual evolution of major liquid fuels used in Brazil, from 1970 to 2012. It is clear that ethanol has shown the largest composed annual growth rate (8.65%) while diesel the one with the larger share has grown at 5.25%. Gasoline, which was partially displaced by ethanol, has grown only 2.87%. This long-term analysis is useful to demonstrate that the rapid growth of one fuel in some short period should not be extrapolated for longer time, since so many political and economic facts must be taken into account. Even so, it is possible to infer that if the past 43 years is a reasonable description of fuel displacement in the future, we can say that ethanol, produced in an amount of 11,500 ktoe, by 2012, will be able to share with gasoline 50% of the Otto engines market by the year 2018, when the production of both fuels will reach 17,500 ktoe. By the year 2040, with a production of 109,000 ktoe, ethanol use would be 5.4 times larger than gasoline. A large number like this one is not infeasible, since the 2012 sugarcane crop area used 8.9 million hectares and does not have to grow in the same proportion than ethanol demand. The reasons for that are (1) by 2012 only 45% of the sugarcane crop area was used for ethanol production, the reaming for sugar [7], and (2) historical sugarcane yield has increased by 1% per year and sugar content by another 0.5% [8]. Assuming technology will continue this trend, total sugarcane crop area for sugar and ethanol, by 2040, has to evolve to 30 Mha, which is 3.4 times bigger than the 2012 area. Such area is available in the country considering the evaluation performed in 2009 by the Ministry of Agriculture, which concludes that more than 34 Mha is available for sugarcane crop expansion over pasture land [9] and that the soya crop, the one occupying the largest plantation area, occupied 26 Mha in 2012 [10].

Nevertheless, there is a significant possibility that due to energy efficiency improvement, total energy demanded by the transportation sector will decline. Probably, even without considering the feasibility of second-generation ethanol, the full Brazilian automobile fleet may be powered with 16 Mha of sugarcane crop for fuel and complemented by electricity, generated from the combustion of sugarcane bagasse and other wastes, in modern sugar mills (see Ref. [11]).

18.2 Major Outcomes from the Success of the Ethanol Program

18.2.1 Currency Savings by Brazilian Society

During the first and second oil shocks (1973–1974 and 1977–1979) due to the strong dependence of the Brazilian energy sector on imported oil significant amount of hard currency was spent in its importation. Figure 18.4 shows the impact of ethanol in the Brazilian trade balance. In some year more than 50% of the total hard currency obtained from products and goods exportation was spent in oil importation. As noted, accumulated net importation of oil has consumed US$$_{2008}$ 245 billion of the total US$$_{2008}$ accumulated commercial trade of the country in the period 1978–2012. It is possible to see that this expenditures where more intense up to 1984, when due to an increase in national oil production and in ethanol fuel replacement of gasoline, the amount of oil importation started to decline. The figure also displays the ethanol contribution in the Brazilian trade balance, by comparing the real accumulated net trade with the possible one without ethanol production. The difference between both scenario reaches a value of US$$_{2008}$ 155 billion, much larger than the cumulative avoided expenses with oil and oil products importation plus gains with ethanol exportation (US$$_{2008}$ 54 billion, due to the interest charged for international loans and return on foreign investments, assumed at 8%/yr, and necessary, since the overall country balance was negative in the period analyzed).

18.2.2 Currency Savings by Passenger Car Drivers

The presence of an alternative to gasoline increases competition and consequently reduces prices. Figure 18.5 shows price of gasoline and diesel relative to imported oil. As seen, diesel has a lower price compared to gasoline and this is a long-term trend set by the government to control price of goods. It is possible to note also that the diesel-to-imported-oil price ratio shows an almost horizontal trend, that is, the ratio is essentially constant over the 39 years period. On the contrary gasoline price decreased significantly with respect to imported oil. From Fig. 18.5 we note that gasoline price decrease by 20.5%, while diesel decreased only by 3.3%, over the full period.

Figure 18.4   Impact of ethanol used internally and exported in Brazilian trade from 1978 to 2012. Prepared by the author on the basis of data from Ref. [3] and use of Ref. [12] for currency calculation in US$$_{2008}$.

Figure 18.6 allow us to see even better the effect of competition in the gasoline market. The oil company, in order to preserve its revenue and minimize market share losses, increased gain margin on diesel while reduced it on gasoline.[3] The conclusion is that alternative sources of energy can bring another advantage—reduction in the cost of energy in case alternatives are available to replace the major fuels used in the transportation sector.

---

[3]Gasoline and diesel prices have been controlled by the government up to the year 2000. Nevertheless, after that date, government decisions were taken after consulting and hearing the government-owned oil company PETROBRAS. Even in more recent years, when PETROBRAS became free to set fuel prices, the company has acted with care, trying to avoid difficulties regarding government policies. On the other hand, ethanol prices are set by the private sector. It is possible to see, in the last 10 years, that negotiation between the government and PETROBRAS concludes that to keep good economic health of the oil company it is more efficient to increase diesel prices faster than gasoline's.

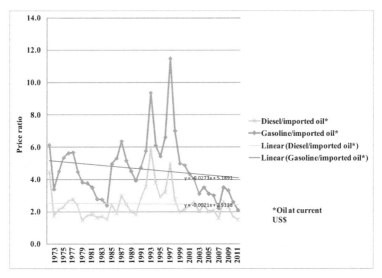

Figure 18.5 Price ratio of gasoline/imported oil to diesel/imported oil from 1973 to 2008. Prepared by the author on the basis of data from Ref. [3].

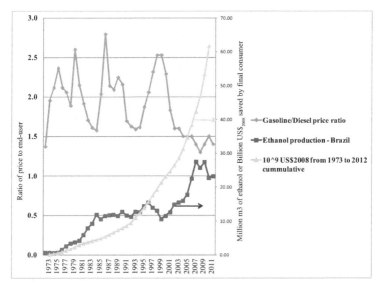

Figure 18.6 Relative price to final consumer for gasoline and diesel, amount of ethanol production, and amount of US$$_{2008}$ saved by gasoline consumers from 1973 to 2008. Prepared by the author on the basis of data from Ref. [3] and use of Ref. [12] for currency calculation in US$$_{2008}$.

18.2.3 Liquid Biofuel and Reduction of GHG Emissions

Production and use of biofuels have been under criticism since 2008 due to some new sustainability indicators. One of them is the biofuel contribution to greenhouse gas (GHG) emissions due to direct land use (LUC) and indirect land use (ILUC). Another source of concern was related with the significant release of N_2O to the atmosphere due to the use of N fertilizers when planting biofuels feedstock. Some authors tried to convince society that not all biofuels are necessarily green [13–15]. But, all these evaluations conclude that sugarcane ethanol is the one with the best capacity to mitigate climate change if properly managed [16, 17].

According to US EPA, ethanol from sugarcane produced in Brazil is able to reduce CO_{2eq} emission by 61% considering LUC and ILUC effects [17]. Pacca and Moreira [8] tried to quantify the overall impact of the PROALCOOL program since its launching (1975) up to 2007. Figure 18.7 shows that in the initial years of the program, the overall effect was negative, increasing GHG emissions, mainly due to C from above- and below-ground biomass lost to the atmosphere when converting earlier vegetation in sugarcane crops. It took 17 years for CO_{2eq} emissions avoided from gasoline, due to its displacement by ethanol, to offset all the initial GHG emissions (see Fig. 18.7). Nevertheless, by 2008, after 32 years of the begin of the PROALCOL program it is possible to see that 125 tCO_{2eq}/ha has been avoided. The relatively long offset time was consequence of the very poor initial efficiency on ethanol production

Figure 18.8 shows how much GHG mitigation could be obtained from the PROALCOOL program if the present surplus electricity generation was performed using modern available technology (high-pressure steam turbines) and carbon capture and storage (CCS) on CO_2 from fermentation since 1975. An accumulated abatement of CO_{2eq} of 400 t/ha would be achieved in 2008. Note that for Fig. 18.8, the historical low ethanol efficiency has been maintained.

For the next 32 years, Pacca and Moreira [8] show that on the basis of the already available technologies and with the present and increasing yield of ethanol and an annual expansion of sugarcane planted area of 4.3% the practice would be always environmentally sound, accumulating CO_{2eq} abatement of 820 t/ha during the next 32 years (up to year 2039).

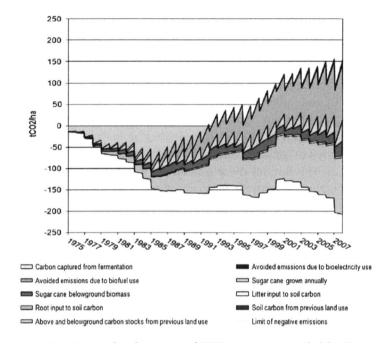

Carbon captured from fermentation

Avoided emissions due to biofuel use

Sugar cane belowground biomass

Root input to soil carbon

Above and belowground carbon stocks from previous land use

Avoided emissions due to bioelectricity use

Sugar cane grown annually

Litter input to soil carbon

Soil carbon from previous land use

Limit of negative emissions

Figure 18.7   Accumulated amount of GHG emissions avoided by the use of ethanol fuel as a replacement of gasoline in Brazil (1975–2008) [8]. See also Color Insert.

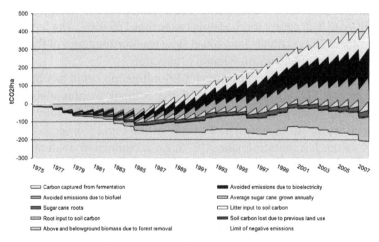

Carbon captured from fermentation

Avoided emissions due to biofuel

Sugar cane roots

Root input to soil carbon

Above and belowground biomass due to forest removal

Avoided emissions due to bioelectricity

Average sugar cane grown annually

Litter input to soil carbon

Soil carbon lost due to previous land use

Limit of negative emissions

Figure 18.8   Accumulated amount of GHG emissions avoided by the use of ethanol as a replacement of gasoline in Brazil if electricity surplus generation and CCS where in effect since the beginning of the program (1975–2008) [8]. See also Color Insert.

18.2.4 Bioelectricity Generation from Sugarcane Residues

Since the late 1990s, and soon after the electricity shortage crisis of 2001, hydroelectric complementation is growing through the use of thermal fossil fuel (natural gas, diesel oil, fuel oil), nuclear energy, and more recently, also through the use of renewable energy (bagasse, wood waste, wind, and small hydropower) [18]. Thermal plants require less investment than hydroelectric, are built in less time, but most of them have a high operating cost due to fuel cost. The thermal fossil fuel power plants are large sources of CO_2 emissions, which imply environmental and social costs which unfortunately still are treated as externalities, and therefore do not enter into the composition of the regular cost of the project, but are paid by the whole society. Renewable power plants such as wind and small hydropower have low operating costs because they do not use fuel, and modest external environmental and social costs compared to fossil fuels, but require high investment cost. Winds have a high investment cost due to low capacity factor utilization and show seasonal variations due to wind speed. Small dams are expensive due to economies of scale, which discourages small-plant investment.

A very interesting complement to hydroenergy input consists of bagasse and other residues of sugarcane. Bagasse is mainly used in thermoelectric plants installed in cane mills, to provide heat and electricity to the industrial process. So, there is tradition and experience in its use and developed technology. In addition, seasonal factors operate in synergy in the case of the thermal bagasse and other waste of sugarcane and hydroelectricity in the southeast and midwest of the country, where much of the country hydroelectricity is generated. In fact the electric network is well integrated over most of the country and thus electricity generated in a region can be transferred to other, within certain limits. Thus, regional electric complementation can have a national character, depending on its amplitude. Cane harvest, throughout that region, including the south, occurs more intensely in the months from April to November, because it is during this period that sugarcane contains lots of sugar. The period of high rainfall in the southeast/midwest occurs between November and March, when the hydroelectric reservoirs have the opportunity to complete. Thus, the supply of electricity generated from sugarcane mills in the region, from the month of April to November, when exported to the grid, replaces hydroelectric generation, preserving water in reservoirs and acting as a virtual expansion of them, at no additional cost. This extraordinary advantage, which has an economic value not considered by the National Electric Energy Agency (ANEEL), is one of the great motivators to exploit this form of bioelectricity. There are numerous other reasons to use bagasse and cane waste to generate surplus electricity in power plants but the most important is related to the efficient use of natural resources for the benefit of the environment and society. This can be understood since cane mills burn bagasse because they need primarily heat to process sugarcane into ethanol and sugar. But, with known technology used in the country for many years, we can simultaneously produce heat and, generate electricity consuming almost the same amount of bagasse used only for heat purpose. This technology known as cogeneration and its improvements still need to be better utilized.

In April 2013 about 367 plants generating electricity from waste sugarcane were enrolled in ANEEL,[4] with an installed capacity of 8,536 MW [19]. This power represents 6.50% of the total installed electric capacity in the country. A share of this power is used for their own purposes, which is known as self-generation, as traditionally the sugarcane sector has always been interested in generating the energy it consumes. This occurs because the processing of sugarcane into sugar and/or ethanol requires a lot of thermal energy and only some mechanical and electrical energy. At the same time, after removal from sugarcane juice there is a so large amount of residual bagasse that no market can absorb it entirety. This process is called cogeneration of heat and electricity and is always fetched because you can convert the energy from bagasse into heat, mechanical power and electricity with an efficiency superior to most advanced systems for electricity generation in use in the world [20]. Unfortunately, the use of cogeneration is often limited in many industrial activities and thermal plants due to lack of demand for heat, because this, unlike electricity, cannot be transported over distances of more than a few miles. However, in sugarcane mills that demand exists on-site, which allows working with cogeneration.

---

[4]ANEEL is the regulatory body for electricity supply.

Until 1985, no mills were selling electricity to the power grid because there was no legislation allowing this operation and because they generated little electricity, enough only for self-consumption. Thereafter appeared a few power plant marketing electricity to the grid, but only after the restructuring of the electricity market in 2004, when it offered a differentiated tariff for that electricity, greater interest has occurred, inducing the generation of electricity at a volume higher than the internal demand of the mills, with the purpose of selling the surplus to the grid. The activity was successful, as shown by the number of mills decided to make money with the opportunity. It is true that success is modest because although there are 367 registered and generating plants installed in mills, the amount of electricity sold to the grid is small. This total, in 2011, was just over 10,000 GWh [21], that is, 2% of electricity generated in Brazil. Considering a sale value of US$ 60/MWh, this represents a turnover of US$ 0.6 billion, which is very little compared with the revenues obtained from the sale of ethanol (US$ 16.9 billion) and sugar (US$ 10.8 billion) [22]. Although modest, there is interest from some plants owners to sell electricity, requiring large investments if their amount is close to the limits of modern technology. The investment to generate electricity is much greater than to make fuel and this fact is well known worldwide [23].

18.3 Future Perspectives on the Use of Ethanol

18.3.1 Very Large Amount of Energy-Built-In Sugarcane Crop in Brazil

The amount of sugarcane planted for sugar and ethanol production, in 2013, occupies a little over 9 million hectares. In almost all mill in operation by that year, only the energy contained in the juice, giving ethanol or sugar as the final product, and the energy contained in the bagasse, which is a by-product of sugarcane after the juice has been removed, are being used. But we know that in the field, cane wastes in the form of leaves and tips (both called straw) are left, when green cane harvest is performed. This biomass, yet has little commercial interest, and is either destroyed due to the traditional practice of burning of sugarcane fields prior to harvest, or is left in the soil when harvesting is mechanized, eliminating the burning. The fact that this biomass is left over the crop area is a demonstration that its transportation cost to the mill has not yet compensates for any gain which may yield from its use or sale. However, there is great expectation of its use as further energy input to fuel boilers to produce more steam, which is used to generate more electricity [24].

Considering the energy contained in the three components discussed in the previous paragraph, which composes the cane primary energy, Fig. 18.9 was prepared to show the amount of such energy over the years and compared with the primary energy of oil produced in Brazil. As can be seen, in the year of the largest production of sugarcane (2010) such primary energy was 80% of oil energy.

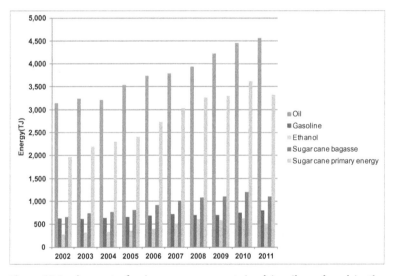

Figure 18.9 Amount of primary energy contained in oil produced in the country, in gasoline consumed in the country, in ethanol, in sugarcane bagasse, and in sugarcane harvested in Brazil between 2002 and 2011. Prepared by the author with data from Ref. [21].

In contrast to this large amount of energy, we see from Fig. 18.9 that ethanol, in the form of final energy, accounted, in 2011, for 17% of sugarcane primary energy, while sugarcane bagasse accounted for 33%. Actually, ethanol represents 35% of primary cane energy; however, we must remember that in Fig. 18.9, only somewhere around 50% of the available cane stock was converted into ethanol, for just over half of it was used to make sugar [21]. Considering this huge amount of primary energy from sugarcane, is understandable why from the economic point of view, there is great interest in improving the efficiency of converting primary energy into final energy. When processing sugarcane for ethanol there is some room for efficiency improvement, but it does not reach 10%. When processing bagasse to transform its primary energy into heat and electricity, there are great opportunities to increase efficiency; one of them depends on the mill operation using less heat (instead of consuming 500 kg of steam/tcane is feasible to consume 300 kg/tcane [24] and thus increase the share of primary energy converted to electricity. Other opportunities are related with the use of high-pressure and high-temperature boilers, and with the use of a share of sugarcane residues left in the field.

By examining Table 18.1 [19] we see the installed electric power capacity of the 367 sugar mills officially enrolled as electricity producers. The installed capacity is listed per Brazilian states and the number of plants per state is shown. Through the MAPA[5] data, we know the amount of cane harvested per state in all years. Thus, assuming that all the available cane was used for electricity generation, we obtain the installed power per unit of cane produced in the 2012/2013 crop. Since power plants operate on average 210 days per year with 80% load factor [24], we can evaluate the power generated in each crop year. Taking, also, into account the phase of intensive recent development of the sugar industry occurred in the period after 2003 and that many new mills, with modern technology, planned during this period started operating from 2005, we can estimate the electricity generation of the new plants, considering the installed power in 2013 and the increased availability of sugarcane between 2005 and the year of greatest harvest after 2005. This volume of sugarcane burned in power plants after 2005 and the year of the greatest harvest roughly corresponds to the use of the cane in new and modern mills, and thus allows estimating the maximum amount of electricity generated by tcane. Figure 18.10 shows the percentage increase in sugarcane production, by state, between the largest crop year and the year 2005. It is seen that the southeast and center west region, where are located the states

---

[5]MAPA: Ministry of Agriculture, Cattle Ranching and Food Supply.

of Goiás, Minas Gerais, and Mato Grosso do Sul, had the largest increase. This increase in the quantity of cane was responsible for generating incremental electricity in the period and, therefore, from the installed power in 2013 we can evaluate the maximum amount of electricity produced by new plants. Figure 18.11 shows the average generation from all plants of each state, as well as the generation from new plants. It is significant to note that the average maximum generation of all power plants is 55 kWh/tcane, while the average of the new plants is 88 kWh/tcane. Remembering that not all new plants have high efficiency, this allows inferring that the most efficient ones may be generating around 100 kWh/tcane.[6] As a check on the validity of our hypotheses, we conclude that the maximum generation of new installed plants would reach 34,000 GWh (55 × 618), which exceeds the approximately 21,000 GWh generated in 2011, for self-consumption and exportation [21]. The difference of almost 40% between the reported generation (21,000 GWh) and the potential generation (34,000 GWh) can be explained by:

(1) The installed capacity data is for 2013, and data from Ref. [21] are for 2011. Numerous plants which began to be installed or planned by 2008, were only completed after 2011, which could mean a significant increase in bioelectricity generation in 2013, compared to 2011.

(2) It is common for plants to have nominal capacity above the average operational capacity since it is not always that the maximum power is being produced [27].

(3) We know that there are around 5%–10% bagasse not consumed in mills but sold to third parties [28].

In conclusion, considering the difficulty in obtaining the data generation for each plant,[7] it seems reasonably valid the approach used here to get them throughout the calculation above described.

---

[6]From the CONAB 2011 [25] study, the mill with the greatest electric plant efficiency generated 99 kWh/tcane and exported 73 kWh/tcane. In Ref. [26], the estimated average exported power is 77 kWh/tcane (thus, a generation above 95 kWh/tcane) for mills which have won the auction for the supply of electricity after 2009.

[7]A study was done by CONAB referring to the 2009/2010 harvest (see Ref. [25]). The values found for the units selling electricity (47 kWh/tcane) are somewhat smaller than those obtained (55 kWh/tcane), but we are evaluating figures for the 2011/2012 harvest. Therefore the values found with our model look quite reasonable. Also, the CONAB 2011 [25] study does not attempt to identify the average efficiency of power plants that started operation after 2005, for which we get the value of 88 kWh/tcane.

Table 18.1 Information on electric power installed in sugarcane mills in Brazil and registered in ANEEL in 2013

	State	Installed power (kW)	Country share (%)	No. of mills
1	Sao Paulo	4,663,208	54.65%	193
2	Goias	955,748	11.20%	26
3	Minas Gerais	949,480	11.13%	37
4	Mato Grosso do Sul	709,287	8.32%	21
5	Parana	369,165	4.32%	22
6	Pernambuco	289,710	3.39%	20
7	Alagoas	238,162	2.79%	20
8	Mato Grosso do Sul	83,332	0.98%	6
9	Rio Grande do Norte	57,000	0.67%	2
10	Sergipe	56,900	0.67%	5
11	Paraiba	55,200	0.65%	3
12	Rio de Janeiro	44,000	0.52%	1
13	Espirito Santo	23,100	0.27%	4
14	Bahia	14,000	0.16%	1
15	Sta. Catarina	11,070	0.13%	3
16	Piaui	8,800	0.10%	1
17	Maranhao	3,200	0.04%	1
18	Para	1,250	0.01%	1
	Total	8,532,612	100.00%	367

Source: Prepared by the author using data from Ref. [19].

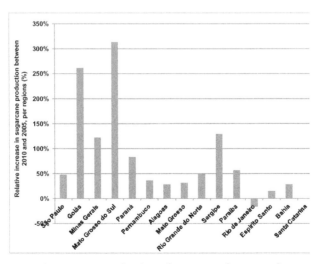

Figure 18.10 Increase in production of sugarcane between the year of the largest crop harvest (2010) and 2005. Prepared by the author on the basis of data from Ref. [25].

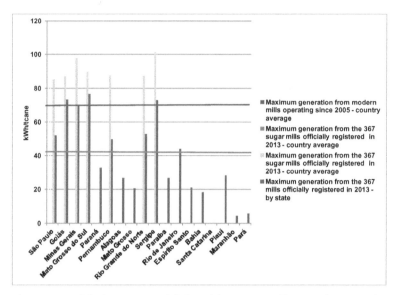

Figure 18.11   Maximum potential electricity generated in crops between the harvest season of 2008/2009 and 2012/2013 if all produced bagasse was used to generate electricity in sugarcane mills. Prepared by the author using data from Refs. [19, 25].

18.3.2   Ethanol as Fuel for Diesel Engines

As shown in Fig. 18.2, major oil product consumed in Brazil is diesel. Diesel oil is used only by trucks since legislation forbids its use in cars. Considering the transportation system is mainly provided by road vehicles due to the modest train system, and the country extension covers 8.5 Mkm^2 it is easy to understand why diesel consumption is 30% higher than gasoline and ethanol, in 2012.

As already discussed, ethanol production can increase significantly in the country while fuels for Otto type engines may observe modest growth in case diesel-type engines are allowed to be used in cars, as is occurring in most European countries. Diesel fuel has been avoided to be used in cities due to its higher local pollution level, but its quality is improving and its efficiency is higher than gasoline or ethanol. The possibility that, in the future, legislation will be modified allowing sales of cars propelled with diesel engines may be a drawback for the ethanol market. Thus, searching space for ethanol to be used in diesel-type engines is a concern of the ethanol producers. Another reason for promoting the use of ethanol in substitution of diesel is GHG emission reduction.

Technologies for replacing diesel fuels by ethanol, in diesel engines, are available. One of them is to blend hydrous ethanol with 5% by volume of a cetane enhancer. A cetane enhancer is already manufactured in the country and a fleet of buses are operational for more than 3 years, at the city of Sao Paulo, showing its technical feasibility [29].

Figure 18.12 shows the fuel cost of several types of buses being tested under the Ecofrota program carried out by the municipality of Sao Paulo, with the main purpose of reducing local and global pollution. Buses powered by diesel, biodiesel, ethanol, and diesel obtained from sugar fermentation are being tested. As noted, ethanol with a cetane enhancer is the least expensive option after diesel. Even so, only with carbon tax and with a larger fleet[8] there is some chance of economic competition with diesel.

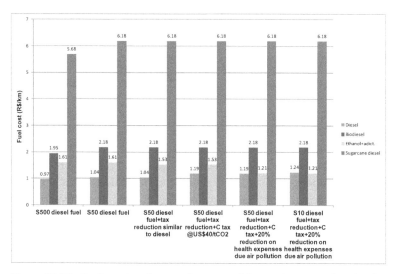

Figure 18.12 Fuel costs of several types of buses in operation in the Ecofrota project in Sao Paulo City (P by the author from data from Ref. [29]). For S10 diesel fuel the value is estimated on the basis of the relative price increase of this fuel and the S50 price quoted in Ref. [29].

---

[8]This is so because a cetane enhancer can be produced at a lower price if its demand increases.

The major barrier is cost, since it is necessary to use 1.6 L of ethanol with a cetane enhancer to displace 1 L of diesel, due to the larger volumetric energy content of diesel. Also, due to political reasons, diesel is less costly than gasoline. Presently, 1 L of diesel is only 26% more expensive than ethanol. Nevertheless, legislation is requiring the use of low sulfur diesel (S10), which increases diesel price and makes it 33% more expensive than ethanol. Thus, it is impossible to compete with diesel, without some form of government policy.

Another aspect has already been discussed in item 2.2. Diesel price is historically increasing faster than gasoline (and consequently ethanol) and when quoted at the international market it is costlier than gasoline. Considering 10% higher price for diesel compared to gasoline and that by 2013 its national price was 17% lower than gasoline, both these changes in value could set diesel price as 71% costlier than ethanol. Under this circumstance ethanol blend with a cetane enhancer would be price competitive with diesel (see Fig. 18.12).

18.3.3 Feasibility to Question Investments in Oil Sources When Sugarcane Production Can Be Expanded

Most papers dealing with ethanol as an alternative to fossil fuels used in the transport sector concludes that due to the higher producer sales price of ethanol compared to the price of oil, ethanol is only cost effective if appropriate policies exist [30, 31]. For Brazil this is an important issue since significant new oil reserves has been found and are already being exploited requiring huge investments and operational costs.

One recent study [32] analyzes from the investor's view, the detailed financial cost of producing oil from traditional and new oil reserves, and from large scale ethanol production for the period 2012–2070. The analysis considers two scenarios for future ethanol production. One of them, named *low scenario*, is based on the future evolution of ethanol production at the same average rate observed in the past (1975 to 2012). The other, *high scenario*, assumes the future growth rate at the average observed in the period 2002–2009, when a significant development of the ethanol production occurred. Figure 18.13 shows both scenarios as well as the assumed future evolution of oil production in the country due to exploitation of well-established traditional reserves and assumption on the total potential of the new reserves, when fully confirmed.

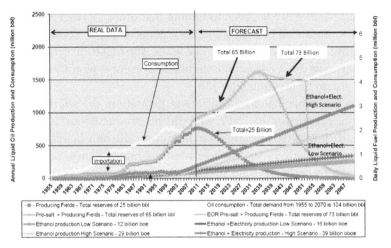

Figure 18.13   Oil consumption, oil production from traditional producing fields and from new oil fields (presalt), ethanol production for two possible scenarios (high and low), ethanol and electricity production for the low and high scenarios, and oil depletion from producing and presalt field reserves. Real values up to 2011 and values calculated on the basis of scenarios after 2011 up to 2070 [32].

The difference between ethanol and oil return on investment (ROI), all over the future period, is very small, and if annual discount rates above 4% are considered, ethanol ROI is always greater than presalt oil's ROI. Regarding oil extracted from the more traditional producing fields its economic return is unequivocally better than ethanol production for both high and low scenarios, in the absence of the CO_2 tax. Nevertheless, half of these reserves have already been used and their contribution for the liquid fuel supply in the future is limited. For the presalt oil, which is presented to Brazilian society as the major future source of oil supply, the evidences pointed out to a different direction. The results show that ethanol in the high scenario, has an economic return slightly lower than the presalt oil for interest rates below 4%, while requiring less investment than the oil scenarios. Furthermore, if a carbon tax of US$ 40/tCO_2 is taken into account, high-scenario ethanol becomes more attractive even for zero discount rate while ethanol in the low scenario presents the same return as the presalt. Thus, ethanol production might be more economically attractive than oil, which is finite. Once the presalt reserve is over, even assuming that other oil reserves could replace presalt oil in the far future, it is necessary to consider that new investments would have to be made, starting from scratch and further costs due to decommissioning of present exploitation would have also to be accounted for [33].

On the other hand for sugarcane plantation, the soil used for ethanol production in 2012 will still be productive in 2070 and beyond. New initial investment in new areas will only be required to expand the planted area, but operation and maintenance costs apply annually to keep the total productive area, sustaining production and jobs. Thus, it is very important that energy policy makers take a wise decision on the allocation of energy investments. Another important finding is that the oil option will not be enough to attend the Brazil's demand beyond 2043. This will threaten the country's energy security, as well as will drag significant amount of hard currency to guarantee importation after these years. Ethanol supply will guarantee partial liquid fuels supply for all the period analyzed. In particular, the high scenario will meet 62% of total fuel demand by 2070 and, even more after that, provided land availability is not a barrier for further sugarcane modest area expansion.[9] It is worthwhile to mention that total ethanol produced in the high scenario between 2012 and 2070 is equivalent to 38.5 billion boe and, for the low scenario, 13.7 billion boe. The high-scenario production capacity, during this period, is almost as big as the presalt inferred reserve (97.5%).

Finally, the CO_2 impact on climate change is not just a matter of cost. Real reduction on CO_2 emission has to be achieved for many countries, mainly the ones with high emission level and some with medium to high economic development. The high scenario reduces emissions up to 2070 by 20.0 billion tons of CO_2, compared to presalt oil, which represents two-thirds of the 2010 CO_2 global emissions.

Appropriate government policies are needed to enhance the interest in ethanol production and slow down oil production

---

[9]The expected sugarcane harvested area by 2070 is 14 Mha, assuming reasonable technology improvement in crop yield and in sugar content of the feedstock [30].

expansion. One of the best measures would be foster energy sector transparency, providing better information to society regarding the physical magnitude of production, economic costs, and prices of energy products. This can be achieved by requiring official and detailed annual reports prepared by experts and independent auditors. The level of uncertainty related to oil cost is significant.

18.4 Conclusion

The text intends to report the evolution of first-generation ethanol derived from sugarcane in Brazil. Together with ethanol, electricity generation from sugarcane bagasse and other wastes also evolved. The ethanol program has already 39 years and has faced many ups and downs due to internal policy changes and international oil and sugar price oscillation. Nevertheless, ethanol production has increased year after year, with few exceptions, while its feedstock— sugarcane—has shown a more steep growth since it is the feedstock for both ethanol and sugar. Sugarcane grown in the country, where soil, water, and low-cost manpower are available has been a success. Since the discovery of potential large oil reserves in the last years, ethanol and bioelectricity have faced new energy competitors, because their costs have increased and climate change issues have lost momentum at the international level. Nevertheless, the recent ethanol production decrease by no means shall be understood as a serious issue, since many past similar conditions occurred and were successfully surpassed. Furthermore, the expectation of a new production record for ethanol may occur by 2014.

The paper also tries to cover some new potentials of the ethanol program which are yet to be confirmed in practice and presents extensive arguments for their possible success in the future.

References

1. União da Industria de Cana-de-Açúcar (2014). http://www.unicadata. com.br/listagem.php?idMn=55.
2. Proalcool (2010). *Programa Brasileiro de Alcool*, http://www. biodieselbr.com/proalcool/pro-alcool.htm.
3. Balanço Enegético Nacional, Ministério de Minas e Energia, Brasilia, 2013.

4. Companhia Nacional de Abastecimento (CONAB) (2013). Presented in Valor Econômico, December 2013, http://www.valor.com.br/agro/3378162/conab-eleva-perspectiva-de-producao-de-cana-em-201314-6598-mit.

5. Moreira, J.R. (2008). *Evaluation of the Ethanol Cost to Brazilian Society.* Internal report, Cenbio, University of Sao Paulo.

6. Poppe, M. (2004). *Biofuels for Latin America and the Caribbean*: Lessons from Brazil, CGEE, Center for Strategic Management and Studies, http://www.eclac.org/dmaah/noticias/paginas/5/23775/poppe.pdf.

7. Datagro (2012). *Historical Ethanol Production in Brazil*, http://www.datagro.com.br/section/315/ethanol?palavra=ethanol+productio&meslnicio=01&anolnicio=2010&mesTermino=09&anoTermino=2012.

8. Pacca, S., Moreira, J.R. (2009). Historical carbon budget of the Brazilian ethanol program, *Energy Policy*, doi:10.1016/j.enpol.2009.06.072.

9. Ministry of Agriculture, Livestock and Supply (MAPA) (2009). Zoneamento Agroecologico da Cana-de Açucar, Brasilia, Brazil, 2009 (in Portuguese).

10. EMATER, Agência Goiana de Assistência Técnica, Extensão Rural e Pesquisa Agropecuária (2012). Goias, Brazil, http://www.emater.go.gov.br/w/4400, accessed in Jan. 2014 (in Portuguese).

11. Pacca, S., Moreira, J.R. (2011). A biorefinery for mobility?, *Environmental Science and Technology*, **45**(22), 9498–9505, doi:10.1021/es2004667 (Epub 2011 Oct. 20).

12. Consumer Price Index (2014). *Consumer Price Index Historical Tables for U.S. Cities Average*, US Department of Labor, http://www.bls.gov/ro3/fax_9160.htm.

13. Timothy Searchinger, T., Heimlich, R., Houghton, R.A., Dong, F., Elobeid, A., Fabiosa, J., Tokgoz, S., Hayes, D., Yu, T.H. (2008). Use of U.S. croplands for biofuels increases greenhouse gases through emissions from land-use change, *Science*, **319**(5867), 1238–1240, DOI: 10.1126/science.1151861.

14. Fargione, J., Hill, J., Tilman, D., Hawthorne, P. (2008). Land clearing and the biofuel carbon debt, *Science*, **319**, 1235–1237.

15. Crutzen, P.J., Mosier, A.R., Smith, K.A., Winiwarter, W. (2008). N_2O release from agro-biofuel replacing fossil fuels, *Atmospheric Chemistry and Physics*, **8**, 389–395.

16. Gibbs, H.K., Johnston, M., Foley, J.A., Holloway, T., Monfreda, C., Ramankutty, N., Zaks, D. (2008). Carbon payback times for crop-based

biofuel expansion in the tropics: the effects of changing yield and technology, *Environmental Research Letters*, **3**(2008), 034001 (10pp), doi:10.1088/1748-9326/3/3/034001.

17. US Environment Protection Agency (2010). *Renewable Fuel Standard Program (RFS2) Regulatory Impact Analysis*, EPA-420-R-10-006, February 2010.

18. Empresa de Pesquisa Energética (2012a). *Anuário Estatístico de Energia Elétrica 2012.*

19. Banco de informações de Geração (2013). *Banco de informações de Geração, disponível no site da ANEEL em,* http://www.aneel.gov.br/area.cfm?idArea=15.

20. Council of Industrial Boiler (2003), www.cibo.org, published in March 2003.

21. Balanço Enegético Nacional, Ministério de Minas e Energia, Brasilia, 2012.

22. Neves, M.F., Trombin, V.G., Consoli, M. (2010). O mapa sucroenergético do Brasil, in *Etanol e Bioeletricidade: A Cana-de-Acucar No Futuro da Matriz Energetica,* prepared for UNICA, Ed. Luc Projeto de Comunicação ltda, Sao Paulo, SP, June 2010.

23. Yeager, K., Dayo, F., Fischer, B., Fouquet, R., Gilau, A., Rogner, H.-H. (2012). Energy and economy, in *Global Energy Assessment (GEA): Towards a Sustainable Future,* Eds. Johansson, T., Patwardhan, A., Nakicenovic, N., Gomez-Echeverri, L., Cambridge University Press, Cambridge, UK and New York; International Institute for Applied System Analysis, Luxenburg, Austria.

24. Olivério, J.L., Do Carmo, V.B., Gurgel, M.A. (2010). *27^0 Congresso da Sociedade Internacional de Tecnologias da Cana de Açúcar,* Vera Cruz, Mexico, 11 de Março de 2010.

25. Ministério da Agricultura, Pecuária e Abastecimento. Estatísticas (2013). *Estatísticas e Dados Básicos de Economia Agrícola,* April 2013, http://www.agricultura.gov.br/vegetal/estatisticas.

26. PDE (2012). *Plano Decenal de Expansão de Energia 2021,* prepared by the National Planning Energy Agency, EPE, Brasilia, Brazil, 2012.

27. Empresa de Pesquisa Energética (2012b). *Plano Decenal 2012–2021.*

28. Companhia Nacional de Abastecimento (CONAB) (2011). A Geração Termoelétrica com a Queima do Bagaço da Cana-de-Açúcar no Brasil-Análise do Desempenho da Safra 2009/10, Março 2011.

29. Ecofrota (2012). Jorge, Eduardo, MEIO AMBIENTE, 2do. Workshop Conjunto dos Programas BIEN-BIOTA-Mudanças Climáticas, FAPESP, SP, 23 de Agosto de 2012.

30. REN21 (2012). *Renewables 2012: Global Status Report, Renewable Energy Policy Network for the 21st Century.*

31. Chum, H., Faaij, A., Moreira, J.R. (2011). Bioenergy, in *IPCC Special Report on Renewable Energy Sources and Climate Change Mitigation*, prepared by Working Group III of the Intergovernmental Panel on Climate Change, Eds. Edenhofer, O., Pichs-Madruga, R., Sokona, Y., Seyboth, K., Matschoss, P., Kadner, S., Zwickel, T., Eickemeier, P., Hansen, G., Schlömer, S., von Stechow, C., Cambridge University Press, Cambridge, UK.

32. Moreira, J.R., Pacca, S.A., Parente, V. (2013). The future of oil and bioethanol in Brazil, *Energy Policy*, http://dx.doi.org/10.1016/j. enpol.2013.09.055.

33. World Bank (2010). *Towards Sustainable Decommissioning and Closure of Oil Fields and Mines: A Tool Kit to Assist Government Agencies*, World Bank multistake-holder initiative, World Bank, Washington, DC, http://siteresources.worldbank.org/EXTOGMC/ Resources/336929-1258667423902/decommission_toolkit3_-full. pdf.

About the Author

José Roberto Moreira is professor of energy at the University of Sao Paulo, Brazil.

Prof. Moreira joined the the University of Sao Paulo in 1971 as professor of physics. He then joined Princeton University in 1979–1980 as an energy expert. In 1987, he moved to the Institute of Energy and Environment at the University of Sao Paulo, where he launched a graduate course in energy/environment. He was associated with the Intergovernmental Panel of Climate Change (IPCC) from 1994 to 2011 and was one of the 2007 Nobel Peace Prize laureate. He has worked for several other United Nations programs and is author of more than 200 books and papers in the areas of nuclear physics, atomic physics, energy conservation, energy planning, and environment.

Prof. Moreira has also been the director of Companhia Energetica de São Paulo (1983–1987), Secretary of Energy at the Ministry of Mines and Energy (1985–1986), and executive director of Biomass Users Network (1992–1997). Currently, he is director of NEGAWATT, a private engineering company.

Chapter 19

Biomass Conversion to Advanced Biofuels: A Short Review on Second Generation Processes to Produce Ethanol

Dario Giordano,[a] Alina Carmen Tito,[b] Valentina Tito,[b] Piero Ottonello,[b] and Stefania Pescarolo[b]

[a]Beta Renewables S.p.A., Mossi Ghisolfi Group, Strada Ribrocca 11, 15057 Tortona (AL), Italy
[b]Biochemtex S.p.A., Strada Savonesa 9 Blocco D, Rivalta Scrivia (AL), Italy
dario.giordano@gruppomg.com

19.1   Introduction

The production of advanced biofuels, starting from lignocellulosic biomasses by second-generation technologies, represents an attracting chance to produce green liquid transportation fuels, biochemicals, and electricity from locally available forestry and agricultural waste.

Currently, the majority of biofuels for gasoline substitution are produced using sugars extracted from agricultural feedstock, like sugarcane juice, or by converting starch into sugars mainly from

Biomass Power for the World: Transformations to Effective Use
Edited by Wim van Swaaij, Sascha Kersten, and Wolfgang Palz
Copyright © 2015 Pan Stanford Publishing Pte. Ltd.
ISBN 978-981-4613-88-0 (Hardcover), 978-981-4669-24-5 (Paperback), 978-981-4613-89-7 (eBook)
www.panstanford.com

edible grains (Fig. 19.1). The sugars are then fermented into ethanol using yeast [1].

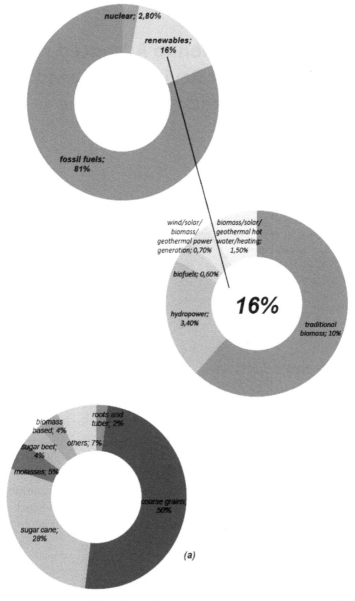

Figure 19.1 Renewable energy share of global energy consumption, 2009, and (a) details about different feedstock used for ethanol production. *Source: Renewables 2011 Global Status Report.*

Nowadays about 85 million barrels of crude are processed and used to meet the energy needs of the world. The demands for crude oil are projected to increase to 116 million barrels by 2030 [2]. Since this may result in depletion of crude oil reserves in the world, it is imperative to consider energy sources alternative to crude oil [3]. So the worldwide interest in biofuels and biochemicals produced using nonedible feedstock such as lignocellulosic biomass is growing.

The biomass conversion technology choice, together with feedstock supply chain development, plays a key role in the commercialization of next-generation biofuels. In fact the production of second-generation biofuels is even more challenging than producing first-generation ones due the complexity of the biomass and issues related to producing, harvesting/picking up, and transporting less dense biomass to centralized biorefineries. In addition to this logistic issue, other challenges with respect to processing steps in converting biomass to liquid transportation fuel, like pretreatment, hydrolysis, fermentation, and fuel separation, still exist and are discussed in this chapter.

The conversion technologies for biobased materials to advanced biofuels can be broadly classified into two major categories, biochemical and thermochemical conversions.

Biochemical conversion describes the transformation of biomass by microorganisms or enzymes into simple sugars and intermediates. Thermochemical conversion describes the thermal breakage of biomass obtained through the use of heat in the biomass and includes torrefaction, pyrolysis, liquefaction, combustion, and gasification [4].

Both technologies are not a "one size fits all"; many possible configurations exist for each conversion approach. In biochemical conversion there are a multitude of pretreatment approaches, as well as fermentation approaches. In thermochemical conversion there are a multitude of gasification, as well as fuel synthesis options, in addition to multiple pyrolysis to fuels options.

The thermochemical biomass conversion process is complex and uses components, configurations, and operating conditions which are more typical of petroleum refining.

The biochemical way, if the process is well designed, has the potential to be scalable, controllable, and economically sustainable. It is in fact important that all the unit operations integrate into an efficient overall conversion process. Whereas performance targets for individual unit operations are defined as levels of conversion at specific conditions, overall integration targets are defined as overall cost, both capital and operating, efficiency, and availability. Ideally, the overall conversion process should be as simple as possible and robust, providing maximum availability.

For any sophisticated conversion process, combining individual unit operations into a complete, integrated, efficient process is a significant challenge. The integration challenge is to design and validate a conversion process which affords maximum efficiencies and operational robustness.

The objective of the present chapter is to illustrate the state-of-art of biochemical conversion technologies which are potentially ready or are already being deployed for large-scale applications, presenting also the process design, integration, and scale-up.

19.2 Thermochemical and Biochemical Processes

The production of liquid and gaseous fuels from lignocellulosic feedstock can be performed both through thermochemical and through biochemical conversion approaches (Fig. 19.2).

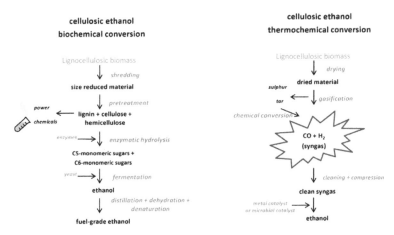

Figure 19.2 The general biomass-to-biofuel (ethanol) conversion pathways.

Biochemical conversion and chemical conversion typically transform the biomass into sugars as intermediates. Thermochemical conversion uses heat to convert the biomass into building blocks, such as carbon monoxide (CO) and hydrogen (H_2), which can be used for the synthesis of fuels. Other thermochemical processes include pyrolysis and liquefaction.

In detail, the *thermochemical processes* convert the organic matter into a mixture of solid, liquid, and gaseous products, whose properties strictly depend on the pretreatment conditions, feedstock, and downstream processing parameters.

The main biomass thermochemical conversion processes are classified as torrefaction, pyrolysis (fast, intermediate, and slow), hydrothermal liquefaction, and gasification.

Torrefaction is a biomass upgrading and energy-densifying pretreatment step in which the lignocellulosic material is processed at 200°C–300°C in the absence of air/oxygen for a certain time. Biomass is thus converted into a hydrophobic product, showing an increased energy density and brittleness (which makes the biomass easy-grindable).

Pyrolysis is a process which decomposes biomass at 300°C–600°C in the absence of air/oxygen. Lowering the process temperature and extending the vapor residence time (*slow pyrolysis*, or *carbonization*) maximize the production of char, the pyrolysis solid product, while increasing both temperature and residence time (*fast pyrolysis*) promotes gas stream production. Thus, varying the operating parameters (including the downstream steps such as vapor condensation), the relative amount of solid (char), liquid (pyrolysis oil or bio-oil), and gaseous products varies, as well as the bio-oil properties. The bio-oil can be upgraded into a transport fuel through catalytic or hydro-de-oxygenation routes.

Hydrothermal liquefaction is a process in which organic material is fed in a wet state to a high-pressure (order of a hundred bars) and high-temperature (300°C–400°C) reactor. The product is less oxygenated than pyrolysis oil and it shows more favorable characteristics for downstream processing and use either as a fuel or chemical precursor, but process conditions are severe, and that is a technological challenge.

Gasification occurs when, in a range of temperature higher than that of pyrolysis (800°C–1500°C or above), the biomass is converted into a H_2- and CO-rich gaseous product with a composition depending on the reactor configuration, process conditions, and gasification medium.

According the final application, the conversion of the produced gas into a syngas fuel can be needed whose composition (e.g., H_2–CO ratio) is suitable for downstream processing (this is always required in the liquid fuel production).

Biochemical conversion of biomass involves use of enzymes, bacteria, and microorganisms to break down biomass into gaseous or liquid fuels, such as biogas or bioethanol.

The most popular biochemical technologies are anaerobic digestion (or biomethanation) and fermentation.

Biofuels can be produced, applying bioprocesses to biobased materials, through first- or second-generation technologies.

Compared to first-generation biofuels from sugars and starch (constituted of C_6 sugars), second-generation production from lignocellulosic materials requires additional processing steps, as the crystalline structure of cellulose makes it highly insoluble and recalcitrant to enzymes action and as the hemicellulose and lignin provide a protective shield around the cellulose, which has to be modified or removed before efficient hydrolysis can occur. Polysaccharides (long carbohydrate molecules) are very stable.

Ethanol production technologies based on biochemical conversion of lignocellulosic material (constituted of C_5 and C_6 sugars and lignin) has been largely studied and developed at different stages. In general, in those technologies, the biomass undergoes a size reduction operation (e.g., from simple cutting or chipping to milling or grinding, depending on the conversion technologies), followed by a pretreatment step. Then, the feedstock is hydrolyzed to fermentable sugars, using both acids and commercial enzymes, and fermented to fuel molecules using different microorganisms or bacteria. Fuel is separated from the components of fermentation broth by a distillation/separation technique (Fig. 19.3).

In contrast to thermal conversion processes, bioconversion processes are less energy intensive and take place at near atmospheric pressure and low temperatures (30°C–80°C). The product can be gaseous and liquid fuels and other chemicals. There is a large scope for improvement in conversion efficiencies through the incorporation of modern processes and technologies such as continuous fermentation, new enzyme cocktails, new genetically modified microorganisms, etc.

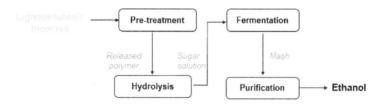

Figure 19.3 Ethanol production from biochemical conversion of lignocellulosic biomass.

Actually, main efforts are focused on the development of advanced biochemical conversion routes able to improve the efficiency and economics of the entire process at a commercial scale.

19.3 Biochemical Process Overview

This section is focused on the biochemical processes for converting lignocellulosic biomass to ethanol in an advanced biorefinery. The processes discussed here occur at the end of the supply chain, when the biomass has been delivered to/stored in the biorefinery.

Some feedstocks have to be washed to remove inorganic and other undesirable materials before pretreatment. Whether washing is needed depends on the source and the manner of storage before the feedstock is delivered to the conversion facility. The biomass is then chopped or ground to the desirable size range to feed into the pretreatment stage. The extent of grinding and size reduction will depend on the type of biomass and the pretreatment technology being used. Lignocellulosic feedstock can be chopped or ground with existing forestry or agricultural techniques. After this feedstock preparation step, pretreatment is needed to release cellulose from the lignin shield, followed by saccharification (breaking down of the cellulose and hemicellulose by hydrolysis to sugars, such as glucose and xylose), fermentation of sugar to ethanol, and distillation to separate the ethanol from the dilute aqueous solution.

19.3.1 Pretreatment

In the *pretreatment* stage the goal is to destroy the lignin shell protecting cellulose and hemicellulose. At the same time the crystallinity of the cellulose is usually decreased. Without breaking lignin, sugar-containing materials would not be accessible for the hydrolysis process. Various side products which act as inhibitors for hydrolysis and/or fermentation steps may be released or generated (e.g., organic acids, aldehydes, or inorganic salts).

The pretreatment process is a crucial process step: depending on its method, the structure of lignocellulosic biomass is broken down and cellulose is made more accessible to enzymatic attack which coverts carbohydrate polymers into fermentable monomeric sugars (Fig. 19.4).

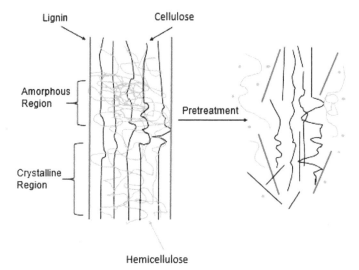

Figure 19.4 Effect of the pretreatment stage on the destructuring biomass. See also Color Insert.

Prominent pretreatment methods include:

(a) *Physical pretreatment*: Size reduction by milling [5] and extrusion at elevated temperature [6]

(b) *Chemical pretreatment under alkaline conditions*: Ammonia fiber expansion (AFEX) [7], ammonia percolation process (ARP) [8], soaking aqueous ammonia (SAA) [9], NaOH [10], lime [11], and alkaline hydrogen peroxide [12])

(c) *Chemical pretreatment under neutral conditions*: Supercritical water, ozonolysis [13], ionic liquids [14], and liquid hot water [15]

(d) *Chemical pretreatment under acidic conditions:* Organosolv under acid conditions [16], organic acid [17, 18], and concentrated acid [19]

(e) *Physicochemical pretreatment*: Supercritical CO_2 [20, 21]

(f) *Biological pretreatment* [22, 23]

Each method generates a different pretreatment stream [24].

Physical pretreatment was found to be effective in reducing cellulose crystallinity, but it requires more cost for power supply, and it gives all the three major compounds in one product stream [25].

Chemical methods are also effective pretreatment procedures, but they need more energy and chemicals than biological processes and may cause secondary pollution problems [26].

Dilute acid hydrolysis has been successfully developed for the pretreatment of lignocellulosic materials.

Dilute sulfuric acid pretreatment can achieve high reaction rates and significantly improve cellulose hydrolysis [27]; in a moderate temperature range, direct saccharification is disadvantaged for low yield because of sugar decomposition, whereas at higher temperature in dilute acid, treatment is favorable for cellulose hydrolysis [28]. Although dilute acid pretreatment can significantly improve the cellulose hydrolysis, its cost is usually higher than some physicochemical pretreatment processes such as steam explosion or autohydrolysis which are not making use of chemicals. Some alkalis can also be used for the pretreatment of lignocellulosic material and the effect of alkaline pretreatment depends upon the lignin content of the materials [28, 29]. The reacting way of alkaline hydrolysis is believed to be saponification of intermolecular ester bonds crosslinking xylan hemicellulosic and other components, for example, lignin and other hemicellulose. The porosity degree of the lignocellulosic material increases with the removal of the crosslinks [30]. Dilute alkaline treatment of lignocellulosic materials causes material swelling, leading to an increase in the internal surface area, a decrease in the degree of polymerization and crystallinity, separation of structural linkages among carbohydrates and lignin, and disruption of the lignin structure [29, 31].

AFEX pretreatment for various lignocellulosic materials shows better performance [33] but ammonia makes the process expensive and it also causes secondary pollution problems. Furthermore an alkali does not attack the hemicellulose structure, requiring a more complex enzymatic cocktail for depolymerization in the subsequent processing phase. Although all these methods, in general, have the potential for cellulose hydrolysis, they usually involve complicated procedures or are economically unfeasible [34].

Among *physicochemical procedures*, steam explosion is considered one of the most cost-effective pretreatment processes for woods and agricultural waste, but it shows limitations due to incomplete disruption of the lignin–carbohydrate matrix and, if the conditions are too severe, generates compounds which may be inhibitory to microorganisms used in downstream processes [32]. To remove the inhibitory products, pretreated biomass needs to be washed by water but washing decreases the overall saccharification yields.

Pretreatment methods having potential commercialization should meet the criteria listed below [35]:

(a) Allow the opening of the cell walls and carrying of lignin to the surface, reduction of cellulose fiber crystallinity, and increase of material porosity and access to hydrolytic agents.

(b) Generate lesser amounts of degradation products, which are toxic for downstream processes (enzyme hydrolysis and microbial fermentation). Processes which generate a large amount of toxic degradation products require a large amount of water to remove the toxins from the pretreated biomass, making the process more expensive.

(c) Scale up to meet the biorefinery needs of handling more than 1000 tons/day or more. The capital cost of a pretreatment reactor could increase depending on the pretreatment operating conditions and the catalyst used. For example, microwave radiation or gamma irradiation requires special reactors and also should undergo several additional safety regulations before this technology can get commercialized. Acid pretreatment requires Hastelloy-made reactors to overcome corrosion, while most other pretreatments require reactors made of stainless steel.

(d) Use less energy and so less process costs and vice versa.

(e) Not foresee the use of chemicals or use cheap chemicals. Using an expensive chemical like ionic liquid requires additional recovery operations, which increases the processing costs.

(f) Conserve the lignin during pretreatment. Part of the lignin can be used as a power source to drive different processing steps in the biorefinery, and in general lignin which can be obtained clean and without functionalization after the processing steps is ideal feedstock for a number of other conversion technologies aimed at producing high-added-value fuels and biochemicals. Pretreatments like ozonolysis and alkaline hydrogen peroxide tend to degrade lignin and hence the energy density of lignin will reduce; strong acid treatment such as sulfuric acid digestion will produce a sulfonated lignin of reduced value.

(g) Require moderate temperatures and pressures. This has implications on the cost of the reactor and the safety measures which companies have to take before these technologies can be implemented. Some of the pretreatment processes which use supercritical fluids (water and CO_2) operate at very high pressure and require additional cost (e.g., reactor construction).

(h) Use less hazardous chemicals. For example, chemicals like hydrofluoric acids should undergo additional safety steps to avoid accidents during biomass processing, which will make high the processing cost.

(i) When a catalyst is required allow catalyst recovery. Although this will slightly increase the processing cost, there will be overall energy saving, since less amounts of chemicals will be used in the process. Moreover, most of the catalysts (acid or base) are soluble in water and end up in the waste stream. They should be neutralized with a base or an acid. Since most of the biorefinery will be reusing the water, removing the neutralization-produced salts from the water will eventually increase the processing cost.

Pretreatment can drastically change the properties of the pretreated material (specific surface area [SSA], cellulose crystallinity index, degree of polymerization, lignin content in biomass, acetyl content in biomass, etc.). Effective pretreatments increase the rate of enzyme hydrolysis and significantly decrease the amount of enzymes needed to convert the biomass into monomeric sugars, which can be utilized by microorganisms.

19.3.2 Hydrolysis

The aim of hydrolysis is to break down the carbohydrate polymers of cellulose and hemicellulose into monomeric sugars. This process is also named saccharification.

During hydrolysis, cellulose and hemicellulose polymers are broken down into monomeric sugar molecules, which can be subsequently fermented. Due to the different structure of the hexose-dominated cellulose and the pentose-dominated hemicellulose, the hydrolysis process is very complicated. Hydrolysis of hemicellulose is usually easier due to its less crystalline structure.

There are two main hydrolysis procedures producing sugar monomers from cellulose for the subsequent fermentation. These comprise acid hydrolysis (with dilute or concentrated acids) and enzymatic hydrolysis [36]. For the time being, the main hydrolysis technology for biofuel production is based on enzymes due to environmental and economic reasons.

In fact acid hydrolysis requires severe conditions (high temperature and low pH), which results in corrosive conditions, so special construction material are needed.

A large number of inhibitors (e.g., different phenolic compounds, aliphatic acids, such as acetic acid, formic acid, and levulinic acid), which may inhibit both microbial growth and product yield, are formed during acid-catalyzed hydrolysis (so an additional inhibitors removal step is needed).

Diluted acid hydrolysis is the most commonly applied method among the chemical hydrolysis methods and it can be performed in either a one-stage or a two-stage process. Dilute acid hydrolysis can be used as a pretreatment process followed by enzymatic hydrolysis or as a method to hydrolyze lignocellulose material to sugars. One of the first established dilute acid hydrolysis processes was the *Scholler process*. It was a batch process in which woody material was kept in 0.5% sulfuric acid at 11–12 bar for about 45 min.

Actually most of dilute acid hydrolysis processes are performed in batch mode with a residence time of a few minutes [37].

Degradation of sugars and formation of unwanted by-products are the main disadvantages of one-stage dilute acid hydrolysis. This brings lower sugar yields and inhibition of ethanol production during the fermentation [38]. Potential inhibitory compounds are furfural, 5-hydroxymethylfurfural (5-HMF), acetic acid, formic acid, uronic acid, 4-hydroxybenzoic acid, vanillic acid, phenol, formaldehyde, etc. Some inhibitors are already present in the biomass (e.g., acetic acid, as acetyl groups); however most of the inhibitors are formed during the hydrolysis process due to sugar degradation. Although several detoxification methods, such as adsorption and lime treatment, have been devised, an appropriate strategy for efficient hydrolysis of lignocellulose to fermentable sugars is still lacking [39].

To avoid formation of inhibitors and degradation of sugars at high temperatures, dilute acid hydrolysis can be carried out in two or more steps. In the first step, hemicellulose is transformed into sugar monomers. As hemicellulose has an amorphous structure, hydrolysis occurs under mild conditions. This step is equivalent to a dilute acid pretreatment step. In general, the maximum yield of pentoses and hexoses recovered from hemicellulose in the first-stage hydrolysis is high (i.e., 80%–95% of the hemicellulose sugars available). However, the yield of cellulose hydrolysis to glucose is not so high—40%–60%). In the second step, the remaining solid portion is hydrolyzed in harsher conditions, which allows cellulose to be hydrolyzed.

The distinct step for hydrolysis of the hemicellulose and cellulose results in higher total sugar yields and less by-product formation since the solid cellulose fraction is separated from the liquid (the hydrolyzed hemicelluloses fraction) before the second hydrolysis step. From the second-stage hydrolysis step a product with high glucose content can be obtained. This stream can then easily be fermented to ethanol or combined with the pentose-rich hemicellulose fraction and then fermented.

Usually batch reactors are used for acid hydrolysis processes.

Hydrolysis of lignocellulosic biomass by concentrated sulfuric or hydrochloric acids has a long history. In 1819, Braconnot discovered that cellulose can be converted to fermentable sugars by concentrated acids. The possibility to dissolve and hydrolyze cellulose by sulfuric acid followed by dilution with water was cited in the literature already in 1883.

In 1948, a concentrated sulfuric acid hydrolysis treatment was commercialized in Japan where a membrane-based system was used to separate sugars from acid.

The *concentrated acid* is able to disaggregate the hydrogen bonds among cellulose chains and convert them into an amorphous form. Once the cellulose has been decrystallized, it generates a homogenous gelatin with the acid; therefore at this point cellulose is susceptible to be hydrolyzed. To avoid high degradation, dilution with water at low temperatures allows complete and quick hydrolysis to glucose.

In general, concentrated acid processes allow higher sugar yield compared to dilute acid ones. In addition, the concentrated acid processes can operated at lower temperatures. However, the concentration of acid is high (30%–70%) and dilution and heating of the acid during the hydrolysis make it corrosive. The process needs special construction materials (ceramic or carbon-brick lining). In addition, environmental impacts confine the application of hydrochloric acid. The concentrated acid has to be recovered after hydrolysis to make the process economically feasible. High investment and maintenance costs limit the potential commercial interest and research about this process.

Enzymatic hydrolysis of cellulose and hemicellulose, performed by fungal or bacterial enzymes (cellulases and hemicellulases), is another alternative to acid hydrolysis to produce sugars monomers.

Unlike acid hydrolysis, enzymatic hydrolysis is carried on at milder conditions (around pH 5.0 and 45°C–55°C). For this reason, enzymatic hydrolysis doesn't cause corrosion problems or generate inhibitors. However, this process has the disadvantage of being a slow process, and coupled with the relatively high cost of enzymes, it has previously been regarded as a bottleneck for biofuel production, but today numerous recent achievements in the enzyme discovery and production techniques are making available new generation of enzymes which can make the lignocellulosic process economically attractive.

Cellulases are within the glycosyl-hydrolase families based on their sequence homology and hydrophobic clusters analysis, and traditionally the degradation of cellulose to glucose has been regarded as a synergistic action of mainly three classes of enzymes (Fig. 19.5):

(a) Endoglucanases (EGs) (EC 3.2.1.4),[1] which randomly cleaves internal β-1,4-glucosidic linkages in the cellulosic chain, forming two new chain ends

(b) Cellobiohydrolases (CBHs), known as exogluconases, (EC 3.2.1.91), which are the most abundant components in most naturally occurring cellulose systems, progress along the cellulose, and cleave off cellobiose units from the ends of the cellulosic chain in a processive manner

(c) β-glucosidases (BGs), known as β-glucoside glucohydrolases, (EC 3.2.1.21),[2] which hydrolyze cellobiose to glucose and cleave off glucose units from cello-oligosaccharides

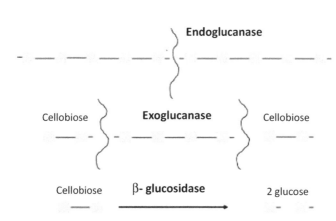

Figure 19.5 Degradation of cellulose to glucose.

---

[1]Enzyme Commission. Enzyme nomenclature of the International Union of Biochemistry and Molecular Biology Accepted name: Cellulose
Reaction: Endohydrolysis of (1→4)-β-D-glucosidic linkages in cellulose, lichenin, and cereal β-D-glucanscellobiohydrolases
[2]Enzyme Commission. Enzyme nomenclature of the International Union of Biochemistry and Molecular Biology Accepted name: β-glucosidase
Reaction: Hydrolysis of terminal, nonreducing β-D-glucosyl residues with the release of β-D-glucose

Recently, though, a new kind of enzyme was identified as an important constituent of efficient enzymatic hydrolysis [42]. The enzyme works, as opposed to the other hydrolytic enzymes, by oxidizing carbon in the glucan chain, and in such a way it generates two new free chain ends for the CBHs to work on. These new enzymes give the chance to break the crystalline cellulose structure and hence enhance the enzymatic hydrolysis rate [43].

19.3.3 Fermentation

Fermentation is the biological process using a wide range of microorganisms to convert sugars into ethanol. The most common fermentation is the conversion of 1 mole of glucose into 2 moles of ethanol and 2 moles of carbon dioxide. However, glucose consumed to generate additional yeast cells (cell mass) does not result in the production of ethanol. Also other hexoses (fructose and galactose) may be converted in a similar way.

Most industrial fermentation processes operate at 90%–95% of the theoretical yield of ethanol from glucose fed to the yeast.

A wide category of yeasts, bacteria, and fungi are reported to produce ethanol through fermentation as the main product. Among fermenting microorganisms, yeasts—and in particular bakery yeast (*Saccharomyces cerevisiae*)—are most common in today's ethanol industry. Ethanol production comes primarily under anaerobic conditions in which a fermentative end product is needed to regenerate the cofactor nicotinamide adenine dinucleotide (NAD+). However, also aerobic formation of ethanol can take place in so-named Crabtree-positive yeasts.

Another well-known ethanol-producing microorganism is the bacterium *Zymomonas mobilis*, which produces this alcohol at high yield, which is a consequence of its use of an alternative pathway in the catabolic production of pyruvate from glucose (Entner–Duodoroff pathway instead of the more common Embden–Meyerhof–Parnas pathway).

The simultaneous conversion of both cellulose and hemicellulose for production of fuel ethanol is being studied intensively with a view to develop a technically and economically viable bioprocess. The fermentation of glucose, the main constituent of cellulose hydrolyzate, to ethanol can be carried out efficiently. On the other hand, although bioconversion of xylose, the main pentose sugar obtained on hydrolysis of hemicellulose, to ethanol presents a biochemical challenge, especially if it is present along with glucose, it needs to be fermented to make the biomass-to-ethanol process economical. A lot of attention therefore has been focused on the utilization of both glucose and xylose to ethanol, and significant developments have been made in the area of simultaneous hexose and pentose sugars to ethanol.

Fermentation for ethanol production at the industrial scale uses one of two major technologies, batch and continuous (or semicontinuous) fermentation. In batch fermentation, the process occurs in a single, large reactor, which is filled with the sugar-rich liquid and living cells (*inoculum*), allowed to ferment, drained, and then cleaned in preparation for the next batch. There are usually multiple reactors in batch operations, each in a different stage of the process (filling, fermenting, emptying, etc.). While the exact size of each fermenter varies between plant designs, usual fermenter size ranges between 1 and 2 million liters each [44]. The major advantage for batch-mode fermentation is that it affords less contamination problem by proper sanitization between trials. Bacteria, especially species of *Lactobacillus*, can infect yeast-driven fermentation and produce organic acids which decrease ethanol yield and interfere with the *S. cerevisiae* activity [45]. In continuous fermentation, the process occurs through a tank cascade where the liquid continuously flows through the process. Fresh fermentation media is continuously added at the front end, and fermented beer is continuously removed from the back end. While continuous fermentation has greater reactor productivity because it is continuously operating with high yeast loadings, much more operating care is needed to prevent contamination.

The following process configurations are applied in the fermentation process (Table 19.1):

(a) *Separate hydrolysis and hermentation* (SHF). This process foresees hydrolysis separately from fermentation. As hydrolysis and fermentation steps are performed in separate reactors, each operation can be carried out under optimal conditions. This configuration therefore should enable more efficient performance of the enzymes. However, a drawback is that the rate of hydrolysis is typically reduced as the activities of some of the cellulases may be inhibited by hydrolysis products (e.g., cellobiose or glucose), thus resulting in a lower yield compared to other approaches.

(b) *Simultaneous saccharification and cofermentation* (SSCF). This process performs hydrolysis and fermentation in one reactor, and sugars (both hexoses and pentoses) produced in hydrolysis are simultaneously fermented. For the cofermentation of pentoses and hexoses, a microorganism capable of both conversions is needed. However, in comparison to SHF, the process is run under nonoptimal conditions, in particular for the enzymatic hydrolysis which takes place at a lower temperature than optimal.

(c) *Consolidated bioprocessing* (CBP), which accomplishes cellulase production, hydrolysis, and fermentation in one reactor by using just one microbial consortium.

Table 19.1 Comparison between different fermentation process configurations

Process	Advantages	Disadvantages
SHF	Hydrolysis and fermentation at optimum conditions	Inhibitory effect Increased contamination
SSCF	Low quality of enzyme input High ethanol yield Reduced foreign contamination Less inhibitory effects Less contamination risk Shorter process time	Either hydrolysis or fermentation under optimal conditions Difficulty in process control
CBP	Cost effective Energy efficient	Lack of suitable organisms Difficulty in process control

19.3.4 Distillation

Distillation is a widely applied process of upgrading ethanol from lower concentration into pure ethanol and is carried out in different steps:

1. Evaporation of ethanol from beer (crude ethanol with ~45% concentration)
2. Purification (ethanol concentration increased to concentration close to the azeotrope, that is, ~96.5%)
3. Dehydration (removal of azeotropic water to reach ~99.8 ethanol concentration)

Distillation is a separation technique which is based upon differences in volatility. If a mixture of ethanol and water is placed in a container at a given temperature and pressure, after some time the mixture will reach equilibrium. At equilibrium, some of the ethanol will be a vapor in the gas above the liquid and some will be in the liquid phase. Similarly, some of the water will be in the vapor phase and some will be in the liquid phase. Because ethanol is more volatile than water (boils at a lower temperature), the ratio of ethanol to water in the vapor phase is greater than the ratio of ethanol to water in the liquid phase. This characteristic allows for the separation of the ethanol from the water. Through subsequent vaporization of the mixture, condensation, revaporization, and recondensation the mixture becomes higher and higher in ethanol content because the vapor at each vaporization step has a higher ethanol concentration than the liquid from which it was vaporized. Thus, multiple fractionation steps can be used to purify ethanol from water. However, the basic principle by which this occurs (difference in boiling points between water and ethanol) ceases to exist when the mixture is 95.6 wt.% ethanol. At this point, named azeotrope, the ethanol and water both vaporize to the same degree and cannot be further fractionated by distillation. Either a third solvent can be introduced to break the azeotrope, or an alternative separation principle can be applied, such as absorption. The most popular method is absorbing water away from ethanol using molecular sieves. Industrial fractional distillation to produce fuel ethanol is one of the major energy inputs for the production of fuel ethanol. A process improvement which captures and recycles energy from the process has greatly reduced the cost of this step. All industrial fuel ethanol production uses continuous-feed distillation column systems.

19.4 Process Design, Integration, and Scale-Up

Large efforts are dedicated to implementing a biochemical conversion pathway for the production of biofuels from lignocellulosic material through a biochemical conversion process.

Demonstration projects for cellulosic biofuel are under development in several countries worldwide. Facilities demonstrate the capability of second-generation technology for continuous production and cover the entire production process. While only few production facilities are operational yet, many projects are under construction. There are a number of biomass feedstocks proposed for these facilities, including cereals residues (e.g., corn stover, corn cobs, wheat straw, rice straw), energy crops (e.g., *Panicum virgatum*, sorghum, energy cane, *Miscanthus x giganteus*, switchgrass), and woody materials (e.g., poplar, eucalyptus, birch).

The goals of the design of an industrial second-generation ethanol plant can be summarized in some essential points:

- Incorporate all component designs into an integrated and technoeconomic sustainable process package.
- Calidate the process technology improvements achieved during the course of R&D operations and realize the scale-up to a commercial plant.
- Confirm the economic viability of the process design.
- Ensure that the environmental, safety, health, and security requirements are fully incorporated and properly implanted into the project's design and execution and can be readily implemented for the plant.

The plant design should follow some general criteria (i.e., sizing safe margins to allow plant improvement and flexibility in operation mode), supported by solid background knowledge through technology scouting activities and experimental trials in the laboratory and the pilot plant to validate the second-generation technology.

The technical key aspects/criteria of design and plant scale-up can be summarized as:

(a) *Mass transfer*. Reactions, mixing, and any physical treatment which entail a change of state should be performed with same efficiency; a smaller size enhances intimate mixing and smaller systems are more effective as concerns mass transfer.

(b) *Heat transfer*. Convective heat transfer as well as heat transfer by direct contact govern the transfer of heat and the process temperature control (conduction and radiation have much smaller effect), and they are both dependent on an efficient mass transfer.

(c) *Mechanical energy transfer and mechanical feasibility.* The size of the machinery and equipment, in general, should stay within proven capability of equipment fabrication in order to ensure effective performance.

(d) *Process stream recycle and presence of undesired components.* Process plants entailing chemical reactions are subject to the presence of trace components which may not be identified as critical on the lab scale or simply may not be detected on the lab scale because of being purged with sampling. If stream recycle is present, trace components will accumulate; if such components are inhibitors, dangerous, or undesirable substances, they may give place to unexpected and undesired process conditions.

(e) *Accumulation of undesired components and fouling.* Lab-scale plants are never run for periods of time of length comparable to industrial plants; they are often stopped and opened for checks ad modifications, and risks of fouling with time are very difficult to appreciate. The process engineer should identify the areas of the plant where components are present which may give place to fouling, and the process design of the industrial plant should account for that.

(f) *Overpressures.* In process plants reactive mixtures may be present (typically, flammable mixtures of gases); the occurrence of overpressure (typically, an explosion) is known to be geometry dependent, that is, it is unlikely in small systems (propagation cannot take place) and increases its probability in larger systems (propagation can take place and presence of sources of ignition may be more likely). The scaled-up plant should be duly protected from the occurrence of overpressures.

(g) *Overheating.* In a lab unit the effect of heat losses to ambient is extremely important, and the occurrence of unexpected overheating may be difficult to be detected; the effect of heat losses is negligible on industrial plants, and the effect of overheating may be critical. The process design should allow for a generous temperature control system and process cooling wherever exothermic reactions may take place.

(h) *Suitability of equipment design.* In several cases the design of special equipment has extensive mechanical references in other industrial fields; in all those cases the adaptation of the features of the equipment to the bioethanol process should be studied in detail with the full cooperation of the equipment manufacturer.

For the plant scale-up the following are also required:

(a) Development of an extensible and sustainable biomass supply system

(b) Optimization of steam consumption in pretreatment and adjustment of stream configuration (recycles, energy integration)

(c) Improvements in biomass constituents separation (hemicellulose, cellulose, lignin) to obtain high-purity streams

(d) Optimization of enzymatic cocktail formulation to limit product inhibition phenomena and shorter hydrolysis residence times

(e) Development of fermenting strain (yeast) able to resist to high inhibitor concentration

(f) Technoeconomic evaluation of the integrated waste water treatment at different discharge limits

A process should be designed to guarantee a balance among all operation units (due to a complete process integration of all key units and an easy setup adaptation to process changes).

Not only the product (biofuel) but also by-products (i.e., lignin, biogas) are high-value-added compounds; for this reason, changes in biofuel production yield, due to different conditions in the biological conversion rate, should be taken into account.

Scale-up of a plant from pilot to demontration to industrial size is considered a fundamental pathway in order to reduce associated techno- and economic scale-up risk and to confirm the fully operability.

As regard, improvements in the entire concerned process unit can provide an optimization of the equipment size and a minimization of chemical and energy demand, with the aim of obtaining better performance and reducing the biofuel production cost. The interrelation between enzymes and process mechanisms is also important.

Further improvements to the cost–performance of the enzymes can and must be achieved as the lignocellulosic industry emerges.

It could be possible to design an enzyme product which is optimized to the process conditions and also identify modifications to process steps which can lead to an optimal use of the enzymes (e.g., adjustments of temperature or pH during the saccharification).

19.4.1   State of the Industrialization Step

Second-generation biofuels produced from lignocellulosic biomass have made so many advances in the last 10 years:

(a) The amount of pilot plants and demos in operation and under construction almost tripled in 2008–2014. Production capacity has increased by a factor of 10 to around 2.4 billion liters (Gl/y).

(b) The technologies for producing these biofuels are approaching maturity. Industrial biotechnology research has made significant progress, especially in the use of enzymes to process biomass. Estimates suggest that in 2008–2012, the cost of enzymes treatment per liter of lignocellulosic bioethanol produced fell by more than 70%.

(c) The global production cost of 0.7 €/liter for second-generation bioethanol has become an achievable near-term target, and it is close to the cost of the historically most economic processes (e.g., sugarcane-based processes) and the market prices seen in recent years.

(d) The first commercial lignocellulosic bioethanol production units came into operation in 2013.

The 2013 commissioning of the first European commercial unit (the Beta Renewables [M&G Group]-owned plant in Crescentino,

Italy), the completion of first US industrial-scale plants (DSM POET in Emmetsburg, Iowa; Abengoa in Hugoton, Kansas; DuPont in Nevada, Iowa), and the start-up of the first industrial installation in Brazil (GranBio, Alagoas, using PROESA™ technology) reflect the nascent maturity of new technologies and mark the initial steps of the second-generation bioethanol industry toward mass production.

Despite the indisputable technological advances achieved, this industry is still young and has more work to do on the long-term economic profitability of its production units and must therefore look beyond possible economic incentives, such as tax breaks, etc.

19.4.2   The Biochemtex Case: The Crescentino Demonstration Plant

The first plant in the world for the industrial production of second-generation bioethanol was started up in Crescentino, a province of Vercelli (Piedmont region, Italy), at the beginning of 2013 by Beta Renewables (M&G Group). At full capacity, the plant will produce 40,000 tons per year.

Crescentino is living proof that it is possible to produce biofuels from nonfood agricultural materials, transforming the cellulose and hemicellulose contained in biomass into simple sugars.

The technological and industrial result seemed out of reach but was turned into reality by the second-generation Beta Renewables **PRO**duzione di **E**tanolo da biomas**SA** (PROESA™) technology (ethanol production from biomass). The bioethanol produced in Crescentino is derived from wheat straw, rice straw, and *Arundo donax*, the common giant reed. This is the first step allowing the global market to open to processes which can produce low-cost ethanol, competitive with fossil fuels.

The PROESA™ technology has been adopted by other industrial projects under start-up, such as the Alogoas GranBio unit, and under development in several industrial sites around the world (which cannot be disclosed at the moment of the editing of this article).

References

1. Mohr, A., Raman, S. (2013). Lessons from first generation biofuels and implications for the sustainability appraisal of second generation biofuels, *Energy Policy*, **63**, 114–122.

2. International Energy Agency (IEA), World Energy Outlook World Energy Outlook, International Energy Agency, Paris, France, 2007.

3. Lee, R.A., Lavoie, J.-M. (2013). From first- to third-generation biofuels: challenges of producing a commodity from a biomass of increasing complexity, *Animal Frontiers*, **3**(2), 6–11.

4. Van Loo, S. (2008). *The Handbook of Biomass Combustion and Co-Firing*, Earthscan, London.

5. Hideno, A., Inoue, H., Tsukahara, K. et al. (2009). Wet disk milling pretreatment without sulfuric acid for enzymatic hydrolysis of rice straw, *Bioresource Technology*, **100**(10), 2706–2711.

6. Yoo, J., Alavi, S., Vadlani, P., Amanor-Boadu, V. (2011). Thermo-mechanical extrusion pretreatment for conversion of soybean hulls to fermentable sugars, *Bioresource Technology*, **102**(16), 7583–7590, 2011.

7. Teymouri, F., Laureano-Perez, L., Alizadeh, H., Dale, B.E. (2005). Optimization of the ammonia fiber explosion (AFEX) treatment parameters for enzymatic hydrolysis of corn stover, *Bioresource Technology*, **96**(18), 2014–2018.

8. Tae, H.K., Lee, Y.Y. (2005). Pretreatment and fractionation of corn stover by ammonia recycle percolation process, *Bioresource Technology*, **96**(18), 2007–2013.

9. Li, X., Kim, T.H., Nghiem, N.P. (2005). Bioethanol production from corn stover using aqueous ammonia pretreatment and two-phase simultaneous saccharification and fermentation (TPSSF), *Bioresource Technology*, **101**(15), 5910–5916.

10. Taherzadeh, M.J., Karimi, K. (2008). Pretreatment of lignocellulosic wastes to improve ethanol and biogas production: a review, *International Journal of Molecular Sciences*, **9**(9), 1621–1651.

11. Xu, J., Cheng, J.J., Sharma-Shivappa, R.R., Burns, J.C. (2010). Lime pretreatment of switchgrass at mild temperatures for ethanol production, *Bioresource Technology*, **101**(8), 2900–2903.

12. Banerjee, G., Car, S., Liu, T., et al. (2012). Scale-up and integration of alkaline hydrogen peroxide pretreatment, enzymatic hydrolysis, and ethanolic fermentation, *Biotechnology and Bioengineering*, **109**(4), 922–931.

13. García-Cubero, M.T., González-Benito, G., Indacoechea, I., Coca, M., Bolado, S. (2009). Effect of ozonolysis pretreatment on enzymatic digestibility of wheat and rye straw, *Bioresource Technology*, **100**(4), 1608–1613.

14. Dadi, A.P., Schall, C.A., Varanasi, S. (2007). Mitigation of cellulose recalcitrance to enzymatic hydrolysis by ionic liquid pretreatment, *Applied Biochemistry and Biotechnology*, **137–140**(1–12), 407–421.

15. Kim, Y., Hendrickson, R., Mosier, N.S., Ladisch, M.R. (2009). Liquid hot water pretreatment of cellulosic biomass, in *Biofuels: Methods and Protocols*, Vol. 581 of *Methods in Molecular Biology*, 93–102, Springer, Berlin, Germany.

16. Sun, F., Chen, H. (2008). Organosolv pretreatment by crude glycerol from oleochemicals industry for enzymatic hydrolysis of wheat straw, *Bioresource Technology*, **99**(13), 5474–5479.

17. Qin, L., Liu, Z.-H., Li, B.-Z., Dale, B.E., Yuan, Y.-J. (2012). Mass balance and transformation of corn stover by pretreatment with different dilute organic acids, *Bioresource Technology*, **112**, 319–326.

18. Gírio, F.M., Fonseca, C., Carvalheiro, F., Duarte, L.C., Marques, S., Bogel-Łukasik, R. (2010). Hemicelluloses for fuel ethanol: a review, *Bioresource Technology*, **101**(13), 4775–4800.

19. Galbe, M., Zacchi, G. (2002). A review of the production of ethanol from softwood, *Applied Microbiology and Biotechnology*, **59**(6), 618–628.

20. Kim, K.H., Hong, J., (2001). Supercritical CO_2 pretreatment of lignocellulose enhances enzymatic cellulose hydrolysis, *Bioresource Technology*, **77**(2), 139–144.

21. Schacht, C., Zetzl, C., Brunner, G. (2008). From plant materials to ethanol by means of supercritical fluid technology, *Journal of Supercritical Fluids*, **46**(3), 299–321.

22. Patel, J., Onkarappa, R., Shobha, K.S. (2007). Comparative study of ethanol production from microbial pretreated agricultural residues, *Journal of Applied Sciences and Environmental Management*, **11**, 137–141.

23. Wan, C., Li, Y. (2012). Fungal pretreatment of lignocellulosic biomass, *Biotechnology Advances*, **30**(6), 1447–1457.

24. Chandrakant, P., Bisaria, V.S. (1998). Simultaneous bioconversion of cellulose and hemicellulose to ethanol, *Critical Reviews in Biotechnology*, **18**(4), 295–331.

25. Cadoche, L., Lopez, G.D. (1989). Assessment of size reduction as a preliminary step in the production of ethanol from lignocellulosic wastes, *Biological Wastes*, **30**, 153–157.

26. Sivers, M.V., Zacchi, G. (1995). A techno-economical comparison of three processes for the production of ethanol from pine, *Bioresource Technology*, **56**, 43–52.

27. Esteghalalian, A., Hashimoto, A.G., Fenske, J.J., Penner, M.H. (1997). Modeling and optimization of the dilute sulfuric acid pretreatment of corn stover, poplar and switchgrass, *Bioresource Technology*, **59**, 129–137.

28. McMillan, J.D. (1994). Pretreatment of lignocellulosic biomass, in *Enzymatic Conversion of Biomass for Fuels Production*, 292–324, Eds. Himmel, M.E., Baker, J.O., Overend, R.P., American Chemical Society, Washington, DC.

29. Fan, L.T., Gharpuray, M.M., Lee, Y.H. (1987). *Cellulose Hydrolysis*, Vol. 3, 1–68, Springer-Verlag, Berlin, Germany.

30. Kumar, R., Singh, S., Singh, O.V. (2008). Bioconversion of lignocellulosic biomass: Biochemical and molecular perspectives, *Journal of Industrial Microbiology*, **35**, 377–391.

31. Saratale, G.D., Chen, S.D., Lo, Y.C., Saratale, R.G., Chang, J.S. (2008). Outlook of biohydrogen production from lignocellulosic feedstock using dark fermentation -a review, *Journal of Scientific and Industrial Research*, **67**, 962–979.

32. Mackie, K.L., Brownell, H.H., West, K.L., Saddler, J.N. (1985). Effect of sulfur dioxide and sulfuric acid on steam explosion of aspenwood, *Journal of Wood Chemistry and Technology*, **5**, 405–425.

33. Holtzapple, M.T., Jun, J.H., Ashok, G., Patibandla, S.L., Dale, B.E. (1991). The ammonia freeze explosion (AFEX) process: a practical lignocellulose pretreatment, *Applied Biochemistry and Biotechnology*, **28–29**, 59–74.

34. Mes-Hartree, M., Dale, B.E., Craig, W.K. (1998). Comparison of steam and ammonia pretreatment for enzymatic hydrolysis of cellulose, *Applied Microbiology and Biotechnology*, **29**, 462–468.

35. Venkatesh, B. (2014). Current challenges in commercially producing biofuels from lignocellulosic biomass, *ISRN Biotechnology*.

36. Saha, C.B., Iten, B.L., Cotta, M.A., Wu, Y.V. (2005). Dilute acid pretreatment, enzymatic saccharification and fermentation of wheat straw to ethanol, *Process Biochemistry*, **40**, 3693–3700.

37. Taherzadeh, M.J., Karimi, K. (2007). Enzymes-based hydrolysis processes for ethanol from lignocellulosic materials: a review, *Bioresources*, **2**, 707–738.

38. Palmqvist, E., Hägerdal, B.H. (2000). Fermentation of lignocellulosic hydrolysates. II: inhibitors and mechanisms of inhibition, *Bioresource Technology*, **7**(1), 25–33.

39. Aden, A. (2008). *Biochemical Production of Ethanol from Corn Stover: 2007 State of Technology Model*, Technical report NREL/TP-510-43205, May 2008.

40. McParland, J.J., Grethlein, H.E., Convers, A.O. (1982). Kinetics of acid hydrolysis of corn stover, *Solar Energy*, **28**, 55–63.

41. Vargas Radillo, J.J., Ruiz-Lòpez, M.A., Rdriguez Macias, R., Barrientos Ramìrez, L., Garcìa-Lopez, P.M., Lopez-Dellamary Toral, F.A. (2011). Fermentable sugars from Lupinus rotundiflorus biomass by concentrated hydrochloric acid hydrolysis, *Bioresources*, **6**, 344–355.

42. Horn, S.J., Vaaje-Kolstad, G., Westereng, B., Eijsink, V.G. (2012). Novel enzymes for degradation of cellulose, *Biotechnology for Biofuels*, **5**, 45.

43. Converse, A.O., Ooshima, H., Burns, D.S. (1990). Kinetics of enzymatic hydrolysis of lignocellulosic materials based on surface area of cellulose accessible to enzyme and enzyme adsorption on lignin and cellulose, *Applied Biochemistry and Biotechnology*, **24/25**, 67–73.

44. Kwiatkowski, J.R., McAloon, A.J., Taylor, F., Johnston, D.B. (2006). Modeling the process and costs of fuel ethanol production by the corn dry-grind process, *Industrial Crops and Products*, **23**, 288–296.

45. Thomsson, E., Larsson, C. (2006). The effect of lactic acid on anaerobic carbon or nitrogen limited chemostatic cultures of Saccharomyces cerevisiae, *Applied Microbiology and Biotechnology*, **71**, 533–542.

About the Authors

Dario Giordano is chief technology officer at Beta Renewables  S.p.A, Mossi Ghisolfi Group, Italy, and Biochemtex S.p.A., Italy. Dario joined Mossi Ghisolfi Group in 1990, where he has held positions of Start-up Manager, Process Engineer, and Technology Transfer Manager in Sinco Engineering S.p.A. and R&D European Director in Sinco Ricerche S.p.A. Currently, he is member of the Mossi Ghisolfi Executive Committee and Corporate Responsible for Research, Technology and Development.

Alina Carmen Tito is R&D Customers Manager at Biochemtex  S.p.A since 2012, and her main activities are the management of pilot-scale trials for biofuels (bioethanol) production for customers and partners. She also takes care of the technical mediation with clients.

Alina worked for Politecnico di Milano University (2005–2006) as scientific collaborator in the field of R&D, working on chemical and energetic valorization of polymers-based waste by pyrolysis technology, and then as researcher (2006–2010) working on carbon nanotubes production by ethylene. In Politecnico, she also led a team of six chemical engineers (PhD researchers), working on hydrogen production (Sulfur–Iodine cycle pilot plant and biomass gasification in a pilot-scale fluidized bed system) and green ethylene production (bioethanol dehydration process), from 2010 to 2012.

She was also Scientific consultant for several companies: Ecological Scrap Industry, Pace de Mela, Messina, Italy (2005–2008); Costech International, Piotello, MI, Italy (2006–2008); Radish Design, Fujian, China (2009–2012); and Turboden srl, Mitsubishi Heavy Industries, Brescia, Italy (2011–2012).

Valentina Tito is project funding leader at Biochemtex S.p.A. She started working for the Mossi Ghisolfi Group as environmental engineer in 2013.

In 2012 Valentina worked for ENEA, Italian National Agency for New Technologies, Energy and Sustainable Economic Development, as Researcher on International Standards of Environmental Certification and Corporate Social Responsibility with a focus on the evaluation of simplified models for the improvement of enterprises environmental performances, the eco-innovation of production systems and the increase of competitiveness in the global market. In 2013 she was volunteer for the United Nations Development Programs (UNDP) of Honduras, Tegucigalpa, for the development of a device for the production of sodium hypochlorite (for water disinfection) in charge of training and introduction to its use in 10 indigenous communities of Honduras. In 2013 she also participated at "Banco de Calidad" project funded by AEA Alianza Energìa Ambiente and finalized to the sustainability assessment of the renewable energy system realized by the institution in the rural communities of Central America (Honduras, Costa Rica, Nicaragua, Belize, and Guatemala).

Her main activities are preparation of projects submission of European and National Funding Programmes in the field of second-generation technologies for the production of biochemicals and bio-fuels from lignocellulosic biomass. She is Responsible of technical projects writing and coordination of administrative activities. She manages the contacts with partners involved in projects in order to develop innovative and structured proposals.

Piero Ottonello is R&D Proesa® Director at Biochemtex S.p.A. and he participated in the PROESA® project to develop technology for the production of second-generation bioethanol. He is also involved in the generation of key inventions on the entire process chain. Piero leads a team of project leaders working in development of processes to produce chemicals for renewable resource based on second- generation technology.

Piero joined the Mossi Ghisolfi Group in 2004 for Cobarr Reseach Center (R&D Application Projects) and was responsible for the commercialization of the process. He managed several projects that involve development of new polyester resins and new packaging solutions in polyester. He was also researcher of innovative solutions for food packaging (bottle and flexible film) with improved barrier against gases.

He also participated in R&D activities with increased level of complexity and integration with other functions inside the company, working in close partnership with the U.S. R&D center of the Group and the marketing team.

In 2006 Piero changed rule and became R&D project leader and his main activities were technical and economical scouting and evaluation of different biodiesel production technologies and preparation of a basic engineering package necessary for the technology transfer activity between the Italian and the foreign offices of the Group. He was also responsible for the commercialization of the process.

From 2007 to 2010 he was involved in the development of novel process for the production of ethanol from ligno-cellulosic material, coordinating the work necessary to bring the core technology from lab scale to pilot scale.

Stefania Pescarolo is head of Public Funding at Biochemtex S.p.A. and she works in the field of R&D Public Funding for Mossi Ghisolfi Group since 2008.

Stefania was General Technical and Scientific Secretary for Federchimica, Italy, and she supported chemical companies in the regulatory, technical, scientific and economic affairs, protecting the interests with public and private institutions, with national and international authorities. She also took care also of the creation of technical committees and round tables on key chemical issues.

Stefania worked for EFfCI (European Federation for Cosmetic Ingredients), Bruxelles, as General Secretary for Communication, and her main activities were dealing with European and international competent authorities (European Commission, EU Parliament, scientific committees), with international associations of various chemical sectors, with opinion leaders, trade fair organizers, and journalists related to the cosmetic sector.

Her main activities are evaluation, preparation, negotiation and management of project funding proposals, coordination of partnership in relation of funding project procedures, public relationship with Italian and international authorities and stakeholders, scouting of funding opportunities. She coordinates the participation in several working groups at European and international level on biofuel and biochemical sector (Biobases Consortium, EBTP, EIA, ESBF, Aviation WG, Cluster Nazionale della Chimica Verde Spring, etc.)

Chapter 20

Biomass to Liquid Fuels via HTU®

Frans Goudriaan and Jaap E. Naber
Biofuel B.V., Rendorppark 30, 1963 AM Heemskerk, the Netherlands
info@biofuel.com

20.1 Thermochemical Conversion of Biomass

The primary aim for the conversion of biomass is the production of transportation fuels as opposed to the direct combustion for the production of electricity and heat. For electricity there are many renewable energy alternatives such as hydroelectricity, solar power (including heat), and wind. This is not the case for transportation fuels, especially for aviation. Therefore much development work has been devoted to biochemical routes, mainly for ethanol manufacture, to the conversion of pure plant oils and to thermochemical routes for the production of diesel and kerosene components. For the latter the main challenge is the removal of oxygen, preferably as CO_2, in view of the required hydrogen content of the product. The main alternatives are:

- *High-temperature pyrolysis plus upgrading*
  - relatively low cost

Biomass Power for the World: Transformations to Effective Use
Edited by Wim van Swaaij, Sascha Kersten, and Wolfgang Palz
Copyright © 2015 Pan Stanford Publishing Pte. Ltd.
ISBN 978-981-4613-88-0 (Hardcover), 978-981-4669-24-5 (Paperback), 978-981-4613-89-7 (eBook)
www.panstanford.com

- requires dry biomass feedstock, necessitating inefficient moisture removal
- limited or no removal of oxygen, resulting in high hydrogen cost for upgrading
- near commercial

- *Gasification plus (Fischer–Tropsch) synthesis (biomass-to-liquid [BtL])*
  - relatively complicated, requiring very large scale, which may be in conflict with biomass collection
  - combination not commercially proven
  - requires dry biomass feedstock, necessitating inefficient moisture removal
  - superior diesel quality

- *Hydro Thermal Upgrading® (HTU®) plus upgrading through hydrodeoxygenation (HDO)*
  - can process all types of biomass, particularly wet organic residues
  - high degree of oxygen removal
  - produces high-quality diesel and kerosene after HDO
  - ready for commercialization

A particular challenge in the development and commercialization of biomass conversion technology is the cost of the initial commercial installations prior to upscaling and multiplication with standardization. It is therefore desirable to utilize the abundantly available and low-cost biomass residues from agriculture, forestry and industrial processing. These residues contain 50–90 wt.% water, which makes drying a no-go.

One must also consider the need to not compete with biomass for food, the environmental concerns for efficient utilization of biomass residues and the ultimate economic viability without subsidies. Taking these factors into account, HTU® plus HDO provides many opportunities for a successful commercial development.

20.2 Activities of Shell

The position of the Royal Dutch/Shell Group concerning fuels derived from biomass was a strategic one. After the oil crisis of the

1970s it was decided to explore alternatives for crude petroleum oil as the energy source. Around 1970 research programs were started to study the promises of conversion routes such as coal gasification, Fischer–Tropsch synthesis, and conversion of oil shale, tar sands, peat, and biomass.

To secure the feedstock for the biomass route Shell acquired large forests in subtropical areas such as Brazil, Thailand, New Zealand, and South Africa. In fact, Shell was one of the world's largest private forest owners (Fig. 20.1). Around 1990 the total forest area owned by Shell was some 250 million hectares. The idea was to operate the forest with "plantation forestry," that is, harvesting a part of the plantation each year and replacing it with new trees on a rotational basis.

It was decided to give the highest priority to the direct conversion of wood to liquid fuels, with the prerequisite that environmental requirements be met.

Figure 20.1   Picture of one of Shell's forests.

Around 1980 an extensive study was made of the open literature on conversion of biomass into liquid fuels. Various laboratories in Europe and North America were visited and international biomass conversion conferences were attended.

The Non Traditional Business (NTB) and Shell Research divisions decided to focus the work on two conversion routes, gasification plus Fischer–Tropsch synthesis and HTU® plus HDO. The gasification route was worked out by NTB (London) together with external parties. The Amsterdam Laboratory started experimental work on the HTU® route.

20.3   Exploratory Work on HTU® by Shell Research

20.3.1   Autoclave Experiments

Earlier, exploratory research work had been started, originally on the beneficiation of peat. This work was now directed to scouting experiments on HTU®. Herman Ruyter, a talented and very enthusiastic inventor, was the project supervisor (Fig. 20.2).

Figure 20.2   Herman P. Ruyter (1949–2007).

In pressure autoclaves of 150 mL or larger, a wide variety of biomass feedstocks were subjected to treatment in water at temperatures of 300°C–350°C and pressures of 150–200 bar.

With all biomass feedstocks it was found that under HTU® conditions a significant amount of the feedstock oxygen is removed both as carbon dioxide and as water. The remaining organic product is referred to as "biocrude."

A few correlations between product distribution and properties with feedstock characteristics and reaction conditions were found:

- Lignin content of the feed:
  - o A higher lignin content in the feed gives a lower gas (CO_2) yield (Fig. 20.3).
  - o A higher lignin content in the feed causes the biocrude to be more aromatic.
  - o The biocrude yield is not affected by lignin.
- The reaction pressure has no significant effect, provided that the water is in the liquid phase.
- In the range of 310°C–350°C a higher reaction temperature gives a higher gas yield.
- The reaction time for an optimal yield of biocrude is 5–15 minutes. At prolonged reaction times the biocrude tends to react away by thermal cracking reactions to heavy char and light gases.

20.3.2 Upgrading to Transportation Fuel

The main purpose of Shell was to produce a transportation fuel. Therefore, from the beginning it was considered to upgrade the biocrude with a process like hydrocracking. Basic chemical considerations indicated that, from the chemical composition of the biocrude, it would seem possible to upgrade it. A difference with traditional hydrocracking feedstocks is the higher oxygen content.

This supposition was tested by performing a few experiments in a small, continuous-microflow reactor with a typical hydrocracking catalyst. Because of the limited availability of biocrude only a few experiments could be done. Therefore high severity was selected. It was shown that biocrude can be converted into light hydrocarbons at a temperature of 400°C. A diesel fraction was isolated from the product by fractionation. Tests in the product laboratory of Shell in Thornton (UK) showed that the ignition properties of this diesel fraction were superior to those of traditional diesel products.

Although the scope of HDO experiments was very limited it was concluded that the principle of upgrading the biocrude by HDO had been demonstrated and that the conversion of biomass to biocrude and subsequent upgrading by HDO is feasible.

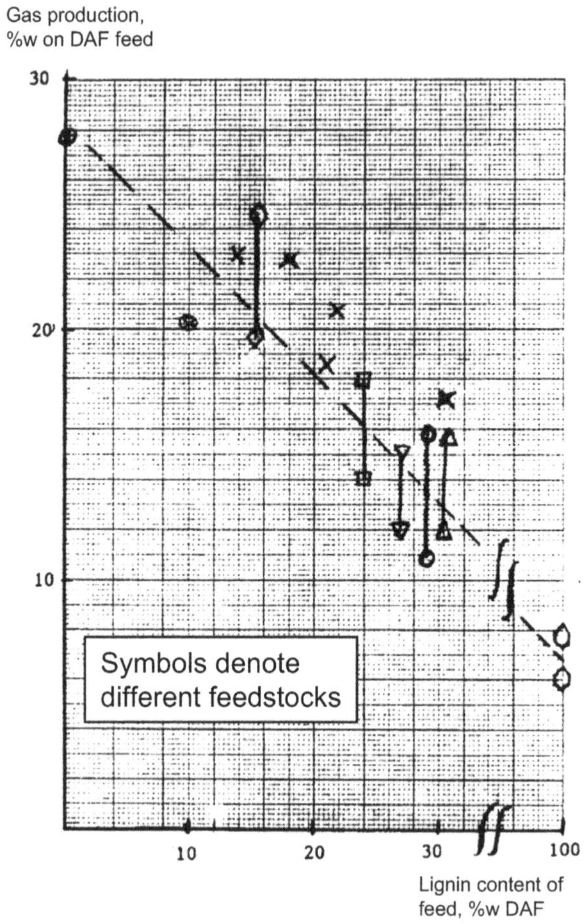

Gas production,
%w on DAF feed

Symbols denote
different feedstocks

Lignin content of
feed, %w DAF

Figure 20.3 Effect of feedstock lignin content on gas (CO_2) production.

20.4 HTU® Process Development by Shell

20.4.1 Technoeconomic Evaluation

In 1984 an evaluation of the HTU® process was performed to assess the technical and economic merits. The base case was a HTU® plant in or near a forest, with an intake of 375 kton/year dry basis (db) of plantation wood. The biocrude product was transported to a refinery where a HDO unit produced 100 kton/a db of C_5^+ product.

No technical obstacles were identified which would make it impossible to construct a large-scale HTU® plant. The problem of bringing the wood chips to the process pressure of 160 bar was solved by first softening them in water at 180°C for 15 minutes. This technique, adopted from the paper industry, converts the wood into a slurry which was judged to be pumpable.

For the base case the economics showed that the process combination HTU®+HDO would be commercially competitive at a crude oil price of US\$ 65 per barrel.

20.4.2 Process Development

The result of the evaluation led to the decision to transfer the project to the process development department. It was decided to construct a modular bench-scale unit in order to demonstrate the HTU® process in continuous operation. The throughput was of the order of 10 kg/h. Figures 20.4 and 20.5 show the process scheme and a picture of the unit.

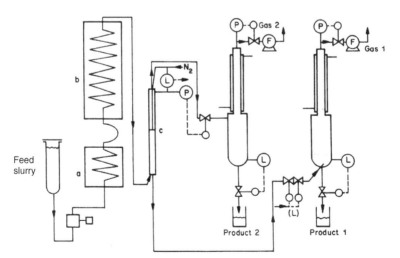

Figure 20.4 Process scheme of the Shell modular continuous bench-scale unit.

Quite some difficulties had to be overcome. First of all the continuous introduction of wood against a pressure of some 150 bar was complicated. In the end this was achieved with a slurry of wood particles (0.1 mm size) and a modified piston pump. Furthermore the operation of the high-pressure gas–liquid separator was troublesome. This was solved with a dip-leg where a small flow of nitrogen gas was introduced into the liquid phase in the separator. By measuring the differential pressure over this dip-leg the level of liquid could be determined (level indicator L in Fig. 20.4).

Figure 20.5 Picture of the Shell modular continuous bench-scale unit.

The temperature and pressure of the unit could be controlled independently, thus allowing the preliminary measurement of the liquid–gas phase equilibrium.

After the teething problems had been solved the unit operated successfully. Mass recoveries were close to 100%. Several wood species as well as peat were used as the feedstock at various process conditions. About 30 experiments were performed. It was shown that with all feedstocks the normal HTU® conversion pattern was obtained. Especially peat was an excellent HTU® feed. On the basis of the results an improved conceptual process design was made. For more details, see Ref. [1].

20.4.3 Termination of Shell Activities

In spite of the promising results the Shell management decided in 1988 to discontinue the development work on HTU®. The main reason was that the crude oil price was as low as $10/bbl and that it was perceived to remain at that level for quite some time.

A study into the chemical reaction mechanism of HTU® [2] was sponsored at the Delft University of Technology. It was intended to provide a basis for later process improvements.

Around 1990 there was in the society at large a growing awareness of the importance of sustainability and protection of the environment (caused by, for example, the Brundtland report [3] and the Rio de Janeiro meeting). It was thought in Shell that the HTU® process might be an important contributor in these matters. Hence, a desk study was made in 1992 to identify process improvements. Also, an independent technoeconomic evaluation was made by Stork Comprimo, a well-known engineering contractor. Both studies indicated that the production of C_5^+ by HTU® plus HDO was competitive at a crude oil price of $55/bbl.

In 1993 Shell decided, for mainly strategic reasons, to put a final halt to the development of the HTU® process. The results were shelved and the Shell patent of 1986 [4] maintained.

20.5 Developments by Biofuel B.V.

In 1996 Frans Goudriaan and Jaap Naber left Shell and founded the consultancy firm Biofuel B.V. They made arrangements with Shell which permitted them to continue the development and commercialization of the HTU® process.

First a development and business plan was made. This was discussed with a number of potentially interested parties.

20.5.1 First HTU® Consortium, 1997–2000

A consortium was formed which had for partners Stork Engineers & Contractors (later Jacobs Engineering), TNO, Shell Netherlands, Biomass Technology Group (BTG), and Biofuel B.V. A comprehensive development project was started at the end of 1997 with a financial support of 2.5 M€ from the Dutch government. The main results of this project which ran until 2000 were:

- The problem of *pumping* was tackled. From a large number of pump manufacturers one was identified, who slightly modified a commercially available pump so that biomass paste could be pumped to 160 bar in one step. This was demonstrated in a few test runs with a commercial pump. Also, they were prepared to downscale the design of this pump to pilot plant scale.

- A *pilot plant* was designed and constructed at the facilities of TNO at Apeldoorn. The process scheme is shown in Fig. 20.6. The intake capacity was around 100 kg/h. Figure 20.7 shows a picture of the plant.
  The plant was officially opened in 1999 (Fig. 20.8). Initial experiments were carried out to test the performance of the plant. In most of these the feedstock was a paste prepared from pellets of sugar beet pulp (SBP). At the end of the project all operational problems had been solved and a number of successful test runs had been done.

- Various methods of *feedstock preparation* were investigated. After quite some trials a set of operations was worked out in cooperation with a specialized company. This is described in Section 20.7.1.

- *Autoclave experiments* were performed with the aim of further investigating the effect of HTU® process conditions and the HTU® performance of different biomass feedstocks. A special feedstock injection device was developed. This allowed experiments at short residence times. Also experiments were now possible to estimate the reaction heat. For more details, see Ref. [5].

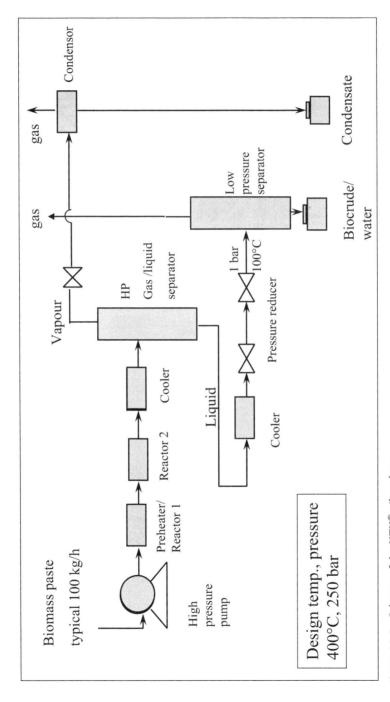

Figure 20.6 Scheme of the HTU® pilot plant.

Figure 20.7 HTU® pilot plant.

Figure 20.8 Minister Jorritsma (Economic Affairs) officially opens the HTU® pilot plant.

- A full professional *basic process design* was made by Jacobs Engineering for a HTU® plant. The intake was 600 kton/a as-received (ar) of SBP (or 130 kton/a db). The total investment cost was estimated as 50 M€. This figure was used in subsequent economic analyses of the HTU® process. A similar design was made for a demonstration plant with 60 kton/a ar of washed municipal waste (or 25 kton/a db) as the feed.

- On the basis of these result a *patent* was obtained for eight countries [6].

20.5.2 Subsequent HTU® Consortia, 2000–2008

20.5.2.1 The DEN project

From 2002 to 2004 the HTU® process development was continued by a consortium with TNO, Shell Research Foundation, the Waste and Energy Enterprise of the City of Amsterdam (Afval Energie Bedrijf [AEB]), and Biofuel B.V. as the partners. Support was obtained from NOVEM, the Dutch Energy Agency under the DEN program.

Major milestones were:

- The *selection and pretreatment* of commercially attractive biomass feedstocks by experiments in autoclaves and in the HTU® pilot plant.
- A *prolonged continuous operation* of the pilot plant was demonstrated in a three weeks campaign with shift personnel from AEB Amsterdam. It was successfully concluded in April 2004, with onion pulp as the feedstock [7]. The on-stream time percentages were 50% and 61% for two different plant configurations. These percentages may seem low but it should be realized that most interruptions were caused by blockages in the plant. In the absence of a flushing system the operators had to wait until the plant was sufficiently cooled down before any actions could be taken. This waiting time considerably reduced the on-stream percentage.

  Considering the fact that until then the pilot plant had only operated in day service, the positive results of the operation and the high on-stream time were beyond expectations.

- Many *improvements* were identified for the pilot plant. Most important was the addition of a flushing system, which allowed, in the case of a plant blockage, to flush the plant sections which were not blocked with hot water, so avoiding these sections being blocked by coking or solidification. Another improvement suggested was to have one uniform pipe diameter throughout the plant. This and other modifications have been applied in later projects and they have increased the plant availability considerably.

- The pilot plant runs provided sufficient amounts of biocrude to verify the *product handling and properties.*

On the basis of the results from this project it was concluded that there were sufficient data to allow the following step, that is, a follow-up with the commercial demonstration of the HTU® process.

20.5.2.2 Other projects

Over the years, Biofuel B.V. has made and maintained many contacts with government and other public bodies, scientific organizations and industrial companies. A specific goal was to find sponsorship or investors for funding the first commercial HTU® demonstration plant. From these contacts quite some smaller projects arose. The most important ones were:

- The NWO-Japan *BIOCON* group with, among others, the Delft and Twente Universities, NWO (scientific organization of the Dutch government), and Shell Research. A visit was paid to a number of Japanese universities and scientific organizations who were active in thermochemical conversion of biomass (1999).
- In the *GAVE-HTU®* project it was worked out how the commercial implementation of HTU®-derived automotive transport fuel could be realized. This included the upgrading of HTU® biocrude, an evaluation program and screening tests to verify compliance of the upgraded transport fuel product with EN 590 specifications, laboratory engine tests, and field tests with trucks, additives, and logistics. This was done in cooperation with TNO automotive, CE Delft (for CO_2 calculations and LCA aspects), AEB, and Van der Sluijs Handelsmij (a fuels logistic firm). The project was financially supported by the government's GAVE program. The result was a complete program with cost and timing (2001–2003).
- The *OTC project* set out the formation of a so-called transition coalition for the commercial realization of the HTU® process. Participants were NOVEM, CE Delft, AEB, North Holland

municipal waste company (HVC), and Tebodin (for drafting a business plan) (2004).

- *Costa Due* was a cooperation between many industrial and public companies and institutes aiming at a plan for the sustainable development of the North Groningen industrial area (2005–2007).

- *Biorefining.* AVEBE, an innovating company originally processing potatoes and now diversifying, had developed a patented process for the extraction of protein from grass. The protein would serve as fodder. By-products of the extraction were the fibers, and the juice containing the nonprotein dissolved organics.

Together with the AFI institute of Wageningen University a technical and economic study was made of the integration of protein extraction and HTU® (see Fig. 20.9). It turned out that there was a large synergy between the two processes. The extraction step had found a good destination for juice and fibers and could get sufficient low-temperature heat from the HTU® step. The HTU® step obtained a low-nitrogen feedstock which was suitable without further pretreatment. Economic analysis showed a good profitability. However the economics were largely dominated by the prices of protein and grass (2005).

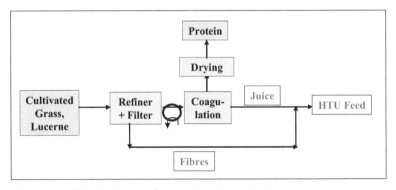

Figure 20.9   Block scheme of extraction of protein from grass.

- In a cooperation project with TOTAL (France) and HVC, development work was done with the aim of:

- investigating the HTU® conversion of GFT (organic domestic waste), for HVC
- showing the technical and economic feasibility of HTU® (for TOTAL)

Quite some experiments were done with different feeds in autoclaves to show trends of conversion and product properties in dependence of feedstock type and process conditions. Also extensive pilot plant runs were performed. The main emphasis here was on GFT pretreatment and conversion.

It was shown that GFT processing is very well possible. However, because of disagreement on the follow-up it was decided to dissolve the cooperation (2006–2008).

20.6 HTU® Chemistry and Product Composition

20.6.1 Hydrothermal Biomass Conversion

Hydrothermal conversion of biomass has been studied for over a hundred years. One of the first investigators who worked in this field was Friedrich Bergius. He pioneered in the development of high-pressure laboratory equipment. Around 1910 he conducted investigations into the reactions of superheated water and its influence on coal in the presence of iron. Also, he demonstrated a simulation of the genesis of bituminous coal from peat, and from plant materials like cellulose or lignin, with water at 200 bar and temperatures up to 500°C [8]. Considerable quantities of carbon dioxide and water were formed. In 1931 he received, together with Carl Bosch, the Nobel Prize for his high-pressure work, particularly on the hydrogenation of coal [9].

Hydrothermal conversion of biomass with the objective of producing liquid fuel has been studied in many institutes starting in the first half of the 20th century. An excellent review is given in Ref. [10]. A number of different sets of reaction conditions are distinguished. Typical examples are:

- Supercritical high-temperature gasification, producing a hydrogen-rich gas at temperatures of typically 600°C–800°C.

- Supercritical moderate-temperature gasification to methane-rich gas at around 400°C, in most cases with the use of a catalyst.
- Supercritical liquefaction such as developed by Aalborg University [11] and now being commercialized by Steeper Energy [12]. This Hydrofaction™ technology operates at 360°C–430°C and 250–350 bar, using a K_2CO_3-dissolved catalyst to convert organic wastes into a raw "base oil" which can be upgraded.
- Subcritical with catalyst such as the Catliq® process [13]. Organic material is converted in water at 280°C–370°C and 250 bar in the presence of a homogeneous alkaline and a heterogeneous zirconia catalyst. A bio-oil with an oxygen content of 6 wt.% (dry and ash free [DAF]) is obtained. The thermal efficiency is 73%.
- Subcritical without catalyst. This is the domain of the HTU® process, described in this chapter.

20.6.2 The Chemistry of HTU®

Ruyter published some considerations on the chemistry of HTU® [14]. He found indications that the chemistry of biomass conversion by HTU® (at 350°C and 15 min.) is similar to that of the reactions of petroleum genesis where biomass is converted at a temperature of some 50°C and reaction times of millions of years. Correlation of the two reactions points to an activation energy of some 30 kJ/gmole, which seems to be a reasonable value.

In chemical terms the key to biomass liquefaction is the removal of oxygen. Biomass contains typically 40–45 wt.% (DAF basis) of oxygen. Oxygen removal increases the heating value, and it leads to a product with more hydrocarbon-like properties causing it to be immiscible with water. In thermochemical liquefaction the oxygen is either removed as water or as carbon dioxide. Removal as water (dehydration) leads ultimately to carbon as the remaining product (e.g., charcoal manufacture, pyrolysis). Removal of carbon dioxide (decarboxylation) leaves a product with a higher H/C ratio and therefore a higher lower heating value (LHV). This is visualized in Fig. 20.10.

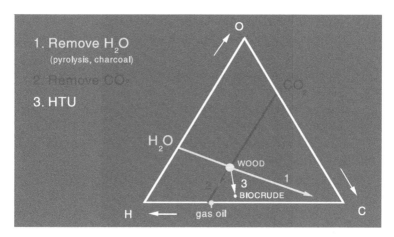

Figure 20.10 Triangular diagram showing pathways for removal of oxygen from biomass.

For HTU® conversion to occur it is essential that sufficient *liquid* water be present and that the temperature be above 300°C. Oxygen is removed both as CO_2 and as water. It appears that the selectivity of oxygen removal is nearly constant for all biomass feeds: the amount of oxygen removed by decarboxylation is almost equal to the amount removed by dehydration. See arrow number 3 in Fig. 20.10.

Various studies have been undertaken to elucidate the reaction mechanism of HTU®. *Physical factors* are important. First of all the pressure should be at least high enough for water to be in the liquid phase. Second, the particle size of the biomass feed should be limited to, typically, 5–10 mm. At larger sizes the transport of water to the centre of the particle is slower than the heat transport. This results in dry pyrolysis in the central part of the particle, and leads ultimately to coal formation. This can also occur if insufficient water is present. A safe rule of thumb is that the mass of water should be at least twice that of dry biomass.

The *chemical* reaction mechanism of HTU® is very complicated. The best one can aim for is a description of reactions of groups of components. One possibility is to use the biomass constituents cellulose, hemicellulose and lignin as model compounds.

An example of the effect of the lignin content on the conversion pattern in HTU® has been given in Section 20.3.1.

Many studies have been published on the conversion of a large variety of model compounds in water at temperatures from 150°C to 500°C with or without catalysts.[1]

Most of the studies deal with cellulose and its derivatives. From literature it is clear that hydrolysis of cellulose (polysaccharides) to monosaccharides like glucose is an important initial step. At Delft University two theses [2, 16] report studies on hydrothermal reactions of carbohydrates like D-glucose and its oligomers, and of supposed reaction intermediates like 5-hydroxymethyl-2-furaldehyde (HMF) and several furan derivatives. Also experiments with eucalyptus wood were reported. Both acid- and base-catalyzed reactions were found, which concurs with the high value of K_w (dissociation constant of water) under near-critical conditions. The main reactions of glucose are bimolecular polycondensations to biocrude-like products.

It is suggested that the base-catalyzed reactions include isomerizations, aldol and retro-aldol reactions, β-eliminations, and α-dicarbonyl cleavages. Acid-catalyzed reactions such as dehydratation/hydratation and alkylation also occur.

At University of Twente HTU®-like studies were done using small quartz capillaries as the reactor [15,17]. As shown in Fig. 20.11, a capillary was filled with water and a splinter of wood. The capillary is heated in an oven and pictures of it are made at regular intervals.

They show a rapid initial increase of the liquid volume as a result of the decrease of its density, and gradual shrinking of the wood splinter. The color of the liquid phase turns brown. The final picture after five minutes, at 340°C, does not allow to judge whether water and biocrude form one liquid phase or two, in the latter case a fine emulsion. The authors analyzed the products and made mass balances. The results indicate that there is always dehydratation next to decarboxylation.

A lumped reaction path model was proposed which covers all experiments with glucose, pyrolysis oil, and wood as the feedstocks. The results (products compositions and ratio dehydratation to decarboxylation) agree with the results found in the HTU® research by Shell and Biofuel B.V. (see Chapters 3–5).

---

[1]A list of some 20 publications is given in the thesis of Luijkx [2] in Table II of Chapter 1.

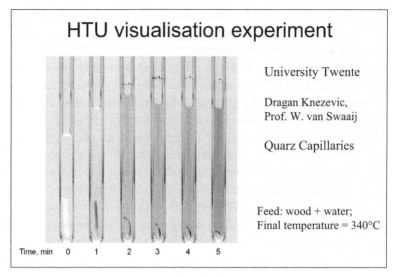

Figure 20.11   HTU in quartz capillary [15]. See also Color Insert.

The effect of the presence of catalysts was also studied. In screening experiments both dissolved salts and heterogeneous catalysts were judged by their rate of enhancement of the formation of CO_2. The best performance was shown by a Ru/TiO_2 catalyst. A substantial increase in the gas formation was found as shown in Table 20.1.

Table 20.1   Effect of catalyst on HTU® of wood (HTU of wood in a quartz capillary, 350°C, 10 min) [17]

	gC/gC in feed		
Catalyst	WSS	Gas	WSIS
None	0.61	0.09	0.25
Ru/TiO_2	0.63	0.24	0.12

WSS = soluble in water + acetone
WSIS = insoluble in water + acetone

In conclusion it can be stated that research into the reaction chemistry has found indications for many types of organic reactions taking place during HTU®. These are limited to cellulose-derived model compounds. The work of Knezevic et al. [15, 17] produced a reaction model with useful indications for process design.

No answer has been given so far to the question if the ratio decarboxylation/dehydratation can be increased. It is not excluded that dehydration is a necessary prerequisite for decarboxylation.

So, the process design has to be made solely on the basis of the experimental results of process development.

There are valid indications that the use of a catalyst can increase the formation of CO_2. Although this is a valuable lead to improvement it is thought that incorporation of a catalyst in the first commercial HTU® reactor is too much of a complication.

20.6.3   Products of HTU®

The product distribution is shown in Table 20.2. It is similar for nearly all biomass feeds, with deviations of a few percent.

Table 20.2   Typical product distribution of HTU® conversion

HTU of sugar beet pulp		
Product	**Wt.% on feed (DAF)**	**% of feed (LHV)**
Biocrude	41	79
Gas	26	2
Organics dissolved (OD)	12	16
Water	21	0
Total products	100	96
Reaction enthalpy [5]		4

Examples of the composition of feed, biocrude, and organics dissolved (OD) are given in Table 20.3. The products are:

- *Biocrude.* The biocrude is a heavy organic material which is solid at room temperature, and becomes liquid above typically 80°C. The oxygen content is 12–20 wt.% (DAF basis) and the heating value is 30–35 MJ/kg (DAF). The atomic H/C ratio is 1.0–1.3. Average molecular weight is about 1000.

- *Water and OD.* OD is the organic fraction dissolved in the process water after cooling the reaction mixture to ambient temperature. It consists mainly of acetic acid and ethanol. A great number of other compounds are found as well. Examples are compounds like acetone, methyl-cyclopentenone, and hydroxypyridine. Depending on the feedstock more nitrogen components and aromatic structures like phenols are also present.

The process water has a pH of 4–5.

Table 20.3 Composition of feed and products

	HTU of sugar beet pulp		
	Feed	Biocrude	OD
C, wt.% DAF	49.7	74.7	46.4
H, wt.% DAF	7.2	7.1	8.3
O, wt.% DAF	41.2	15.3	42.7
N, wt.% DAF	1.9	2.9	2.6
Total	100.0	100.0	100.0
LHV, MJ/kg DAF	17.5	33.3	22.8
Minerals, wt.% db	13.5	6.0	

- *Gases.* Apart from carbon dioxide, some carbon monoxide is formed and traces of methane and hydrogen.
- *Minerals.* The minerals content of biomass feedstocks shows a wide variation from 2 wt.% (db) for wood to 50 wt.% (db) for digestate. In the HTU® products the water-soluble minerals end up in the water phase and the insoluble part ends up in the biocrude. The minerals content of biocrude is therefore between 5 and 50 wt.% (db).

20.7   HTU® Process Design

A conceptual design for a commercial HTU® plant is described in this chapter. Many designs have been worked out by Biofuel B.V. As an example a plant with intake capacity of 100 kton/a db is described. A block scheme of the plant is presented in Fig. 20.12. The individual process steps are discussed below.

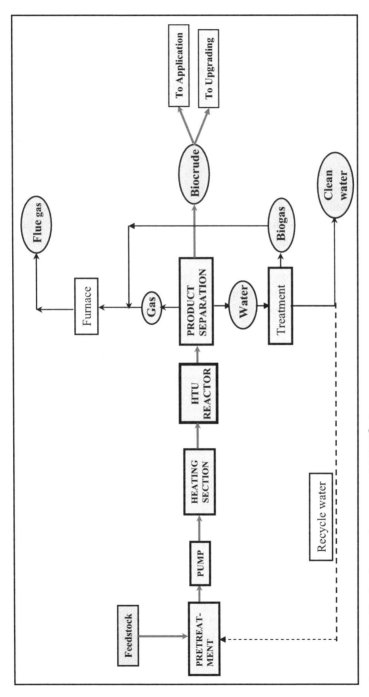

Figure 20.12 Block scheme of a commercial HTU® plant.

20.7.1   Feed Pretreatment

Biomass feeds have widely different physical structures and minerals contents. The challenge for the feedstock pretreatment is to transform them into a paste which can be pumped and easily flows through pipes of, for example, heat exchangers. The flow properties of such a paste depend on the nature of the feed such as its minerals content, presence of fibers, etc. For a given feed the flow properties are mainly determined by the particle size and the dry matter content.

Different feeds need different pretreatment. Easy feeds are, for instance, SBP, which is a homogeneous material right from the sugar mill. An example of a difficult one is household waste where large parts and sand have to be removed first. After that a specific combination of unit operations such as sieving, washing, and milling has to be found for each biomass feed to arrive at the particle size and dry matter content which is desired for that feed.

Many feedstocks of different nature were successfully transformed into a paste suitable for handling in the pilot plant. Examples are SBP pellets, road side grass, various wood species, municipal waste, onion pulp, peat, etc.

For most feedstocks a paste with a dry matter content between 20 and 25 wt.% gives good results in the pilot plant. Particle sizes are usually 1–2 mm.

For a commercial application the requirements are likely to be less strict. A larger pump and pipe diameter will allow larger particles. This will reduce the energy requirement of the feed preparation. However, too large a particle size can lead to nondesired reactions as pointed out in Section 20.6.1.

20.7.2   High-Pressure Pump

In Section 20.5.1 the cooperation with a pump manufacturer was mentioned. A design and a cost quotation for a commercial-scale pump were made.

Depending on the nature of the feed and the plant size, one or more pumps will be used in parallel. Also, at least one spare pump will have to be installed.

20.7.3   Heating Section

20.7.3.1   Heat transfer

A major uncertainty in the HTU® process design is the heat transfer to a flow of biomass feed paste. To get a first impression of the heat transfer characteristics a series of experiments was conducted in the pilot plant. The preheater/reactor was an electrical heater. It consisted of six inconel tubes of 6 m in length in series, connected by insulated bends (shown in Fig. 20.7 at the left hand of the operator).

First, experiments with water were done to calibrate the system. This was done for a number of values for the water flow and the temperature. Repeated measurements showed a deviation in heat transfer coefficient from the theory [18] by no more than a factor of 2. This result was judged to be satisfactory.

Then the measurements were repeated with biomass paste. The result was that the heat transfer coefficient for biomass paste was about two times lower than that for water.

In the commercial situation the Reynolds number will be much larger, so the pilot plant gives conservative results for the rate of heat transfer.

20.7.3.2   Heat exchangers

In the heating section the feed paste flow has to be heated from room temperature to the reaction temperature of 330°C. This is accomplished in a series of heat exchangers, the design of which is made using the results from Section 20.7.3.1.

20.7.3.3   Process furnace and hot-oil system

To heat the feed stream to 320°C[2] the required heat (5 MW) is supplied by the process furnace via the circuit of hot oil (see Fig. 20.13). The furnace fuel is biogas (see Section 20.7.6).

The flue gases from the furnace will most likely have to be cleaned by deNOxing.

---

[2]It is thought that the heat of reaction (see Section 20.7.4) will take care of the last 10°C heating.

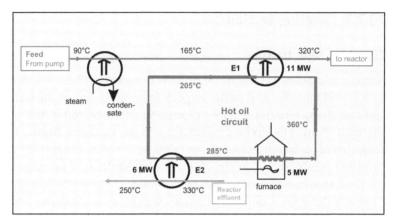

Figure 20.13 Hot-oil circuit and main heat exchangers.

20.7.4 Reactor Section

20.7.4.1 Reaction enthalpy

An estimate of the reaction enthalpy was made from autoclave and from pilot plant experiments. Both point to a slightly exothermic reaction with ΔH_r = –0.65 MJ/kg DAF feed. For details, see Ref. [5]. For the enthalpy balance of the reaction, see Table 20.2.

20.7.4.2 Reactor conditions

On the basis of autoclave and pilot plant experiments a HTU® reactor temperature of 330°C is selected, with a liquid residence time of 15 minutes. The reaction pressure does not have much of an effect on the kinetics, provided sufficient water is in the liquid phase. The decisive factor in choosing the reactor pressure is the extent of vapor phase formation which is desired in the reactor. The gas/liquid phase equilibrium at the reactor outlet can be estimated with sufficient accuracy from that in the CO_2–water system [19]. This has been verified in pilot plant experiments.

Originally, on the basis of autoclave results of Shell Research, it was thought that the biocrude would be distributed over the gaseous and liquid phases, and a considerable part of it would be present as "light oil" in the gas phase. This would allow the isolation of biocrude with a premium quality. However, the results obtained in the pilot plant have shown that only a very limited amount of biocrude passes to the gas phase.

20.7.4.3 Reactor type

For a liquid residence time of 15 minutes one calculates that the required reactor volume is about 30 m³. The limited knowledge of the reaction kinetics indicated that the residence time of 15 minutes should not be exceeded, because this would lead to consecutive reactions causing degradation of the biocrude. So, the residence time distribution should be as narrow as possible. Hence, the reactor flow pattern should approximate plug flow.

20.7.5 Product Separation Section

20.7.5.1 Phase behavior biocrude water

Separation depends on the liquid/liquid phase equilibrium in the system biocrude and water. From this system it is known that at temperatures up to 100°C a perfect phase separation occurs.

Knezevic [17] reported visual observation of the phase behavior in his quartz capillaries. At temperatures of 180°C and lower there are clearly two phases. Between 180°C and 220°C the situation is unclear. Above 220°C there is either only one phase or an emulsion of two phases.

If it would be desired to have a biocrude–water separation at temperatures above 180°C more data on the phase behavior are required. The presently available data from pilot plant and autoclave research pertain to phase separation at up to 100°C.

20.7.5.2 Separation of the products

The product separation section should isolate the individual product streams. Here only a few considerations are given. For the overall process heat economy it is desirable to utilize the enthalpy above, say, 250°C for heating up the feed stream. The cooling to this temperature has to be done at reactor pressure.

The further cooling and depressurizing can be done in many ways. In general a combination of flashers and coolers will be applied. It requires a trade-off between heat economy and equipment cost. The final product streams are:

- *Water.* This is discussed in Section 20.7.6.
- *Biocrude.* Depending on the intended application a conditioning step may be required. For instance, if upgrading by HDO is intended it is useful to remove light ends by flashing.
- *Gas.* On a dry basis some 5 wt.% of the gas is CO, H_2 and CH_4. It represents 1% to 2% of the heating value of the HTU® feed. In the process design one has the choice of either passing the off-gas through an incinerator or utilizing the combustion heat in the furnace.

20.7.6 Water Treatment

The waste water from the HTU® process contains some 3 wt.% of OD. They represent some 20% of the heating value of the HTU® feed. The composition of the OD (see Section 20.6.2) originally cast some doubt on the possibility to convert them via anaerobic digestion. Phenols, for instance, are known to be hard to convert. Therefore it was decided already in an early stage to verify this conversion experimentally. Samples were supplied to a well known company specializing in anaerobic waste water treatment. They performed small scale digestion experiments with HTU® water samples. They found that the conversion into biogas proceeded with a thermal efficiency up to 60%, which was higher than anticipated. A commercial digester can be designed without problems. So, the production of biogas is incorporated in the HTU® process design. The biogas is the main fuel for the process furnace.

Posttreatment by aerobic digestion is required in most cases to comply with water discharge regulations. Also, the minerals dissolved in the effluent of the water treater are removed to a level corresponding to the regulations.

20.7.7 Process Enthalpy Balance

The heat required for heating of process streams is obtained from heat exchange against products and from the furnace. The biogas and combustible product gases provide sufficient combustion heat to generate the required 5 MW in the furnace. So, no external fuel is required.

The electricity consumption of the plant is about 1.5 MW_e. This is mainly used for the feedstock pump(s) and, to a lesser extent, for the feed preparation section. For other feeds the consumption for the latter can be higher than for the present SBP feed.

The *thermal efficiency* (heating value in biocrude product related to that in feed) is 79%, as shown in Table 20.2. If the electricity consumption is taken into account, the thermal efficiency drops by 3%.

20.7.8 Materials of Construction

From the beginning of the HTU® research at Shell much attention has been given to corrosion aspects. On the basis of the chemical composition of the process streams an assessment was made of corrosion hazards by the materials department. The result was that under HTU® conditions the potential corrosion problems when using austenitic stainless steel of the AISI 300 series were:

- pitting and crevice corrosion by chlorides
- stress corrosion cracking (SCC) by chlorides and/or high-temperature (>200°C) water

Chloride SCC can occur on stainless steel at temperatures of 80°C and higher if the chloride concentration is 10 ppmw or higher and in the presence of minor amounts of oxygen. SCC is a particularly insidious form of corrosion because it can lead to material failure without prior warning.

The initial experiments were done in stainless-steel autoclaves in a bunker with remote operation of the installations. So any danger caused by potential corrosion was excluded.

One autoclave was cut in two after several hundreds of operation hours, with heating and cooling before and after each experiment. Inspection by the materials department did not show any corrosion.

The Shell continuous HTU® bench-scale unit was constructed in 300 series stainless steel. In normal operation conditions the chloride content of the HTU® reaction mixture is 100–500 ppmw. Also, oxygen cannot be excluded. Hence, operation is in the window where SCC can occur. To prevent this, the feed (a slurry with wood flour, particles of <0.1 mm) was washed in a household washing machine with demineralized water. The chloride content was reduced to <10 ppmw.

During the development work by Biofuel B.V. there was a permanent advice by a materials consultancy company. The pilot plant was constructed in stainless steel, with the exception of the preheater/reactor, which was in inconel. Corrosion coupons in a variety of materials were placed at selected points in the plant. Regular inspections of the coupons gave valuable insight into the corrosion behavior.

20.8   Opportunities for HTU®

The main objectives for the commercial application of the HTU® process are:

- No competition with biomass for food
- Economic viability without subsidy
- Solution for environmental concerns
- Sizable impact on the utilization of biomass for energy, notably for transportation fuels

Therefore, prospective feedstocks include residues from primary processing. Especially those which contain a high percentage of water, for which there are almost no viable alternatives. The following feedstocks present excellent opportunities for a rapid commercial development:

- Empty fruit bunches (EFBs) and other residues from palm oil plantations
- Bagasse
- Forestry waste
- Wet organic fraction of municipal solid waste (MSW). Wet separation (washing) of the MSW provides great opportunities for fast growing cities in view of the ever increasing global urbanization
- Manure
- Peat

All these feedstocks are available in great quantities and (except may be peat) at low cost or even at a premium at the gate. The initial commercial applications will most likely aim at the utilization of the biocrude for electricity production. After demonstrating the commercial viability there are excellent opportunities for multiplication and/or upscaling. The latter is a prerequisite for the development of the upgrading to transport fuels, especially for aviation.

The challenge is to establish the first commercial demonstration in a difficult economic climate.

References

1. Goudriaan, F., Peferoen, D.G.R. (1990). Liquid fuels from biomass via a hydrothermal process, *Chemical Engineering Science*, **45**(8), 2729–2734.

2. Luijkx, G.C.A. (1994). *Hydrothermal Conversion of Carbohydrates and Related Compounds*, PhD thesis, Delft University of Technology.

3. World Commission on Environment and Development (1987). *Our Common Future,* Oxford University Press.

4. Annee J.H.J., Ruyter H.P. (1986). *Process for Producing Hydrocarbon-Containing Liquids from biomass.* European Patent 0204354, to Shell Internationale Research Maatschappij B.V.

5. Goudriaan, F., Van de Beld, B., Boerefijn, F.R., Bos, G.M., Naber, J.E., van der Wal, S., Zeevalkink, J.A. (2000). Thermal efficiency of the HTU® process for biomass liquefaction, *Proceedings of the Conference Progress in Thermochemical Biomass Conversion*, 1312–1325, Ed. Bridgwater, A.V., Tyrol, Austria, 18–21 September 2000, ISBN 0-632-05533-2.

6. Beld, L., van de, Boerefijn, F.R., Bos, G.M., Goudriaan, F., Naber, J.E., Zeevalkink, J.A. (2001). *Process for the Production of Liquid Fuels from Biomass*, US Patent 7.262.331, to Biofuel B.V.

7. Berends R.H., Zeevalkink J.A., Goudriaan F., Naber J.E. (2004). Results of the first long duration run of the HTU® pilot plant at TNO-MEP, *Proceedings of the 2nd World Biomass Conference: Biomass for Energy, Industry and Climate Protection*, Rome, 535.

8. Bergius, F.K.R. (1912). *Die Anwendung hoher Drucke bei chemischen Vorgängen und eine Nachbildung des Entstehungsprozesses der Steinkohle*, Verlag Wilhelm Knapp, Halle a.S., Germany.

9. Bergius, F.K.R. (1932). Nobel lecture, http://www.ava-co2.com/web/media/downloads_DE/dokumente/Bergius_Lecture.pdf.

10. Peterson, A.A., Vogel, F., Lachance, R.P., Fröling, M., Antal, Jr., M.J., Tester, J.W. (2008). Thermochemical biofuel production in hydrothermal media: a review of sub- and supercritical water technologies, *Energy and Environmental Science*, **1**, 32–65.

11. Hoffmann, J. (2013). *Upgrading of Bio-crude from Hydrothermal Liquefaction*, presented at TCBiomass 2013.

12. http://www.steeperenergy.com.

13. Toor, S.S., Rosendahl, L., Nielsen, M.P., Glasius, M., Rudolf, A., Iversen, S.B. (2012). Continuous production of bio-oil by catalytic liquefaction from wet distiller's grain with solubles (WDGS) from bio-ethanol production, *Biomass and Bioenergy*, **36**, 327–332.

14. Ruyter, H.P. (1982). Coalification model, *Fuel*, **61**, 1182.

15. Knežević, D., Rep, M., Kersten, S.R.A., Prins, W., van Swaaij, W.P.M., (2004). Hydrothermal liquefaction: visualization and screening experiments, paper P069 in *Science in Thermal and Chemical Biomass Conversion*, Victoria, Vancouver Island, 30 August–2 September 2004.

16. Srokol, Z.W. (2009). *Reaction Pathways during Hydrothermal Upgrading of Biomass*, Thesis, TU Delft.

17. Knežević, D. (2009). *Hydrothermal Conversion of Biomass*, Thesis, University of Twente, the Netherlands.

18. Beek, W.J., Muttzall, K.M.K., van Heuven, J.W. (1999). *Transport Phenomena*, 2nd edition, John Wiley & Sons.

19. Tödheide, K., Franck, E.U. (1963). *Z. Phys. Chem. (Frankfurt)*, **394**, 387–401; also in Gmelin *Handbuch der Anorganischen Chemie*, 6th edition (1973), Vol. 3, 54–57, Verlag Chemie, Weinheim/Bergstr.

About the Authors

Frans Goudriaan was born on January 15, 1941, in Ouderkerk aan den IJssel, the Netherlands. He did chemical engineering from the Delft University of Technology, the Netherlands (1996), and his PhD in chemical engineering from the University of Twente, the Netherlands (1974).

From 1966 to 1971 Dr. Goudriaan was a university teacher in chemical process technology and economic analysis of chemical processes at the University of Twente.

From 1971 to 1996 he was at Shell Research, mainly at the Amsterdam laboratory (KSLA), the Netherlands, where his activities included exploratory research and process development. He was the project leader for process development of Hydro Thermal Upgrading® (HTU®) technology for the conversion of biomass into fuels. The last position he held was manager operations of hydrocarbon process research.

In 1996, together with Ir. J. E. Naber, Dr. Goudriaan started BIOFUEL B.V., the Netherlands, and developed the HTU® process in the period 1997–2007 in various consortia. Since 2008 he is involved in commercial development programs for the HTU® process for the conversion of biomass to transport fuels.

Jaap E. Naber was born on June 17, 1938, in Blangkedjeren, Indonesia. He did chemical engineering from the Delft University of Technology (1963).

After graduating he joined Royal Dutch/Shell Laboratory Amsterdam (KSLA). He was transferred to Shell Oil Woodriver Refinery, USA, in 1975, to a business center in Houston, USA, in 1977, and to Pernis Refinery, the Netherlands, until 1980. He subsequently worked in the Shell International Petroleum Company, The Hague, the Netherlands, in various functions, with an intermediate period (1985/1986) as manager of hydrocarbon processing at KSLA. He was appointed as director of process research at KSLA on December 1, 1989, and as director of downstream oil and gas research in 1993. He retired on April 1, 1996.

Naber started Biofuel B.V. with Dr. Ir. Frans Goudriaan in 1996 and developed the Hydro Thermal Upgrading® (HTU®) biomass conversion process in the period 1997–2012 in various consortia. He is now involved in commercial development programs for the HTU® process for the conversion of biomass to transport fuels.

Chapter 21

Catalytic Hydrothermal Gasification

Douglas C. Elliott

Chemical and Biological Process Development Group, Pacific Northwest National Laboratory, P. O. Box 999, MSIN P8-60, Richland, Washington 99352, USA
dougc.elliott@pnnl.gov

Catalytic hydrothermal gasification is an innovative method to thermochemically convert wet biomass into a gas product consisting essentially of methane and carbon dioxide and recover a clean water by-product.

The term "hydrothermal" used here refers to the processing of biomass in water slurries at elevated temperature and pressure to facilitate the chemical conversion of the organic structures in biomass into useful fuels. The process is meant to provide a means for treating wet biomass materials without drying and to access ionic reaction conditions by maintaining a liquid water-processing medium. Typical hydrothermal processing conditions are 520–640 K of temperature and operating pressures from 4 to 22 MPa of pressure. The temperature is sufficient to initiate pyrolytic mechanisms in the biopolymers, while the pressure is sufficient to maintain a liquid water-processing phase.

Biomass Power for the World: Transformations to Effective Use
Edited by Wim van Swaaij, Sascha Kersten, and Wolfgang Palz
Copyright © 2015 Pan Stanford Publishing Pte. Ltd.
ISBN 978-981-4613-88-0 (Hardcover), 978-981-4669-24-5 (Paperback), 978-981-4613-89-7 (eBook)
www.panstanford.com

Hydrothermal gasification is accomplished at the upper end of the process temperature range. It can be considered an extension of the hydrothermal liquefaction (HTL) mechanisms which begin at the lowest hydrothermal conditions with subsequent decomposition of biopolymer fragments formed in liquefaction to smaller molecules and eventually to gas. Typically, hydrothermal gasification requires an active catalyst to accomplish reasonable rates of gas formation from biomass. Supercritical water processing is an extension of hydrothermal gasification. In fact, the earliest publications suggested that supercritical water conditions were a prerequisite for effective gasification of biomass, but later work has shown that effective catalysts can allow lower temperature gasification operations.

When operating in a system which reaches thermodynamic equilibrium, the resulting gas product composition will be determined by the pressure, temperature, and concentration and composition of organics. Operation at subcritical temperature results in a product gas high in methane and low in hydrogen [1], while operations at supercritical temperatures will produce more hydrogen and less methane. A confounding factor is that the relative amount of water in the system will also affect the gas product composition in that lower biomass concentration in the reactor system (and therefore higher water content) will move the equilibrium toward hydrogen and away from methane by known steam-reforming mechanisms. The use of low temperature will also impact the mechanical systems for containing the reaction. Lower-temperature operation allows lower operating pressures, which result in lower capital costs with lower requirements for containment structure and less severe corrosive attack on the reactor walls, which allows the use of less costly alloys.

At subcritical conditions the vapor pressure of water is a direct function of the temperature. To maintain liquid water in the processing environment, the operating pressure must be maintained above the vapor pressure. If the operating pressure is allowed to drop below the vapor pressure, the water will boil to regenerate sufficient water vapor to increase the pressure back to the vapor pressure. In this manner a hydrothermal process system can "boil dry" if allowed to operate at too low a pressure.

Near the critical temperature changes in vapor pressure, liquid density, dielectric constant, and solvating power happen quickly with small changes in temperature. For example, with a temperature increase from 573 to 647 K, the operating pressure must be increased by 13.5 MPa in order to maintain the liquid phase [2]. In addition, the volume of the liquid water will have expanded by 230% because of the drop in density of the liquid phase. Although the actual solubilities of inorganic materials in water have not been extensively determined near the critical point of water, it is clear from the available data, for example, that for sodium carbonate [3], they will have only limited solubility in water near the critical point. In the case of sodium carbonate, its solubility drops significantly over the range from 512 to 621 K, from 18.7 wt.% to <2.0 wt.%.

An advantage of the use of hydrothermal conditions is that it reduces the dewatering requirement of wet biomass. Drying of wet biomass before use in a thermochemical conversion process can have a large negative impact on the overall process efficiency. The drying process is inherently an energy sink and, even if accomplished with low-temperature excess process heat, will be a capital cost and will reduce overall energy efficiency.

The use of water in hydrothermal systems also allows ionic reaction conditions. The ion product of liquid water near its critical temperature is much higher than at ambient conditions. Ionizable compounds will be present as ions and able to react via ionic mechanisms. The ionic medium facilitates mass transfer. Hydroxyl and hydronium ions are present for reacting with the substrates such that both acid-catalyzed and base-catalyzed reactions can be facilitated. Siskin and Katritzky [4] have provided examples of many organic molecules previously considered unreactive in liquid water which undergo chemical reactions when the water temperature was increased from 523 to 623 K. The hydrothermal transformation of biopolymers in biomass into a range of oxygenated fragments invariably produces organic acids, such as acetic acid and formic acid. Carboxylic acid products generate a low pH medium in the water, which can have deleterious effects when considering corrosion of the reactor metallurgy and structure. The pressurized operating environment requires a high-pressure reactor system, typically constructed of steel. Because of corrosion concerns, stainless steel (typically 300 series) appears to be minimally required for this processing environment.

The use of water as the processing medium in hydrothermal processing results in a large water-handling requirement. The process input of water is often met by the water in the wet biomass feedstock, but additional water may also be required. Recycle and reuse of the water becomes a major consideration in the design of hydrothermal processes. Depending on the effectiveness of the process for organic transformation to fuel products and their recovery, a significant waste or recycle water treatment load may result from the process.

To perform hydrothermal processing, a pressurized system is required to minimize the vaporization of the water and the resulting energy requirement. Pumping of the biomass slurries at these pressures is a key technical challenge to utilization. The feeding of the biomass is different from other biomass-processing systems, which more typically include dry particulate (solid) feeding systems of biomass at low or near-atmospheric pressure. To achieve slurry pumping, the biomass material is needed in small-sized particles. Size reduction can be accomplished in a wet medium in order to minimize energy requirements for drying. To feed at high pressure, these size reduction requirements become more stringent. Although it is generally considered that high-pressure slurry feeding can be accomplished more easily at a larger scale, most process development to date has been at a smaller laboratory scale [5]. High-pressure pumping of biomass slurries required for hydrothermal processing at a large scale is not well developed [6].

Hydrothermal processing is envisioned as a means to process high-moisture biomass without preliminary drying of the biomass. Hydrothermal processing is suitable for processing very wet feedstocks such as algae or water hyacinth. Wet wastes from food processing or other agricultural processing systems can be utilized efficiently and effectively in hydrothermal systems.

For additional details of hydrothermal processing of biomass see also reviews by Elliott [5] and Tester's group [7]. A useful update of the work in several laboratories using conditions on both sides of the critical point of water has been published in 2005 demonstrating the extent of interest in the concept [8]. The specific use of catalysts in hydrothermal gasification of biomass is the subject of Elliott's review [9]. Another review by Azadi and Farnood [10] covers the field of catalytic gasification, broadly lumping together subcritical and supercritical processing and discussing biomass and chemical waste also.

21.1   Bench-Scale R&D in Hydrothermal Processing

Hydrothermal processing of biomass has been an active research topic since the first Arab oil embargo in the mid-1970s. Work in both liquefaction and gasification process development began at that time [11].

Initial efforts in hydrothermal gasification reported the use of catalysts. These early tests at the Massachusetts Institute of Technology (MIT) were predicated on the need for supercritical water conditions for effective gasification, without which significant char formation resulted [12]. However, later studies at the Pacific Northwest National Laboratory (PNNL) showed that with adequate catalyst activity the gasification could be accomplished in hot, pressurized water at conditions less severe than supercritical [13]. The PNNL work demonstrated the high activity, but long-term shortcomings, of nickel metal as a catalyst [14] as well as the superior qualities of ruthenium metal as a catalyst in this system [15]. Subsequent studies in Japan [16] and at the Paul Scherrer Institute (PSI) in Switzerland [17] have confirmed the catalytic gasification under subcritical hydrothermal conditions. This work has been reviewed recently [18].

21.1.1   Pacific Northwest National Laboratory

Catalytic hydrothermal gasification (CHG) of biomass R&D began at the PNNL in the late 1970s with initial batch reactor tests. Those initial studies were aimed at the elucidation of the kinetics and mechanisms of catalytic gasification of biomass [13]. In that work, attempts were made to recover the intermediates in catalytic gasification of biomass by using a batch reactor fed with a biomass/water slurry, in which a remotely controlled high-pressure and high-temperature sampling system was employed to recover condensed-phase products, expected to be the gasification intermediates [19]. The work focused on wood as the biomass feedstock and also tested wood fractions, including microcrystalline cellulose, hollocellulose (lignin-free wood), and lignin, using nickel metal and sodium carbonate as catalysts. In that research it was concluded that the rates of catalytic gasification at low temperatures in a pressurized water medium (hydrothermal processing) were similar to those seen at higher temperatures at atmospheric pressure in conventional steam gasification.

That research was further extended to the examination of other biomass feedstocks, including kelp, Napier grass, sorghum, spent grain, grape pomace, potato-processing wastes, apple pomace, cheese whey, and unconverted residue from anaerobic digestion. A continuous-flow processing system was constructed, which used a fixed bed of nickel catalyst with the sodium carbonate catalyst dissolved in the feedstock water slurry of finely ground biomass. These tests validated the process application to a wide range of biomass feedstocks but identified the catalyst stability issues inherent in hydrothermal processing [20].

Development of a catalyst with long-term stability in the hydrothermal gasification environment was accomplished through extensive research in which a range of catalyst metals and supports were tested at the PNNL. Using a batch reactor method, many different types of potential catalyst support material were tested for physical and chemical stability at hydrothermal conditions of 523 K for up to 70 hours. In this way a large number of silica and alumina supports and those of other oxides in combination with them were shown to be unstable. Metal oxides determined to have long-term stability at hydrothermal conditions were titania and zirconia, although only in the crystalline forms of lower surface area, specifically rutile titania and monoclinic zirconia [15]. Carbon was also identified as a stable support, despite the logic that it would react with steam in the presence of a catalytic metal [14]. Among the catalytic metals, ruthenium was identified as the most active. Rhodium and osmium appeared to also have significant activity. Nickel was identified as the best base metal for catalysis of CHG. Other metals with limited activity, such as iron and cobalt, readily oxidized [14]. Copper and tin were found to remain reduced but had very limited catalytic activity. Although nickel remained reduced at hydrothermal conditions it was found to sinter at a rapid rate sufficient to deactivate the catalyst within 24 hours of operation. A method to inhibit the sintering was developed by doping the nickel catalyst with another stable metal such as ruthenium, copper, silver, or tin. However, the inherent solubility of nickel metal in the hydrothermal environment remains, and the long-term operation with nickel has been found to result in measureable loss of nickel into the aqueous system. This effect is not noticeable with ruthenium metal catalyst [15].

21.1.1.1 Catalyst lifetime in model compound tests

Validation of catalyst formulations for CHG of wet biomass has been done with model compound feedstocks and chemical wastewaters, which were free of the inorganic contaminants which complicate biomass processing. Long-term test results have been reported in a 30 mL reactor with ruthenium on a number of supports [15]. In those tests a 10 wt.% phenol in a water stream was converted at over 99.9% to clean water for up to 3190 hours on-stream. More recently other materials have been tested at the PNNL for stability at hydrothermal gasification conditions and to determine if they could serve as a catalyst support for ruthenium in hydrothermal gasification.

Useful catalyst support materials identified in batch reactor testing in water were loaded with ruthenium and tested with phenol as a model for hydrothermal gasification. These catalysts were tested in the microscale continuous-flow reactor in extended period tests to validate catalyst physical stability and catalytic activity stability. The results of these tests are summarized in Table 21.1. All experiments were operated at a nominal temperature of 623 K and a pressure of 21 MPa. Some of the early tests were performed with an 8 wt.% phenol in water solution, while later tests used a 5 wt.% feedstock. The different catalysts were kept on-stream for different periods of time. Those noted by "plugged" were terminated when the catalyst pellets disintegrated and particulate collected downstream of the reactor and made pressure control difficult. The other catalysts were operated for extended periods of time and were shut down for reasons other than catalyst deactivation.

The Hyperion nanocarbon, SiCat silicon carbide, and BASF carbon granule catalysts all showed good activity and good stability. They are reasonable substitutes for the previously reported ruthenium on carbon extrudate catalyst, C3610 from BASF. An obvious conclusion is that the level of ruthenium loading directly translates to the acceptable rate of operation, that is, liquid hourly space velocity (LHSV). A full understanding of the limits of operation is not presented here, as the tests were maintained at levels of high conversion in most cases in order that extended time operations could be demonstrated. A loading of 1 to 2 wt.% ruthenium per unit of LHSV (liter of 5% phenol in water solution per liter of catalyst bed per hour) appears appropriate as a design parameter. No limitations for the activity resulting from overloading metal onto the support can be seen from these data.

Table 21.1 Results with ruthenium catalysts on different supports

Catalyst number	Material manufacturer and description	Feedstock, wt.% phenol	Time on-stream (hour)	COD conversion	LHSV
1	Norit, ROX carbon extrudate (8 wt.% Ru)	8	2.4 (plugged)	NA	2
2	Hyperion, CS-02C-063 XD, nanocarbon 59894-129-1 (8 wt.% Ru)	8 8	70 142	100 99.8	2 2
3	SGL, grade 33G graphite (8 wt.% Ru)	5	171 (plugged)	49 56	2 1
4	SICAT, SB0681A silicon carbide (8 wt.% Ru)	5 5	201 808	100 98	4.0 7.6
5	SICAT, SB0681A silicon carbide (1 wt.% Ru)	5	250	100	0.5
6	Jacobi, Eco sorb BX-Max carbon PNNL graphitized (8 wt.% Ru)	5	<1 (plugged)	NA	2
7	Johnson–Matthey, AcCarbon Type 482 carbon granules (2 wt.% Ru)	5	10 (plugged)	100	1
8	BASF, carbon granules (2 wt.% Ru)	5 5	91 1305	70 100	2.4 1.0
9	BASF, C3610 carbon extrudates (7.8 wt.% Ru)[*]	10 10	504 1008	100 100	1.8 3.1

Note: Operating parameters for experiments: 623 K and 21 MPa.
[*]From Ref. [15].

Table 21.2 provides data to evaluate the catalyst activity over a longer time for better catalysts. Catalysts with good hydrothermal stability are included along with comparative data from our earlier test with C3610, the BASF catalyst with 7.8 wt.% ruthenium on a proprietary carbon extrudate.

Table 21.2 Results with ruthenium catalysts on stable supports in long-term tests

Catalyst number	3–4 days		1 week		3 weeks		5–6 weeks		7 weeks	
	Conv.	LHSV	Conv.	LHSV	Conv.	LHSV	Conv.	LHSV	Conv.	LHSV
5	99.99	0.50	99.99	0.50	–	–	–	–	–	–
4	99.994	4.00	99.997	4.00	99.6	5.0	97.75	7.6	–	–
8	82.38	2.42	99.89	1.00	99.99	1.00	83.95	1.55	99.98	1.00
2[a]	99.96	2.00	94.4	2.03	–	–	–	–	–	–
9[b]	–	–	99.99	1.15	99.99	1.83	99.99	3.14	–	–

Note: Tests at 623 K and 21 MPa with 5% phenol feedstock.
[a]8% phenol feedstock.
[b]From Ref. [15] with 10% phenol feedstock.

In the case of the 2 wt.% ruthenium on granulated carbon, the data for chemical oxygen demand (COD) in the aqueous effluent is plotted in Fig. 21.1. In this test, which extended over a two-month period, a range of LHSVs was used to determine long-term operability of the catalyst at a high conversion level. The high conversion level was defined as a reduction by >99.9% of the initial COD of the 5% phenol in water at 119,000 ppm. At this high conversion level, the gas product consisted of 63% methane, 35% carbon dioxide, 1% hydrogen, and <1% of ethane and higher hydrocarbon gases. The operation was maintained near the flow rate limit to verify that operation in an overload condition (too high LHSV) could be remedied by continued operation at lower LHSV. On the basis of these results, an LHSV of 1.0 can be used for the regeneration/cleaning of the catalyst bed. Operations at LHSVs of 2, 1.75, 1.55, and 1.5 all led to a condition of reduced COD conversion (overload) after a period of time, which varied on the basis of the flow rate. It appears

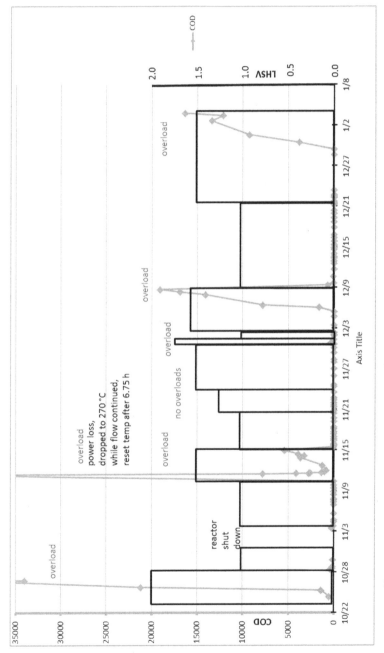

Figure 21.1 Effluent COD over a range of LHSVs using 2 wt.% Ru on C granule catalyst. COD is plotted in blue and refers to the left axis; LHSV levels are indicated by horizontal lines and refer to the right axis.

that operation at an LHSV between 1.0 and 1.5 would be the limit for high conversion of the phenol (>99.9%).

In contrast to earlier reported results suggesting solubilization/ sintering for nickel metal catalyst under hydrothermal conditions, ruthenium has been found to be quite stable. Nickel is typically found at about 1 ppm in the aqueous effluent when using a highly loaded nickel catalyst. An initial peak of 16 ppm nickel found in the aqueous by-product recovered in the early stage of one test [15]. Ruthenium results reported for a 1 wt.% ruthenium catalyst were less than 0.08 ppm in the aqueous effluent throughout an eight-month CHG test of that ruthenium-stabilized nickel catalyst. In the tests reported here, a more careful examination at the start-up conditions of long-term tests with ruthenium catalysts was undertaken. When using the silicon carbide–supported catalyst (7.8 wt.% ruthenium), it was found that ruthenium spiked to 19 ppm in the first hour before falling to 1 ppm after 3.5 hours and was still detectable at 0.8 ppm after 21 hours on-stream. Later in the same test, after a malfunction caused a shutdown at 79 hours on-stream and a subsequent restart, the ruthenium was measured at 0.15 ppm in the aqueous effluent in the first 8 hours of the restart before falling below the level of detection (0.08 ppm) for the balance of the test. Similarly, when using the BASF carbon granule–supported catalyst (2 wt.% ruthenium), at 9 hours after start-up there was no detectable (<0.15 ppm) ruthenium in the aqueous effluent, and that level was maintained throughout 111 hours on-stream. The amount of ruthenium lost in the aqueous effluent is trivial in these tests and likely represents a small amount of physical removal of dust from the catalyst bed. In contrast, using the SGL porous graphite-supported catalyst (7.8 wt.% ruthenium), which had poor activity from the start, 1270 ppm ruthenium was found in the aqueous effluent at 2 hours into the test and the ruthenium level remained at 45 ppm after 9.5 hours and 5 ppm at 26 hours. Clearly, the choice of support material has a dramatic effect on the stability of the ruthenium metal catalyst.

21.1.1.2   Catalyst lifetime tests with industrial wastewater feedstock

The condensate from chemical manufacturing served as the feedstock for long-term catalyst lifetime tests of actual wastewater feedstock. Chemical analysis on the condensate confirmed the presence (by chromatography) of three organic acids, fumaric, maleic, and phthalic, as well as propylene glycol, propanal, and dioxolanes. The stream was nearly free of inorganic contaminates, which is important for long-term processing over a metal catalyst. Neither sulfate nor phosphate was detectable. The COD was measured initially at 121,800 ppm, and the pH at 1.5. The tests were around-the-clock, continuous-flow operation in the 30 mL reactor system.

The experimental program evaluated a range of processing flow rates and catalyst formulations. High conversion of the organic materials to gas products was achieved with the catalysts. The effluent goal of <500 ppm COD was attained with a range of processing rates and catalysts. The product gas was primarily methane and carbon dioxide, with a low level of hydrogen and ethane in some cases. Carbon monoxide and olefins were undetectable.

Three nickel catalysts were tested. These were versions of the BASF G1-80 catalyst, which was not designed for use in hydrothermal gasification, but earlier work had shown that it had reasonable stability in the processing environment [15]. The lifetime of the catalyst had been directly correlated with the nickel crystallite size, and stabilized forms of the catalyst included dopants of either silver or copper. Three versions, as-produced G1-80, silver-doped G1-80, and copper-doped G1-80, were tested at typical CHG conditions. None was found to be suitable in long-term operation in this application.

Two ruthenium catalysts were also tested, one from Engelhard (now BASF) on a carbon extrudate and one from Degussa on a titania extrudate. The titania-supported catalyst has been patented for use in aqueous processing systems and has a demonstrated lifetime of at least three months, depending on an earlier test [15]. The carbon-supported catalyst contains a higher level of ruthenium and has shown high activity in extended operations of CHG [15].

The carbon-supported catalyst expressed very high activity initially, as shown in Table 21.3. The high activity was maintained for only a few days before the effluent COD increased, indicating a deactivation of the catalyst. With decreasing catalyst activity, the methane content of the product gas decreased and the hydrogen and ethane content increased. The flow rate was reduced from an LHSV of 5 to 4 and the effluent COD came back into range, that is, <500 ppm. After an additional week of operation, the LHSV was reduced again to 3.5 and then to 3 to maintain the desired level of gasification. As the effluent COD again went high the feed was stopped and the catalyst bed washed with water. After 10 days of subsequent operation, catalyst deactivation again became apparent. After two additional weeks of continuing decreasing activity the test was terminated.

Table 21.3   Bench-scale process results with ruthenium catalysts

Catalyst	Ru/C	Ru/TiO$_2$	Ru/TiO$_2$	Ru/C	Ru/C
On-stream, active	4 weeks	First 3 weeks	Balance of test	First 3 weeks	Balance of test
Reactor 1, K	603–633	626–628	NA	Only one	Only one
Reactor 2, K	593–623	605–623	625–631	618–625	616–625
Pressure, MPa	19.3–20.9	19.7–20.6	19.8–20.1	20.8	20.3–20.8
LHSV, L/L/h	5–3	1.5–1.05	0.77	4	4
pH of feedstock	1.5	6.7	6.7	6.7	6.7
Gas, L/g organic	0.85–0.78	0.78–0.88	0.85	0.91	0.79
COD conv., %	99.9–98.6	99.8–99.3	99.7	99.9	99.9
Effluent COD, ppm	164–1650	276–832	405–444	172	134–153
CH$_4$ vol.%	58–25	63	65–63	63	63
CO$_2$, vol.%	38–44	32–33	31–33	32.5	32
H$_2$, vol.%	3–9	0.8–2.8	0.9–1.1	2.0	1.8–2.1
Ethane, vol.%	0–17	0.4–1.2	0.7	0.4	0.5–0.8
Back-flush, vol.%	1–3	0.3–1.5	0.4	0.3	0.6–0.3

Initial samples of the titania-supported catalyst were damaged in situ because of over-temperature exposure during pump stoppages. The stability of this catalyst is such that it must be reduced at temperatures below 573 K and should not be exposed to high partial pressures of hydrogen gas at higher temperatures because of a strong metal support interaction identified as a partial reduction of the titania support and its migration to the ruthenium surface [21].

The long-term test with the ruthenium on titania catalyst was started with neutralized feedstock. Sodium hydroxide pellets were added with stirring to the condensate to bring the feedstock to near-neutral conditions of about 6.7 pH. The sodium content of the resulting feedstock was about 4000 ppm. The equipment operated consistently for the three-month test. This test was begun with an LHSV of 1.5 and over the first three weeks of operation the LHSV was reduced in steps to 0.75, where it remained for the balance of the test. The effluent COD was about 300 ppm initially and the processing rate reduced subsequently in order to maintain it at <500 ppm (see Fig. 21.2). At one week short of the three-month goal, the effluent COD exceeded 500 ppm with an LHSV of 0.75. At three months the effluent COD was at 800 ppm, with the LHSV maintained at 0.75. The test was then terminated. However, it is likely that a water wash at that point may have returned the catalyst to an active form, which would have allowed further processing.

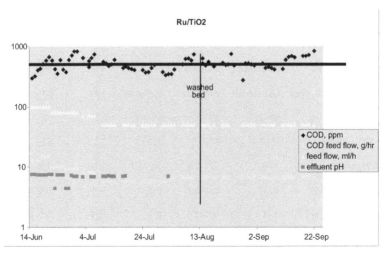

Figure 21.2 Process data during a three-month test processing industrial condensate wastewater feedstock using Ru/TiO$_2$ catalyst.

A second long-term test was performed with the ruthenium on carbon catalyst using the neutralized feedstock and it was successfully operational for over three months at an LHSV of 4 (see Fig. 21.3). The gas composition remained stable throughout the three-month test with high methane content and low hydrogen and ethane and undetectable carbon monoxide and ethylene. There was no trend indicating loss of activity and the full lifetime of the catalyst is expected to be greatly in excess of three months. Neutralization of the organic acids in the wastewater was an important discovery and a key to long-term operation.

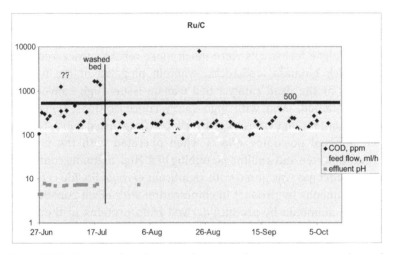

Figure 21.3   Process data during a three-month test processing industrial condensate wastewater feedstock using Ru/C catalyst.

21.1.1.3   CHG of algal HTL aqueous by-product

A higher level of complication for CHG is its application to the aqueous by-product from HTL of biomass. In this case, algal biomass (*Nannochloropsis salina*) was processed [22]. CHG of the HTL aqueous by-product, which can carry a COD of 80,000 ppm or more, resulted in nearly complete gasification of the remaining organic components. COD of the water was reduced by 98.8% to 99.8%. The typical high methane and carbon dioxide gas was produced with little hydrogen or higher hydrocarbons. The results presented are for only limited times on-stream ranging from 8.1- to 42.2-hour operation. The catalyst composition was 7.8 wt.% ruthenium metal on a partially graphitized carbon extrudate.

These tests showed that high-methane-content gas was produced by CHG of the HTL aqueous by-product. The removal of the organic material should facilitate the recycle and reuse of the dissolved nutrients (N and K) in the aqueous stream.

21.1.1.4 CHG of algal biomass

The application of CHG to actual biomass is complicated by the mineral content of the feedstock. Formation of insoluble phosphates and sulfates can plug reactor systems. Sulfur can poison the catalyst system. Recent results with algal biomass have been reported by the PNNL [23], in which several species were tested.

The algae feedstocks were much more reliably processed than lignin-rich biomass feedstocks, wherein plugging of the feeding systems or the fixed catalyst bed was an issue. High conversions were obtained, even with high slurry concentrations. Consistent catalyst operation in these short-term tests suggested good stability and minimal poisoning effects when operated with the mineral separation step and a sulfur-scrubbing bed. High methane content in the product gas was noted with significant carbon dioxide captured in the aqueous by-product in combination with alkali constituents and the ammonia by-product derived from proteins in the algae. High conversion of algae to gas products was found with low levels of by-product water contamination and minimal loss of carbon in the mineral separation step.

21.1.2 Paul Scherrer Institute

Research at the PSI has progressed along a pathway similar to the PNNL, except for the PSI's focus on operation at supercritical water conditions. Significant developments at the PSI include evaluation of Ni as a catalyst [24] and subsequent testing of a ruthenium catalyst. The PSI has also confirmed the catalytic gasification under subcritical hydrothermal conditions [25]. At supercritical conditions, it evaluated sulfur poisoning of the ruthenium catalyst by sulfate and found it to be a significant problem with the formation of a sulfate Ru^{3+} complex [26]. Through studies of mineral component solubilities, a salt brine separation system was developed for mineral contaminants, which allows operation at supercritical water conditions without mineral precipitation and plugging in the system piping [27]. Although Modell's efforts at the MIT were diverted into supercritical water oxidation, since then the MIT has returned to the gasification arena in collaboration with the PSI [28]. The collaboration with Tester's group provided a useful link in the development of these technologies. More recently the Vogel group has investigated the supercritical water catalytic gasification of algae. In batch reactor tests, they validated the method and also considered the effect of metals (from reactor wall or catalyst) solubilized into the aqueous by-product, which would be recycled to the algae growth ponds [29]. Most recently the Vogel group has returned to the issue of catalyst formulation and has reported tests of a suite of titania- and zirconia-supported ruthenium catalysts in a batch hydrothermal stability test followed by hydrothermal gasification testing in a continuous-flow reactor of the most promising catalysts. Their results were similar to those at subcritical water conditions—monoclinic zirconia and rutile titania were found to be stable [30].

21.1.3 University of Twente

The University of Twente in the Netherlands has also played a role in the development of CHG. Its work was initially focused on supercritical water gasification [31] and without a catalyst, but more recently it has considered CHG at subcritical temperatures [18]. Knezevic's research focused effort on noncatalytic hydrothermal processing and the development of chemical mechanisms for glucose conversion [31]. Subsequently, he reported hydrothermal processing of biomass (wood) and pyrolysis bio-oil as well as glucose, both with and without a catalyst [33]. The catalysts included alkali salts and ruthenium catalysts. When using the heterogeneous catalysts (ruthenium on titania, alumina or carbon) important differences in the reaction rate (greater conversion to gas with ruthenium) and the mechanism (more methane with ruthenium and less char) are reported.

21.1.4 Others

The Karlsruhe Institute of Technology in Germany has also investigated supercritical water gasification (see Kruse's review

[34]) in continuous-flow reactor systems, both at the laboratory scale [35] and in a pilot plant operation [36]. The catalyst of choice for its work was an aqueous potassium base, such as carbonate or bicarbonate.

The University of Hawaii has developed the use of carbon catalysts for gasification at supercritical water conditions [37].

All other researchers in the field have undertaken batch reactor tests. Savage's group at the University of Michigan has recently investigated supercritical water gasification of biomass with [38] and without a catalyst [39], although the catalyst was only a nickel wire, which would have a low surface area and not be expected to be as active as a supported nickel catalyst.

Another recent participant in the field is the Farnood group at the University of Toronto. The research has evaluated glucose as a model in comparing a heterogeneous Raney nickel catalyst with homogeneous nickel, iron, or cobalt catalysts in subcritical water gasification and finding the homogeneous catalysts to be only slightly effective [40]. Further they have compared nickel and ruthenium catalysts in CHG and compared to atmospheric pressure steam reforming. It was concluded that the steam reforming was more energy efficient at low concentration (2 wt.%) but that the efficiency of CHG would be improved at higher concentration [41]. Tests using Raney nickel for gasification of activated sludge suggested little difference in activity over the range of 623 to 683 K, from below to above the critical point of water. It was reported that in supercritical water the nickel sintered and became deactivated, while also being poisoned by sulfur [42]. Similar evaluations at subcritical temperature were not done. More recent research by the group has focused only on supercritical operation, testing nickel and ruthenium catalysts with lignocellulosics and model compounds [43].

Onwudili and Williams, the University of Leeds, entered the field of CHG by evaluating sodium hydroxide catalysis using a municipal solid waste feedstock as well as carbohydrate models [44], following their initial research in non-CHG. Although they report no liquefaction (no tar formation), the base clearly catalyzed the formation of gas and dissolution of the feedstock into the water as no solids were formed. The group also performed subsequent research on the base-catalyzed gasification of food wastes, glucose, and dairy wastes in subcritical water, with the emphasis on evaluating the catalytic effects on hydrogen production by a water–gas shift [45]. Model tests with glucose and glutamic acid were also used to compare nickel catalysts, on either alumina or silica, along with NaOH. As expected, the base catalysis improved the hydrogen yield, while the nickel catalysis improved the methane yield. An added effect was noted that the NaOH appeared to decrease the amount of carbon deposited on the nickel catalyst surface [46]. Onwudili et al. [47] have also reported supercritical water gasification with nickel catalysts, with and without sodium hydroxide, for hydrogen production from algae feedstocks, and additional studies of ruthenium-catalyzed gasification biomass also at supercritical conditions [48].

Researchers in Japan have also investigated many of these process issues. Studies at the Agency for Industrial Science and Technology (AIST) by Minowa [49] began with supercritical water processing with a nickel catalyst. Later studies with cellulose feedstock were performed at subcritical temperatures and showed the gasification activity in a batch reactor to be directly related to the amount of the nickel catalyst and with hydrogen as the primary gas product [50]. In tests with a number of different oxide supports for the nickel catalyst, the batch tests produced a more carbon-concentrated product gas, and methane was at a higher concentration than hydrogen in some cases but still less than CO_2 [51]. Minowa also concluded that the gasification was only effective for the water-soluble products of HTL and not the biocrude product [52].

Osada et al. compared catalysts (nickel, ruthenium, sodium hydroxide) for lignin and cellulose gasification at subcritical and supercritical water conditions in batch reactors [53]. They report a clear dependence of gas yield on temperature over the range from 523 K to 673 K and a clear advantage of ruthenium over the other catalysts for both gas yield and methane production. Osada's group has made an extensive study of ruthenium catalysis of lignin gasification in supercritical water. Osada et al. [54] have reported results at supercritical water conditions, wherein γ-alumina was transformed to α-alumina (as opposed to boehmite found at subcritical temperatures) with dramatic loss of surface area and activity. They tested an anatase titania and found, after three 3-hour batch tests, it remained active and the crystal form had not transitioned to rutile (as has been reported in subcritical water).

A carbon-supported Ru catalyst became deactivated in the same period. The deactivation was attributed to loss of active catalyst surface area due to sintering of the metal. Further, they report on sulfur poisoning of the ruthenium catalysts by the addition of elemental sulfur and showed that the ruthenium was converted into sulfide, sulfite, and sulfate [55], and they report that sulfur-poisoned (by sulfuric acid treatment) catalyst is deactivated toward carbon–carbon bond scission and methanation at supercritical water conditions [56]. In the case of the ruthenium on titania catalyst, they have reported the ability to regenerate the sulfur-poisoned catalyst by subcritical water treatment, while supercritical water treatment was not so effective [57].

The AIST (Shirai's group) has returned to CHG in recent years and reported comparative tests of metal catalysis for cellulose. The carbon gasification activities of the metals were reported such that they can be ranked Ru > Rh >> Pt >> Pd at 523 K [58]. A companion effort evaluated ruthenium salts deposited on carbon or titania. They concluded that Ru (III) nitrosyl nitrate formed a more active catalyst than Ru (III) chloride because the ruthenium metal formed in the process of lignin gasification at 673 K was of a smaller particle size (providing more active sites) [59].

21.2 Process Studies in Scaled-Up Reactor Systems

There have only been limited reports of CHG at a larger than laboratory scale of operation. From the PNNL there is a report of gasification tests carried out in a mobile scaled-up reactor system (MSRS), depending on the bench-scale continuous-flow design [60]. The MSRS was designed at a scale of 10 L/hour of aqueous feed for obtaining engineering data for further scale-up [61]. Design working conditions for the reactor systems were 623 K at 24 MPa.

In the MSRS, the wet biomass feedstock could be loaded into the feed tank equipped with an electrically driven paddle stirrer to agitate the contents. The feed stream was pumped with a high-pressure reciprocating plunger pump. In earlier bench-scale tests the initial portion of the tubular reactor acted as the preheater as well. The addition of a continuous-flow stirred-tank reactor (CSTR) was required to prevent the buildup of solids at the opening of the reactor, as was found when using many biomass slurry feedstocks [62]. In the scaled-up system the feedstock was pumped directly from the feed tank through the tube side of the heat exchanger. The heat exchanger was a double-tube heat exchanger constructed of 316SS tubing. With a total length of 17 m, the heat exchanger could bring the feedstock to within 373 K of the final operating temperature when using biomass feedstocks. The final heating of the feed occurred in the coiled tubular preheater. The catalytic gasification reactors were of a tubular fixed-bed design. After leaving the reactor(s), the product stream was routed through a heat exchanger, which provided heat for preliminary heating of the feed stream. Downstream of the exchanger, the process pressure was reduced to ambient over a back-pressure regulator. The product stream then entered a liquid/gas separator tank, where process water was reclaimed and combustible gases were sampled for analysis, measured, and then vented.

21.2.1 CHG of Corn Ethanol Fermentation Sludge

A limited amount of scaled-up testing was completed processing distillers dried grains and solubles (DDG&S), the unconverted residue from corn ethanol fermentation, in the MSRS engineering demonstration unit [60]. Engineering issues related to feeding the slurry to the high-pressure reactor were the focus of much of the work. As in the bench-scale unit, a CSTR was installed between the pump and the catalytic reactor beds. Solids build-up at the entrance to the catalyst bed and the resulting flow stoppage were to be avoided by the liquefaction caused in the preheating by the CSTR. The feed slurry was processed by wet grinding in a Union Process Attritor stirred ball mill to achieve a pumpable slurry of the DDG&S feedstock. Plugging at the entrance to the reactor was also a problem and was not relieved by using the CSTR, that is, it occurred whether the CSTR was on line or not. When the reactor entrance was reconfigured to allow better flow into the reactor bed the system was run only without the CSTR. The tube-in-tube heat exchanger provided all the required liquefaction of the biosolids and also provided heat recovery, preheating the feed from 298 K to

523 K, while effectively cooling the product from 628 K to 303 K. Following the test, an examination of the inside of the tube-in-tube heat exchanger showed only a light powder coating on the tube wall and no significant fouling of the surface. Consistent pumping was less of a problem in this test, but by eight hours on-stream, there was an indication of plugging at the front end of the reactor, as evidenced by a 2 MPa pressure drop.

Samples from the plugging materials and the catalysts were analyzed by instrumental methods, scanning electron microscopy (SEM), X-ray photoelectron spectroscopy (XPS), and X-ray diffraction (XRD), to determine changes in the catalyst, in addition to inductively coupled plasma (ICP) and X-ray fluorescence (XRF) elemental analysis. These analyses clearly showed that certain biomass trace components precipitated and plugged the catalyst bed entrance, that a crust of trace components from the biomass was deposited onto the catalyst pellets, and that some of the biomass components passed through the catalyst bed reacting with and poisoning the bed during processing.

A combination of ICP and XRF elemental analysis and XRD showed that the plugging precipitate at the entrance to the catalytic bed in the reactor was composed primarily of hydroxyapatite, $Ca_5(PO_4)_3OH$, and iron–chrome stainless steel from the wet-grinding media. Sulfur contamination only was found within the catalyst pellet and was highly associated with the ruthenium but that contamination of the bulk of the material was found throughout all four reactor beds.

The wet gasification of biomass was demonstrated in continuous-feed, fixed-bed catalytic reactor systems in a scaled-up engineering development system. The system was operated at conditions of 603–633 K and 21 MPa at processing rates up to 10 L/ hour despite complications related to the slurry nature of the feed and the inorganic components in the feedstocks. Aqueous effluents with low residual COD (as low as 100 ppm) and a product gas of medium-heating value were produced from DDG&S. These results have shown that careful monitoring and control of feedstock trace components (e.g., calcium, magnesium, sulfur, and phosphorus) are critical for maintaining long-term operability and catalyst activity. Clearly, more development work was needed to provide longer-term operation without mineral precipitation in the catalyst bed.

21.2.2 Catalytic Hydrothermal Gasification of Wet Biosludge

Tests were also performed in the MSRS using dewatered wastewater treatment biosludge from a chemical manufacturing facility. The tests were undertaken to develop a mineral matter separation system to prevent blockage of the catalyst bed by the precipitate. In addition, a sulfur-scrubbing system was developed to protect the catalyst bed from sulfur poisoning. An initial round of bench-scale continuous-flow reactor tests appeared to validate the mineral matter separation and the sulfur-scrubbing methods, so development could proceed to the scaled-up reactor work. As a result of analysis, it was determined that a sulfate separation method was needed, in addition to the sulfur-scrubbing system. The sulfate precipitation step utilized the addition of barium solution to the feedstock, which effectively allowed separation of precipitated barium sulfate in the mineral separation step. The sulfur-scrubbing system was designed as, and appeared to function only as, a capture method for reduced sulfur forms.

The MSRS was modified on the basis of the results of the bench-scale demonstration of the technology. The system was then operated in multiday tests to evaluate the modifications and refine them. Some results are presented in Table 21.4.

The final test of the MSRS, depicted schematically in Fig. 21.4, incorporated the best of the processing ideas. It demonstrated the operability of the process with wet biosludge using a high-pressure syringe pump. High conversion of COD in the product water with stable catalyst operation was shown. Loss of carbon through the mineral matter separation step was identified as a process deficiency. The conclusions are provided below:

- A high yield of medium-heating-value gas and a clean water stream was produced from a range of biosludge samples using the ruthenium on carbon catalyst and operations at 623 K and 20 MPa. However, there was a significant loss of carbon from the biosludge feed into the separated solids, which were intended to be a concentration of the mineral component of the biosludge.

- Sulfur contaminants in biosludge must be addressed with sulfur capture systems to protect the activity of the gasification catalyst. Whereas traditional sulfide capture systems could be used to protect the catalyst in aqueous processing systems, separation of sulfate by these systems was not effective. Sulfate precipitation was achieved with barium to reduce its concentration in the aqueous medium to less than 10 ppm. The precipitation was done before heat up to gasification because carbon dioxide generated would compete with the sulfate in precipitating the barium.

Table 21.4 Summary of process results with biosludge in the mobile scaled-up reactor system

Feedstock	1	2	3
Ba addition to hot feed	Yes		
Ba addition to cold feed		Yes	Yes
Feed dry solids, %	3.4	3.0	3.0
Feed COD, ppm	50900	45100	44400
Effluent COD, ppm	233	8	20
COD conversion, %	99.7	99.98	99.96
Separator temperature, K	627	632	629
Reactor temperature, K	609	624	623
Feed rate, mL/h	6600	6600	2200
Gas yield, L/g DS	0.65	0.58	0.65
C gasification, %	84.1	84.5	85.0
Solid yield, g/g DS	0.24	0.28	0.41*
Solid ash content, %	50.0	56.9	60.0
Solid C content, %	39.6	32.9	30.5
C loss in solids, %	20.0	21.6	29.7

*May indicate holdup of solids from earlier higher-flow-rate operation.

21.2.3 Catalytic Hydrothermal Gasification of Industrial Condensate Wastewater

An example of further development of CHG outside the biomass arena is the application to condensate wastewater, which was

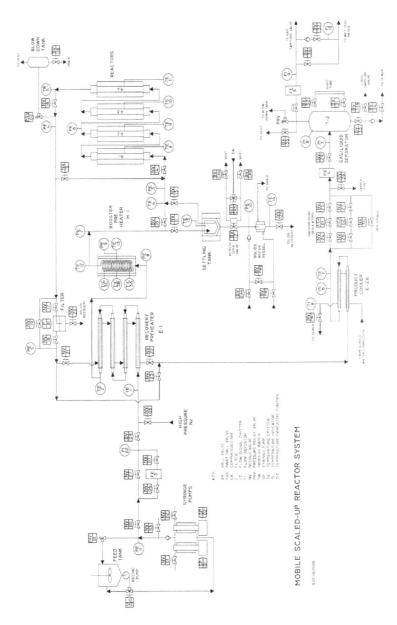

Figure 21.4 Schematic of the mobile scaled-up reactor system (MSRS) at the PNNL.

demonstrated in the MSRS with operations of over 50 hours and effluents at or below the 500 ppm COD limit at optimized conditions of 5 LHSV at 603–623 K and 19.3 MPa. The MSRS was used to demonstrate CHG with condensate as the feedstock at a scale ranging from 9 to 32 L/hour and temperatures from 513 to 638 K. The three-day test period included 63 hours on-stream with 52 hours on-stream with condensate feed. The test was an around-the-clock operation of the MSRS.

The experimental program evaluated a range of processing flow rates and processing temperatures. Table 21.5 summarizes the several data windows evaluated in this program at the lower flow rates. High conversion of the organic materials to gas products was achieved in the CHG system. The effluent goal of <500 ppm COD was attained with a range of processing rates and temperatures. The product gas was primarily methane and carbon dioxide with a low level of hydrogen. As processing conditions were adjusted to achieve lower levels of conversion, measurable levels of ethane, propane, and

Table 21.5   Process results for CHG of industrial condensate wastewater in the scaled-up reactor operated at flows of 2.3 to 6.6 LHSV

Reactor 1, K	583	601	615	627	625	638	628	628
Reactors 2–3, K	513	598	601	582	608	621	604	605
Reactor 4, K	540	602	605	557	610	624	601	607
LHSV, L/L/h	2.6	2.8	2.8	2.7	3.0	2.3	5.4	6.6
Gas, L@0°C	644	708	659	738	873	687	1579	1766
Gas, L/g organic	0.79	0.77	0.74	0.87	0.91	0.88	0.91	0.84
COD conversion, %	99.70	99.76	98.15	99.98	99.99	99.98	99.98	99.87
Effluent COD, ppm	396	332	2521	26	10	27	22	171
Product gas								
CH_4/CO_2, %	54/37	47/37	40/38	59/36	56/36	58/35	60/36	53/36
H_2/ethane, %	0.06/7.6	1.4/13	1.8/17	0.25/3.8	1.7/5.2	3.8/2.1	1.3/2.2	1.5/8.6
Back-flush, %	1.4	1.6	3.6	0.65	1.1	0.8	0.4	1.2

even butanes and higher hydrocarbons were found in the product gas. Carbon monoxide and olefins were undetectable. Lower-temperature operation (<620 K) at low flow results in moderately good conversion; high conversion requires higher-temperature operation (>620 K) and can be accomplished even with higher flow.

The operation of the scaled-up reactor demonstrated the tube-in-tube heat exchanger efficiency in this application. The process data shown in Table 21.6 have heat recovery ranging from 70% to 92% for the flows and temperatures tested. It appears that 90% heat recovery should be achieved at the demonstrated appropriate operation for the condensate stream of 5 LHSV and 623 K.

Table 21.6   Heat exchanger results in the scaled-up reactor

Feed temp., K	293	290	291	293	294	296	296	293	291	293
Preheat feed, K	475	554	561	569	576	546	560	523	526	572
Product temp., K	557	601	605	607	611	609	619	601	606	624
Cooled product, K	299	297	304	305	305	297	295	293	293	293
LHSV, L/L/h	2.70	5.40	7.24	6.62	6.82	2.98	2.73	2.82	2.76	2.27
Heat recovery, %	70.5	86.8	89.7	91.4	92.2	80.1	81.5	74.7	75.1	84.3

21.3   Current Trends for CHG Treatment

Application of CHG is currently focused on treatment of aqueous by-products from biomass thermochemical conversions. Its utilization for wet biomass also remains as a reasonable application. Studies have been completed to evaluate the use of CHG as an alternative to anaerobic digestion for waste or by-product treatment. A study of its use in swine manure processing suggests that the capital and operating costs for a first-generation CHG system would be significantly higher than those of an anaerobic lagoon system. However, there are many significant environmental advantages of the CHG option, such as greatly reduced plant footprint, production of relatively clean water, odor elimination, pathogen destruction, and the potential for valuable by-product recovery for nutrient recycle [63].

CHG is currently being evaluated for use in wastewater treatment related to fast-pyrolysis bio-oil upgrading to hydrocarbon fuels [64]. Tests are being completed with various stream compositions in order to understand the impacts of various levels of hydrotreating severity relative to the organic components remaining in the aqueous stream and their potential for conversion in CHG. The intention is to use organics in water by-products as a source of hydrogen for bio-oil upgrading via CHG and reforming of the methane product gas. The bench-scale work currently underway at the PNNL should lead to a scaled-up demonstration of the application in the MSRS at the integrated biorefinery being constructed in Hawaii.

An extensive study has been completed on CHG of process streams related to algae conversion to fuels as part of the National Alliance for Advanced Biofuels and Bioproducts (NAABB). The process research was undertaken at the PNNL and included both CHG of whole algae and lipid-extracted algae [22] but also the aqueous by-product from HTL of algae [21].

An important development is the licensing of the PNNL technology for CHG to Genifuel Corporation for use in biomass processing. Genifuel is currently designing a small demonstration plant for CHG of algae with the participation of fellow NAABB partner, Reliance Industries Ltd. (RIL) of India. Construction of the plant was completed in September 2014 with commissioning tests underway in late 2014. The plant was to be installed and operated at the RIL industrial site in Jamnagar, India.

21.4 Conclusions

Hydrothermal processing of biomass to gaseous fuels requires expanded process development to take the technology to a scale for industrial demonstration. Technical challenges associated with hydrothermal processing of biomass include the issues associated with defining the properties of the by-products, which are highly dependent on the feedstock composition; optimizing the gasification process variables; and demonstrating the effectiveness of separation techniques to remove impurities before catalyst poisoning. Recycle of nutrients from the recovered by-products (phosphorus in solids and nitrogen, potassium, and carbon in aqueous) is a potential area for process cost savings and improved sustainability. It is clear that CHG has many potential advantages over biological conversion for water by-product cleanup. Construction and operation of a scaled-up demonstration of the technology is needed to validate its utility and confirm a capital cost basis for commercial implementation.

References

1. Butner, R.S., et al. (1986). Effect of catalyst type and concentration on thermal gasification of high-moisture biomass feedstocks, *Biotechnology and Bioengineering*, XXVIII (Biotechnology and Bioengineering Symposium 17), 169–177.

2. Walter Barker (1968–1969). *Handbook of Chemistry and Physics*, 49th Ed., Chemical Rubber Company, Cleveland, Ohio.

3. Seidell, A. (1953). *Solubilities of Inorganic and Metal Organic Compounds*, 3rd Ed., Vol. 1, 1194.

4. Siskin, M., Katritzky, A.R. (1991). Reactivity of organic compounds in hot water: Geochemical and technological implications, *Science*, **254**, 231–237.

5. Elliott, D.C. (2011). Hydrothermal processing, in *Thermochemical Processing of Biomass: Conversion into Fuels, Chemicals and Power*, 200–231, Ed. Brown, R.C., John Wiley and Sons, Chichester, UK.

6. Berglin, E.J., Enderlin, C.W., Schmidt, A.J. (2012). *Review and Assessment of Commercial Vendors/Options for Feeding and Pumping Biomass Slurries for Hydrothermal Liquefaction*, PNNL-21981 Pacific Northwest National Laboratory, Richland, Washington, November 2012.

7. Peterson, A.A., et al. (2008). Thermochemical biofuel production in hydrothermal media: a review of sub-and supercritical water technologies, *Energy and Environmental Science*, **1**, 32–65.

8. Matsumura, Y., et al. (2005). Biomass gasification in near- and super-critical water: status and prospects, *Biomass and Bioenergy*, **29**, 269–292.

9. Elliott, D.C. (2008). Catalytic hydrothermal gasification of biomass, *Biofuels, Bioproducts and Biorefining*, **2**, 254–265.

10. Azadi, P., Farnood, R. (2011). Review of heterogeneous catalysts for sub- and supercritical water gasification of biomass and wastes, *International Journal of Hydrogen Energy*, **36**, 9529–9541.

11. Sealock, Jr., L.J., et al. (1993). Chemical processing in high-pressure aqueous environments: 1. Historical perspective and continuing development, *Industrial and Engineering Chemistry Research*, **32**, 1535–1541 (1993).

12. Modell, M. (1985). Gasification and liquefaction of forest products in supercritical water, in *Fundamentals of Thermochemical Biomass Conversion*, 95–120, Eds. Overend, R.P., Milne, T.A., Mudge, L.K., Elsevier Applied Science, London.

13. Elliott, D.C., Sealock, Jr., L.J. (1985). Low-temperature gasification of biomass under pressure, in *Fundamentals of Thermochemical Biomass Conversion*, 937–950, Eds. Overend, R.P., Milne, T.A., Mudge, L.K., Elsevier Applied Science, London.

14. Elliott, D.C., Sealock, Jr., L.J., Baker, E.G. (1993). Chemical processing in high-pressure aqueous environments: 2. Development of catalysts for gasification, *Industrial and Engineering Chemistry Research*, **32**, 1542–1548.

15. Elliott, D.C., Hart, T.R., Neuenschwander, G.G. (2006). Chemical processing in high-pressure aqueous environments. 8. Improved catalysts for hydrothermal gasification, *Industrial and Engineering Chemistry Research*, **45**(11), 3776–3781.

16. Minowa, T., Zhen, F., Ogi, T. (1998). Cellulose decomposition in hot-compressed water with alkali or nickel catalyst, *Journal of Supercritical Fluids*, **13**, 253–259.

17. Vogel, F., and Hildebrand, F. (2002). Catalytic hydrothermal gasification of woody biomass at high feed concentrations, *Chemical Engineering Transactions*, **2**, 771–777.

18. van Rossum, G., Potic, B., Kersten, S.R.A., Van Swaaij, W.P.M. (2009). Catalytic gasification of dry and wet biomass, *Catalysis Today*, **145**, 10–18.

19. Barrows, R.D., Elliott, D.C. (1984). Analysis of chemical intermediates from low-temperature steam gasification of biomass, *Fuel*, **63**, 4–8.

20. Elliott, D.C., Sealock, Jr., L.J., Butner, R.S., Baker, E.G., Neuenschwander, G.G. (1989). *Low-Temperature Conversion of High-Moisture Biomass: Continuous Reactor System Results*, PNL-7126 Pacific Northwest National Laboratory, Richland, Washington, October 1989.

21. Riley, C.C.A., et al. (1991). Investigation of the strong metal support interaction state of Ru/TiO$_2$ by ^{1}H nuclear magnetic resonance, *Catalysis Today*, **9**, 121–127.

22. Elliott, D.C., Hart, T.R., Schmidt, A.J., Neuenschwander, G.G. Rotness, L.J., Olarte, M.V., Zacher, A.H., Albrecht, K.O., Hallen, R.T., Holladay, J.E. (2013). Process development for hydrothermal liquefaction of algae feedstocks in a continuous-flow reactor, *Algal Research,* **2**, 445–454.

23. Elliott, D.C., et al. (2012). Chemical processing in high-pressure aqueous environments: 9. Process development of catalytic gasification of algae feedstocks, *Industrial and Engineering Chemistry Research,* **51**, 10768–10777.

24. Waldner, M.H., Vogel, F. (2005). Renewable production of methane from woody biomass by catalytic hydrothermal gasification, *Industrial and Engineering Chemistry Research,* **44**, 4543–4551.

25. Vogel, F., Hildebrand, F. (2002). Catalytic hydrothermal gasification of woody biomass at high feed concentrations, *Chemical Engineering Transactions,* **2**, 771–777.

26. Waldner, M.H., Krumeich, F., Vogel, F. (2007). Synthetic natural gas by hydrothermal gasification of biomass: selection procedure towards a stable catalyst and its sodium sulfate tolerance, *Journal of Supercritical Fluids,* **43**, 91–105.

27. Schubert, M., Regler, J.W., Vogel, F. (2010). Continuous salt precipitation and separation from supercritical water. Part 1: type 1 salts, *Journal of Supercritical Fluids,* **52**, 99–112; Part 2: type 2 salts and mixtures of two salts, *Journal of Supercritical Fluids,* **52**, 113–124; Schubert, M., Aubert, J., Müller, J.B., Vogel, F. (2012). Continuous salt precipitation and separation from supercritical water. Part 3: interesting effects in processing type 2 salt mixtures, *Journal of Supercritical Fluids,* **61**, 44–542.

28. Peterson, A.A., et al. (2008). In situ visualization of the performance of a supercritical-water salt separator using neutron radiography, *Journal of Supercritical Fluids,* **43**(3), 490–499.

29. Stucki, S., et al. (2009). Catalytic gasification of algae in supercritical water for biofuel productin and carbon capture, *Energy and Environmental Science,* **2**, 535–541; Haiduc, A.G., et al. (2009). SunCHem: an integrated process for the hydrothermal production of methane from microalgae and CO_2 mitigation, *Journal of Applied Phycology,* **21**, 529–541.

30. Zohrer, H., Mayr, F., Vogel, F. (2013). Stability and performance of ruthenium catalysts based on refractory oxide supports in supercritical water conditions, *Energy and Fuels,* **27**, 4739–4747.

31. Kersten, S.R.A., et al. (2006). Gasification of model compounds and wood in hot compressed water, *Industrial and Engineering Chemistry Research,* **45**, 4169–4177.

32. Knežević, D., Van Swaaij, W.P.M., Kersten, S.R.A. (2009). Hydrothermal conversion of biomass: part I; glucose conversion in hot compressed water, *Industrial and Engineering Chemistry Research,* **48**(10), 4731–4743.

33. Knežević, D., Van Swaaij, W.P.M., Kersten, S.R.A. (2010). Hydrothermal conversion of biomass: part II; conversion of wood, pyrolysis oil, and glucose in hot compressed water, *Industrial and Engineering Chemistry Research,* **49**(1), 104–112.

34. Kruse, A. (2008). Supercritical water gasification, *Biofuels, Bioproducts and Biorefining,* **2**, 415–437.

35. Kruse, A., et al. (2005). Influence of proteins on the hydrothermal gasification and liquefaction of biomass. 1. Comparison of different feedstocks, *Industrial and Engineering Chemistry Research,* **44**, 3013–3020; Kruse, A., Maniam, P., Spieler, F. (2007). Influence of proteins on the hydrothermal gasification and liquefaction of biomass. 2. Model compounds, *Industrial and Engineering Chemistry Research,* **46**, 87–96.

36. Boukis, N., et al. (2008). Biomass gasification in supercritical water. Experimental progress achieved with the VERNA pilot plant, *15th European Conference and Exhibition,* 1013–1016, May 7, 2007, Berlin, Germany.

37. Antal, Jr., M.J., et al. (2000). Biomass gasification in supercritical water, *Industrial and Engineering Chemistry Research,* **39**, 4040–4053.

38. DiLeo, G.J., Neff, M.E., Kim, S., Savage, P.E. (2008). Supercritical water gasification of phenol and glycine as models for plant and protein biomass, *Energy and Fuels,* **22**, 871–877.

39. Resende, F.L.P., Neff, M.E., Savage, P.E. (2007). Noncatalytic gasification of cellulose in supercritical water, *Energy and Fuels,* **21**, 3637–3643; Guan, Q., Savage, P.E., Wei, C. (2012). Gasification of alga *Nannochloropsis sp.* in supercritical water, *Journal of Supercritical Fluids,* **61**, 139–145.

40. Azadi, P., Khodadadi, A.A., Mortazavi, Y., Farnood, R. (2009). Hydrothermal gasification of glucose using Raney nickel and homogeneous organimetallic catalysts, *Fuel Processing Technology,* **90**, 145–151.

41. Azadi, P., et al. (2010). Hydrogen production by catalytic near-critical water gasification and steam reforming of glucose, *International Journal of Hydrogen Energy,* **35**, 3406–3414.

42. Afif, E., Azadi, P., Farnood, R. (2011). Catalytic hydrothermal gasification of activated sludge, *Applied Catalysis B: Environmental,* **105**, 136–143.

43. Azadi, P., et al. (2012). Hydrogen production from cellulose, lignin, bark, and model carbohydrates in supercritical wter using nickel and ruthenium catalysts, *Applied Catalysis B: Environmental,* **117–118**, 330–338; Azadi, P., et al. (2013). Catalytic reforming of activated sludge model compounds in supercritical water using nickel and ruthenium catalysts, *Applied Catalysis B: Environmental,* **134–135**, 265–273.

44. Onwudili, J.A., Williams, P.T. (2007). Hydrothermal catalytic gasification of municipal solid waste, *Energy and Fuels,* **21**, 3676–3683.

45. Muangrat, R., Onwudili, J.A., Williams, P.T. (2010). Reactions products from the subcritical water gasification of food wastes and glucose with NaOH and H_2O_2, *Bioresource Technology,* **101**, 6812–6821); Influence of alkali catalysts on the production of hydrogen-rich gas from the hydrothermal gasification of food processing wastes, *Applied Catalysis B: Environmental,* **100**, 440–449.

46. Muangrat, R., Onwudili, J.A., Williams, P.T. (2010). Influence of NaOH, Ni/Al2O3 and Ni/SiO2 catalysts on hydrogen production from subcritical water gasification of model food waste compounds, *Applied Catalysis B: Environmental,* **100**, 143–156.

47. Onwudili, J.A., Lea-Laangton, A.R., Ross, A.B., Williams, P.T. (2013). Catalytic hydrothermal gasification of algae for hydrogen production: composition of reaction products and potential for nutrient recycling, *Bioresource Technology,* **127**, 72–80.

48. Onwudili, J.A., Williams, P.T. (2013). Hydrogen and methane selectivity during alkaline supercritical water gasification of biomass with ruthenium-alumina catalyst, *Applied Catalysis B: Environmental,* **132–133**, 70–79.

49. Minowa, T., et al. (1994). Methane production from cellulose by catalytic gasification, *Renewable Energy,* **5**(11), 813–815.

50. Minowa, T., Ogi, T., Yokoyama, S.-Y. (1995). Hydrogen production from wet cellulose by low temperature gasification using a reduced nickel catalyst, *Chemistry Letters,* 937–938.

51. Minowa, T., Ogi, T. (1998). Hydrogen production from cellulose using a reduced nickel catalyst, *Catalysis Today,* **45**, 411–416 (1998).

52. Fang, Z., et al. (2004). Liquefaction and gasification of cellulose with Na_2CO_3 and Ni in subcritical water at 350°C, *Industrial and Engineering Chemistry Research,* **43**, 2454–2463.

53. Osada, M., Sato, T., Watanabe, M., Adschiri, T., Arai, K. (2004). Low-temperature catalytic gasification of lignin and cellulose with a ruthenium catalyst in supercritical water, *Energy and Fuels*, **18**, 327–333.

54. Osada, M., Sato, O., Arai, K., Shirai, M. (2006). Stability of supported ruthenium catalysts for lignin gasification in supercritical water, *Energy and Fuels*, **20**, 2337–2343.

55. Osada, M., Hiyoshi, N., Sato, O., Arai, K., Shirai, M. (2007a). Effect of sulfur on catalytic gasification of lignin in supercritical water, *Energy and Fuels*, **21**, 1400–1405.

56. Osada, M., Hiyoshi, N., Sato, O., Arai, K., Shirai, M. (2007b). Reaction pathway for catalytic gasification of lignin in the presence of sulfur in supercritical water, *Energy and Fuels*, **21**, 1854–1858.

57. Osada, M., Hiyoshi, N., Sato, O., Arai, K., Shirai, M. (2008). Subcritical water regeneration of supported ruthenium catalyst poisoned by sulfur, *Energy and Fuels*, **22**, 845–849.

58. Yamaguchi, A., et al. (2010). Gaseous fuel production from nonrecyclable paper wastes by supported metal catalysts in high-temperature liquid water, *ChemSusChem*, **3**, 737–741.

59. Yamaguchi, A., et al. (2008). Lignin gasification over supported ruthenium trivalent salts in supercritical water, *Energy and Fuels*, **22**, 1485–1492.

60. Elliott, D.C., et al. (2004). Chemical processing in high-pressure aqueous environments: 7. Process development of catalytic gasification of wet biomass feedstocks, *Industrial and Engineering Chemistry Research*, **43**, 1999–2004.

61. Elliott, D.C., Neuenschwander, G.G., Phelps, M.R., Hart, T.R., Zacher, A.H., Silva, L.J. (1999). Chemical processing in high-pressure aqueous environments. 6. Demonstration of catalytic gasification for chemical manufacturing wastewater cleanup in industrial plants, *Industrial and Engineering Chemistry Research*, **38**(3), 879–883.

62. Elliott, D.C., Hart, T.R. (1996). *Low-Temperature Catalytic Gasification of Food Processing Wastes: 1995 Topical Report*, PNNL-11246, Pacific Northwest National Laboratory, Richland, Washington, 1996.

63. Ro, K.S., Cantrell, K., Elliott, D.C., Hunt, P.G. (1999). Catalytic wet gasification of municipal and animal wastes, *Industrial and Engineering Chemistry Research*, **38**(3), 879–883.

64. Lupton, S. (2013). *Sustainable Transport Fuels from Biomass and Algal Residues Via Integrated Pyrolysis and Catalytic Hydroconversion*, 2013 DOE BETO IBR Project Peer Review, Alexandria, Virginia, May 21, https://www2.eere.energy.gov/biomass/peer_review2013/Portal/presenters/public/InsecureDownload.aspx?filename=UOP IBR Peer Review rev 1.pdf.

About the Author

Douglas C. Elliott has over 40 years of experience in research and project management at the Pacific Northwest National Laboratory. His work has mainly been directed toward development of fuels and chemicals from biomass and waste. His experience is primarily in high-pressure batch and continuous-flow processing reactor systems. This research has also involved him in extensive study of catalyst systems. Mr. Elliott's research has involved such subject areas as biomass liquefaction and hydroprocessing of product oils, catalytic hydrothermal gasification of wet biomass and wastewaters and chemicals production from renewable sources.

Mr. Elliott did his MBA in operations and systems analysis from the University of Washington in 1980 and B.S. in chemistry (departmental honors) from Montana State University in 1974.

He has authored or coauthored over 80 peer-reviewed papers and book chapters. He is listed as inventor or coinventor on 23 US patents with numerous foreign filings and was recognized as a Battelle Distinguished inventor in 2004. He is an active participant in IEA Bioenergy and currently serves as task leader for Task 34 on Pyrolysis, which involves him with technology leaders from six European countries and the United States.

Mr. Elliott is a member of the American Chemical Society and its Energy and Fuels Division as well as the Richland Section. He has won the following awards:

- Special Award for Excellence in Technology Transfer, 1989, US Federal Laboratory Consortium
- R&D 100 Award, TEES Biomass Gasification System, 1989, *Research & Development Magazine*

- R&D 100 Award, Petroleum Sludge Treatment, 1991, *Research & Development Magazine*
- Presidential Green Chemistry Challenge Award, Economic Conversion of Cellulosic Biomass to Chemicals, 1999, US Environmental Protection Agency

Chapter 22

Biomass to Fuels or Rather to Chemicals?

Jacob A. Moulijn, Jianrong Li, and Michiel Makkee

Department of Chemical Engineering, Delft University of Technology, Julianalaan 136, 2628BL Delft, the Netherland

j.a.moulijn@tudelft.nl

Biomass is an important raw material for energy and a variety of applications. Besides utilization as material it is a popular energy carrier. However, it will be shown that the majority of biomass has a structure that has the potential for producing attractive products, provided that dedicated processes are developed. As an example it is shown that for the two main classes, namely, (i) fats/oils and (ii) lignocellulosic biomass the production of valuable products by dedicated processes is in harmony with the structure of the biomass in contrast with the more robust but not very selective thermal processes. From an evaluation of the resources it is concluded that the contribution of biomass can be significant, provided that an adequate agricultural policy is implemented. In general, from an economic point of view, for favorable biomass resources production of chemicals is preferred above energy. It should be noted that for the

Biomass Power for the World: Transformations to Effective Use
Edited by Wim van Swaaij, Sascha Kersten, and Wolfgang Palz
Copyright © 2015 Pan Stanford Publishing Pte. Ltd.
ISBN 978-981-4613-88-0 (Hardcover), 978-981-4669-24-5 (Paperback), 978-981-4613-89-7 (eBook)
www.panstanford.com

widely available diluted aqueous waste systems, not the production of chemicals but the production of biogas for energy is optimal. A concrete example, the conversion of lignocellulosic biomass into sugar-based platform molecules, is discussed in some detail.

Biomass is the oldest source of energy for consumers and currently provides roughly 10% of total energy demand. Traditionally, biomass in the form of fuel wood, used for heating and cooking, is the main source of bioenergy, but liquid biofuel production has shown rapid growth during the last decade. Similar to crude oil, biomass can be and is processed in several ways. Figure 22.1 gives an overview showing the main approaches and processes for biomass conversion [1, 2].

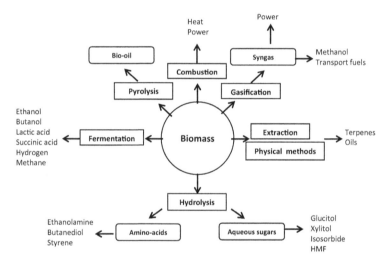

Figure 22.1 Main routes and their products in biomass conversion processes [1, 2].

The processes range from high-temperature thermochemical processes, viz., combustion, gasification, and pyrolysis, to more subtle (bio)chemical processes in the liquid phase, viz., hydrolysis and fermentation. The former category is very robust in the sense that the detailed structure of the biomass plays only a minor role and the complete organic part of the biomass is converted into a large pool of various chemical compounds. The latter category involves the selective conversion routes under milder conditions. For this class, the biomass structure offers the potential of efficient processes with high yields of target products.

Biomass can be gasified at high temperature in the presence of a substoichiometric amount of oxygen and the produced synthesis gas [a mixture of CO (carbon monoxide) and H_2 (hydrogen)] can be further processed to obtain the "normal" product spectrum, including, for example, methanol and Fischer–Tropsch liquids [3]. During pyrolysis, which takes place at intermediate temperature in the absence of oxygen, biomass is converted into a mixture of gas, solid material, and liquid, referred to as bio-oil [4]. This robust process can be used with a large variety of feedstocks, but the bio-oil produced is a low-quality fuel with low stability and the yield is modest.

All these processes are analogous to major processes applied in the oil refinery. However, because of the completely different structure of biomass compared to crude oil, fundamentally different processes are possible.

22.1 Structure of Biomass

In a simplified analysis, biomass resources can be divided in three major groups, viz., oils and fats, sugars, and lignocellulosic biomass. In addition, specific biomass materials contain large amounts of vegetable oils (palm oil) and proteins [5]. This is in particular the case for palm oil, sunflower, and algae. Algae and especially microalgae draw a lot of attention as a third-generation biofuel [6, 7], which fall outside the scope of this chapter.

Animal fats and vegetable oils are primarily composed of triglycerides, esters of fatty acids with glycerol. Figure 22.2 shows an example of a typical vegetable oil triglyceride.

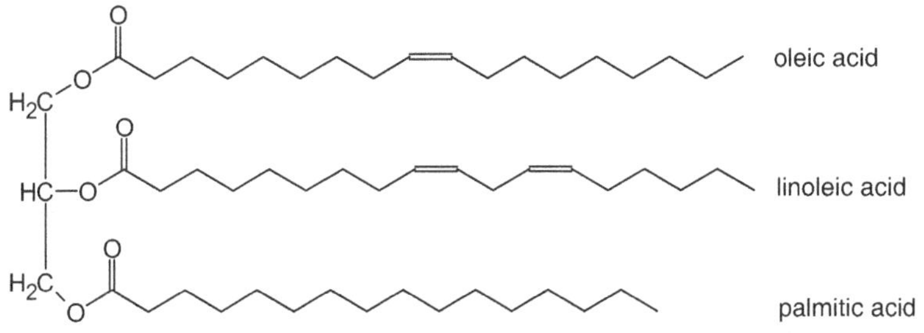

Figure 22.2 Typical chemical structure of vegetable oil.

Lignocellulosic biomass consists mainly of three components: cellulose (35–50 wt.%), hemicellulose (15%–25%), and lignin (15%–30%). Plant oils, proteins, different extractives, and ashes make up the rest of the lignocellulosic biomass structure (Fig. 22.3).

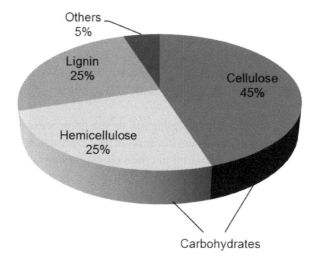

Figure 22.3 Average composition of lignocellulosic biomass.

The macrostructure of lignocellulosic biomass is complex. A schematic representation is given in Fig. 22.4.

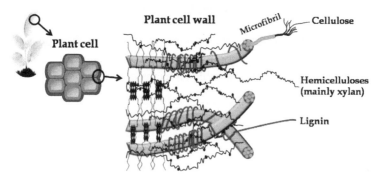

Figure 22.4 Structure of lignocellulosic biomass [8].

The cell walls are built up from cellulose and hemicellulose, held together by lignin. The hemicellulose in turn is a matrix containing cellulose fibers.

Cellulose is the most abundant organic polymer on earth and its chemical structure, which is largely crystalline, is remarkably simple. It consists of linear polymers of cellobiose, a dimer of glucose (see Fig. 22.5). The multiple hydroxyl groups of the glucose molecule form hydrogen bonds with neighbor cellulose chains making cellulose microfibrils of high strength and crystallinity.

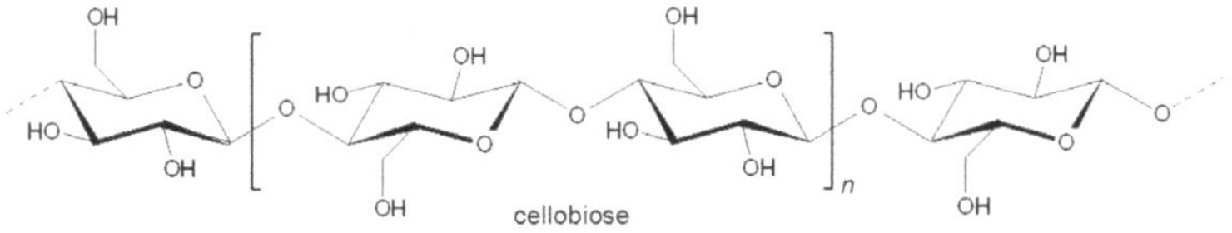

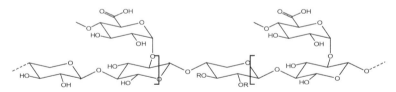

cellobiose

Figure 22.5 Chemical composition of cellulose; n = 2500–5000.

Hemicellulose is chemically related to cellulose in the sense that it is comprised of a carbohydrate backbone. However, due to its random and branched structure hemicellulose is amorphous. It has a more complex composition than cellulose. Figure 22.6 shows the structure of xylan, a polymer representative of hemicellulose. Whereas cellulose is completely built up from glucose monomers, hemicellulose consists of a mixture of five-carbon sugars (xylose, arabinose), six-carbon sugars (glucose, mannose, galactose), and uronic acids (e.g., glucoronic acid) (see Fig. 22.7). In hemicellulose xylose is the most abundant monomer.

Figure 22.6 Chemical composition of xylan, a typical hemicellulose; n = 3–40.

Lignin, the third main component (see Fig. 22.8) is an amorphous three-dimensional polymer which fills the spaces in the cell wall between cellulose and hemicellulose. It is aromatic and hydrophobic in comparison with cellulose and hemicellulose. The complexity and variability of the lignin composition and its chemical resistance make its conversion to base chemicals difficult, but because of its high energy density direct combustion is often favorable. A comprehensive review on the catalytic valorization of lignin has been published [9].

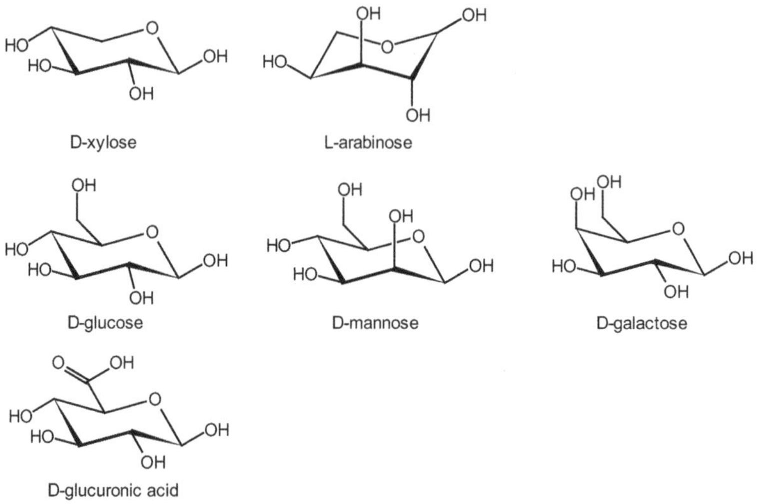
D-xylose
L-arabinose
D-glucose
D-mannose
D-galactose
D-glucuronic acid

Figure 22.7 Monomers present in hemicellulose.

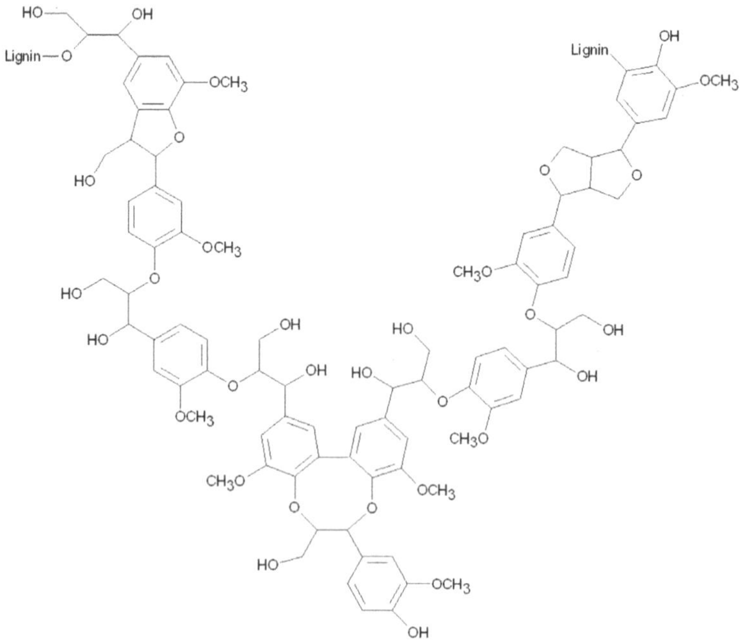
Lignin

Figure 22.8 Chemical composition of (a small piece of) lignin polymer.

22.2 Biomass Resources: Which Are the Most Attractive?

Which biomass is the most attractive feedstock? For an answer we need numbers for resources and markets, including the desired product(s), and both economical and ethical considerations should be taken into account.

Obviously, there is a danger of energy based on biomass competing with the food chain. Ethically, it is preferable to give priority to food and feed applications. There are several biomass resources where such a competition is not the case, for example, waste streams and in principle the large amounts of lignocellosic feedstocks such as wood. When we want to substitute mineral oil by biomass, some kind of agricultural policy is needed. Figure 22.9 gives the composition of major crops. Clearly, cellulose, hemicellulose, and lignin dominate, but not surprisingly, proteins and vegetable oils are main components in specific resources (palm oil, soybeans, for example). In essence, the attractiveness of biomass is based on the fact that they continuously capture solar energy. Compared to fossil fuels providing *photons from the past*, biomass offers *photons from yesterday*. The question arises, which crops harvest photons most efficiently?

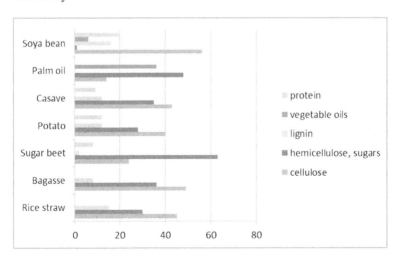

Figure 22.9 Composition of some major lignocellulosic crops. Data taken from Ref. [5].

The productivity in harvesting solar energy can be estimated from the dry weight per hectare per year (see Table 22.1). Sugarcane shows the highest productivity followed by sugar beet, while soybeans have a very low productivity [11]. Later it will become clear that the vegetable oil–rich biomass (soybeans, palm oil) have the advantage of simple engineering leading to low production costs, compensating for the low to modest harvest production numbers.

Table 22.1 Productivity of selected crops

Crop	Largest producer	Production* (t/ha/a)
Cassava	Nigeria	50
Grass	Netherlands	14
Jatropha seeds	Zimbabwe	8
Maize	USA (Iowa)	14
Rice straw	Egypt	11
Palm oil	Malaysia	25
Potato	Netherlands	65
Soybean	USA (Illinois)	4
Sugar beet	Germany	100
Sugarcane	Brazil	125
Sunflower	France	6

Note: Data from Refs. [5, 11].
*Best-practice technology.

Figure 22.10 gives a clear message with respect to the practical availability of biomass when we compare the numbers with the current usage of crude oil, viz., 170 EJ/a [1, 10]. Thus, waste streams are sizable but not sufficient. Biomaterials are a sizable production but they should be used as such and use as feedstock for biomass conversion is in general not wise. Clearly, use of degraded land and surplus land is needed for a major impact. The good news is that in principle this land is available [12], although the real availability will still be subject to a political debate. Also the availability of water, fertilizers, groundwater, transportation, etc., has to be taken into account.

When biomass is to be used as feedstock for the chemical industry, the amounts required are much lower than for global energy: we estimate that for the year 2050 50 EJ/a are sufficient for producing the organic feedstock for the chemical industry [1]. The data in Fig. 22.10 show that plenty of biomass is available for such a production.

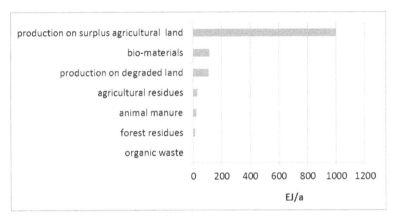

Figure 22.10 Biomass resources which can be made available. Adapted from Ref. [11].

22.3   Optimal Products: Energy or Chemicals?

Biomass has a highly functionalized structure and contains a large amount of oxygen, in sharp contrast with crude oil. The highly functionalized structure has resulted in an extensive biotechnological industry, the most important process being the production of ethanol by fermentation of sugars obtained from sugarcane and corn. Another category of potential processes involves dissolution and depolymerization (by hydrolysis) of lignocellulosic biomass into the corresponding monomers. Such chemical processes are hot topics in R&D. Obvious monomers are sugars, the most important being glucose produced from either starch (in direct competition with the food chain) or cellulose and xylose from hemicellulose. The sugars may subsequently be fermented or converted to useful chemicals by (bio)catalytic means.

One might wonder whether it would not be advisable to process biomass in the existing industrial infrastructure such as oil refineries. The simplest way would be to defunctionalize the biomass, so the stoichiometry of the products approaches that of crude oil, and subsequently utilize these products as feedstock for the refinery. It has been proposed to defunctionalize by hydrodeoxygenation (HDO), similar to other hydrotreating processes such as hydrodesulfurization (HDS), a major process in the oil refinery. However, as will be shown later, for lignocellulosic biomass the economics of such a process are questionable.

Apart from the cost factor, there is the question what the optimal products of biomass conversion are. A priori there is no need to copy the hydrocarbon industry. In principle, it is better to produce more oxygen-containing products. Firstly, when less oxygen atoms need to be removed, less process steps are needed and consumption of hydrogen (typically used for removing oxygen) is reduced. Secondly, the product quality might be better with oxygen atoms present. For instance, in the case of diesel fuel, oxygen-containing molecules often result in smokeless combustion and, as a result, in less pollution [13]. Thirdly, when less oxygen is removed less mass is lost in the conversion steps.

A fundamental point is the following: Intuitively, it is clear that it would be attractive to produce products which have a structure similar to that of the feedstock monomers. The consequence of making such products would be that only a minimal number of bond breaking and forming steps would be required. Usually this leads to an efficient process, as far as the chemistry is concerned. Therefore, in deciding upon the desired products, it makes sense to take into account the structure of the feedstock. One of the great challenges in biomass applications is to identify those attractive molecular structures.

22.4 Biomass Utilization: From Combustion and Anaerobic Fermentation to a Biorefinery

At present the combustion of biomass is the most important technology, probably mainly because of its simplicity and easiness of scaling down. The conversion of "leftover" biomass in biogas, a mixture of CH_4 and CO_2 (1,2), by anaerobic fermentation is the other important route. In wastewater treatment this technology is widely applied. The technology is robust with respect to the raw material, there is no product separation problem (products in the gas phase) and it is easy to scale down. Obviously, the value of the product is limited, but in diluted aqueous streams, the production of specific products does not make sense.

Biomass can be used in a variety of applications where attractive products are generated. In general, multiple products will be produced and the collected biomass should be processed completely. In analogy with the oil refinery the term "biorefinery" is used for a facility combining the production of transport fuels, chemicals, and other energy carriers. Several ideas have been put forward for a design of the biorefinery.

An attractive option would be to give the biomass a kind of pretreatment enabling using the existing industrial infrastructure, in particular oil refineries. Why not defunctionalize the biomass so that the stoichiometry of the product approaches that of crude oil and subsequently process them further in (existing) oil refineries? However, it is easily shown that such process is not economical.

Taking glucose a model for the cellulose and hemicellulose fraction (corresponding roughly to 75% of the lignocellulose biomass) and $(CH_2)_6$ as a typical oil refinery fuel component:

$$C_6H_{12}O_6 + 6H_2 \rightarrow C_6H_{12} + 6H_2O \tag{22.1}$$

This equation is very informative: in an ideal case of a full conversion with a 100% efficiency on a mass basis the yield is less than 50% (see Table 22.2). According to present day sugar prices the value of the product is less than that of the reactant. When we substitute glucose by lignocellulosic feedstock at first sight it seems economical. However, for several reasons such a conclusion is too optimistic. Firstly, the cost of H_2 is considerable, making the economic margin small. In addition, the experience has learned that the price of lignocellulosic feestock is highly dependent on the local situation and tends to increase with time. Secondly, processing costs of lignocellulosic biomass into glucose are relatively high (see later in this chapter). Perhaps the scene has changed with the shale gas/oil boom. The hydrogen price considerably decreased caused by the development of shale gas in the U.S. This development might make (temporally?) extensive hydroprocessing of biomass economically feasible.

Table 22.2 The hydrodeoxygenation of glucose into hexane [10]

	Amount* (t)	Price (kEuro/t)	Raw materials (kEuro/$t_{C_6H_{12}}$)
$C_6H_{12}O_6$	180	0.5	1.1
H_2	12	2.2	0.3
C_6H_{12}	84	0.7	0.7
Lignocellulosics	220	0.1	0.25

*In agreement with Eq. 22.1.

The pyrolysis process is based on thermal decomposition, and as a consequence, no hydrogen is consumed. This process is robust and the bio-oil produced is relatively easily separated. Cons are the poor quality of the bio-oil (generally requiring extensive hydroprocessing) and the low yields because of the intrinsically low selectivity of the chemical reactions occurring.

On the basis of Fig. 22.1 we can distinguish two main types of biorefineries, (i) a thermochemical refinery with a syngas and/ or a pyrolysis platform and (ii) a refinery based on a fermentation platform and/or a sugar platform. Depending on the local situation and the desired product spectrum a combination might be optimal.

Figure 22.11 shows an often cited example of a conceptual biorefinery scheme based on a sugar platform and a syngas platform. A sugar platform breaks down lignocellulosic biomass into different types of sugars for fermentation or other (bio)chemical processing into various fuels and chemicals. In a syngas platform, biomass is transformed into synthesis gas by gasification. The synthesis gas can then be converted into a range of fuels and chemicals by established catalytic chemical processes. The residues mainly consist of lignin and cogeneration of heat and power (electricity) suggests itself.

This design is very general. We believe that an optimal design primarily is dominated by the kind of feedstock and the desired products. Earlier we made the remark that it makes sense to minimize the number of bond-breaking and bond-making steps. When we convert biomass into syngas and subsequently fuels, we certainly do not follow this concept! Thus, we prefer subtle processes in harmony with the structure of the biomass.

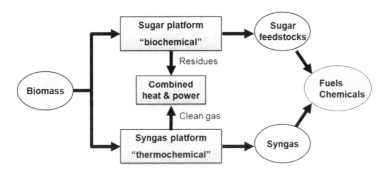

Figure 22.11 Concept of a biorefinery based on a combination of sugar and a syngas platform [14].

When vegetable oil is the feedstock the obvious application is the use as food or cattle feed. When a conversion in order to replace fossil fuels is aimed at, then the structure of the vegetable oil suggests that a minor conversion leads to diesel (see Fig. 22.12).

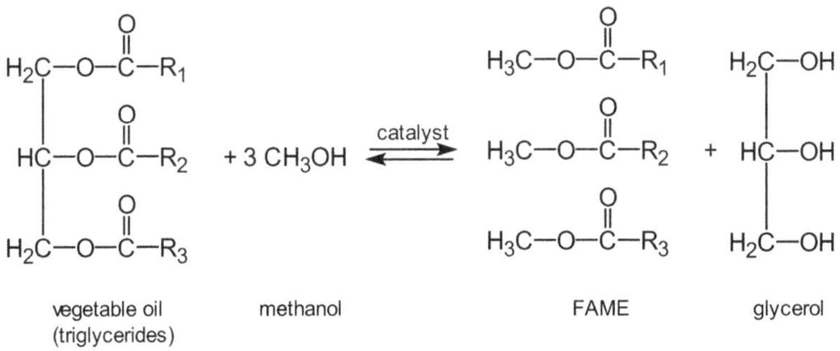

$$
\begin{array}{c}
\text{H}_2\text{C}-\text{O}-\overset{\overset{\text{O}}{\|}}{\text{C}}-\text{R}_1 \\[2mm]
\text{HC}-\text{O}-\overset{\overset{\text{O}}{\|}}{\text{C}}-\text{R}_2 \quad +3\ \text{CH}_3\text{OH} \underset{}{\overset{catalyst}{\rightleftharpoons}} \\[2mm]
\text{H}_2\text{C}-\text{O}-\overset{\overset{\text{O}}{\|}}{\text{C}}-\text{R}_3
\end{array}
\qquad
\begin{array}{c}
\text{H}_3\text{C}-\text{O}-\overset{\overset{\text{O}}{\|}}{\text{C}}-\text{R}_1 \qquad \text{H}_2\text{C}-\text{OH} \\[2mm]
\text{H}_3\text{C}-\text{O}-\overset{\overset{\text{O}}{\|}}{\text{C}}-\text{R}_2 \quad + \quad \text{HC}-\text{OH} \\[2mm]
\text{H}_3\text{C}-\text{O}-\overset{\overset{\text{O}}{\|}}{\text{C}}-\text{R}_3 \qquad \text{H}_2\text{C}-\text{OH}
\end{array}
$$

vegetable oil   methanol   FAME   glycerol
(triglycerides)

Figure 22.12 Transesterification of vegetable oil with methanol. FAME stands for fatty acid methyl esters. R_1, R_2, and R_3 are long-chain hydrocarbons.

The process is simple and the yields will be high. Note that the conversion steps in terms of chemistry are very limited. Only a few bonds are broken and made. Without going in detailed calculations, it is to be expected that this process is an example of a combination of an expensive raw material and a low-cost engineering process. This process is at present the most important process for the production of biodiesel [15]. An alternative is hydroprocessing of the vegetable oils, resulting in a product ("green diesel") without oxygen [16]. Compared with the carbohydrate case discussed before the oxygen content of green diesel is low, and as a consequence, the costs of the hydrogen in this case are not prohibitive.

Thus, in designing a biorefinery for lignocellulosic feedstocks, different reaction schemes suggest themselves. The structure of biomass suggests a key position for sugar-based platform chemicals such as glucose. Glucose is an excellent platform molecule allowing the production of a large variety of biobased chemicals such as sorbitol (by reduction) and hydroxymethylfurfural (HMF) (by dehydration of the fructose intermediate).

HMF has a large potential in the production of substitutes for oil-based monomers in the synthesis of large-scale polymers including polyesters, polyamides, and polyurethane (see Fig. 22.13). A notable example is furan dicarboxylic acid (FDCA), which has a chemical structure and properties similar to terephthalic acid (TPA). It can substitute terephthalate acid in the production of polyesters, giving poly(ethene furanoate) (PEF) instead of poly(ethene terephthalate) (PET). The success of this technology depends on the design of efficient routes for HMF production. Usually HMF is indirectly formed by the dehydration of glucose, catalyzed by mineral acids. An issue in this reaction is the instability of HMF under acidic conditions. The company Avantium has developed an approach avoiding this problem of unstable HMF by using an alcohol or organic acid forming a relatively stable HMF ether derivative, which then undergoes catalytic oxidation to form FDCA [17,18].

Thus, glucose is an excellent platform molecule. However, although numerous studies have been published showing its high potential, not so much is known about the first step, the conversion of biomass into glucose. Acid catalyzed hydrolysis is extensively used [19, 20]. Most common is using diluted HCl or H_2SO_4 at temperatures of 100°C–150°C. Not surprisingly, compared to cellulose, hemicellulose is converted at a lower temperature. Higher acid concentration results in higher sugar yields.

The company Roquette has pioneered producing base chemicals starting from starch from corn (competition with the food chain) [21, 22]. They developed a biorefinery where starch is hydrolyzed in an enzymatic process, resulting in, among others, glucose. The glucose produced is the base chemical for a large number of biotechnological and chemical processes, giving a variety of products including polyols, sorbitol, and isosorbide, for which compound they are presently the market leader.

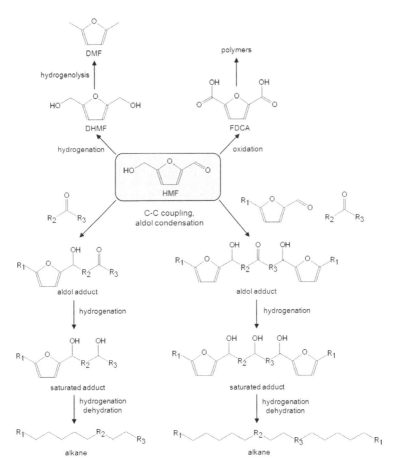

Figure 22.13 HMF as a platform molecule for the production of chemicals and biofuels. HMF, hydroxymethylfurfural; DHMF, dihydroxymethylfurfural; DMF, dimethylfuran; FDCA, furan dicarboxylic acid.

Dumesic et al. report a good example of a process based on subtle chemistry, resulting in valuable products. They did not limit themselves to exploratory chemistry but they included in their research the development of a separation strategy resulting in optimized concentrations in the chemicals steps. In their process the carbohydrate fraction of lignocellulosic biomass is converted into butene via, respectively, livulinic acid and γ-valerolactone; the latter also functions as solvent in the acidic hydrolysis step [23]. The butene produced is subsequently converted into butane oligomers, which can be used as transport fuel.

22.5 Biorefinery Based on Depolymerization of Cellulose and Hemicellulose

In a recent process the carbohydrate fraction of lignocellulosic feedstock was dissolved and depolymerized followed by hydrogenation into sorbitol. Subsequently, the polarity of the product was reduced by defunctionalization of the sorbitol produced in order to facilitate separation [24–28].

Cellulose is the most refractory part of the carbohydrate fraction of biomass. In literature it is well documented that cellullose can be depolymerized in the appropriate solvents. Several solvents have been reported, including mineral acids, aqueous HCl, and H_2SO_4, ionic liquids, and salt hydrates, for instance, $ZnCl_2$. The critical aspects of such processes are undesired degradation reactions and the difficult separation of the products from the solvent system. $ZnCl_2$ has been selected in a recent process development because of its good performance and low cost.

Figure 22.14 shows the superb performance of $ZnCl_2$: at room temperature the cellulose fibers are clearly visible and upon heating they quickly disappear: at 60°C cellulose is dissolved, corresponding with a residence time within seven minutes!

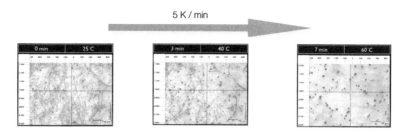

Figure 22.14   Dissolution of cellulose in $ZnCl_2$ hydrate in a temperature-programmed experiment under a microscope. The images at the left, the middle, and at the right represent, successively, the initial situation at room temperature, the situation after 3 min (40°C), and the situation after 7 min (60°C).

In the same solvent hydrolysis takes place resulting in glucose as main product. Careful analysis showed that to a limited extent isomerization and oligomerization takes place besides some dehydration (to HMF). This phenomenon reflects the fact that glucose is a rather unstable molecule. Sorbitol, the monohydrogenated product of glucose is more stable.

An innovative scheme is based on a combination of hydrolysis and hydrogenation [26]. When these reactions are carried out simultaneously under well-established reaction conditions, it might be expected that sequential side reactions of glucose are suppressed. This appears to be the case, as is demonstrated the simultaneous hydrolysis and hydrogenation of cellobiose (Fig. 22.15).

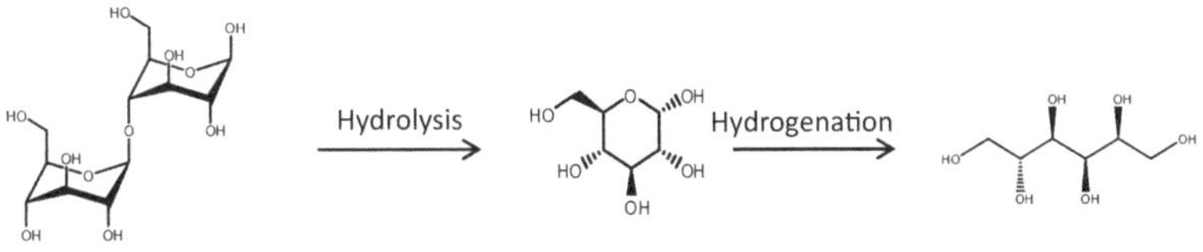

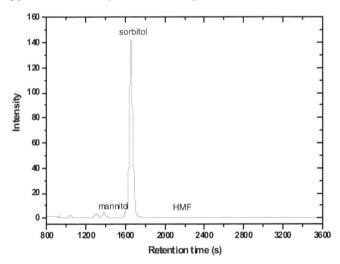

Figure 22.15 Reaction of cellobiose into sorbitol.

Typical results are presented in Fig. 22.16.

Figure 22.16 Simultaneous hydrolysis and hydrogenation of cellobiose ($125°$, 75 min, Ru/C hydrogenation catalyst) in $ZnCl_2$ hydrate. Reproduced from Ref. [26] with permission from The Royal Society of Chemistry.

Figure 22.16 shows a clean reaction with a yield > 95%. However, this result does not prove that this technology can be used in practice. Economic separation of sorbitol from this solvent will be very difficult, because of the high polarity of both the product and the solvent. Thus, it was decided to attempt to reduce the polarity. It is unavoidable to do this by removing functional groups. Several reactions are possible. In particular the following reactions suggest themselves: hydrogenation, dehydration, alkylation, esterification, and dehydration. The main criteria for a choice are the costs and value of the product(s). As discussed before, extensive hydrogenation is not the most promising route. We have chosen a reaction scheme where no other reactants are involved, viz., dehydration into cyclic ethers. The solvent used for the hydrolysis shows also activity in the dehydration. Figure 22.17 shows results of the dehydration of sorbitol, illustrating the serial kinetic scheme and the high yields (>85%) obtained. The main by-products are anhydrosorbitols ("sorbitan," a commercial product). The underlying reaction mechanism is discussed in Ref. [28].

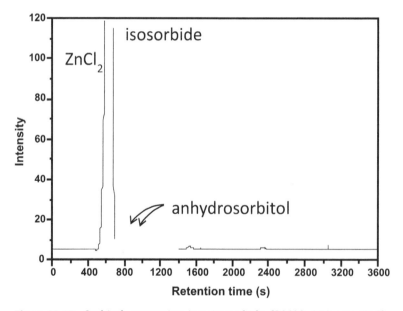

Figure 22.17 Sorbitol conversion into isosorbide (200°C, 270 min, $ZnCl_2$ hydrate solvent). Reproduced from Ref. [27] with permission from The Royal Society of Chemistry.

For cellulose conversion this leads to the scheme in Fig. 22.18.

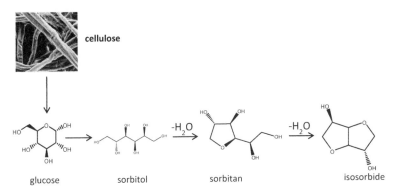

cellulose

glucose   sorbitol   sorbitan   isosorbide

Figure 22.18 Reaction scheme for the conversion of cellulose into isosorbide.

On the basis of this technology a dedicated biorefinery can be designed, as illustrated in Fig. 22.19.

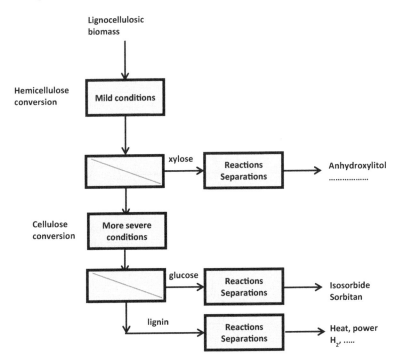

Figure 22.19 Conceptual design biorefinery based on the hydrolysis and reactions scheme.

The lignocellulose can be fed after drying/cleaning. In the first step, hemicellulose is removed, being the most reactive towards hydrolysis. In the second step cellulose is dissolved and hydrolyzed, forming glucose; lignin is inert and removed as a solid. In this scheme the lignin is used as an energy source or it is gasified, giving H_2 as a product. The glucose (and oligomers) are converted (hydrogenation and two dehydration steps) in a serial-reaction network, generating sorbitan and isosorbide.

22.6 Applications of Isosorbide

Isosorbide is a molecule with two functional groups and can act as monomer in polymerization processes (Fig. 22.20). Presently it is produced from starch and it has already a significant market share [21].

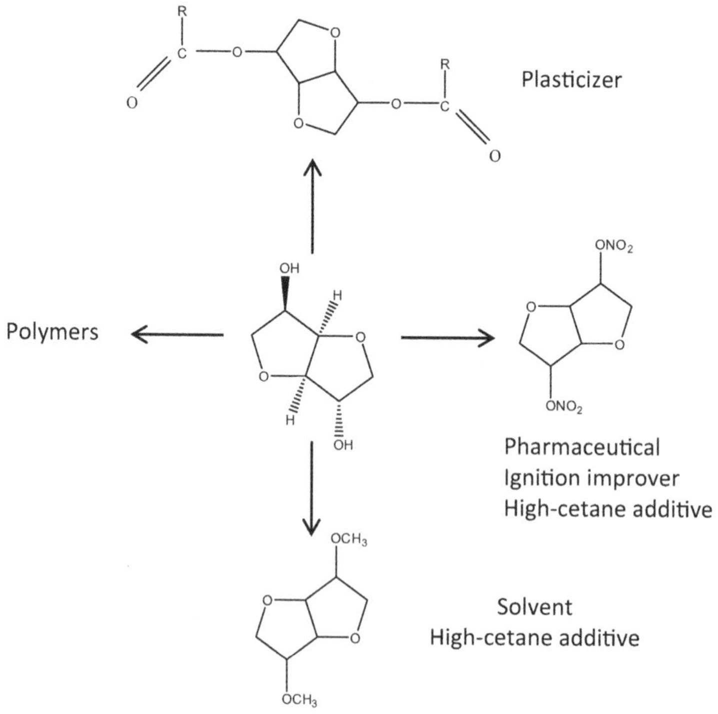

Figure 22.20 Isosorbide, a platform chemical in a variety of applications.

22.7   Concluding Remarks

An optimal process design is related to the characteristics of raw material and products. In Fig. 22.21 the products of biomass conversion are listed, roughly ordered according to the value. In addition criteria for the "ideal" plant are given.

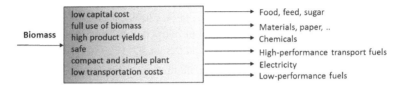

Figure 22.21   The ideal plant for the utilization of biomass with its products.

Lignocellulosic biomass is abundantly available and it is the preferred biomass feedstock. The carbohydrate fraction (75%) of lignocellulosic biomass has a highly functionalized structure, and, as a consequence, it is an ideal raw material for several O-containing base chemicals. Subtle hydrolysis-based technology suggests itself instead of high-temperature processes such as pyrolysis and gasification.

It should be noted that such subtle technology is not always preferred. For instance, in urban wastewater treatment robust, nonspecific technology is required: in anaerobic treatment biogas is produced, a product low in the hierarchy of Fig. 22.21, but the process is economical. When fuels are explicitly the desired product, conventional technologies such as pyrolysis and gasification can be preferred because they are robust and often product separation is relatively straightforward, but in terms of economy they are problematic with respect to yields and product value.

In general, for an economic process the production of well-selected chemicals has more potential than the production of fuels.

References

1. Sanders, J.P.M., Clark, J.H., Harmsen, G.J., Heeres, H.J., Heijnen, J.J., Kersten, S.R.A., van Swaaij, W.P.M., Moulijn, J.A. (2012). Process intensification in the future production of base chemicals from biomass, *Chemical Engineering and Processing*, **51**, 117–136.

2. Moulijn, J.A., Makkee, M., van Diepen, A.E. (2013). *Chemical Process Technology 2nd edition*, Wiley, West Sussex.

3. Puig-Arnavat, M., Bruno, J.C., Coronas, A. (2010). Review and analysis of biomass gasification models, *Renewable Sustainable Energy Reviews*, **14**, 2841–2851.

4. Bridgwater, A.V. (2012). Review of fast pyrolysis of biomass and product upgrading, *Biomass and Bioenergy*, **38**, 68–94.

5. B. Brehmer, Boom, R.M., Sanders, J.P.M. (2009). Maximum fossil fuel feedstock replacement potential of petrochemicals via biorefineries, Chemical Engineering *Research and Design*, **87**, 1103–1119.

6. Brennnan, L., Owende, P. (2010). Biofuels from microalgae: A review of technologies for production, processing, and extractions of biofuels and co-products, *Renewable and Sustainable Energy Reviews*, **14**, 557–577.

7. Mata, M.M. (2010). Microalgae for biodiesel production and other applications: a review, *Renewable and Sustainable Energy Reviews*, **14**, 217–232.

8. Ratanakhanokchai, K., Waeonukul, R., Pason, P., Tachaapaikoon, C., Kyu, K.L., Sakka, K., Kosugi, A., Mori, Y. (2013). INTECH, http://creativecommons.org/licences/by/3.0.

9. Zakzeski, J., Bruijnincx, P.C.A., Jongerius, A.L., Weckhuysen, B.M. (2010). The catalytic valorization of lignin for the production of renewable chemicals, *Chemical Reviews*, **110**, 3552–3599.

10. Moulijn, J.A., Babich, I.V. (2011). The potential of biomass in the production of clean transportation fuels and base chemicals, In *Production and Purification of Ultraclean Transportation Fuels*, 1088, American Chemical Society, 65–77.

11. Hoogwijk, M., Faaij, A., Eickhout, B., de Vries, B., Turkenburg, W. (2005). Potential of biomass energy out to 2100, for four IPCC SRES land-use scenarios, *Biomass and Bioenergy*, **29**, 225–257.

12. European Environment Agency. *Land Available for Biomass Production for Energy*, http://www.eea.europa.eu/data-and-maps/figures/land-availablefor-biomass-production-for-energy.

13. de Almeida, R.M., Klotz Rabella, C.R., US patent US20100064574A1.

14. NREL (2009). *What Is a Biorefinery?* http://www.nrel.gov/biomass/biorefinery.html.

15. Lestari, S., Mäki-Arvela, P., Beltramini, J., Max Lu, G.Q., Murzin, D.Y. (2009). Transforming triglycerides and fatty acids into biofuels, *ChemSusChem*, **2**, 1109–1119.

16. Kalnes, T.N., Marker, T., Shonnard, D.R., Koers, K.P. (2008). *Green diesel production by hydrorefining, Biofuels Technology*, 7–11, www.Biofuels-tech.com, addressed 14-04-2014.

17. Gruter, G.J.M., Dautzenberg, F. (2007). EU patent 1834950A1.

18. Eerhart, A.J.J.E., Huijgen, W.J.J., Grisel, R.J.H., van der Waal, J.C., de Jong, E., Dias, A. de Sousa, Faaij, A.P.C., Patel, M.K. (2014). Fuels and plastics from lignocellulosic biomass via the furan pathway: a technical analysis, *RSC Advances*, **4**, 3536–3549.

19. Liu, S., Wang, Y., Buyondo, J.P., Wang, Y., Garver, M. (2013). Biochemical conversion, in *Integrated Biorefineries*, Eds. Stuart, P.R., El-Halwagi, M.M., Taylor and Francis, 590–649.

20. Katzen, R., Schell, D.J. (2010). Lignocellulosic feedstock biorefinery: history and plant development for biomass hydrolysis, in *Biorefineries: Industrial Processes and Products*, Eds. Kamm, B., Gruber, P.R., Kamm, M., Wiley-VCH Verlag.

21. Rupp-Dahlem, C., ec.europa.eu/research/energy/pdf/gp/gp_events/biorefinery/09_rupp-dahlem_en.pdf.

22. http://www.mitsubishi-chemical.de/no_cache/products/bio-polymers/durabio/index.html.

23. Jeehoon, H., Murat Sen, S., Alonso, D.M., Dumesic, J.A., Maravelias, C.T. (2014). A strategy for the simultaneous catalytic conversion of hemicellulose and cellulose from lignocellulosic biomass to liquid transportation fuels, *Green Chemistry*, **16**, 653–661.

24. de Almeida, R.M., Li, J., Nederlof, C., O'Connor, P., Makkee, M., Moulijn, J.A. (2010). Cellulose conversion to isosorbide in molten salt hydrate media, *ChemSusChem*, **3**(3), 325–328.

25. Makkee, M., Moulijn, J.A., de Almeida, R.M., O'Connor, P. *Converting polysaccharide-containing material to fuel additive in inorganic molten salt medium involves dissolving the material then hydrolyzing, hydrogenating monosaccharides, dehydrating sugar alcohols, and derivatizing (di)anhydro sugars.* WO2010106053-A2; WO2010106053-A3.

26. Li, J., Soares, H.S.M.P., Moulijn, J.A., Makkee, M. (2013). Simultaneous hydrolysis and hydrogenation of cellobiose to sorbitol in molten salt hydrate media, *Catalysis Science and Technology*, **3**(6), 1565–1572.

27. Li, J., Spina, A., Moulijn, J.A., Makkee, M. (2013). Sorbitol dehydration into isosorbide in a molten salt hydrate medium, *Catalysis Science and Technology*, **3**(6), 1540–1546.

28. Li, J. Buijs W., Berger, R.J., Moulijn, J.A., Makkee, M. (2014). Sorbitol dehydration in a $ZnCl_2$ molten salt hydrate medium: molecular modeling, *Catalysis Science and Technology*, **4**, 152.

About the Authors

Jacob A. Moulijn is a professor emeritus of chemical engineering  at the Delft University of Technology, the Netherlands (1990–2007) and the University of Amsterdam (1986–1990). He was director R&D of a start-up company in biomass conversion (2010–2012). At present he is a distinguished research professor at the Cardiff University (2007–present) and founder/director of a consultancy company.

His research interests include catalysis engineering, structured reactors, petroleum conversion, environmental catalysis (diesel soot abatement, N_2O removal), selective hydrogenation, coal conversion ([catalytic] gasification, pyrolysis, combustion), and biomass conversion (liquefaction, chemicals, and fuel production).

He has supervised over 300 master and 70 PhD students. He is a co-author of over 750 technical papers and 2 books, editor of 7 books, and holder of several patents (reactor design, zeolitic membranes, catalyst development, biomass conversion).

Jianrong Li was born (1981) in Tianjin, China. She did her PhD  research in the catalysis engineering group in the Department of Chemical Engineering at the Delft University of Technology (2007–2011), under supervision of Prof. Dr. Jacob A. Moulijn and Prof. Dr. Ir. Michiel Makkee. She received her doctor's degree in 2012 on the topic "Lignocellulosic Biomass Conversion into Platform Chemicals in Molten Salt Hydrate Media" (ISBN 978-90-9027073-9). At present she is working at the Van Swinden Laboratory (VSL), the Dutch Metrology Institute, as project leader in the fields of metrology in health, energy, and environment. She is the work package and/or task leader of several European joint research projects. She is a co-inventor of 1 patent, a co-author of 6 journal articles, and a co-author of over 15 poster and oral presentations at international conferences.

Michiel Makkee (1954) did his master and PhD studies on catalysis and organic chemistry at the Delft University of Technology. After more than six years at Exxon Chemicals, he was nominated as associate professor in the field of catalysis engineering and process development at the Delft University of Technology (1990). He was nominated as special part-time professor at Politecnico di Torino, Italy (2011). He has supervised more than 150 master and 25 PhD students. Prof. Dr. Ir. Makkee is a co-inventor of 25 patents/patent applications and a co-author of 200 peer-reviewed papers with an H-factor of 47. He has lectured in several courses on chemical and reactor engineering and is a co-author of the book *Chemical Process Technology* (2nd edition).

Index

Color Insert

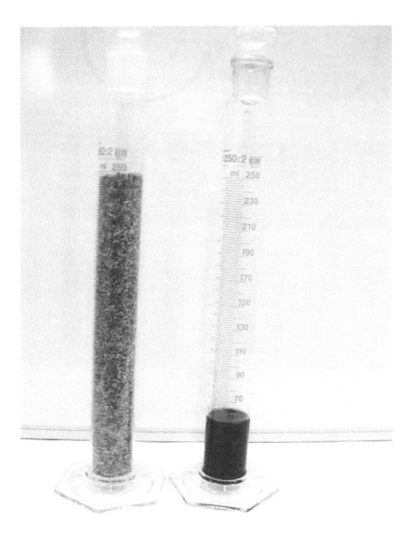

Figure 1.49

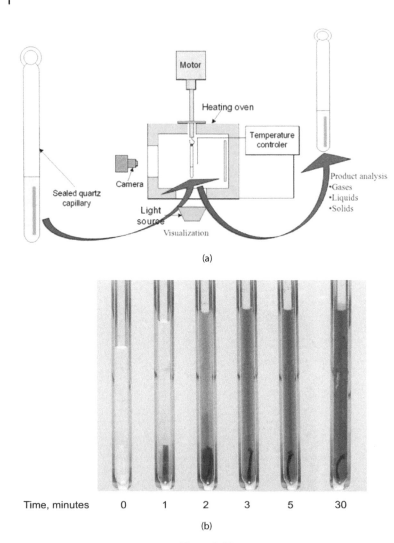

(a)

Time, minutes   0   1   2   3   5   30

(b)

Figure 1.61

Figure 5.9

Figure 6.9

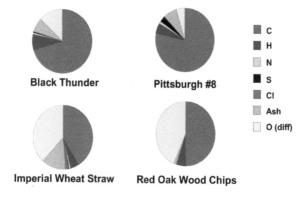

Black Thunder **Pittsburgh #8**

- C
- H
- N
- S
- Cl
- Ash
- O (diff)

Imperial Wheat Straw **Red Oak Wood Chips**

Figure 7.7

Figure 8.1

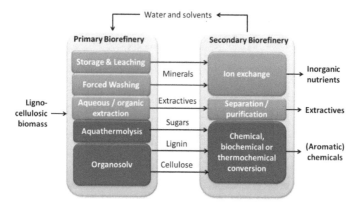

Figure 9.1

Figure 10.3

Figure 11.3

Figure 14.8

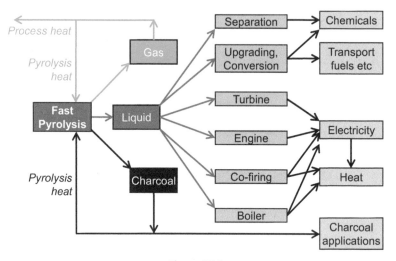

Figure 15.6

(a)

(b)

(c)

(d)

Figure 17.1

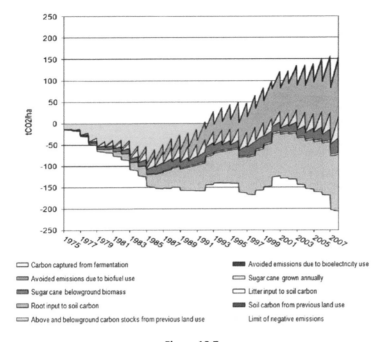

Figure 18.7

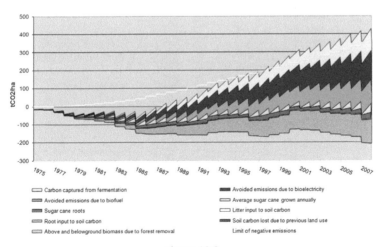

Figure 18.8

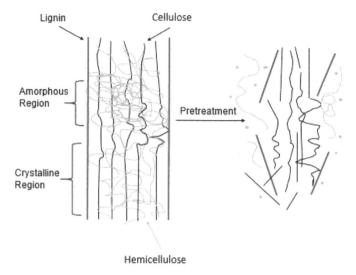

Figure 19.4

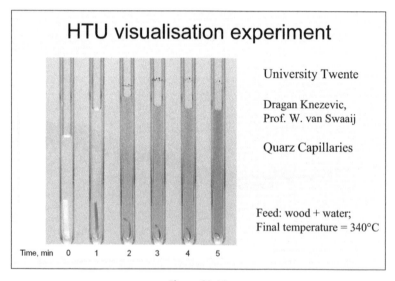

Figure 20.11